T0203929

STATISTICAL METHODS

METHODS

for

ENGINEERS and SCIENTISTS

STATISTICS: Textbooks and Monographs

A Series Edited by

D. B. Owen, Founding Editor, 1972–1991

W. R. Schucany, Coordinating Editor
Department of Statistics
Southern Methodist University
Dallas, Texas

R. G. Cornell, Associate Editor
for Biostatistics
University of Michigan

W. J. Kennedy, Associate Editor
for Statistical Computing
Iowa State University

A. M. Kshirsagar, Associate Editor
for Multivariate Analysis and
Experimental Design
University of Michigan

E. G. Schilling, Associate Editor
for Statistical Quality Control
Rochester Institute of Technology

Additional Volumes in Preparation

STATISTICAL METHODS
for
ENGINEERS and SCIENTISTS

THIRD EDITION, REVISED AND EXPANDED

ROBERT M. BETHEA
Chemical Engineering Department
Texas Tech University
Lubbock, Texas

BENJAMIN S. DURAN
Department of Mathematics
Texas Tech University
Lubbock, Texas

THOMAS L. BOULLION
Department of Mathematics and Statistics
University of Southwestern Louisiana
Lafayette, Indiana

CRC Press
Taylor & Francis Group
Boca Raton London New York

CRC Press is an imprint of the
Taylor & Francis Group, an **informa** business

CRC Press
Taylor & Francis Group
6000 Broken Sound Parkway NW, Suite 300
Boca Raton, FL 33487-2742

First issued in paperback 2019

ISBN-13: 978-0-8247-9335-7 (hbk)
ISBN-13: 978-0-367-40182-5 (pbk)

Library of Congress Cataloging–in–Publication Data

Bethea, Robert M.
    Statistical methods for engineers and scientists / Robert M.
Bethea, Benjamin S. Duran, Thomas L. Boullion. -- 3rd. ed., rev. and
expanded.
        p. cm. -- (Statistics, textbooks, and monographs ; v. 144)
    Includes bibliographical references and index.
    ISBN 0-8247-9335-8
    1. Mathematical statistics. 2. Probabilities. I. Duran,
Benjamin S. II. Boullion, Thomas L. III. Title.
IV. Series.
QA276.B425 1995
519.5--dc20   95-3642
                                                                      CIP

Visit the Taylor & Francis Web site at
http://www.taylorandfrancis.com

and the CRC Press Web site at
http://www.crcpress.com

Preface

The first two editions of the book have been used numerous times, mostly in teaching statistical methods to students of engineering and science, where the purpose was to provide the student with an understanding of the statistical techniques as well as practice in using them. Students, colleagues, and other users have provided us with ample constructive comments to make the third edition both a better medium for prospective users of statistics and better for classroom instruction or self-study. The text should prove useful for professional engineers and scientists because of its emphasis on regression, experimental design, analysis of variance, and the use of computer software to aid in data analysis related to these topics. This text is intended as a basic introductory course in applied statistical methods for students of engineering and the physical sciences at the undergraduate level. However, the text can also be used by graduate students who have very little or no statistical training. Theoretical developments and mathematical treatment of the principles involved are included as needed for understanding of the validity of the techniques presented. More intensive treatment of these subjects should be obtained, if desired, from the many theoretical statistics texts available.

The major changes in this edition are the deletion of Chapter 10, "Orthogonal Polynomials in Polynomial Regression," from the second edition, and its replacement with the new Chapter 10, "Statistical Process Control and Reliability." Several nonparametric techniques have been added at appropriate places, and numerous additional changes have been made throughout the text. The material on subsampling and three- and four-way analysis of variance, including the corresponding problems, has been moved from Chapter 11, "Experimental Design," to Chapter 8, "Analysis of Variance." Thirty problems have been added at appropriate places throughout the text. In addition to the major modifications, numerous changes have been made throughout the text to improve readability and presentation.

The material in this text can be arranged for either two 3-credit quarter courses or else one 3-credit semester course at the option of the instructor. For the former case, the material would be covered *in toto*, with additional time allowed throughout for applications in the discipline of the students involved. The most logical place for division is at the end of Chapter 7, on statistical inference. For the latter case, the material in the first four chapters is presented in three weeks. Chapter 5 is allotted one week; Chapters 6 through 8, two weeks each; Chapters 9 and 11 are covered in three weeks each; and Chapter 10, in one week. Instructors desiring only one three-credit quarter course in methodology would probably use only Chapters 6 through 9, provided the students have already been introduced to the basics.

Although many of the example problems are oriented towards chemical engineering and chemistry, this text is by no means limited to those areas. The examples merely illustrate general statistical principles as applicable to all fields, but particularly engineering and the related physical sciences. The problems at the end of each chapter are graded in difficulty for the convenience of the instructor. They are arranged to logically follow the outline of the material presented in the text.

The student should have access to some type of electronic calculator. There are many such calculators on the market that have some built-in statistical capability. Utilization of such calculators allows the student to concentrate on the interpretation of the analysis instead of the mechanical calculations. Those analyses involving larger data sets will be handled in terms of existing statistical software such as the Statistical Analysis System (SAS*), now available for some small computers in addition to the mainframe version.

We are indebted to many students and associates who have contributed to this text by way of suggestions for improvements, problem statements, experimental data, and consultation.We are also grateful to the American Mathematical Society for permission to reprint parts of the tables from "Exact Probability Levels for the Kruskal-Wallis Test" from *Selected Tables in Mathematical Statistics*, Volume 3, 1975, and parts of the tables from "Critical Values and Probability Levels for the Wilcoxon Signed-Rank

---

* SAS is the registered trademark of SAS Institute Inc., Cary, North Carolina 27511-8000.

Test" from *Selected Tables in Mathematical Statistics*, Volume 1, 1973. Our gratitude also goes to John Wiley & Sons for permission to reprint Table A7 from *Practical Nonparametric Statistics*, by W. J. Conover.

We are grateful to SAS Institute Inc. for permission to use and reproduce the SAS programs and other material that are their property. We appreciate all the comments we have received from all the individuals who used the first two editions. Many of those comments and suggestions have been incorporated into the third edition. We also thank all those individuals who have allowed us to use the data from their research and laboratory experiments for examples and problems.

In addition to these, our thanks go to The Iowa State University Press, Addison-Wesley Publishing Co., Prentice Hall, Inc., and Gulf Publishing Co. for permission to reprint data for use in examples and problems. Thanks are also due to the editors of the American Institute of Chemical Engineers, the Committee of Editors of the American Chemical Society, the American Society for Quality Control, the American Society for Testing and Materials, and the American Society of Mechanical Engineers for permission to reprint data extracted from articles in their respective publications.

Last, but not least, we wish to thank our families for their patience and support during the work of completing the third edition.

Robert M. Bethea
Benjamin S. Duran
Thomas L. Boullion

Contents

List of Worked Examples

List of Tables

Table No.   *Table Illustrating*

1

Introduction

The subject matter of the field of statistics is often defined as the scientific study of techniques for collecting, analyzing, and drawing conclusions from data. The engineer and scientist use statistics as a tool, which, when correctly applied, is of enormous assistance in the study of the laws of physical science. This is why an introduction to statistical methods is useful to students who are preparing themselves for careers in these areas. There are no statistical procedures that are applicable only to specific fields of study. Instead, there are general statistical procedures that are applicable to any branch of knowledge in which observations are made. Statistical procedures now constitute an important part of all branches of science. Procedures that have been developed for use in one field are inevitably adapted for use in a number of other fields. However, some procedures are used more frequently in one group of related disciplines than in another. In this book we concentrate on those procedures that are most widely used by engineers and scientists.

The scientific discipline of statistics has been used for describing, summarizing, and drawing conclusions from data. We, as scientists and engineers, have made enormous use of this most important and highly versatile discipline.

We use statistics principally to aid us in four ways. The first of these is to assist us in designing experiments and surveys. We desire that our experiments yield adequate answers to the questions that prompted their performance. The experiments should also be efficient ones. That is, they should provide the desired answers with maximum precision and accuracy with a minimum expenditure of time and other resources.

Another way in which the engineer and scientist use statistics is in describing and summarizing experimental data. This is properly termed *descriptive statistics*. The person's first concern with a body of data is whether it is to be considered as all possible data or as part of a larger body. It is extremely important to make this distinction, since failure to do so often results in loose thinking and erroneous results being obtained. To be perfectly clear about this distinction, we define a *population* as the entire set or collection of similar values, attributes, and so on, which are characteristic of a given experiment. Measurements or characteristics that are not constant, but show variability upon repeating the experiment under the same conditions, are called *values* of random variables. Thus a population could be defined as the set of all possible values of a random variable. These values need not all be different or finite in number.

A *sample* is a part of a population. In some situations a sample may include the entire population. Usually, we desire to use sample information to make an inference about the population. For this reason, it is particularly important to define the population under consideration and to obtain a representative sample from the population. To obtain a representative sample, we embody the principle of randomness in the rules for drawing the sample items.

Randomly selected samples guarantee us that the mathematical laws of probability will hold. Thus we are able to make statements about the population from which a sample is taken and give a quantitative measure of chance that these statements are true for the population.

A third use made of statistics is in testing *hypotheses*. You would not have performed an experiment or collected a set of data about some phenomenon without a definite purpose in mind. Your purpose may simply have been to satisfy your curiosity. It may have been an attempt to predict the behavior of a process or group of individuals. In any event, you will have had some idea, feeling, or belief that a population possessed a particular attribute. This, we say, is a hypothesis concerning the population. This is where statistics becomes invaluable. Through its proper use, we can plan experiments to evaluate any of these hypotheses. This is done by determining whether or not the results are reasonable and likely to be valid or are probably due strictly to random variation.

Thus far, the uses of statistics described above have been qualitative in nature. In many cases it is of equal, if not greater importance to obtain quantitative relationships between variables. This is the last of the principal ways in which we use statistics. You have probably done this quite often in the past, many times subconsciously. Perhaps some of your most conscious efforts at quantification in this regard started with analytic geometry or the calculus. There you were confronted with the problem of describing a line, curve, surface, or other shape which itself defined one or more variables or occurrences in terms of others. Your most simple effort would have been to draw a line through a collection of data points and from it determine the equation relating the independent and dependent variables. You undoubtedly would have drawn that line so as to minimize the variation of data points around it. But the line you drew and the line drawn by someone else to describe the same data will almost invariably be different. We therefore say that the estimates obtained from these "eyeball" curve fits are empirical and quite subjective. A better method would have been for you both to use some consistent manner of determining the probable relation, if any, between the variables involved. Statistics here has provided us with just such procedures. The most common is the method of least squares.

In this book we begin by studying the basic concepts of simple and compound probabilities of events occurring. From this you should obtain a feeling for the role played by chance in the outcome of investigations. These concepts are then applied to discrete and continuous distributions. You should then learn how to recognize the type of distribution involved so that you can handle sample data effectively.

After studying the ways in which data are distributed and the probabilities affecting those distributions, we turn to ways of describing data. These descriptions are given in terms of measures of location and variability. The chapter on descriptive statistics is followed by a chapter on expectations. In that portion of the book we delve into theory to a sufficient degree so it will become obvious why statistics calculated from samples can be used to describe populations adequately.

We next direct our attention to the methods involved in estimation and testing hypotheses about populations by the use of sample data. Estimation includes procedures for constructing confidence intervals for the poulation parameters. In hypothesis testing the necessary statistical techniques for evaluating the hypotheses are developed. As a continuation of these topics we then study the techniques of analysis of variance. These procedures are used for comparing several things at a time or several estimates of the same thing whenever certain assumptions are satisfied. They can also be used to aid us in making qualitative judgments concerning our observations.

The statistical methods involved in estimating quantitative relations between variables are the subject of the chapter dealing with regression and correlation analysis. In regression analysis we try to find the best estimates of the relationships between population variables from sample data and to test the validity of those estimates by suitable procedures. The technique for

Suppose that we have such an experiment. If all possible outcomes of the experiment can be described prior to its performance, and if it can be repeated under the same conditions, it will be referred to as a *random experiment*. The set of all possible outcomes will be referred to as the *sample space*. For instance, the toss of a coin, assuming that the toss may be repeated under the same conditions, is an example of a random experiment having as its sample space the set {H, T}, where H denotes heads and T denotes tails. We will take a *set* to be a collection of objects and use braces to enclose the elements of a set.

Since the idea of a set allows us to discuss probability expeditiously, we will define several set operations that will be used later. The set of outcomes contained in either A or B (or both), called the *union of A and B*, is denoted by $A \cup B$. The set of outcomes contained in *both* A and B, called the *intersection of A and B*, is denoted by $A \cap B$ (or AB). In the language of probability we say that the events A and B are *mutually exclusive* if $A \cap B$ contains no outcomes; that is, $A \cap B$ is the empty set, denoted by $\emptyset$. If every outcome in A is an outcome in B, then A is said to be a *subset* of B, denoted by $A \subset B$. An event consisting of a single outcome will be called a *simple event*.

Let S denote the set of all possible outcomes of a random experiment; that is, S is the sample space. The outcomes in S will on occasion be referred to as *sample points*. We desire to define a function $\mathscr{P}$ such that if E is a subset of S, hereafter called an *event*, then $\mathscr{P}(E)$ is the probability that the outcome of the random experiment is an element of E. Since we are only considering experiments which are such that all possible outcomes can be described prior to their performance and which can be repeated under the same conditions, we will take $\mathscr{P}(E)$ to be that number about which the relative frequency of the occurrence of E tends to stabilize after a long series of repetitions of the experiment. This phenomenon is usually referred to as *statistical regularity*. For instance, in the coin-tossing experiment we would take $\mathscr{P}(H) = \frac{1}{2}$ and $\mathscr{P}(T) = \frac{1}{2}$, since we would expect the relative frequency of each to stabilize around the same value after a large number of performances of the experiment. Table 2.1 illustrates the results of the coin-tossing experiment. The number of heads is recorded for:

1.   Every 20 tosses for 15 times
2.   Every 200 tosses for 15 times
3.   Every 2000 tosses for 15 times

From each of these groups a relative frequency is calculated.

Each value of the experimental relative frequency of occurrence of H is only approximately equal to the probability of occurrence of H. As the number of tosses is increased, the relative frequency values approach the

TABLE 2.1 Number of Heads per Group in Coin-Tossing Experiment

	Number of tosses					
	20		200		2000	
Group	Frequency	Relative frequency	Frequency	Relative frequency	Frequency	Relative frequency
1	14	0.70	104	0.520	1010	0.5050
2	11	0.55	91	0.455	990	0.4950
3	13	0.65	99	0.495	1012	0.5060
4	7	0.35	96	0.480	986	0.4930
5	14	0.70	99	0.495	991	0.4955
6	10	0.50	108	0.540	988	0.4940
7	11	0.55	101	0.505	1004	0.5020
8	6	0.30	101	0.505	1002	0.5010
9	9	0.45	101	0.505	976	0.4880
10	9	0.45	110	0.550	1018	0.5090
11	9	0.45	108	0.540	1021	0.5105
12	6	0.30	103	0.515	1009	0.5045
13	6	0.30	98	0.490	1000	0.5000
14	10	0.50	101	0.505	998	0.4990
15	13	0.65	109	0.545	988	0.4940

Source: Data from R. Lowell Wine, *Statistics for Scientists and Engineers,* copyright 1964, p. 105. Adapted by permission of Prentice Hall, Englewood Cliffs, NJ.

defined or expected value of $\frac{1}{2}$. It should be noted that even after 30,000 tosses, the relative frequency (0.4998) is still only approximately equal to the expected value of 0.5.

We now summarize the relative frequency definition (or interpretation) of probability.

Definition 2.1 If a random experiment is repeated a large number of times, n, the observed relative frequency of occurrence, n_E/n, of the event E will tend to stabilize at an unknown constant $\mathscr{P}(E)$, which is called the *probability* of E or the *theoretical relative frequency* of E.

Another definition of probability, which preceded the frequency definition, is the so-called *classical definition.*

Definition 2.2 If a random experiment can result in n equally likely and mutually exclusive outcomes, and if n_E of these outcomes possess attribute E, the probability of E is the ratio n_E/n.

A certain amount of caution regarding the terms *equally likely* and *mutually exclusive* must be exercised in using the classical definition. That the outcomes are equally likely cannot be proved but must be assumed. This assumption is reasonable in some problems. The classical definition also requires n to be finite. The frequency definition is more useful than the classical definition in that one generally does not know the true probability structure associated with the sample space.

The frequency and classical definitions of probability both suggest some of the properties that we would want $\mathscr{P}(E)$ to have. Since the relative frequency is never negative, we would want $\mathscr{P}$ to be a nonnegative function. If S is the entire sample space, its relative frequency is always 1, and we would want $\mathscr{P}(S) = 1$. Finally, if $E_1, E_2, \ldots$ are mutually exclusive subsets of S, that is, no two of these subsets have a point in common, the relative frequency of the union of these sets is the sum of the relative frequencies of the sets, so we would want

$$\mathscr{P}(E_1 \cup E_2 \cup E_3 \cup \cdots) = \mathscr{P}(E_1) + \mathscr{P}(E_2) + \mathscr{P}(E_3) + \cdots.$$

The classical definition suggests the same properties, although it is restricted to a finite number of outcomes.

Thus we are led to the following axiomatic definition of probability.

Definition 2.3   Let S be a sample space. Assume that $\mathscr{P}$ is a function defined on subsets (events) of S satisfying the following:

(a) $\mathscr{P}(E) \geqslant 0$, for every event E.
(b) $\mathscr{P}(E_1 \cup E_2 \cup \cdots) = \mathscr{P}(E_1) + \mathscr{P}(E_2) + \cdots$, where the events E_i, $i = 1, 2, \ldots$, are such that no two have a point in common.
(c) $\mathscr{P}(S) = 1$.

Then $\mathscr{P}$ is called a *probability function*. The number $\mathscr{P}(E)$ will be called the probability that the outcome of the random experiment is an element of the set E, or simply the *probability of the event E*.

2.3   POSSIBLE WAYS FOR EVENTS TO OCCUR

Many problems in elementary probability involve use of the classical definition of probability; that is, the sample space has a finite number of equally likely and mutually exclusive outcomes. In such cases, evaluating the probability of a given event E consists of determining or counting the number of outcomes, n_E, favorable to the event E and then computing $\mathscr{P}(E) = n_E/n$. In some problems the determination of n_E can become quite tedious. Consequently, it is advantageous to develop some enumeration procedures that will be less time consuming and more efficient than total enumeration procedures.

We now state a counting principle that will be useful in determining the number of ways in which events can occur.

Counting Rule   If an operation can be performed in m_1 ways, and if for each of these a second operation can be performed in m_2 ways, the two operations can be performed in $m_1 m_2$ ways. In probability terms, if an event E consists of the occurrence of an event B, followed by the occurrence of an event C, we can obtain the number of ways that E can occur by taking the product of the number of ways that B can occur times the number of ways that C can occur.

We now consider permutations and combinations, two ideas that will help in determining the number of ways in which certain events can occur.

2.3.1 Permutations

Given a set of n distinguishable objects, all ordered sets formed with r objects chosen from the n given objects in any manner are called *permutations* of the n objects taken r at a time. The number $P(n, r)$ of such permutations is given in most algebra books to be

$$P(n, r) = \frac{n!}{(n - r)!}. \tag{2.1}$$

In particular, the number of ways of rearranging n objects is $n!$.

Actually, the result of Eq. (2.1) follows from the counting rule of Section 2.3 since there are n ways of choosing the first object, $n - 1$ ways of choosing the second object, $\dots, n - (r - 1)$ ways of choosing the rth object. Consequently,

$$\begin{aligned} P(n, r) &= n(n - 1)\cdots(n - r + 1) \\ &= \frac{n(n - 1)\cdots(n - r + 1)(n - r)!}{(n - r)!} \\ &= \frac{n!}{(n - r)!}. \end{aligned}$$

Example 2.1   A fellowship, a research assistantship, and a teaching assistantship are available in this department this semester. In how many different ways may they be awarded to nine applicants?

Solution

$$n = 9, \qquad r = 3$$

Number of ways $= P(9, 3) = \dfrac{9!}{6!} = 504.$

Permutations can be extended quite simply to the case where k groups of objects are involved, provided that the objects within each group are alike and that they are different from the objects of all other groups. Suppose that there is a total of N objects of which n_1 are alike, n_2 are alike, ..., n_k are alike. Then the number of permutations of these N objects taken N at a time is given by

$$P(N; n_1, n_2, n_3, \ldots, n_k) = \frac{N!}{n_1!\, n_2!\, n_3! \cdots n_k!}. \qquad (2.2)$$

Example 2.2 How many different seating arrangements are possible for a materials science class composed of seven chemical engineers, three textile engineers, nine civil engineers, and one petroleum engineer?
Solution

$$N = 20, \quad n_1 = 7, \quad n_2 = 3, \quad n_3 = 9, \quad n_4 = 1$$

Number of arrangements $= \dfrac{20!}{7!\, 3!\, 9!\, 1!} = 221{,}707{,}200.$

2.3.2 Combinations

Suppose that we select r objects from among n objects without regard to the order of arrangement of the objects among themselves. Any such selection is called a *combination* of n objects taken r at a time or a combination of n objects of order r. The number of combinations of n objects taken r at a time, $\binom{n}{r}$, is given as

$$\binom{n}{r} = \frac{n!}{(n - r)!\, r!}. \qquad (2.3)$$

Example 2.3 In how many different ways can a pair of students choose two calculators from the six machines available?
Solution

$$n = 6, \qquad r = 2$$

Number of ways $= \dbinom{6}{2} = \dfrac{6!}{4!\, 2!} = 15.$

To tell whether a permutation or combination is involved, remember that the order of selection is important in permutations. The order of selection is immaterial for combinations. If we consider the letters u, v, and w, we have the following two-letter permutations: uv, uw, vw, vu, wu, and wv. There are, however, only three combinations of pairs of letters. We see that uv and vu, uw and wu, and vw and wv represent the pairs of permutations possible for each two-letter combination. From this example one should realize that the number of permutations of *n* objects taken *r* at a time is equal to the number of combinations of these *n* objects taken *r* at a time multiplied by the number of permutations of the *r* objects taken *r* at a time, which is *r*!. Thus

$$P(n, r) = \binom{n}{r} r!.$$

2.4 PROBABILITY COMPUTATION RULES

If the probabilities of simple events are known, the probabilities of compound (i.e., nonsimple) and related events may be computed in terms of the simple event probabilities. For clarity, certain terms, some of which have been defined previously, associated with such compound and related events are defined as follows (*A* and *B* are considered to be arbitrary events of an experiment):

1. If *A* is an event of an experiment, then the event in which *A* does not occur, denoted by $\bar{A}$, is called the *complementary event* of *A*, or simply the *complement* of *A*.
2. The event "*A* and *B*" is just $A \cap B$.
3. The event "*A* or *B*" or "$A \cup B$" is that event in which at least one of *A* or *B* occurs.
4. Two or more events, *A*, *B*, *C*, and so on, are mutually exclusive if the occurrence of any one event precludes the occurrence of each of the others.
5. The probability of occurrence of event *A*, given that event *B* has occurred, is called the *conditional probability* of event *A* given *B* and is denoted $\mathscr{P}(A|B)$.
6. A collection of events is *exhaustive* if the collection includes all possible outcomes of the experiment.
7. The event E_1 is independent of the event E_2 if the probability of occurrence of event E_1 is not affected by the prior occurrence of event E_2, and vice versa, so that

$$\mathscr{P}(E_1|E_2) = \mathscr{P}(E_1) \quad \text{and} \quad \mathscr{P}(E_2|E_1) = \mathscr{P}(E_2). \tag{2.4}$$

Rule 1: Multiplication

If $E_1, E_2, E_3, \ldots, E_n$ are independent events having respective probabilities $\mathscr{P}(E_1), \mathscr{P}(E_2), \mathscr{P}(E_3), \ldots, \mathscr{P}(E_n)$, the probability of occurrence of E_1 and E_2 and E_3 and $\cdots E_n$ is

$$\mathscr{P}(E_1 \cap E_2 \cap E_3 \cap \cdots \cap E_n) = \mathscr{P}(E_1 E_2 E_3 \cdots E_n)$$
$$= [\mathscr{P}(E_1)][\mathscr{P}(E_2)][\mathscr{P}(E_3)]\cdots[\mathscr{P}(E_n)]$$
$$= \prod_{i=1}^{n} \mathscr{P}(E_i). \tag{2.5}$$

This is immediate from the definition of independence of two events given in item 7 above.

Example 2.4 If color, solution viscosity, and percent ethoxyl are independent characteristics of ethyl cellulose resin and the probabilities of any given batch being off-grade with respect to these characteristics are, respectively, 0.03, 0.05, and 0.02, what is the probability of a batch being on-grade with respect to all three characteristics?

Solution The event of the batch being on-grade is the complementary event of its being off-grade. Thus the probabilities of a batch being on-grade with respect to the three characteristics are 0.97, 0.95, and 0.98, and the probability of it being on-grade with respect to all three is

$$\mathscr{P}(E_1 \cap E_2 \cap E_3) = 0.97(0.95)(0.98) = 0.9031.$$

Rule 2: Addition

(a)   If E_1, E_2, and E_3 are events, the probability of at least one of these events is

$$\mathscr{P}(E_1 \cup E_2 \cup E_3) = \mathscr{P}(E_1) + \mathscr{P}(E_2) + \mathscr{P}(E_3)$$
$$- \mathscr{P}(E_1 E_2) - \mathscr{P}(E_1 E_3) - \mathscr{P}(E_2 E_3)$$
$$+ \mathscr{P}(E_1 E_2 E_3).$$

$$\tag{2.6}$$

(b)   If only two events E_1 and E_2 are considered,

$$\mathscr{P}(E_1 \cup E_2) = \mathscr{P}(E_1) + \mathscr{P}(E_2) - \mathscr{P}(E_1 E_2). \tag{2.7}$$

Although no proof of these relationships is given here, they are clarified by observation of the Venn diagrams in which areas represent probabilities. Diagrams for two and three events are shown in Fig. 2.1, in which the circles represent the probabilities of single events. The areas included in the intersecting circles represent the probabilities of the compound events indicated.

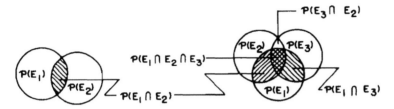

FIGURE 2.1 Venn diagrams for two events (E_1 and E_2) and three events (E_1, E_2, and E_3).

Equation (2.6) may be generalized for more than three events by adding all odd-membered probabilities and subtracting all even-membered probabilities.

(c) If the events $E_1, E_2, \ldots, E_n$, are *mutually exclusive*, the addition rule becomes

$$\mathscr{P}(E_1 \cup E_2 \cup E_3 \cup \cdots \cup E_n) = \sum_{i=1}^{n} \mathscr{P}(E_i). \tag{2.8}$$

Example 2.5 From the data given in Example 2.4, calculate the probability of a given batch being off-grade with respect to any of the characteristics color, solution viscosity, and percent ethoxyl.

Solution The event E_1 is a batch not meeting the color, E_2 the solution viscosity, and E_3 the percent ethoxyl specification. Thus

$$\mathscr{P}(E_1) = 0.03, \quad \mathscr{P}(E_2) = 0.05, \quad \text{and} \quad \mathscr{P}(E_3) = 0.02,$$
$$\mathscr{P}(E_1 \cup E_2 \cup E_3) = 0.03 + 0.05 + 0.02 - 0.03(0.05)$$
$$- 0.03(0.02) - 0.05(0.02) + 0.03(0.05)(0.02)$$
$$= 0.0969.$$

It will be noted that the event above is the complementary event of that in Example 2.4 and that the sum of the probabilities of the two events is 1.0.

Rule 3: Conditional Probability

The probability of the compound event $E_1 \cap E_2$ is equal to the probability of one event times the conditional probability of occurrence of the other event, given that the first event has occurred. That is,

$$\mathscr{P}(E_1 \cap E_2) = \mathscr{P}(E_1)\mathscr{P}(E_2|E_1)$$
$$= \mathscr{P}(E_2)\mathscr{P}(E_1|E_2), \tag{2.9}$$

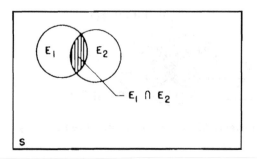

Figure 2.2 Sample space S and Venn diagram for events E_1 and E_2.

The proof of Rule 3 is not given here: however, the rule can be clarified by making reference to Fig. 2.2. If event E_1 has occurred, the probability that E_2 occurs is given by

$$\mathscr{P}(E_2|E_1) = \frac{n_{E_1 \cap E_2}}{n_{E_1}},$$

where $n_{E_1 \cap E_2}$ and n_{E_1} denote the number of sample points in $E_1 \cap E_2$ and E_1, respectively. Thus

$$\mathscr{P}(E_2|E_1) = \frac{n_{E_1 \cap E_2}/n}{n_{E_1}/n} = \frac{\mathscr{P}(E_1 \cap E_2)}{\mathscr{P}(E_1)}$$

or

$$\mathscr{P}(E_1 \cap E_2) = \mathscr{P}(E_1)\mathscr{P}(E_2|E_1).$$

If the events E_1 and E_2 are independent, there is no contingency of one on the other. Thus

$$\mathscr{P}(E_2|E_1) = \mathscr{P}(E_2) \tag{2.10}$$

and

$$\mathscr{P}(E_1|E_2) = \mathscr{P}(E_1), \tag{2.11}$$

so

$$\mathscr{P}(E_1 \cap E_2) = \mathscr{P}(E_1)\mathscr{P}(E_2), \tag{2.12}$$

which is a proof of Rule 1 in the case of two events.

Example 2.6 In a batch manufacturing sequence, the material is tested after each processing step to determine its suitability for further processing. Batches not meeting test specifications are reprocessed. Past records indicate that the percentages of on-grade material produced by steps A and B, based on the input to the particular step, are 97.0 and 95.0, respectively. What is the probability of a batch passing the requirements for both steps A and B?

Solution Since input to step B is contingent on satisfactory material being produced in step A, the 0.95 probability of success in the second step is a conditional probability. Thus

$$\mathscr{P}(A \cap B) = \mathscr{P}(A)\mathscr{P}(B|A) = 0.97(0.95) = 0.9215.$$

Example 2.7 The contents of two boxes used for storing hydrometers in the stockroom have become mixed. Box A, which is easier to reach and thus gone into by two-thirds of the students, contains six API hydrometers and four Baumé hydrometers. Box B contains 10 hydrometers, of which five are Baumé and five are API hydrometers.

(a) If you remove one from each box, what is the probability that the two thus selected will be of the same type?

(b) What is the probability that a second hydrometer removed from one of the boxes will be of the same type as the one initially removed from that box?

(c) To conclude this example, suppose that one of the seniors asked for a Baumé hydrometer before you started to sort the contents of the two boxes. What is the probability that if one hydrometer is selected from each box, you will get exactly one Baumé hydrometer?

Solution (a) They will be of the same type if they are both API hydrometers or if they are both Baumé hydrometers. Thus

$$\mathscr{P}(2 \text{ alike}) = \mathscr{P}(2 \text{ Baumé or 2 API}) = \mathscr{P}(2 \text{ Baumé})$$
$$+ \mathscr{P}(2 \text{ API}) = 0.4(0.5) + 0.6(0.5)$$
$$= 0.5.$$

(b) Note that since two-thirds of the students select box A, the probability of box A being selected is $\frac{2}{3}$ and for box B it is $\frac{1}{3}$. The desired event can occur by selecting a box and then selecting two hydrometers of the same type from it. Hence the desired probability is

$$\mathscr{P}(\text{box } A \text{ and 2 hydrometers alike or box } B \text{ and 2 hydrometers alike})$$
$$= \mathscr{P}(\text{box } A) \, \mathscr{P}(2 \text{ hydrometers alike}|\text{box } A)$$
$$+ \mathscr{P}(\text{box } B) \, \mathscr{P}(2 \text{ hydrometers alike}|\text{box } B)$$
$$= \tfrac{2}{3}\mathscr{P}(2 \text{ Baumé or 2 API}|\text{box } A) + \tfrac{1}{3}\mathscr{P}(2 \text{ Baumé or 2 API}|\text{box } B)$$
$$= \tfrac{2}{3}[(0.4)\tfrac{3}{9} + (0.6)\tfrac{5}{9}] + \tfrac{1}{3}[(0.5)\tfrac{4}{9} + (0.5)\tfrac{4}{9}]$$
$$= \tfrac{62}{135}.$$

(c) $\mathscr{P}(\text{exactly 1 Baumé}) = \mathscr{P}(\text{Baumé from } A \text{ and API from } B \text{ or API}$
$$\text{from } A \text{ and Baumé from } B)$$
$$= 0.6(0.5) + 0.4(0.5) = 0.5.$$

Consider the situation where the probability of an event E_1 occurring in a single performance of an experiment is $p = \mathscr{P}(E_1)$ and this probability remains the same for n independent trials of the experiment. If we desire the probability that the event E_1 will occur r times out of n trials of the experiment, we reason as follows: The probability that E_1 occurs the first r times the experiment is performed is $p \cdots p = p^r$ and does not occur the next $n - r$ times is $(1 - p) \cdots (1 - p) = (1 - p)^{n-r}$. Letting $q = 1 - p$, we see that the probability that E_1 occurs on each of the first r trials and does not occur on the last $n - r$ trials is given by $p^r q^{n-r}$. Now, since we are only interested in E_1 occurring r times, we are not concerned about the order in which E_1 occurred and did not occur. Hence since there are $\binom{n}{r}$ rearrangements of n objects of which r are alike and $n - r$ are alike, yet different from the other r objects, we conclude that there are $\binom{n}{r}$ ways for us to get E_1 occurring r times and not occurring $n - r$ times. Since these are all independent of each other, the probability of E_1 occurring r times and not occurring $n - r$ times is $\binom{n}{r} p^r q^{n-r}$, where r is any integer between 0 and n.

Example 2.8   If n copper elbows are tossed into k bins in the storeroom so that each elbow is equally likely to land in any of the bins (sophomores often seem to return materials in this haphazard manner), what is the probability that a particular bin, say bin A, will contain m of the n elbows?
Solution

$$\mathscr{P}(\text{any elbow landing in bin } A) = \frac{1}{k},$$

$$\mathscr{P}(m \text{ elbows landing in bin } A) = \binom{n}{m}\left(\frac{1}{k}\right)^m\left(1 - \frac{1}{k}\right)^{n-m}.$$

For this situation, let us assume that $n = 12$ elbows are involved, that only $k = 4$ bins are in use, and that we desire to obtain the probability that three elbows land in bin A.

$$\mathscr{P}(3 \text{ elbows land in bin } A) = \binom{12}{3}\left(\frac{1}{4}\right)^3\left(1 - \frac{1}{4}\right)^{12-3}$$

$$= \left(\frac{12!}{9!\,3!}\right)\left(\frac{1}{4}\right)^3\left(\frac{3}{4}\right)^9$$

$$= 220\left(\frac{1}{64}\right)\left(\frac{3}{4}\right)^9$$

$$= 0.2581.$$

This rule finds frequent applications in quality control where batches of material are repeatedly sampled for defects.

2.4.1 Applications

Fault-tree analysis (FTA) is a formalized deductive approach for the analysis of hazards. It can be used to identify potential accidents in a system design and can help eliminate costly design changes. It is a technique of system reliability analysis and is useful in all branches of engineering and science. For example, the technique can be applied to a variety of systems, such as large chemical processing plants, transportation systems, and light-water reactor-safety systems. The fault tree itself is a graphical representation of the Boolean failure logic associated with the development of a particular "undesired event" (the TOP event) from basic events. The term *event* denotes a dynamic change of state that occurs to a system component. A basic inherent failure of a system component is referred to as a *basic event*. A basic failure (event) is also referred to as a *primary failure* (event) and is considered to be a generic failure. A *fault tree* is a model that graphically and logically represents the various combinations of possible events, both fault and normal, occurring in a system that lead to the top event. A fault event is an abnormal system state. A *normal event* is an event that is expected to occur. In many engineering applications, FTA is used to evaluate the impact of alternate processes, controls, or preventive measures.

Fault-tree analysis describes in vertical graphical form the specific chain of events or conditions necessary for the undesired (top) event to occur [2,5,6,9,10]. Three principles are associated with FTA. They are that:

1. Each actual/potential top event is the result of multiple, interacting causes.
2. Each cause or interaction can be separated into basic events or component failures.
3. Solutions can be found to control each cause or interaction.

FTA is retrospective in nature and is used in risk analysis to determine what could happen or must have happened to result in the occurrence of the top event being considered. Contributing causes and interactions are followed down the tree until the primary events (root causes) have been identified. A basic event is one that cannot be further decomposed into other events. All events between the top event and the basic events are either intermediate, external, or undeveloped.

If the probabilities are propagated logically, situations may be encountered where a primary event is counted more than once, resulting in a fault tree that does not consist of minimal cut sets. A *cut set* is a basic event or combination of basic events that will cause the top event to occur. A *minimal cut set* is defined as a set of events that is not a subset of any other set of events. That is, a cut set is minimal if it cannot be reduced and still ensure the occurrence of the top event. All superfluities in a fault tree must be eliminated to produce a unique group of sets of basic events leading to the top event. Such groups of sets are referred to as minimal cut sets. They may be determined by methods developed [3] using Boolean algebra.

One of the advantages of FTA is that it begins with the event of interest as selected by the user. FTA provides insight into interactions between multiple events by examining the various ways in which the top event can occur. FTA is also adaptable to computer applications. Perhaps the chief disadvantage of FTA is that the developer or user of the fault tree can never be certain that all failure modes (basic events) have been included. Another disadvantage is that the fault tree quickly expands, becoming almost too cumbersome to handle without a computer. Other problems associated with FTA arise from its two basic assumptions: (1) that failures are complete, not partial, and (2) that all failures (basic events) are independent of each other. Other aspects of FTA are that fault trees for the top event of interest prepared by different persons may not be the same, due to differences in their experience, available data, and so on. Finally, for FTA to be quantitative, failure probabilities must be available for all basic events in the fault tree. Those probabilities are developed based on experience with the same or similar items or actions, use of reliability tables [2,3], manufacturers' estimates, and so on.

Most of the items in the fault tree will be connected by AND or OR gates. The *AND gate* is a process involving the probabilities of multiple events $[\mathscr{P}(E_1)\mathscr{P}(E_2)\cdots\mathscr{P}(E_n)]$ and follows the multiplication rule of Eq. (2.5). Use of the *OR gate* involves a situation in which at least one of several events E_i occurs and is described by the addition rule of Eq. (2.6). Note that the convention for the algebraic signs of the product terms as shown by the addition rule and by Venn diagrams (Fig. 2.1) is "plus odds, minus evens." In that way, no component of the probability expression will be counted more than once. The symbols used in the construction of fault trees are shown in Fig. 2.3. Sample fault trees are shown in Fig. 2.4.

Event-tree analysis (ETA) begins with an indentifiable but undesired initiating event and proceeds forward (prospectively) to construct a horizontal logic tree while considering the response of systems and personnel [4–8]. From the initiating event, a sequence of corrective responses is postulated as a series of binary steps (top = success, bottom = failure). The probabilities of the sequential outcomes are calculated (as in FTA) to determine the probability of the final outcome. The completed event tree

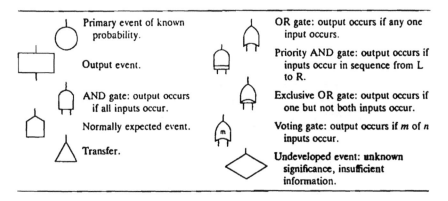

FIGURE 2.3 Some common fault-tree symbols.

shows the process or system in alternative failure states. Unfortunately, ETA has no way to identify multiple initiating events and interactions between events.

Common problems with FTA and ETA are critical omissions, resulting in underestimation of the true hazards. These problems may be categorized as:

1. Failure to understand causes of operator error
2. Failure to consider an unanticipated change or state of the surroundings outside the system or process being considered
3. Failure to realize how different components or subsystems react and interact
4. Insensitivity to the possibility of "common-mode failures"
5. Uncertainties in the estimates of failure rates or probabilities associated with new/untried equipment, systems, or processes
6. Uncertainties in estimating the rates or probabilities of successful operator action/intervention associated with a critical processing step that could lead to potential system failure
7. Oversimplification or overspecification of the available degrees of freedom
8. Incorrect specification of the sequence of events

Other topics related to reliability are considered in Chapter 10. A good reference for concepts related to fault-tree analysis is Barlow et al. [1].

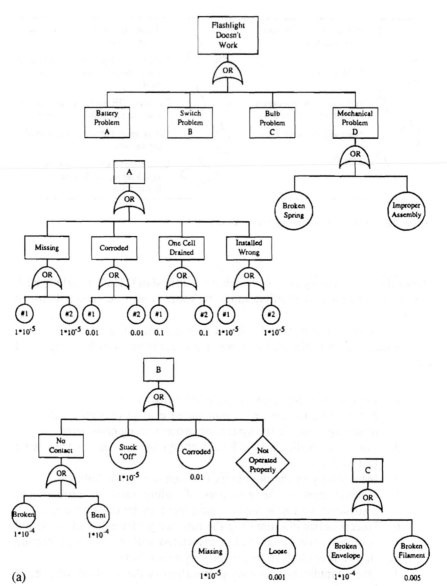

FIGURE 2.4   Fault trees for (a) inoperative flashlight.

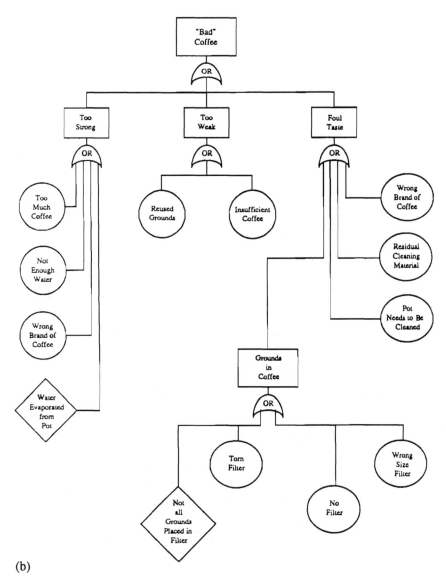

(b)

FIGURE 2.4 (*continued*)   (b) "Bad" coffee at office. (*Continued*)

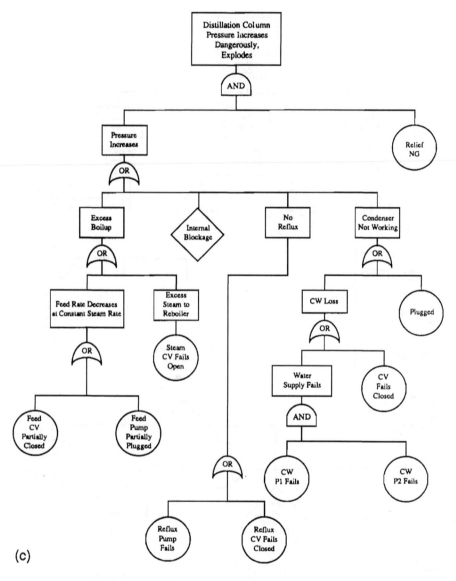

Figure 2.4 (*continued*)   (c) Increase in pressure of a distillation column. CV, control valve; CW, cold water; P1, P2, pumps 1 and 2.

2.5 *A POSTERIORI* PROBABILITY

In the preceding section we assumed that the events would happen, that is, that the probabilities were known *a priori* and that they could be predicted or calculated directly. It is not uncommon, however, for the opposite situation to exist. In that case, we wish to determine the probability that a particular set of conditions or circumstances existed from the results already obtained. These *a posteriori* or "after the fact" probabilities can be determined from Bayes' theorem, which we now derive.

Suppose that we have a two-stage experiment. Let $B_1, B_2, \ldots, B_n$ be the mutually exclusive and exhaustive events in the first stage of the experiment. In the second stage of the experiment, an event E can occur only in conjunction with one or more of the events $B_1, B_2, \ldots, B_n$. Thus we have

$$\mathscr{P}(E) = \mathscr{P}[(E \cap B_1) \cup (E \cap B_2) \cup \cdots \cup (E \cap B_n)]$$
$$= \mathscr{P}(E \cap B_1) + \mathscr{P}(E \cap B_2) + \cdots + \mathscr{P}(E \cap B_n)$$
$$= \sum_{i=1}^{n} \mathscr{P}(E \cap B_i)$$
$$= \sum_{i=1}^{n} \mathscr{P}(B_i)\mathscr{P}(E|B_i).$$

Given that the event E has occurred, if one is interested in obtaining the probability that event B_k occurred in the first stage, it may be calculated as follows:

$$\mathscr{P}(B_k|E) = \frac{\mathscr{P}(B_k \cap E)}{\mathscr{P}(E)}$$
$$= \frac{\mathscr{P}(B_k)\mathscr{P}(E|B_k)}{\sum_{i=1}^{n} \mathscr{P}(B_i)\mathscr{P}(E|B_i)}. \qquad (2.13)$$

Formula (2.13) is known as *Bayes' formula.*

Example 2.9 Consider the case of two bottles of reagents. One of these bottles, which we will call B_1, is filled with KOH pellets. The other bottle, which we will call B_2, is filled with an equal number of KOH and NaOH pellets. A pellet is drawn at random from one bottle and is found to be KOH. What is the probability that the pellet came from B_1?

Solution Let E be the event that the pellet drawn is KOH.

$$\mathscr{P}(B_1) = \mathscr{P}(B_2) = \tfrac{1}{2}, \qquad \mathscr{P}(E|B_1) = 1, \qquad \mathscr{P}(E|B_2) = \tfrac{1}{2}$$

$$\mathscr{P}(B_1|E) = \frac{\mathscr{P}(B_1)\mathscr{P}(E|B_1)}{\mathscr{P}(B_1)\mathscr{P}(E|B_1) + \mathscr{P}(B_2)\mathscr{P}(E|B_2)}$$
$$= \frac{\tfrac{1}{2}(1)}{\tfrac{1}{2}(1) + \tfrac{1}{2}(\tfrac{1}{2})} = \frac{2}{3}.$$

Example 2.10   In the production of Orlon–cotton fabrics for consumer use, Orlon spinning solution is forced through hundreds of tiny orifices in a spinneret that is mounted in a spinning machine. The filaments so produced are dried by a rising airstream, collated into fibers, and then combined into *tow*. Tow is a fairly ordered arrangement of parallel fibers. The next step is to process the tow in a crimping machine in which the fibers are uniformly stretched. Samples of the finished tow are then routinely inspected before baling.

Among the tests performed on each sample is one to measure stretch resistance. If the tow has been insufficiently stretched, fabrics made from it will very likely be deformable in use. If the tow has been stretched too much, there will be too many broken or weakened filaments to use in making top-quality garments.

As an Orlon production engineer, your responsibilities include all aspects of filament and tow production, starting with the arrival of the spinning solution at the spinning heads and ending with the tow-baling operation. The hourly tow report has just come in from the lab and indicates that the sample has poor stretch resistance. How can you decide quickly which mechanical operation is probably at fault?

Solution   The event E is the occurrence of poor stretch resistance. The probability of crimper malfunction is known to be $\mathscr{P}(B_1) = 0.04$; for clogged spinnerets, $\mathscr{P}(B_2) = 0.41$; and for ineffective filament collation, $\mathscr{P}(B_3) = 0.55$.

From past experience we know that poor stretch resistance occurs half the time when the crimper malfunctions. So $\mathscr{P}(E|B_1) = 0.5$. Drawing in a like manner on previous observations, we have $\mathscr{P}(E|B_2) = 0.1$ and $\mathscr{P}(E|B_3) = 0.8$.

Let us utilize Bayes' formula to calculate the equipment malfunction probabilities, $\mathscr{P}(B_i|E)$, for the various mechanical operations.

$$\mathscr{P}(B_1|E) = \frac{0.04(0.5)}{0.04(0.5) + 0.41(0.1) + 0.55(0.8)}$$

$$= \frac{0.02}{0.501} = 0.0399$$

$$\mathscr{P}(B_2|E) = \frac{0.41(0.1)}{0.501} = 0.0818$$

$$\mathscr{P}(B_3|E) = \frac{0.55(0.8)}{0.501} = 0.8782.$$

Thus we conclude that we are probably getting ineffective filament collation.

PROBLEMS

2.1 In how many ways could this department's two $500 and two $250 freshman scholarships have been awarded to the 21 applicants last spring?

2.2 If you are taking three of the 11 departmental courses offered this semester, in how many ways could you have selected your course load in this curriculum?

2.3 In how many ways can a student select an engineering major from the seven engineering sciences on our campus and a minor from the remaining eight sciences?

2.4 In how many ways can the eight experiments in Ch.E. 3351 be assigned to the two groups in each section if no group can do more than one experiment in the allotted time and if equipment limitations are such that the groups cannot do the same experiment at the same time?

2.5 Our sophomore laboratory course in physical property measurement techniques has four groups in each section. In how many ways can the 10 experiments be arranged so that only one group is assigned the same experiment at any given lab period? There are four students in each group.

2.6 A single die is tossed. What is the probability that the outcome will be a 1 or a 2?

2.7 The die is tossed again. For this toss, what is the probability that an even number or a 1 or 2 will be on the upper face?

2.8 This time, two dice are cast simultaneously. What is the probability that the sum of the numbers on their upper faces will be 7 or 8?

2.9 The same die is tossed twice.

(a) What is the probability of getting 7 by rolling, in order, a 2 followed by a 5?

(b) How does this compare with the probability of rolling 7 by any combination if two dice are thrown together?

2.10 A balanced die is fairly tossed. Let the event A denote the occurrence of an odd number, B the occurrence of an even number, C the occurrence of a number greater than 3, D the occurrence of a number less than 3, and E the occurrence of the number 3.

(a) Find $\mathscr{P}(A)$, $\mathscr{P}(B)$, and $\mathscr{P}(D)$.

(b) Which of the events are equally likely?

(c) Which of the events are mutually exclusive?

(d) Find the probability of "event C or event D."

2.11 Cut out the letters, one per small card, of the word ENGINEERING, place them in an empty ball mill, and tumble thoroughly for 5 min. Open the mill and draw out seven letters, one at a time. What is the probability of drawing, in the correct order, the letters of the word GINNING?

2.12   In your class of 12, there will be 3 A's, 3 B's, 4 C's, 1 D, and 1 F at the end of the semester. Assuming that the grades are not curved in any way, what is the chance of your earning:

(a)   A B or a C?
(b)   At least a C?

2.13   The results of the rigorous examination of a dozen automobile tires showed the following: one tire was perfect, three had only slight flaws in appearance, two had incompletely formed treads, one had a serious structural defect, and the rest had at least two of these defects. What is the probability that the next set of four tires you buy of this particular brand:

(a)   Will be perfect?
(b)   Will have at most only undesirable appearances?
(c)   Will have fewer than two defects?

2.14   The storeroom has received 19 dozen disposable pipettes and 1 dozen class A pipettes (definitely not disposable). The new stockboy, not realizing that the pipettes were of two different varieties, yesterday opened all 20 dozen packets, mixed them, and placed all the pipettes in the same supply bin. What are the chances that you will receive one, two, or five class A pipettes when you check out a dozen tomorrow?

2.15   The probability of our ancient mass spectrometer operating satisfactorily for a month is 0.4. The probability of the oscilloscope performing adequately during that period is 0.9. What is the probability of:

(a)   Both instruments needing a service call this month?
(b)   One needing a service call?

2.16   The probability of a microsyringe being broken by a student in our analytical instrumentation course is $\frac{1}{8}$. What is the probability of survival of any particular syringe for the current semester? Assume that 30 students are enrolled and that they are arranged into three lab sections each having two groups of five students.

2.17   The chances of making a flawed pellet for the infrared examination of solid samples is 0.8. What is the probability of making four usable pellets in one 3-h lab period? Approximately 15 min is needed to make each pellet.

2.18   A testing organization wished to rate a particular brand of table radios. Five radios are selected at random from stock and the brand is judged satisfactory if nothing is wrong with any of the five.

(a) What is the probability that the brand will be rated as satisfactory if 10% of the radios are actually defective?

(b) What is the probability of getting one or fewer defective from the sample if 10% of the stock radios are defective?

2.19 A company manufactures a lubricant that must pass an accelerated heat stability test. Past records indicate that the normal percentage of failures is 5%. During one month of operation, 120 batches were produced and 11 were found defective from the heat stability standpoint. What do you conclude after comparing the proportions of failures?

2.20 A lot contains 1400 items. A sample of 200 items is taken and the lot accepted if no more than two defective items appear in the sample. If the lot actually contains 28 defectives, what is the probability of its acceptance? (*Hint*: Use the binomial distribution.)

2.21 The following numbers are obtained from a mortality table based on 100,000 people.

Age	Deaths per 1000 during year
17	7.688
18	7.727
19	7.765
20	7.805
21	7.855

If these numbers are used to define probabilities of death for the corresponding age group and if *A*, *B*, and *C* denote persons of ages 17, 19, and 21, respectively, calculate the probability that during the year:

(a) *A* will die and *B* will live.

(b) At least one of *A*, *B*, and *C* will die

(c) *A* and *B* will both die.

2.22 When buying flasks for the storeroom, the clerk has the habit of checking three flasks out of each case of 24 for chipped lips, cracks, and so on. If any defects are found in this sample of three, the entire case is rejected. Assuming that each possible combination of three flasks had the same chance of being selected as the sample, what is the probability that a case of flasks containing three defective units will be accepted?

2.23   Box *A* contains three white and four black balls. Box *B* contains two white and three black balls. Balls are drawn (at random) without replacement.

    (a)   If two balls are chosen, one from each box, what is the probability that they will be of the same color?

    (b)   If two balls are chosen from box *A*, what is the probability that both balls will be white?

    (c)   If a box is selected (at random) and two balls drawn from it, what is the probability that they will be of the same color?

2.24   Two balls are to be drawn, without replacement, from an urn containing two white, three black, and four green balls.

    (a)   What is the probability that both balls will be green?

    (b)   What is the probability that both balls will be of the same color?

2.25   A relatively new batch process is still producing 10% defective (off-grade) material.

    (a)   What is the probability of producing no off-grade material in the next eight batches to be produced?

    (b)   What is the probability of more than one defective batch in the next eight produced?

    (c)   What is the expected number of defective batches in 30, assuming that the batches are independent?

    (d)   Would you say that zero off-grade batches in a successive series of 30 production runs indicate a change in the process? What is the probability of being wrong in so stating?

2.26   A sample of 20 independent pieces is taken from a production line and none of the pieces is defective.

    (a)   What can be said about the percentage defective in the process?

    (b)   If the actual percentage defective in the process is $p = 2\%$, what is the probability of the foregoing event?

    (c)   What is the probability if $p = 4\%$?

2.27   In the manufacture of a modified natural resin, the three key product properties are melting point (*A*), color (*B*), and acid number (*C*), each of which is independent of the others. Marketing suggests sales specifications of minimum melting point of 130°C, maximum color of 3 (arbitrary scale), and maximum acid number of 12. Analyses of available data indicate that the probabilities of material being off-grade with respect to each of these properties are $\mathscr{P}(A) = 0.03$, $\mathscr{P}(B) = 0.05$, and $\mathscr{P}(C) = 0.02$. Assuming no process changes:

(a) What percentage of the resin produced will be off-grade?
(b) For those customers for which color is unimportant, what percentage off-grade is to be expected?
(c) If off-grade material is valued at 20 cents/lb and first-grade material (on specification with respect to all properties) is priced at 50 cents/lb, what reduction in price can be given to non-color-conscious customers?

2.28 In the manufacture of plastic objects by an injection-molding process, it is found that on the average 5% are oversize as to length, 91% satisfy the length specification, and 4% are undersize as to length. With regard to thickness specification, 2% are undersize, 96% are on specification, and 2% are oversize. Assume that length and thickness are independent. In a random sample of 10 independent pieces, what is the probability of finding:

(a) All good pieces?
(b) Exactly one piece oversize as to length?
(c) Eight pieces or fewer on specification with respect to both length and thickness?

REFERENCES

1. Barlow, R. E., J. B. Fussell, and N. D. Singpurwalla, *Reliability and Fault Tree Analysis*, Society for Industrial and Applied Mathematics, Philadelphia (1975).
2. Center for Chemical Process Safety, *Guidelines for Process Equipment Reliability Data*, American Institute of Chemical Engineers, New York (1989).
3. Center for Chemical Process Safety, *Guidelines for Chemical Process Quanatitative Risk Analysis*, American Institute of Chemical Engineers, New York (1989).
4. Crowl, D. A., and J. F. Louvar, Hazards Identification, Ch. 11 (pp. 338–365), in *Chemical Process Safety: Fundamentals with Applications*, Prentice Hall, Englewood Cliffs, NJ (1990).
5. Greenberg, H. R., and B. Salter, Fault tree and event tree analysis, Ch. 9 (pp. 127–166) in H. R. Greenberg and J. J. Cramer (eds.), *Risk Assessment and Risk Management for the Chemical Process Industry*, Van Nostrand Reinhold, New York (1991).
6. Gressel, M. G., and J. A. Gideon, An overview of process hazard evaluation techniques, *Am. Ind. Hyg. Assoc. J. 52*(4): 158–163 (1991).
7. Hammer, W., Safety analyses, Ch. 26 (pp. 551–566), in *Occupational Safety Management and Engineering*, 4th ed., Prentice Hall, Englewood Cliffs, NJ (1989).

8.   Lees, F. P., Hazard assessment, Ch. 9 (pp. 175–210), in *Loss Prevention in the Process Industries*, Vol. 1, Butterworth, Boston (1980; reprinted with corrections, 1989).

9.   Prugh, R. W., Application of fault tree analysis, *Chem. Eng. Prog.* 76(7): 59–66 (1990).

10. Rankine, J. E., and G. O. Tolley, *Fault Tree Analysis*, MSHA Safety Manual 8, U.S. Dept. of Labor, Washington, DC (repr. 1986).

3

Distributions

3.1 INTRODUCTION

Most statistical methods useful to engineers and scientists deal with the collection, organization, analysis, and presentation of data. Such analyses of experimental data are used in making reasonable decisions that are based at least partly on the data.

Application of the methods of statistics requires an understanding of data and its characteristics, the most common of which are average and variation. These characteristics are reflected by distributions, either experimental (empirical) or theoretical. Most statistical methods are based on theoretical distributions that approximate the actual distributions. Thus the user of statistical methods must be familiar not only with the theoretical distributions but also with the distribution characteristics of the population(s) under consideration.

3.2 DEFINITIONS

Certain terms relative to distributions have special meanings in statistical usage. The most common of these terms are:

Population: a collection of objects that have at least one characteristic (attribute) in common
Sample: a subset of a population

Observation: a recording of information on some characteristic of an object

Measurement: a numerical value indicating the extent, intensity, or size of a characteristic of an object

Sample space: the set of all possible outcomes of an experiment, that is, all possible results of a process that generates data

Random variable: a function from a sample space to the real line, that is, a function that assigns a real value to each outcome in a sample space

Discrete variable: a variable that can assume only isolated values, that is, values in a finite or countably infinite set

Continuous variable: a variable that can assume any value between two distinct numbers

Cumulative frequency: the sum of the frequencies of all values less than or equal to a particular value

All of these terms will be used in the ensuing sections, where their meanings will become increasingly clearer to the reader. However, since the terms *population* and *sample* are quite often confused with each other, a word of clarification is in line here. A population can be considered to be the set of *all* objects that have at least one characteristic in common: for example, the set of all engineers in the world. A sample is a subset of the population: for example, all the engineers living in Texas.

It should also be noted that a random variable is just a "rule" that associates a real number with each outcome in the sample space. Thus in tossing two coins, $S = \{(H,H), (H,T), (T,H), (T,T)\} = \{s_1, s_2, s_3, s_4\}$. Let X denote the number of heads so that $X(s_1) = 2$, $X(s_2) = X(s_3) = 1$, $X(s_4) = 0$. Uppercase letters such as X, Y, Z will denote random variables, and the corresponding lower case letters x, y, z will denote values of the random variables.

3.3 THEORETICAL DISTRIBUTIONS

Frequency distributions are useful largely as a means of indicating which type of theoretical distribution best describes the statistical properties of the population under consideration. For many scientific and engineering applications, the normal (or Gaussian) distribution (discussed later) is an appropriate model. However, many cases can be expected in which the frequency distribution is strongly skewed. Application of statistical methods based on the Gaussian distribution to such cases will lead to erroneous results and conclusions. Often, careful study of the nature of the physical situation from which the data are obtained will indicate the type distribution followed by the data. Before the statistical analysis of any experimental data is attempted, the data should be plotted appropriately. Often, this will give a strong indication of the nature of the relationship involved.

The three most commonly used theoretical distributions are the binomial and Poisson, both of which are discrete distributions, and the normal (or Gaussian) distribution, which is continuous. Before discussing some important theoretical distributions, we define two characteristics associated with a distribution. In general, a characteristic associated with a distribution is called a *parameter*, a numerical value associated with a theoretical distribution. The two parameters we consider here are the *mean* and *variance*, the former being a measure of location and the latter a measure of variation. Since the theoretical probability distribution gives a complete description of the corresponding random variable, we will call the two parameters the mean and variance of the random variable X or, equivalently, of the probability distribution. The parameters of a probability distribution, be it discrete or continuous, will be denoted by Greek letters, for example μ and σ.

Suppose that a random variable X can take on the values of a discrete (finite) set $S = \{x_1, x_2, \ldots, x_n\}$ according to the probability function $f(x_i) = 1/n$, $i = 1, 2, \ldots, n$. The mean, denoted by μ, of the discrete random variable X is defined by

$$\mu = \frac{\sum_{i=1}^{n} x_i}{n} = \sum_{i=1}^{n} x_i f(x_i), \tag{3.1}$$

since $f(x_i) = 1/n$. If X can assume the values of the set $S = \{x_1, x_2, \ldots, x_n\}$, which may be countably infinite, according to the probability function with values $f(x_1), f(x_2), \ldots, f(x_n)$, the mean of X is given by

$$\mu = \sum_{i=1}^{n} x_i f(x_i). \tag{3.2}$$

It should be noted that the mean in (3.1) is the usual arithmetic average of the numbers $x_1, x_2, \ldots, x_n$. The mean in (3.2) is the usual weighted arithmetic average of $x_1, x_2, \ldots, x_n$. The probability function in each case is theoretical, which is not known in general. This prompts use of the term *expectation* or *expected value*. We say the mean μ is the expectation of X or the expected value of X, and denote this operation by

$$\mu = E(X) = \sum_{i=1}^{n} x_i f(x_i). \tag{3.3}$$

The second parameter of interest is the variance of X, denoted by σ^2.

The variance of the discrete random variable X is the expected value of $(X - \mu)^2$, or

$$\sigma^2 = E[(X - \mu)^2] = \sum_{i=1}^{n} (x_i - \mu)^2 f(x_i). \tag{3.4}$$

The variance of X is simply the weighted average of $(x_1 - \mu)^2, (x_2 - \mu)^2, \ldots, (x_n - \mu)^2$, with respect to the weights $f(x_1), f(x_2), \ldots, f(x_n)$.

Since a continuous variable may assume any value between two distinct numbers, the function describing a continuous distribution must be continuous and the geometric representation of the relationship between the probability density function $f(x)$ and x must be a continuous curve. While measurement of a continuous variable may yield data that appear discrete because of the manner in which data are recorded, the data should be handled as a sample from a continuous distribution. For example, temperature is a continuous variable even though the thermometer readings are recorded only to the nearest degree, giving a series of recordings that appear discrete. The probability density function of a continuous random variable has the following properties:

(a) $f(x)$, the probability density function is single valued and non-negative, that is, $f(x) \geq 0$ for all x.

(b) $\displaystyle\int_{-\infty}^{+\infty} f(x)\, dx = 1.$ $\tag{3.5}$

(c) $\displaystyle\int_{x=a}^{x=b} f(x)\, dx = \mathscr{P}(a < X \leq b).$ $\tag{3.6}$

Thus the probability of a value x falling between the values a and b is geometrically represented as the area under the probability density function between the limits a and b as shown in Fig. 3.1. The mean μ and variance

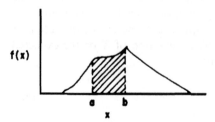

f(x)

FIGURE 3.1 Probability density function.

σ^2 of a continuous variable X are given by

$$\mu = E(X) = \int_{-\infty}^{\infty} xf(x)\, dx \tag{3.7}$$

and

$$\sigma^2 = E[(X - \mu^2] = \int_{-\infty}^{\infty} (x - \mu)^2 f(x)\, dx. \tag{3.8}$$

The positive square root of σ^2 is called the *standard deviation* of X and is denoted by σ. The definitions (3.7) and (3.8) are suggested by similar considerations that lead to the definition of the Riemann integral.

The parameters μ and σ^2 are referred to as the population mean and variance, respectively. They are generally unknown and information about them, in the form of inferences, is obtained by considering a sample from the appropriate population. This leads to the sample analogs of μ and σ^2, which are called the *sample mean* $\bar{X}$ and *sample variance* S^2 or S_x^2. These ideas are discussed in more detail in Chapter 4. However, we note in passing that $\bar{X}$ and S^2 are computed on the basis of the observations in the sample, which represents the corresponding population, according to (3.1) and (3.4), with $f(x_i) = 1/n$ and $f(x_i) = 1/(n-1)$, respectively. The reason for this will be pointed out later. The quantities $\bar{X}$ and S^2 are used in Section 3.5.1, which deals with the normal distribution.

3.4 DISCRETE DISTRIBUTIONS

Discrete distributions are used to describe count or enumeration data. Examples of such data are the number of attendees at the next Sigma Xi luncheon, the score of the next Super Bowl game, and the number of heads in n flips of a coin. All data of this type have one characteristic in common: the number of values in each case is finite. If one considers the random variables associated with the sample spaces of these examples, there are measurable step changes between values of the random variables involved. Consequently, there is a specific, fixed probability associated with each value of the random variable. The function relating these probabilities to the values of the random variable is called the *probability function*.

Assume that the random variable X may take on any value of the finite, ordered, discrete set of values $S = \{x_1, x_2, x_3, \ldots, x_n\}$, where n is finite or countably infinite. Let E be any subset of S; that is, let E be an event. The probability of E, $\mathscr{P}(E)$, is a real, nonnegative number. In the case of a finite sample space S with equally likely x_i's, if f_i is the frequency of occurrence of x_i and N is the sum of the f_i, the probability function, also called the *point probability function*, $f(x_i)$, can be expressed as

$$\mathscr{P}(X = x_i) = f(x_i) = \frac{f_i}{N} = \frac{f_i}{\sum_{j=1}^n f_j} \tag{3.9}$$

and will have the following properties:

(a) $f(x_i)$ is positive and real for each x_i.

(b) $\sum_{i=1}^{n} f(x_i) \equiv 1.$

(c) $\mathscr{P}(E) = \sum f(x_i)$ where the sum is taken over all those x_i that are in E.

Any function $f(x_i) = \mathscr{P}(X = x_i)$ satisfying properties (a), (b), and (c) is said to be a *probability function*. A random variable corresponding to a point probability function is called a *discrete random variable*. For a die, the probability $f(x_i)$ for getting any particular value on a single toss is $\frac{1}{6}$. A plot of the point probability function is shown in Fig. 3.2.

Corresponding to each probability function is a function, $F(x)$, which is called the *cumulative distribution function*. $F(x)$ can be obtained from the probability function according to the following definition. For $x_1 < x_2 < x_3 < \cdots < x_n$, $F(x)$ is defined by

$$F(x) = \mathscr{P}(X \le x) = \begin{cases} 0 & \text{for } x < x_1 \\ \sum_{i=1}^{r} f(x_i) & \text{where } x_r \le x < x_{r+1} \\ 1 & \text{for } x \ge x_n. \end{cases} \tag{3.10}$$

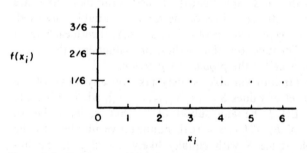

FIGURE **3.2** Point probability function for tossing one die.

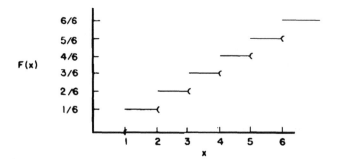

Figure 3.3 Cumulative distubution function for tossing one die.

The cumulative distribution function is a nondecreasing function with the following properties:

(a) $0 \le F(x) \le 1$.

(b) $\lim_{x \to -\infty} F(x) = 0$.

(c) $\lim_{x \to \infty} F(x) = 1$.

The cumulative distribution function $F(x)$ corresponding to the probability function $f(x_i)$ illustrated above is graphed in Fig. 3.3 It should be noted that $F(x)$ takes a jump at $x = 1, 2, \ldots, 6$. This is indicated by a small tick on the graph. For instance, $F(2) = \frac{2}{6}$ and $F(x) = \frac{1}{6}$ for $1 \le x < 2$.

It follows from the definition of the cumulative distribution function that

$$\mathscr{P}(x_1 < X \le x_2) = F(x_2) - F(x_1). \tag{3.11}$$

In the next section we consider an extension of the cumulative distribution function which is useful when the outcomes of an experiment are not countable.

3.4.1 Binomial Distribution

When an observation is placed in one or the other of two mutually exclusive categories, as is often the case in practice, a discrete distribution called the *binomial* or *dichotomous distribution* results. Although the variable is usually considered qualitative (yes or no, pass or fail, dead or alive), the distribution may be made quantitative by assigning the values 0 and 1 to the two categories. It makes no difference how these are assigned. The two categories are usually labeled *success* and *failure* and are assigned the values 1 and 0, respectively.

Let p denote the probability of success (the event occurring) and q the probability of failure (the event not occurring) in one trial of an experiment. Since these events are mutually exclusive and exhaustive, $q = 1 - p$. If the probability of a success is the same for each independent trial of the experiment, the probability of exactly x successes in n trials is given by the expression

$$f(x) = \frac{n!}{x! \, (n - x)!} \, p^x q^{n-x}, \qquad x = 0,1,2, \ldots ,n, \qquad (3.12)$$

where $n!/x! \, (n - x)!$ is the number of combinations of n things taken x at a time. It is recognized that Eq. (3.12) is the general term of the familiar binomial expansion in which $x = 0,1,2, \ldots ,n$, and the factor $n!/x! \, (n - x)!$ is the binomial coefficient of the xth term.

Table I of Appendix B contains values of the binomial cumulative distribution function

$$\mathcal{P}(X \leq x) = \sum_{k=0}^{x} \binom{n}{k} p^k (1 - p)^{n-k}, \qquad x = 0,1,2, \ldots ,n,$$

for selected values of n and p. By means of this table one can compute other probabilities, such as $\mathcal{P}(x_1 \leq X \leq x_2)$, since

$$\mathcal{P}(x_1 \leq X \leq x_2) = \mathcal{P}(X \leq x_2) - \mathcal{P}(X \leq x_1 - 1)$$

$$= \sum_{k=0}^{x_2} \binom{n}{k} p^k (1 - p)^{n-k} - \sum_{k=0}^{x_1 - 1} \binom{n}{k} p^k (1 - p)^{n-k}.$$

If we plot $f(x)$ for a fixed number of observations, say $n = 8$, the shape of the resulting graphs changes with p and q as shown in Fig. 3.4, where each $f(x)$ is associated with a fixed value of n. The first of these, (a), is markedly skewed, the second, (b), is only slightly skewed, and the third, (c), is symmetric. As we will see, as n becomes large while the value of p stays fixed, the binomial distribution corresponding to (3.12) approaches the normal distribution and then reverts to skewness as p increases.

The mean and variance of the binomial distribution are given by

$$\mu = np = E(X) \qquad (3.13)$$

and

$$\sigma^2 = npq = E[(X - \mu)^2]. \qquad (3.14)$$

Example 3.1 The probability that a compression ring fitting will fail to seal properly is 0.1. What is the expected number of faulty rings and their variance if we have a sample of 200 rings?

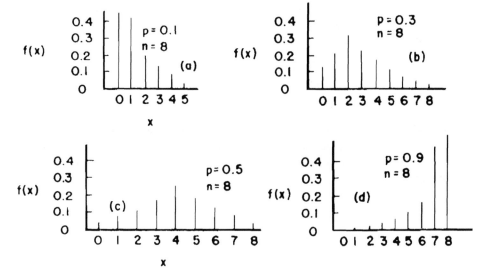

FIGURE **3.4** Binomial probability functions.

Solution Assuming that we have a binomial distribution, we have $\mu = np = 0.1(200) = 20$ and $\sigma^2 = npq = 200(0.1)(0.9) = 18$.

The special case of the binomial distribution when $n = 1$ is commonly called the *point binomial* or *Bernoulli distribution* and is completely defined by one parameter p (or $q = 1 - p$). Obviously, for the point binomial, $\mu = p$ and $\sigma^2 = pq$.

Specific binomial probabilities, such as the probability that there will be exactly b successes in n trials, may be calculated from Table I by considering

$$\mathscr{P}(X = b) = \mathscr{P}(X \le b) - \mathscr{P}(X \le b - 1). \tag{3.15}$$

Example 3.2 Past history indicates that the probability of a given batch of chemical being on-grade is 0.90. Let X denote the number of off-grade batches. What is the probability of getting exactly one off-grade batch in 10 runs?

Solution From Table I, with $n = 10$ and $p = 1 - 0.90 = 0.10$, we obtain

$$\mathscr{P}(X = 1) = \mathscr{P}(X \le 1) - \mathscr{P}(X = 0)$$
$$= 0.736 - 0.349$$
$$= 0.387.$$

Example 3.3   For the case of Example 3.2, what is the probability of getting more than two off-grade batches in 10 runs?

Solution   From Table I we obtain

$$\mathscr{P}(X > 2) = 1 - \mathscr{P}(X \le 2)$$
$$= 1 - 0.930$$
$$= 0.070.$$

Example 3.4   The probability that an entering freshman at one of the Ivy League schools will graduate is 0.4. Of any group of six entering freshmen, what are the probabilities that (a) one will graduate and (b) at least one will graduate?

Solution   Let X denote the number that will graduate. From Table I we get

(a) $\mathscr{P}(X = 1) = \mathscr{P}(X \le 1) - \mathscr{P}(X = 0)$
$$= 0.233 - 0.047$$
$$= 0.186.$$

(b) $\mathscr{P}(X \ge 1) = 1 - \mathscr{P}(X = 0)$
$$= 1 - 0.047$$
$$= 0.953.$$

Example 3.5   Twenty percent of the items produced in a certain stamping process are defective. Here $p = 0.2$ since we associate success with obtaining a defective. Let X denote the number of defectives in a randomly chosen sample of 8.

    (a)   What is the probability that zero, one, or two items in the randomly chosen sample will be defective?

    (b)   What is the probability of obtaining at least two defective items?

    (c)   What is the probability of obtaining at most two defective stampings?

Solution

    (a)   From Table I we obtain

$$\mathscr{P}(X = 0, 1, \text{ or } 2) = \mathscr{P}(X \le 2)$$
$$= 0.797.$$

    (b) $\mathscr{P}(X \ge 2) = 1 - \mathscr{P}(X \le 1)$
$$= 1 - 0.503$$
$$= 0.497.$$

    (c) $\mathscr{P}(X \le 2) = 0.797.$

3.4.2 Poisson Distribution

The relation between the Poisson and binomial distributions is quite simple. To begin with, let $\lambda = np$, where n is the number of observations (occurrences) and p is the probability of success. Thus $p = \lambda/n$ and $q = 1 - \lambda/n$. The binomial distribution can then be written as

$$p(x) = \binom{n}{x}\left(\frac{\lambda}{n}\right)^x\left(1 - \frac{\lambda}{n}\right)^{n-x}. \tag{3.16}$$

If n approaches infinity and p approaches zero in such a way that np is constant, the binomial distribution approaches the Poisson distribution with probability function $f(x)$ given by

$$f(x) = \frac{\lambda^x e^{-\lambda}}{x!}, \qquad x = 0,1,2,\ldots, \tag{3.17}$$

where $f(x)$ approximates the probability of x successes and e is the base of the natural logarithm, 2.71828. The cumulative Poisson distribution function $F(x)$ is given by

$$F(x) = \sum_{k=0}^{x} \frac{e^{-\lambda}\lambda^k}{k!}, \tag{3.18}$$

where the parameter λ is the mean and also the variance of the Poisson. Small values of λ yield skewed probability functions as shown in Fig. 3.5. As in the case of the binomial distribution, the Poisson approaches the normal distribution as λ increases. The Poisson distribution function $F(x)$ given by (3.18) is given in Table II of Appendix B for selected values of x and λ.

From the way the Poisson distribution was introduced as a limiting case of the binomial, we see that for small p and large n it could be used as a relatively good approximation to the binomial distribution. In general, if $p \leq 0.1$ and $n \geq 20$, the Poisson distribution can be used to approximate the binomial without introducing any significant errors. This is illustrated in the examples that follow.

Example 3.6 In the spray coating of aluminum test panels with paint for use in weathering tests, 8% of the panels are unusable because of improper degreasing prior to paint application. For a sample of 20 such panels, what is the probability of having two defectives? Let X denote the number of defectives.

Solution A binomial distribution is appropriate as a model since the panels are either defective or suitable for use, and the degreasing of each panel can be assumed to be independent of the degreasing of every other panel. Hence

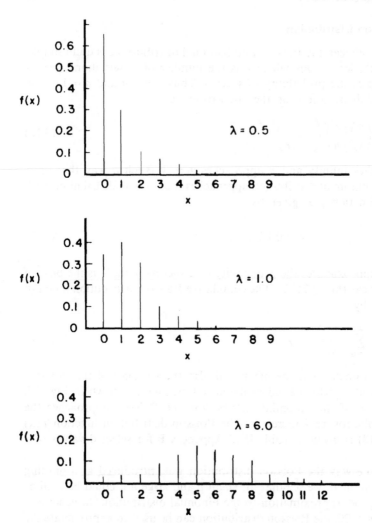

FIGURE 3.5   Poisson probability functions.

the desired probability, obtained from the binomial with $p = 0.08$ and $n = 20$, is

$$\mathscr{P}(X = 2) = \frac{20!}{2! \ 18!} (0.08)^2 (0.92)^{18}$$

$$= 190(0.0064)(0.92)^{18} = 0.2711.$$

[Table I does not cover the value of $p = 0.08$. In cases such as this, one has to resort to more extensive tables or evaluate the desired quantities such as $(0.92)^{18}$ by means of logarithms or with a pocket calculator.] However, since p is small and n is large, the Poisson with $\lambda = np = 20(0.08) = 1.6$ can be used to get

$$\mathscr{P}(X = 2) = \mathscr{P}(X \leq 2) - \mathscr{P}(X \leq 1)$$
$$= 0.783 - 0.525$$
$$= 0.258.$$

Thus we see that the error involved in using the Poisson to approximate the binomial is about 4.8%, which may be significant or not, depending on the requirements of the experiment.

Example 3.7 It is known that three people out of every 2000 have an adverse reaction to a particular drug. What are the probabilities of (a) three adverse reactions and (b) fewer than two adverse reactions in a group of 1000 test subjects? Let X denote the number of adverse reactions.

Solution (a) The appropriate model is again the binomial with $p = 0.0015$ and $n = 1000$. Table I does not cover these values of p and n. The probability of three adverse reactions is given by

$$\mathscr{P}(X = 3) = \frac{1000!}{997!\ 3!} (0.0015)^3 (0.9985)^{997} = 0.12555.$$

Since calculations of this type are tedious using logarithms, and since the Poisson is applicable, we can obtain a close approximation to the probability above by calculating $\mathscr{P}(X = 3)$ from the Poisson with $\lambda = np = 1000(0.0015) = 1.5$. From Table II we have

$$\mathscr{P}(X = 3) = \mathscr{P}(X \leq 3) - \mathscr{P}(X \leq 2)$$
$$= 0.934 - 0.809$$
$$= 0.125.$$

(b) Since the error in using the Poisson approximation is minute in this case, we can estimate the probability of getting fewer than two adverse reactions using the Poisson as follows:

$$\mathscr{P}(X < 2) = \mathscr{P}(X \leq 1)$$
$$= 0.558.$$

Another approach to the Poisson distribution that motivates its use in physical problems follows.

Let the probability that a certain event E occurring in a time interval $(t, t + dt)$ be $k\ dt$, where k is a constant. Also assume that the length of the time interval $(t, t + dt)$ is so small that the probability of the event occurring more than once in the time interval is very small and this probability may be neglected. In addition, assume that the occurrence of E in nonoverlapping time intervals are independent events. Now the problem is to determine the probability $\mathscr{P}_x(t)$ that the event E will occur x times within a time t. Note that $\mathscr{P}_0(t + dt)$ is the probability that event E will not occur in the time period $(0, t + dt)$. This means that the event E does not occur in the time interval $(0, t)$, nor in the interval $(t, t + dt)$. The probability of the former is $\mathscr{P}_0(t)$ and of the latter is $(1 - k \, dt)$. Since we assumed that these probabilities are independent, we have

$$\mathscr{P}_0(t + dt) = \mathscr{P}_0(t)(1 - k \, dt) = \mathscr{P}_0(t) - k\mathscr{P}_0(t) \, dt,$$

$$\frac{\mathscr{P}_0(t + dt) - \mathscr{P}_0(t)}{dt} = -k\mathscr{P}_0(t).$$

The limit of the left-hand side as dt approaches 0 is just the derivative of $\mathscr{P}_0(t)$ with respect to t. Thus a differential equation is obtained, namely,

$$\frac{d\mathscr{P}_0(t)}{dt} = -k\mathscr{P}_0(t),$$

whose solution is $\mathscr{P}_0(t) = e^{-kt}$, as can easily be checked, noting that $\mathscr{P}_0(0) = 1$.

Now to determine the probability that the event E will occur x times in the interval $(0, t + dt)$, we know that this can happen by the event E occurring $x - 1$ times during $(0, t)$ and once during $(t, t + dt)$ or E occurring x times during $(0,t)$ and not occurring during $(t, t + dt)$. This is illustrated in Fig. 3.6.

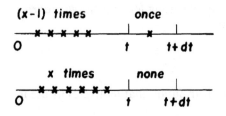

Figure 3.6 Possible ways event E can occur x times during time interval $(0, t + dt)$.

Since these two possibilities are mutually exclusive events, we have

$$\mathscr{P}_x(t + dt) = \mathscr{P}(E \text{ occurs exactly } x \text{ times within time } t \text{ and } 0 \text{ times in } dt)$$
$$+ \mathscr{P}(E \text{ occurs } x - 1 \text{ times within time } t$$
$$\text{and } 1 \text{ time in time } dt)$$
$$= \mathscr{P}(E \text{ occurs } x \text{ times within time } t)\mathscr{P}(E \text{ occurs } 0 \text{ times in } dt)$$
$$+ \mathscr{P}(E \text{ occurs } x - 1 \text{ times within time } t)$$
$$\times \mathscr{P}(E \text{ occurs } 1 \text{ time in } dt)$$
$$= \mathscr{P}_x(t)(1 - k\,dt) + \mathscr{P}_{x-1}(t)k\,dt.$$

Rearranging yields

$$\frac{\mathscr{P}_x(t + dt) - \mathscr{P}_x(t)}{dt} = -k\mathscr{P}_x(t) + k\mathscr{P}_{x-1}(t).$$

The limit of the left-hand side is just the derivative of $\mathscr{P}_x(t)$, and the resulting differential equation is

$$\frac{d\mathscr{P}_x(t)}{dt} = -k\mathscr{P}_x(t) + k\mathscr{P}_{x-1}(t).$$

A solution, subject to the conditions $\mathscr{P}_0(0) = 1$ and $\mathscr{P}_x(0) = 0$, for $x > 0$, is

$$\mathscr{P}_x(t) = \frac{(kt)^x e^{-kt}}{x!},$$

as can readily be verified. Therefore, the probability function of the number of times the event E occurs in the time interval $(0, t)$ is given by $\mathscr{P}_x(t)$ with parameter kt.

The derivation above describes a general physical situation that leads to the Poisson distribution. From the approach above we say that if the probability of each radioactive atom in a mass disintegrating is constant, the number of atoms disintegrating in a time period t has the Poisson distribution. Also, the Poisson may apply to the number of phone calls per time interval during a given part of a day, to the number of flaws per yard of insulated wire, to the number of misprints per page, to the number of blood cells on individual squares on a hemocytometer, and so on. Another specific example is the distribution of the number of failures, x, of electronic equipment within a time t assuming independent failures and a constant failure rate.

3.4.3 Negative Binomial Distribution

Assume that the assumptions necessary for a binomial distribution are satisfied; that is, each performance of an experiment results in one of two mutually exclusive outcomes S and F, the outcome of each performance is independent of other performances, and the probabilities of S and F remain fixed at p and q, respectively, for each performance.

In Section 3.4.1 we observed that when the foregoing assumptions were satisfied, the binomial probability function yielded the probability of having S occur exactly x times out of n performances of the experiment. We now consider the situation where n is not fixed. Since the number of performances of the experiment can vary, we will call it x. We wish to obtain an expression for the probability that x performances of the experiment will be required to produce exactly r of the outcomes S. This expression can be obtained from the binomial by reasoning as follows. The only way that exactly r outcomes will be S, out of x performances of an experiment, is for exactly $(r - 1)$ outcomes to be S, out of the first $(x - 1)$ performances of the experiment, and the last performance to be an S. Therefore, letting X be the number of trials needed to produce r of the outcomes S and setting $\mathscr{P}(X = x) = \mathscr{P}_r(x)$, we have

$$\mathscr{P}_r(x) = 0 \qquad \text{for } x < r,$$
$$\mathscr{P}_r(x) = \mathscr{P}(r - 1 \text{ out of } x - 1 \text{ are } S \text{ and the } x\text{th outcome is } S)$$
$$= \mathscr{P}(r - 1 \text{ out of } x - 1 \text{ are } S) \cdot \mathscr{P}(x\text{th outcome is } S).$$

Since the assumptions necessary for a binomial are satisfied, we have

$$\mathscr{P}(r - 1 \text{ out of } x - 1 \text{ are } S) = \binom{x - 1}{r - 1} p^{r-1} q^{x-r}.$$

Also, $\mathscr{P}(x\text{th outcome is } S) = p$, so that

$$\mathscr{P}_r(x) = \binom{x - 1}{r - 1} p^r q^{x-r}, \qquad x \geq r. \tag{3.19}$$

The probability function $\mathscr{P}_r(x)$ is called the *negative binomial*.

Example 3.8   Suppose that light bulbs are being tested for instant failure. The bulb is turned on, and if it burns out immediately, the bulb is called *defective*, which we choose to call success. Otherwise, it is classified as *nondefective*. It is realistic to assume that the performance of each bulb is independent of the performance of the others whenever the production process is under control. Suppose further that it has been established over a long period of time that the manufacturing process produces 1% defective bulbs. What is the probability that 10 bulbs need to be tested to get exactly one defective?

Solution   In this case $x = 10$, $r = 1$, and $p = 0.01$. Thus, letting X denote the number of trials needed, we have

$$\mathcal{P}(X = 10) = \mathcal{P}_1(10) = \binom{9}{0}(0.99)^9(0.01)^1 \simeq 0.08.$$

3.4.4 Hypergeometric Distribution

The hypergeometric distribution is often the model used to obtain probabilities where the sampling is done without replacement. The ensuing
example of its use will make it clear when one should consider it as the
appropriate model.

Example 3.9 Suppose that five radiator caps are selected at random from
each lot of 50 caps and tested to determine if they hold pressure to within
the prescribed limits, as part of a manufacturer's quality-control program.
Assume also that if more than one out of the five caps does not hold the
proper pressure and is thus classified as defective, all 50 caps are tested to
ensure that they are all right. If in a certain lot of 50 caps, 10 are defective,
what is the probability that all 50 will end up being tested?

Solution   This seems to be closely related to the binomial distribution if we
let the testing of a radiator cap be a performance of an experiment, and let
a nondefective cap be termed a success and a defective cap be termed a
failure. We are then just asking for the probability of more than one failure
out of five, which is very much like the general problem solved by the
binomial distribution. However, there is one basic difference that should be
noted. The probability of a success differs from test to test. For instance, for
the first test the probability of a success is $p_1 = \frac{40}{50}$, but for the second test
it will be $p_2 = \frac{40}{49}$ if the first was a failure, or $p_2 = \frac{39}{49}$ if the first was a success,
and so on.

   To obtain the desired probability we have to enumerate. The number
of ways of selecting x defectives from the 10 that are defective is given as
$\binom{10}{x}$ and the number of ways of selecting $5 - x$ nondefectives from the 40
nondefectives is $\binom{40}{5-x}$. Hence the number of ways of selecting x defectives
and $5 - x$ nondefectives is $\binom{10}{x}\binom{40}{5-x}$. Since a sample of size 5 can be
selected in $\binom{50}{5}$ ways from the lot of 50 caps, we have, by letting X denote
the number of defectives,

$$\mathcal{P}(X = x) = f(x) = \frac{\binom{10}{x}\binom{40}{5-x}}{\binom{50}{5}}, \qquad x = 0,1,\ldots,5.$$

Since all 50 caps will be tested if more than one defective is found, the probability that more than one defective is obtained is given by the following probability:

$$\mathcal{P}(X \geq 2) = \sum_{x=2}^{5} \frac{\binom{10}{x}\binom{40}{5-x}}{\binom{50}{5}}$$

$$= \frac{\binom{10}{2}\binom{40}{3}}{\binom{50}{5}} + \frac{\binom{10}{3}\binom{40}{2}}{\binom{50}{5}} + \frac{\binom{10}{4}\binom{40}{1}}{\binom{50}{5}} + \frac{\binom{10}{5}\binom{40}{0}}{\binom{50}{5}}$$

$$= 0.251959.$$

In general, let N be the lot size, n the sample size, D the number of defectives in the lot, and x the number of defectives in the sample; then

$$f(x) = \frac{\binom{D}{x}\binom{N-D}{n-x}}{\binom{N}{n}}, \tag{3.20}$$

where it is understood that $\binom{a}{b}$ is zero whenever $b > a$. The probability function above is called the *hyerpgeometric distribution*.

Example 3.10   In the manufacture of precision tube fittings for high-vacuum service, a sample of 5 is selected from every production run of 100 fittings of any given type and size. If any of the five randomly selected samples is found to be defective (i.e., it will not hold an applied vacuum equivalent to 0.01 psia), all 100 fittings in that particular run are declared suspect. All fittings are then checked individually for vacuum service. The severity of this quality control criterion is indicative of two things: a company's desire to uphold its high reputation and management's belief that their production and testing techniques are excellent. If, out of the next group of 100 fittings to come off the line, three are actually defective, what is the probability that the defect in the processing will be identified and corrected? As we can assume that the defective processing step will be rectified if it can be found, the real question is: What is the probability that one or more of the faulty fittings will be in the quality control sample of 5?

Solution   For this problem, $N = 100$, $n = 5$, and $D = 3$. The probability is found from

$$\mathscr{P}(X \geq 1) = \sum_{x=1}^{5} \frac{\binom{3}{x}\binom{100-3}{5-x}}{\binom{100}{5}}$$

$$= \frac{\binom{3}{1}\binom{97}{4}}{\binom{100}{5}} + \frac{\binom{3}{2}\binom{97}{3}}{\binom{100}{5}} + \frac{\binom{3}{3}\binom{97}{2}}{\binom{100}{5}} + \frac{\binom{3}{4}\binom{97}{1}}{\binom{100}{5}} + \frac{\binom{3}{5}\binom{97}{0}}{\binom{100}{5}}$$

$$= 0.144,$$

for which the last two terms are zero because $\binom{3}{4}$ and $\binom{3}{5}$ are both zero.

We now discuss some continuous distributions, the first of which is somewhat difficult to work with without tables and therefore will be discussed only briefly. This fact should not lead the reader to believe, however, that it is not used frequently as a probability model, only that it will not be dealt with at great length in this book.

3.5 CONTINUOUS DISTRIBUTIONS

If there is a function $f(t)$ such that

$$F(x) = \mathscr{P}(X \leq x) = \int_{-\infty}^{x} f(t)\, dt, \tag{3.21}$$

we say that $F(x)$ is a continuous distribution function. The function $f(t)$ is then referred to as a *probability density function*. From (3.21) we see that the probability density function can be obtained from the distribution function by taking a derivative, since

$$f(x) = \frac{d}{dx}[F(x)]. \tag{3.22}$$

A random variable X such that $\mathscr{P}(X \leq x) = F(x)$ with F as given in (3.21) is referred to as a *continuous random variable*. Distribution functions of continuous random variables differ from those of discrete random variables in that the former do not show step changes with changing values of x as do the latter. Indeed, the variable x cannot make step changes but may assume any value within an interval. This was not allowed for discrete random variables. It should be noted that if X is a continuous random variable, then $\mathscr{P}(X = a) = 0$ for any real number a.

Some examples of variables with continuous distributions are flow rate, pressure drop, and temperature. The common factor in all these variables is that they cannot physically undergo abrupt step changes but change continually with time.

3.5.1 Normal Distribution

The most widely used of all continuous distributions is the normal distribution, or the common bell-shaped error curve. The basic assumption underlying this type of distribution is that any errors in experimental observation are due to variation of a large number of independent causes, each of which produces a small disturbance or error. Many experimental situations are subject to random error and do yield data that can be adequately described by the normal distribution, but this is not always true.

Regardless of the form of distribution followed by the population, the *central limit theorem* allows the normal distribution to be used for descriptive purposes subject to the limitation of a finite variance. This law states that if a population has a finite variance σ^2 and a mean μ, the distribution of the sample mean, denoted by $\bar{X}$, approaches the normal distribution with variance σ^2/n and mean μ as the size of the sample, n, increases. The proof is left to advanced statistical theory texts.

This theorem is all the more remarkable for its simplicity and profound impact on the treatment and description of data. The only requirement for having the sample mean behave normally is that the population variance be finite. One of the chief problems is to tell when the sample size is large enough to give reasonable compliance with the theorem.

The probability density function $f(x)$ for the normal distribution is given by

$$f(x) = \frac{1}{\sigma\sqrt{2\pi}} e^{-(x-\mu)^2/2\sigma^2} \qquad \text{for } -\infty < x < \infty. \tag{3.23}$$

Inspection of the equation reveals that two parameters, μ the mean and σ the standard deviation, are required to describe the normal distribution completely. For a constant mean, the spread of the curves varies with the standard deviation and increases with increasing values of σ. Differences in the means at constant σ are reflected merely by changes in the location on the x axis with no change in the shape of the curve. These characteristics are shown in Figs. 3.7 and 3.8.

The normal distribution is a two-parameter distribution. The cumu-

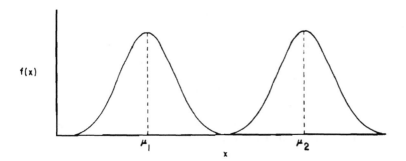

FIGURE 3.7   Distributions with equal variance but different means.

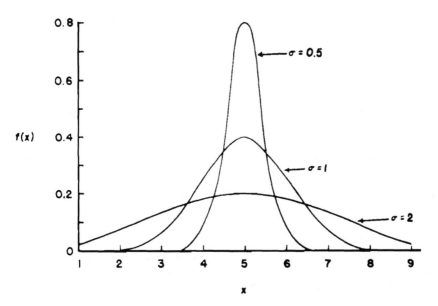

FIGURE 3.8   Normal density curves for constant mean ($\mu = 5$) with varying σ.

lative distribution function for this probability density function is

$$F(x) = \mathscr{P}(X \leq x) = \frac{1}{\sigma\sqrt{2\pi}} \int_{-\infty}^{x} e^{-(x-\mu)^2/2\sigma^2} \, dx. \qquad (3.24)$$

The probability of X falling within the range x_1 to x_2 is

$$\mathscr{P}(x_1 \leq X \leq x_2) = F(x_2) - F(x_1) = \frac{1}{\sigma\sqrt{2\pi}} \int_{x_1}^{x_2} e^{-(x-\mu)^2/2\sigma^2} \, dx. \qquad (3.25)$$

Since the two parameters μ and σ are required to establish a particular normal curve, it is convenient to consider a particular member from this family of distributions. The cumulative distribution functions are not easily evaluated and tables are handled more readily if they are made for the particular case of $\mu = 0$ and $\sigma = 1$. Once the integrals have been evaluated for this case, they can easily be converted to those corresponding to different values of μ and σ.

The *standard normal deviate* is defined as the number of standard deviations by which the variable differs from the mean. The standard normal deviate Z, defined by

$$Z = \frac{X - \mu}{\sigma},$$
(3.26)

is normally distributed with a mean of 0 and a variance (or standard deviation) of 1. Noting that $dx = \sigma\, dz$, the cumulative distribution function $F(z)$ of the random variable Z becomes

$$F(z) = \mathscr{P}\left(\frac{X - \mu}{\sigma} \le z\right) = \frac{1}{\sqrt{2\pi}} \int_{-\infty}^{z} e^{-t^2/2}\, dt.$$
(3.27)

Thus the distribution functions of all normal variables are indentical when expressed in terms of the standard normal deviate.

The standard normal distribution is only the particular case of the normal distribution having a mean of 0 and a variance of 1. Values of the cumulative distribution function $F(z)$ depend only on z. Table III of Appendix B is an abbreviated tabulation of these values. Figure 3.9 shows graphically some of the areas under the standard normal curve.

Let us now consider how Table III can be used. This is best done by way of illustrations. We first show how the table can be used to provide standard or normalized scores on examinations so that they can be compared with scores from other exams having different scales.

Example 3.11   On a particular physics exam, the class average was 72 and the standard deviation was 15. Three students taking a makeup exam (assumed to be the same as the original exam) made actual scores of 60, 93, and 72. What were their normalized scores?

Solution   Here $\bar{x} = 72$, $s_x = 15$. Replacing μ and σ by their "estimates," we have z approximately equal to $(x - \bar{x})/s_x$. Hence $z_{60} = (60 - 72)/15 = -0.8$, $z_{93} \doteq 1.4$, and $z_{72} = 0$. We could also use z with a knowledge of $\bar{x}$ and s_x to determine the actual test grades from the normalized scores since $x = \bar{x} + z s_x$. Hence a z score of -1.0 corresponds to $x = 72 + (-1.0)15 = 57$.

Example 3.12   If on a different physics exam the Z scores of two class members were 0.8 and -0.4, corresponding to grades of 88 and 64, respectively, what were the mean and standard deviation?

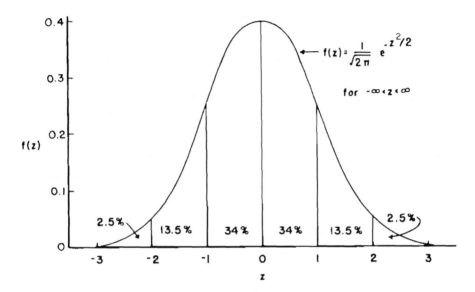

FIGURE 3.9 Standard normal: $N(0,1)$ (with approximate areas under the curve).

Solution Using the relationship $x = \bar{x} + zs_x$, we have two simultaneous equations: $8 = \bar{x} + 0.8s_x$ and $64 = \bar{x} - 0.4s_x$ to solve for $\bar{x}$ and s_x, from which we get $\bar{x} = 72$ and $s_x = 20$.

We now illustrate how probabilities for any member of the family of normal distributions can be obtained from the table for the standard normal. Table III gives the probability of getting a z value between $-\infty$ and some given value of z, where Z ranges from -4.0 to $+4.0$. The use of the table is illustrated below.

Example 3.13 Using Table III for the standard normal distribution, find the probabilities associated with the ranges of Z given below. This is really done by finding the area under the probability density function for the given z ranges.

  (a) $0 \le z \le 1.4$. This is obtained by finding $\mathcal{P}(Z \le 1.4) - \mathcal{P}(Z < 0)$, which is the area under the curve between $z = 0$ and $z = 1.4$. From the table we get this to be 0.4192.
  (b) $-0.78 \le z < 0$. This probability can be obtained by finding $\mathcal{P}(Z < 0) - \mathcal{P}(Z < -0.78) = 0.5 - 0.2177 = 0.2823$.
  (c) $-0.24 \le z \le 1.9$. The result is obtained by taking

$$\mathcal{P}(-0.24 \le Z \le 1.9) = \mathcal{P}(Z \le 1.9) - \mathcal{P}(Z \le -0.24)$$
$$= 0.9713 - 0.4052 = 0.5661.$$

(d) $0.75 \le z \le 1.96$. To get the probability of Z lying in this interval, we find that $\mathscr{P}(Z \le 1.96) - \mathscr{P}(Z \le 0.75) = 0.9750 - 0.7734 = 0.2016$.

(e) $-\infty < z \le -0.44$. The probability can be obtained directly from the table as $\mathscr{P}(-\infty < Z \le -0.44) = 0.32997$.

(f) $-\infty < z < 1.2$. For this interval we desire the entire area to the left of 1.2, which is read from the table to be 0.8849.

From the above it should be clear that the probability of Z being in any interval can be obtained from Table III.

It is sometimes necessary to find values z_1 and z_2 for some given value of $\mathscr{P}(z_1 \le Z \le z_2)$.

Example 3.14

(a) Find z_1 if the probability that Z is between 0 and z_1 is 0.4.

(b) Find z_1 if $\mathscr{P}(-\infty < Z \le z_1) = 0.92$.

(c) Find z_1 if $\mathscr{P}(-1.6 \le Z \le z_1) = 0.03$.

Solution (a) This merely requires us to find the value of z_1 for which the area between $z = -\infty$ and $z = z_1$ is 0.9 or 0.1. From Table III, $z_1 = \pm 1.282$ (using linear interpolation between adjacent tabular values).

(b) From Table III we see that z_1 is between 1.40 and 1.41 and by interpolation we obtain $z_1 = 1.405$.

(c) Since $\mathscr{P}(-1.6 \le Z \le z_1) = \mathscr{P}(Z \le z_1) - \mathscr{P}(Z \le -1.6) = \mathscr{P}(Z \le z_1) - 0.05480 = 0.03$, the area to the left of z_1 must be 0.08480. The value of z_1 is thus found to be -1.3733.

Example 3.15 A sample of 36 observations was drawn from a normally distributed population having a mean of 20 and variance of 9.

(a) What portion of the population can be expected to have values greater than 26?

(b) If the population were sampled repeatedly in groups of 36 observations, what portion of the population of means can be expected to have a value greater than 22?

Solution (a) From Eq. (3.26), we have for the standard normal deviate

$$z_{calc} = \frac{x - \mu}{\sigma_x} = \frac{26 - 20}{\sqrt{9}} = 2.$$

From Table III we find that $\mathscr{P}(Z \le 2) = 0.97725$. Therefore, the probability of obtaining a $Z_{calc} > 2$ is $1 - 0.97725$ or 0.02275. Therefore, 2.28% of the population can be expected to have values greater than 26.

(b) The standard error of the mean is defined as

$$\sigma_{\bar{x}} = \frac{\sigma_x}{\sqrt{n}},$$

which for this problem is $3/\sqrt{36} = 0.500$. The z value in this case is

$$z = \frac{\bar{x} - \mu}{\sigma_{\bar{x}}} = \frac{22 - 20}{0.5} = 4.$$

From Table III we see that the area under the standard normal curve reaches 0.9999 when z reaches 3.9. Therefore, the probability of another sample of 36 (drawn from this population) having a mean greater than 22 is nil.

3.5.2 Exponential Distribution

From the foregoing derivation of the Poisson probability function in Section 3.4.2, one may wonder about the distribution of the time, T, required for the first occurrence of the event E. We assumed that the probability of the event E occurring in the time interval $(t, t + dt)$ was $k\,dt$ and found $\mathscr{P}_0(t) = e^{-kt}$. Hence we can write

$\mathscr{P}$(time for the first occurrence of $E > t$)

$$= \mathscr{P}[E \text{ occurs } 0 \text{ times in } (0, t)] = \mathscr{P}(T \le t) = e^{-kt}.$$

Thus

$\mathscr{P}$(time for the first occurrence of $E \le t$)

$$= 1 - e^{-kt}.$$

The left-hand side is in the form of a cumulative distribution; therefore, differentiating yields the probability density function of the time interval between successive occurrences of the event E. This yields

$$p(t) = ke^{-kt}, \qquad t > 0$$

which is exponential or *negative exponential* probability density function.

In conclusion, if the number of times an event E occurs per interval of time t is distributed as the Poisson with parameter kt, the distribution of time between successive occurrences of E is distributed as the negative exponential with parameter k.

The exponential (or negative exponential) distribution is widely used for modeling time to failure and is very popular in the area of reliability. It is also the distribution of the time *between* successive failures for events occurring according to the postulates of the Poisson distribution in Section 3.4.2.

This distribution is also applicable to sites in evaluations of mirrors, prepared or painted surfaces, and so on. In this case, the negative exponential distribution is appropriate if the quality control test cannot distinguish between an originally damaged or blemished site and subsequent damage to the same site. The exponential distribution also describes the decay of radioisotopes, the attenuation of all forms of electromagnetic radiation (light, noise, etc.) through matter with uniform properties (e.g., air, thermal, or sound insulation), failure rate of light bulbs or transistors, and the residence-time distribution of particles in a continuously stirred tank.

The exponential distribution can be used to model those situations where the items fail at random according to a Poisson process. It was shown above that the time to the first failure has an exponential distribution. It can be shown that the time between successive failures also has the same exponential distribution. The probability density function is

$$f(x) = \alpha e^{-\alpha x}, \qquad 0 \le x < \infty, \quad \alpha > 0 \tag{3.28}$$

and the cumulative distribution function is

$$F(x) = 1 - e^{-\alpha x}, \qquad x \ge 0. \tag{3.29}$$

The probability density function when $\alpha = 0.7$ is shown in Fig. 3.10. The mean and variance of the exponential distribution are

$$\mu = \frac{1}{\alpha} \tag{3.30}$$

and

$$\sigma^2 = \frac{1}{\alpha^2}. \tag{3.31}$$

The continuous random variable X may have any units. The units on μ and σ will be the reciprocal of those of X. Therefore, α represents the fraction of failures or events occurring per unit of time or space.

Example 3.16 A new material that forms a protective oxide film is being considered for use in a marine environment. Although the corrosion rate is constant, the amount of corrosion experienced by the test specimens at any time is a function of the remaining uncorroded surface. After several replications of a salt-spray corrosion test, the average surface degradation is 0.168% for a standard 8-h exposure. Prior work with related systems has indicated that 8 h is approximately equivalent to 30 days in a marine environment. Is this material suitable if parts made from it must withstand at least 4 years (48 months) of exposure in the service environment with less than 3% material lost due to corrosion?

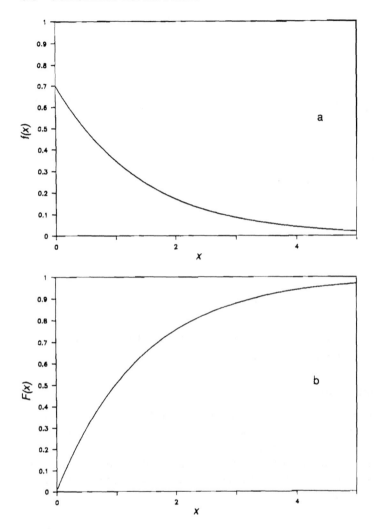

FIGURE 3.10 Exponential distribution: (a) probability density function; (b) cumulative distribution; both for $\alpha = 0.7$.

Solution From Eq. (3.29),

$$0.168\% = 0.00168 = F(t) = 1 - e^{-\alpha t} = 1 - e^{-\alpha(1)},$$

which gives $\alpha = 0.001681413$ for a simulated 1-month exposure. Continuing, $0.03 = F(t) = e^{-0.001681413t}$, which gives $t = 18.1$ months of service life before the maximum loss specification is exceeded. This material is probably not suited for the service environment.

3.5.3   Chi-Square Distribution

Suppose that $Y_1, Y_2, \ldots, Y_v$ are independent random variables each distributed normally with mean 0 and variance 1. Let χ^2 (chi-square) be the sum of their squares. The random variable

$$\chi^2 = Y_1^2 + Y_2^2 + Y_3^2 + \cdots + Y_v^2 \tag{3.32}$$

has probability density function

$$f(\chi^2) = \frac{1}{2^{v/2}\Gamma(v/2)} e^{-\chi^2/2}(\chi^2)^{v/2-1} \qquad \text{for } 0 \le \chi^2 \le \infty, \tag{3.33}$$

where $\Gamma(\alpha)$ is the *gamma function*, defined as

$$\Gamma(\alpha) = \int_0^\infty x^{\alpha-1} e^{-x} \, dx \qquad \text{for } \alpha > 0. \tag{3.34}$$

The parameter v, called *degrees of freedom* (discussed later), distinguishes the members of the family of χ^2 distributions.

3.5.4   Student's *t*-Distribution

Student's *t*-distribution was originally developed by W. S. Gosset, who referred to himself as "Student." It is the basis for the simple *t*-test so widely used in the analysis of scientific and engineering data.

  If Z has a normal distribution with mean 0 and variance 1, and V is distributed as χ^2 with v degrees of freedom and Z and V are independent, the random variable

$$t = \frac{Z}{\sqrt{V/v}} \tag{3.35}$$

has the probability density function

$$f(t) = \frac{1}{\sqrt{v\pi}} \frac{\Gamma((v+1)/2)}{\Gamma(v/2)} \left(1 + \frac{t^2}{v}\right)^{-(v+1)/2} \qquad \text{for } -\infty < t < \infty. \tag{3.36}$$

It is noted that this distribution depends only on the value of v that distinguishes between members of the family of *t*-distributions.

3.5.5 *F*-Distribution

The F-distribution is the distribution of the random variable F, which is defined as

$$F = \frac{U/v_1}{V/v_2}, \tag{3.37}$$

where U and V are independent variables distributed as χ^2 with v_1 and v_2 degrees of freedom, respectively. The probability density function of F is given by

$$f(F) = \frac{\Gamma((v_1 + v_2)/2)}{\Gamma(v_1/2)\Gamma(v_2/2)} \left(\frac{v_1}{v_2}\right)^{v_1/2} \frac{F^{(v_1 - 2)/2}}{[1 + (v_1/v_2)F]^{(v_1 + v_2)/2}}. \tag{3.38}$$

This family of distributions is a two-parameter, v_1 and v_2, family, and thus tables and charts based on the F-distribution are inherently more complex than those based on the single-parameter χ^2- and t-distributions. Equation (3.38) is restricted to the region where $F \geq 0$. A very important application of the F-distribution occurs in analysis of variance, which is treated in Chapter 8.

3.5.6 Weibull Distribution

A distribution function that is being used more and more as a time-to-failure probability model is the Weibull probability density function, given by

$$f(t) = \frac{\alpha}{\lambda}\left(\frac{t}{\lambda}\right)^{\alpha - 1} \exp\left[-\left(\frac{t}{\lambda}\right)^{\alpha}\right], \qquad t \geq 0, \quad \alpha > 0, \quad \lambda > 0. \tag{3.39}$$

One of its most frequent uses has been in studies on the service life of electron tubes. Some plots of the distribution are given in Fig. 3.11. Note that for $\alpha = 1$ the Weibull reduces to the negative exponential probability density function.

3.5.7 Gamma Distribution

The gamma probability density function is

$$f(x) = \begin{cases} \dfrac{x^{\alpha - 1}e^{-x/\beta}}{\Gamma(\alpha)\beta^{\alpha}}, & x > 0 \\[2ex] 0, & x \leq 0. \end{cases} \tag{3.40}$$

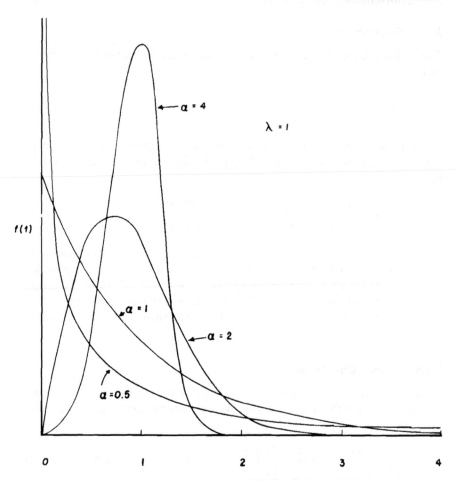

FIGURE 3.11   Weibull probability density functions with $\lambda = 1$.

This distribution has been used very extensively in statistics. The parameters α and β are both positive. If $\alpha = 1$, the gamma density function reduces to the exponential density function,

$$f(x) = \frac{1}{\beta}\, e^{-x/\beta}, \qquad x > 0.$$

If $\alpha = v/2$ and $\beta = 2$, the gamma distribution becomes the chi-square probability density function in Eq. (3.33).

3.5.8 Log-Normal Distribution

Many processes, such as crystals in solution, grain growth due to annealing, and crushing and grinding operations, yield a continuous distribution that is skewed to the right. In these examples, the random variable, X, is the grain or particle size. Empirical research has shown that the distribution of $\ln X$ (or $\log X$) has the familiar bell shape of the normal distribution. That is, if X has a log-normal distribution, then $Y = \ln X$ has a normal distribution, as shown in Fig. 3.12. This approach is also useful in representing particle size in air pollution emissions. As a result of the logarithmic transformation, the random variable involved cannot have negative values.

The log-normal probability density function is

$$f(x) = \begin{cases} \dfrac{1}{\sigma x \sqrt{2\pi}} \exp\left[-\dfrac{(\ln x - \mu)^2}{2\sigma^2} \right], & 0 < x < \infty \\ 0, & \text{otherwise.} \end{cases} \tag{3.41}$$

The parameters μ and σ^2 are the mean and variance of $Y = \ln X$, which has a normal distribution with mean μ and variance σ^2, where

$$\mu \approx \overline{\ln x} = \frac{1}{n} \sum_{i=1}^{n} \ln x_i \tag{3.42}$$

and

$$\sigma^2 \approx S^2 = \frac{1}{n-1} \sum_{i=1}^{n} (\ln x_i - \overline{\ln x})^2. \tag{3.43}$$

Although Eq. (3.41) is the most commonly used form of the log-normal distribution, other forms have also been found useful. When describing particle distributions in industrial hygiene, the form

$$f(x) = \frac{1}{(\ln \sigma_g) x \sqrt{2\pi}} \exp\left[-\frac{(\ln x - \ln D_{nmd})^2}{(\ln \sigma_g)^2} \right], \tag{3.44}$$

where D_{nmd} is the number-median diameter and σ_g is the geometric standard deviation, is often used. The relation between the number-median diameter and the mass-median diameter is

$$\log D_{mmd} = \log D_{nmd} + 6.909(\log \sigma_g)^2. \tag{3.45}$$

With this relation, the log-normal distribution can easily be used to describe particulate matter in air pollution control studies.

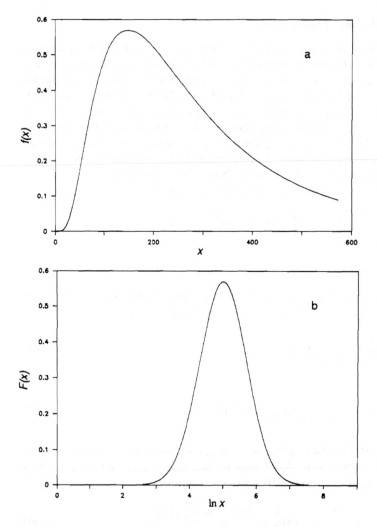

Figure 3.12 Log-normal distributions: (a) original data; (b) transformed data.

To use the log-normal distribution, calculate the mean and variance of ln x_i from Eqs. (3.42) and (3.43) and use the standard normal distribution to test hypotheses on the log-transformed data. The results may be meaningless due to the approximations involved. The log-normal distribution, however, is a convenient graphical tool for visual comparisons of distributions.

3.6 EXPERIMENTAL DISTRIBUTIONS

The distributions defined in Sections 3.4 and 3.5 are theoretical distributions that correspond to some populations. For example, $S = \{1,2,3,4,5,6\}$ is the population of values corresponding to the outcomes in rolling a die. The probability function $\mathcal{P}(X = x_i) = \frac{1}{6}$ for $x_i = 1,2,3,4,5,6$ gives the theoretical distribution for this population. However, the die may not be fair, so that the theoretical distribution is given by some other unknown probability function. One may, however, by rolling the die a large number of times and using the frequency definition of probability in Chapter 2, arrive at an empirical probability function that is close to the theoretical distribution. For example, suppose that 1000 rolls of the die yielded the results summarized in Table 3.1. The resulting function $f(x_i)$, $x_i = 1,2,3,4,5,6$, in Table 3.1 satisfies all the conditions of a theoretical distribution but is, in actuality, an experimental distribution.

Before considering experimental distributions, we define a few terms pertinent to such distributions.

Classes: groups into which data are distributed

Class boundary: that numerical value which divides two successive classes

Class length: the numerical difference in the boundaries of a class

Class mark: the midvalue of a class

Class frequency: the frequency with which values of observations occur

Relative frequency: the frequency expressed as a fraction of the total number of observations

Example 3.17 The methods commonly used in arriving at and displaying the distribution characteristics of data are illustrated using the raw data given in Table 3.2, which give the melting points by the ring and ball method of batches of modified resin produced during a 6-month period. It should be noted that the values in Table 3.2 comprise a sample from the "conceptual" population of all batches that would be obtained henceforth.

TABLE 3.1 Experimental Distribution of Die Outcome Data

	Outcome, x_i					
	1	2	3	4	5	6
Frequency	142	200	58	328	122	150
Relative frequency, $f(x_i)$	0.142	0.200	0.058	0.328	0.122	0.150

TABLE 3.2   Ring and Ball Melting Point Data (Melting Point, °C)

108.4	108.2	108.9	106.2	107.1	108.5
105.5	107.2	107.4	107.7	108.4	107.6
107.4	107.0	109.2	107.1	108.8	109.7
111.5	109.4	107.2	107.6	108.8	109.2
110.9	108.8	106.2	108.5	108.4	108.9
107.1	108.3	108.5	107.3	108.3	107.7
109.1	107.3	107.0	108.3	106.9	106.9
108.7	108.4	108.2	107.6	108.6	110.0
108.9	107.7	106.5	106.2	109.4	109.9
108.1	108.4	108.5	108.0	109.3	107.8
108.8	106.3	108.3	108.3	108.3	108.4
108.4	109.0	109.1	109.0	108.2	109.1
108.2	106.9	108.3	107.6	107.1	108.3
109.2	108.0	109.1	107.1	107.4	109.6
108.3	108.6	106.0	106.8	108.2	106.3
106.6	106.4	106.8	107.4	109.1	107.4
107.3	107.7	108.9	109.2	108.4	108.4
107.9	108.9	108.6	107.6	108.5	106.3
106.7	107.6	106.1	107.3	106.8	107.1
109.1	109.3	106.3	107.7	107.9	108.2
106.2	107.7	108.8	107.8	108.6	107.3
109.2	107.6	109.5	107.9	107.5	108.3
109.2	108.3	108.1	106.8	108.8	108.1
109.5	108.4	108.3	108.1	108.1	107.8

In summarizing large quantities of raw data, grouping is commonly employed. For convenience, the values may be placed in rank order as in Table 3.3. In grouping data, a decision must be made as to how many classes should be used. The related question of class length must also be answered. Too many classes make the population characteristics difficult to picture; too few classes destroy the details of the picture. It is convenient, but not absolutely necessary, that all class lengths be equal. Generally, 10 to 20 classes are adequate even for large sets of observations.

Some trial and error may be required in deciding on the class boundaries that result in the best picture of the data. The class length chosen should be a convenient one to work with. The classes should cover the range of the population with no empty intervals; however, this is not always possible.

For the data under consideration, a class length of 0.5°C was selected,

TABLE 3.3 Rank Order for Melting Point Data

Rank	Value	Rank	Value	Rank	Value
1	111.5	41	108.6	81	108.0
2	110.9	42	108.5	82	107.9
3	110.0	43	108.5	83	107.9
4	109.9	44	108.5	84	107.9
5	109.7	45	108.5	85	107.8
6	109.6	46	108.5	86	107.8
7	109.5	47	108.4	87	107.8
8	109.5	48	108.4	88	107.7
9	109.5	49	108.4	89	107.7
10	109.4	50	108.4	90	107.7
11	109.4	51	108.4	91	107.7
12	109.3	52	108.4	92	107.7
13	109.3	53	108.4	93	107.7
14	109.2	54	108.4	94	107.6
15	109.2	55	108.4	95	107.6
16	109.2	56	108.4	96	107.6
17	109.2	57	108.3	97	107.6
18	109.2	58	108.3	98	107.6
19	109.2	59	108.3	99	107.5
20	109.1	60	108.3	100	107.5
21	109.1	61	108.3	101	107.5
22	109.1	62	108.3	102	107.4
23	109.1	63	108.3	103	107.4
24	109.1	64	108.3	104	107.4
25	109.1	65	108.3	105	107.4
26	109.0	66	108.3	106	107.3
27	109.0	67	108.3	107	107.3
28	108.9	68	108.3	108	107.3
29	108.9	69	108.2	109	107.3
30	108.9	70	108.2	110	107.3
31	108.9	71	108.2	111	107.2
32	108.8	72	108.2	112	107.2
33	108.8	73	108.2	113	107.2
34	108.8	74	108.2	114	107.1
35	108.8	75	108.1	115	107.1
36	108.8	76	108.1	116	107.1
37	108.7	77	108.1	117	107.1
38	108.6	78	108.1	118	107.1
39	108.6	79	108.1	119	107.1
40	108.6	80	108.0	120	107.0

(continued)

TABLE **3.3**   *(Cont.)*

Rank	Value	Rank	Value	Rank	Value
121	107.0	129	106.7	137	106.3
122	106.9	130	106.6	138	106.2
123	106.9	131	106.6	139	106.2
124	106.9	132	106.5	140	106.2
125	106.8	133	106.4	141	106.2
126	106.8	134	103.3	142	106.1
127	106.8	135	106.3	143	106.0
128	106.8	136	106.3	144	105.5

and since the range of the values is 105.5 to 111.5°C, 13 classes are required. To assure that there is no doubt as to which class a value belongs, class boundaries are set at 105.05, 105.55, 106.05, . . . ,111.55°C. Results of the data grouping are shown in Table 3.4.

The data may conveniently be presented graphically in the form of frequency diagrams. The most commonly used diagram is the *frequency histogram*. The data of Table 3.4 are shown graphically in Fig. 3.13 in the form of a histogram. In the construction of the histogram, class marks and

TABLE **3.4**   Grouped Frequencies of Melting Point Data with 0.5°C Class Length

Class	Class mark	Class frequency	Cumulative frequency	Relative frequency	Cumulative relative frequency
105.05–105.55	105.3	1	1	0.00694	0.00694
105.55–106.05	105.8	1	2	0.00694	0.01388
106.05–106.55	106.3	11	13	0.07639	0.09028
106.55–107.05	106.8	12	25	0.08333	0.17361
107.05–107.55	107.3	21	46	0.14583	0.31944
107.55–108.05	107.8	19	65	0.13194	0.45137
108.05–108.55	108.3	38	103	0.26388	0.71528
108.55–109.05	108.8	16	119	0.11111	0.82639
109.05–109.55	109.3	19	138	0.13194	0.95833
109.55–110.05	109.8	4	142	0.02778	0.98611
110.05–110.55	110.3	0	142	0.00000	0.98611
110.55–111.05	110.8	1	143	0.00694	0.99306
111.05–111.55	111.3	1	144	0.00694	1.00000

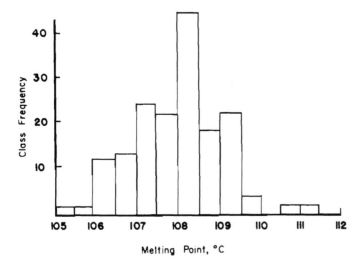

FIGURE **3.13** Histogram of melting point data.

boundaries are put on the horizontal axis and frequency and/or relative frequency on the vertical axis of a rectangular coordinate system. Thus each of the group frequencies is represented as a rectangle, the height of which is equal to or proportional to the corresponding group frequency. This is true for the example here since the class intervals are of constant size. For the general case it is the areas of the rectangles that relate to the group frequencies.

The effect of class length on the distribution "picture" obtained is illustrated by a comparison of Figs. 3.13 and 3.14 in which the class length is changed from 0.5°C to 1.0°C. By decreasing the number of groups, the irregularities in the histogram are eliminated. It is unreasonable, considering the nature of the operation on which the data are based, that the irregularities are significant. Hence it is believed that in this case a class length of 1.0°C gives a better picture of the data as a whole than does a 0.5°C length. The grouped frequencies associated with the longer class length are given in Table 3.5.

Grouped data may also be depicted by means of the *frequency polygon* or *cumulative frequency polygon*. The upper class boundary and cumulative frequency of the corresponding group locate points that are connected by straight lines to form the cumulative frequency polygon. The same data from Table 3.5 are shown in Fig. 3.15, which is a cumulative frequency polygon. Since the cumulative frequency polygon shown in Fig. 3.15 depicts the total

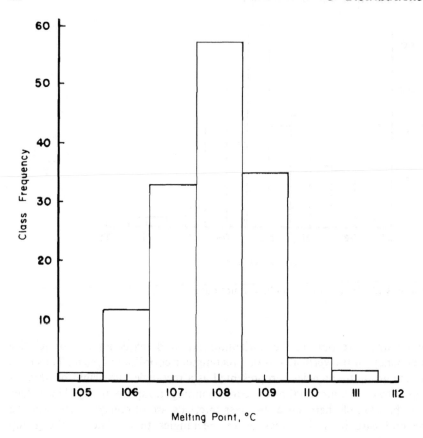

Figure 3.14 Histogram of Table 3.5 data.

Table 3.5 Grouped Frequencies of Melting Point Data with 1°C Class Length

Class	Class mark	Class frequency	Cumulative frequency	Relative frequency	Cumulative relative frequency
104.55–105.55	105.05	1	1	0.00694	0.00694
105.55–106.55	106.05	12	13	0.08333	0.09028
106.55–107.55	107.05	33	46	0.22917	0.31944
107.55–108.55	108.05	57	103	0.39583	0.71528
108.55–109.55	109.05	35	138	0.24306	0.95833
109.55–110.55	110.05	4	142	0.02778	0.98611
110.55–111.55	111.05	2	144	0.01388	1.00000

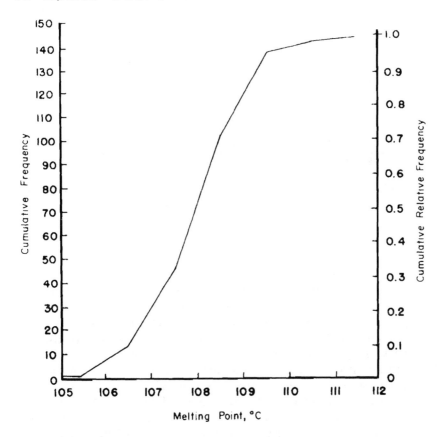

FIGURE 3.15   Cumulative frequency polygon for melting point data.

of frequencies having values less than a particular x_i value, the x_i value plotted is not the class mark (midpoint) but the upper class boundary.

Since the major purpose of frequency polygons and diagrams is to depict distributions, little is gained by drawing smooth curves through the vertices of the polygons to give frequency curves for predictive purposes. For such purposes it is preferable to fit an appropriate theoretical function to the relative frequency histogram and use this function for predictions.

In the handling of data representing observations of discrete random variables (count data), frequency is represented by a line on the diagram and the cumulative frequency polygon is a step polygon. These differences between continuous and discrete distributions arise from the fact that discrete variables can take only isolated values. Figure 3.16 shows the frequency

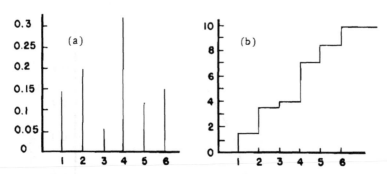

Figure 3.16 (a) Frequency histogram and (b) cumulative frequency polygon for data of Table 3.1.

histogram and cumulative frequency polygon for the die outcome data given in Table 3.1 obtained in rolling a die 1000 times.

Now that you have learned to handle experimental distributions, let us examine the use of commonly available computer programs for obtaining histograms and relative frequency polygons. Two such commonly used packages are BMD (registered trademark, UCLA) and SAS (registered trademark, SAS Institute, Cary, NC). Although we feel that the former is better documented, the latter is, in our opinion, easier to use. From here on, we include appropriate example problems using SAS programs. Before you begin to read this example, study the material of Appendix A carefully so that you know how to set up your data into the formats required for executing SAS programs. We emphasize that the material there and in the SAS programs illustrated in this book is presented in abbreviated form. The serious student should obtain copies of all the SAS Institute publications to which we refer.

Example 3.18 An experiment entitled "Friction Losses in Valves and Fittings" is currently being performed in the unit operations laboratory. The experimental apparatus is equipped with a small orifice that is used to determine the flow rate through the system. For purposes of comparison, several values of the orifice coefficient which have been calculated by different lab groups are presented below. Prepare an appropriate frequency table, histogram, and frequency polygon from these data. Using the PROC PRINT, PROC FREQ, and PROC CHART programs (*SAS Introductory Guide*, SAS Institute, Cary, NC, 1978, 83 pp.), the data are printed (not shown again), arranged in ascending order with the relative frequency (percent column) and cumulative relative frequency (cum percent column), and displayed in histogram form (Fig. 3.17). As the job control language

(JCL) and control cards vary with the computer, we will only show the programs and the results

(put initial JCL cards/statements here)

```
DATA CO;
INPUT @@;
LABEL LABORATORY ORIFICE COEFFICIENT VALUES;
CARDS;
```

0.626	0.670	0.500	0.648
0.615	0.579	0.517	0.898
0.625	0.627	0.495	0.883
0.640	0.605	0.488	0.882
0.623	0.628	0.526	0.964
0.625	0.588	0.610	0.830
0.608	0.559	0.630	0.654
0.615	0.624	0.633	0.603
0.633	0.615	0.610	0.621
0.622	0.616	0.650	0.621
0.624	0.579	0.640	0.584
0.615	0.670	0.773	0.583
0.616	0.517	0.641	0.632

```
PROC PRINT DATA = CO;
PROC FREQ DATA = CO;
TITLE 'RESULTS OF USING PROC FREQ';
PROC CHART DATA = CO;
VBAR VALUES/TYPE=FREQ NOSPACE MIDPOINTS= 0.4875 TO 1.005 BY 0.01;
TITLE 'RESULTS OF USING PROC CHART';
```

(put final JCL cards/statements here)

SAS

RESULTS OF USING PROC FREQ FOR CO DATA

CO	FREQUENCY	CUM FREQ	PERCENT	CUM PERCENT
0.488	1	1	1.923	1.923
0.495	1	2	1.923	3.846
0.5	1	3	1.923	5.769
0.517	2	5	3.846	9.615
0.526	1	6	1.923	11.538
0.559	1	7	1.923	13.462
0.579	2	9	3.846	17.308
0.583	1	10	1.923	19.231
0.584	1	11	1.923	21.154
0.588	1	12	1.923	23.077
0.603	1	13	1.923	25.000
0.605	1	14	1.923	26.923
0.608	1	15	1.923	28.846
0.61	1	16	1.923	30.769
0.615	4	20	7.692	38.462
0.616	2	22	3.846	42.308
0.621	2	24	3.846	46.154
0.622	1	25	1.923	48.077
0.623	1	26	1.923	50.000
0.624	2	28	3.846	53.846
0.625	2	30	3.846	57.692
0.626	1	31	1.923	59.615
0.627	1	32	1.923	61.538
0.628	1	33	1.923	63.462
0.63	1	34	1.923	65.385
0.632	1	35	1.923	67.308
0.633	2	37	3.846	71.154
0.64	3	40	5.769	76.923
0.641	1	41	1.923	78.846
0.648	1	42	1.923	80.769
0.65	1	43	1.923	82.692
0.654	1	44	1.923	84.615
0.67	2	46	3.846	88.462
0.773	1	47	1.923	90.385
0.83	1	48	1.923	92.308
0.882	1	49	1.923	94.231
0.883	1	50	1.923	96.154
0.898	1	51	1.923	98.077
0.964	1	52	1.923	100.000

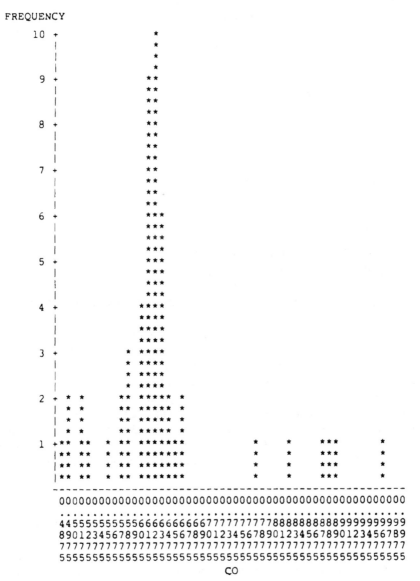

FIGURE 3.17 Results of using PROC CHART: frequency bar chart for orifice coefficients.

Use either the output from **PROC FREQ** or the text-editing capabilities of your computer to save and enter the values of the orifice coefficient CO and the corresponding cumulative relative frequency CRF into a new SAS routine, **PROC PLOT** (*SAS Introductory Guide*). Having saved the output of PROC FREQ under the retrieval name ORIFICE, the procedure to generate the CRF graph is shown below. The CO data were in columns 13 to 18 and the CRF values in columns 57 to 63 of the output from PROC FREQ.

(put initial JCL cards/statements here)

```
DATA ORIFICE;
INPUT CO 13-18 CRF 57-63;
CARDS;
```

(type in CO and CRF values here or call them in from storage)

```
PROC PLOT DATA = ORIFICE;
PLOT CRF*CO;
TITLE 'DISTRIBUTION OF ORIFICE COEFFICIENT';
```
(put final JCL cards/statements here)

The result of this SAS procedure is the cumulative relative frequency polygon shown in Fig. 3.18.

As an illustration of the application of a theoretical distribution to experimental data, Fig. 3.19 shows a relative frequency histogram of the data given in Table 3.5. Superimposed on this histogram is a plot of the normal distribution curve using values of $\bar{x}$ and s calculated from the raw data of Table 3.2. The mean of the data, which is a good estimate of the population mean μ, was calculated to be 108.05°C. The standard deviation of the population was estimated from the data to be 0.960°C. In calculating points on the x vs. $f(x)$ curve in Fig. 3.19, it is necessary to convert from normal deviate values z to the corresponding values of x. From the definition of z in Eq. (3.26),

$$x = \mu + z\sigma = 108.05 + 0.960z,$$

the x value corresponding to each value of z is calculated. Plotting these values vs. $f(x)$ results in the symmetrical normal distribution curve.

There is no evidence that the distribution of the data, as pictured in the relative frequency histogram, deviates significantly from the normal curve. Thus the normal distribution may be used with confidence in describing the melting point population.

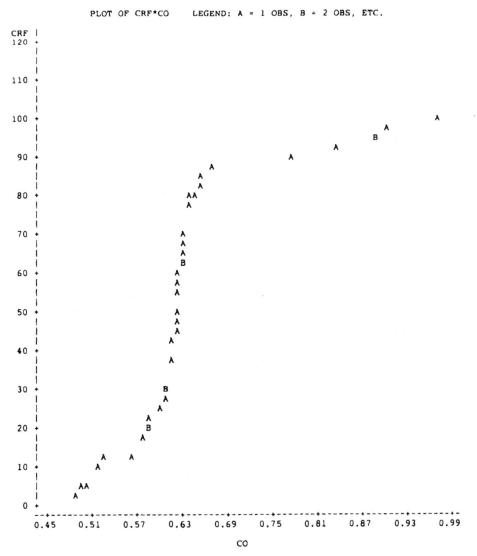

FIGURE 3.18 Distribution of orifice coefficient.

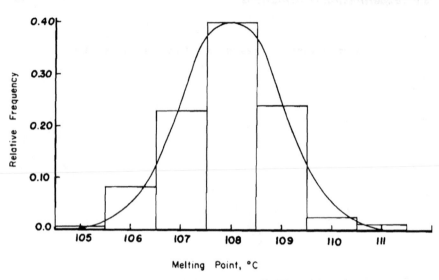

Melting Point, °C

FIGURE 3.19   Relative frequency and normal curve for melting point data.

PROBLEMS

3.1   The average particulate concentration in micrograms per cubic
meter for a station in a petrochemical complex was measured every 6 h for
30 days. The resulting data are as follows:

5	7	9	12	13	16	17	19	23	24	41
18	24	6	10	16	14	23	19	8	20	26
15	6	11	16	12	22	9	8	15	18	13
7	13	14	8	17	19	11	21	9	55	72
23	24	12	220	25	13	8	9	20	61	48
565	65	10	43	20	45	27	20	72	12	115
130	82	55	26	52	34	66	112	40	34	89
85	95	28	110	16	19	61	67	45	34	32
103	72	67	30	21	122	42	125	50	57	56
25	15	46	30	35	40	16	53	65	78	98
80	65	84	91	71	78	58	26	48	21	

For these data, prepare a table similar to Table 3.4 using appropriate class
lengths and boundaries. Plot the class frequency and relative frequency on
one graph as in Fig. 3.16 with a double ordinate. Plot the cumulative
frequency and cumulative relative frequency polygons as in Fig. 3.15.

3.2 On a diagnostic examination given recently to a group of 56 high-school science students, the following scores were obtained.

65	76	81	25	36	103	144	73	184	64	143
84	92	96	67	94	158	203	121	186	40	155
97	106	108	89	118	196	234	162	147	123	161
114	120	150	103	164	246	238	46	111	151	155
134	136	159	148	213	183	145	90	105	55	132
200	218	230	206	236	205	245	147	126	129	148
124	177	137	194	173	159	156	151	157	137	
187	200	214	237	213	149	185	132	160	168	

Prepare appropriate tables of data and diagrams as in Problem 3.1 for these test scores. The maximum possible score was 250 points.

3.3 The following numbers are the final averages for three courses: Ch.E. 3351, Ch.E. 3311, and Ch.E. 4312–5341 (as the last pair are mass-transfer courses, they are grouped together).

Ch.E. 3351:

89	72	89	81	89	85
87	83	91	64	83	69
90	82	91	91	90	88
91	76	91	78	85	75
88	90	84	70	82	86
75	87	89	77	82	75
89	80	91	82	71	85
74	69	78	94		

Ch.E. 3311:

75	59	91	71	70	76
77	80	28	70	73	48
68	71	76	48	65	78
54	76	30	29	62	60
69	85	84	64	66	31
28	46	74	71	74	54
76	72	24	68	61	53
82	80	83	61	66	40
58	59				

Ch.E. 4312–5341:

93	97	71	75	90	94
93	96	77	93	95	87
96	79	99	98	98	68
99	76	99	91	96	97
81	88	88	61	87	94
90	85	83	79	95	77
90	90	93	79	96	91
82	84	88	82	90	87
85	84	78	86	82	83
85					

For each set, plot class frequency, relative frequency, cumulative frequency, and cumulative relative frequency.

3.4   Air pollution control and monitoring in the city of Lubbock falls under the responsibility of the Lubbock City–County Department of Health. Air samples are drawn from various locations within the city limits and analyzed for both naturally occurring and industrial contaminants. The results of analyses for sulfate (SO_4^{2-}) content in air over a 72-day period are as follows (concentrations are expressed as parts per million in air):

3.14	1.87	1.00	1.12	1.03	5.88	0.91	1.86
0.08	1.95	6.38	2.61	0.10	1.96	0.00	0.00
2.00	5.57	2.96	2.68	2.44	2.62	7.61	0.80
2.11	6.99	9.02	3.24	1.25	0.00	8.06	0.00
3.55	7.80	8.90	1.95	1.85	2.07	0.00	6.85
3.51	10.83	14.71	1.24	1.23	4.11	1.75	0.00
3.69	7.17	1.95	0.63	1.24	1.84	1.78	0.00
0.98	1.93	3.30	0.00	1.66	3.60	0.00	0.00
2.01	2.58	2.01	1.24	2.38	0.00	1.83	1.14

Prepare a table of grouped frequencies for the data employing suitable class lengths and boundaries. Plot the class frequency and relative frequency on one graph. Also plot the cumulative frequency and cumulative relative frequency polygons.

3.5   The Department of Chemical Engineering has received a request from the company that manufactured our chain balances for the distribution data on the sensitivity of these balances to compare with the sensitivity of newly manufactured balances. Use the following data (sensitivity, mg/division) from several Ch.E. 3111 laboratory reports.

0.765	1.012	0.810	0.600
0.0721	0.485	2.125	0.711
0.791	1.500	0.790	0.711
0.7725	1.000	0.7725	0.990
0.680	0.702	0.818	0.870
0.715	0.711	0.714	
0.435	3.92	0.560	

(a) Prepare a table of classes, class frequency, cumulative frequency, relative frequency, and cumulative relative frequency.

(b) Prepare graphs of double ordinates of class frequency and relative frequency vs. classes and of cumulative frequency and cumulative relative frequency vs. classes.

3.6 One of the limiting factors for cooling tower design is the ambient dry-bulb temperature. As part of your design procedure, you have accumulated the average monthly temperature (°F) from 1897 to 1966 for July for the proposed plant location. (*Note*: Two years were not recorded.)

77.8	78.5	76.8	76.1	78.0	78.4	78.0
77.1	77.8	76.2	80.7	79.4	79.3	80.5
77.3	77.6	78.4	80.1	79.4	82.7	79.8
81.7	79.2	76.2	78.0	81.1	80.3	78.9
81.1	79.5	79.4	77.9	79.8	79.6	79.2
78.1	79.1	79.2	77.2	76.8	79.8	80.1
79.2	80.1	81.0	79.1	79.9	79.4	77.0
78.2	80.7	79.3	76.8	78.7	80.7	80.9
78.0	80.2	75.8	78.8	79.8	81.8	
77.4	78.9	81.9	80.5	80.4	78.3	

(a) Prepare a table similar to Table 3.4 and submit a plot of the class frequency and relative frequency using appropriate class lengths.

(b) Plot the cumulative frequency and cumulative relative frequency polygons as illustrated in Fig. 3.15.

3.7 One of the demands agreed to by a certain oil company after a recent labor strike was to locate a plant in an area that would be suitable for workers with respiratory diseases. The following data give annual precipitation values (inches) for Tombstone, Arizona from 1898 to 1961.

Year	Precipitation	Year	Precipitation
1898	13.50	1931	23.82
1901	14.84	1932	16.35
1904	11.82	1933	13.54
1905	27.84	1934	11.21
1906	12.14	1935	15.11
1907	19.31	1936	11.07
1909	14.91	1937	12.97
1910	11.77	1938	16.62
1911	19.20	1941	12.29
1912	13.67	1942	9.07
1913	15.97	1943	11.99
1914	19.78	1944	15.05
1915	11.55	1945	9.38
1916	15.99	1946	14.78
1919	21.50	1948	14.07
1920	11.18	1950	10.84
1921	18.55	1954	14.88
1922	10.90	1955	18.77
1923	15.09	1956	8.89
1924	7.36	1957	16.49
1925	14.40	1958	20.44
1927	16.40	1959	14.71
1928	11.90	1960	8.82
1929	14.21	1961	13.32
1930	20.92	1962	12.31

From these data prepare a table similar to Table 3.3 using appropriate class lengths and boundaries. Plot the class frequency and relative frequency on one graph as in Fig. 3.16 with a double ordinate. Plot the cumulative frequency and cumulative relative frequency polygons as in Fig. 3.15.

3.8 Contained in the results of the dissertation of H. R. Heichelheim* are values for the compressibility factors for carbon dioxide at 100°C, over the pressure range from 1.3176 to 66.437 atm.

---

* H. R. Heichelheim, The compressibility of gaseous 2,2-dimethyl propane by the Burnet method, p. 26, Ph.D. dissertation, Library, The University of Texas, Austin, Texas, 1962, Data reproduced by permission of the author.

0.9966	0.9969	0.9971
0.9956	0.9957	0.9960
0.9936	0.9938	0.9940
0.9913	0.9912	0.9915
0.9873	0.9874	0.9980
0.9821	0.9823	0.9829
0.9747	0.9750	0.9758
0.9648	0.9646	0.9657
0.9507	0.9500	0.9515
0.9447	0.9440	0.9454
0.9380	0.9371	0.9388
0.9306	0.9292	0.9314
0.9223	0.9215	0.9234
0.9130	0.9122	0.9141
0.9029	0.9022	0.9042
0.8916	0.8908	0.8930
0.8793	0.8783	0.8805
0.8656	0.8650	0.8657
0.8506	0.8501	0.8510
0.8343	0.8345	0.8354

Prepare a table and plot the data as in Problem 3.1.

3.9 In the experimental evaluation* of a distillation tower for the production of ethanol, the following data were recorded for the steam rate (lb/hr).

1170	1350	1640	1800	1800
1260	1440	1730	1710	1710
1350	1440	1800	1530	1530
1440	1170	1260	1350	1350
1530	1440	1350	1170	1170
1620	1170	1440	1800	1800
1260	1260	1170	1710	1710
1350	1440	1530	1530	1530
1440	1495	1620	1350	1170
1260	1540	1170	1170	1440

---

* Data from R. Katzen, V. B. Diebold, G. D. Moon, Jr., W. A. Rogers, and K. A. Lemesurier, *Chem. Eng. Prog.* 64(1): 79–84 (1968). Copyright 1968 by the American Institute of Chemical Engineers. Reprinted by permission of the editor and publisher.

(a)   Prepare a frequency table and plot class frequency and relative frequency on one graph.

(b)   Plot class frequency and cumulative relative frequency on another graph.

3.10   The diameters of MnO_2 particles used for SO_2 removal in a postcatalytic adsorber for the purification of spacecraft atmospheres were measured in 1/1000 of an inch, yielding the following particle data:

49	68	55	29	10	51
60	49	45	31	62	58
69	40	72	60	56	45
33	39	51	50	60	63
35	42	52	59	59	67
33	31	42	52	60	68
35	39	52	37	67	52
44	45	47	36	65	65
39	62	46	40	61	74
37	70	51	34	61	58
32	62	62	61	84	56
44	61	67	51	65	39
35	52	56	54	54	35
31	58	65	61	32	45
37	29	56	42	70	59
43	40	29	59	64	36
40	76	24	56	53	26
30	74	58	56	26	43
75	54	65	61	64	70
73	65	50	58	55	43

Plot the frequency histogram and cumulative frequency polygon.

3.11   An experiment entitled "Friction Losses in Valves and Fittings" is currently being performed in the unit operations laboratory. The experimental apparatus is equipped with a small orifice that is used to determine the flow rate through the system. For purposes of comparison, several values of the orifice coefficient which have been calculated by different lab groups are presented as follows:

0.626	0.670	0.500	0.648
0.615	0.579	0.517	0.898
0.625	0.627	0.495	0.883
0.640	0.605	0.488	0.882
0.623	0.628	0.526	0.964
0.625	0.588	0.610	0.830
0.608	0.559	0.630	0.654
0.615	0.624	0.633	0.603
0.633	0.615	0.610	0.621
0.622	0.616	0.650	0.621
0.624	0.579	0.640	0.584
0.615	0.670	0.773	0.583
0.616	0.517	0.641	0.632

Prepare the appropriate frequency tables of data and diagrams from these values.

3.12 A chemical engineer desires to estimate the evaporation rate of water from brine evaporation beds for the month of July. She obtains the following daily evaporation data (in./day) from local weather bureau records (1964–1968).

0.36	0.34	0.35	0.55	0.33
0.33	0.11	0.30	0.49	0.26
0.29	0.35	0.33	0.37	0.33
0.25	0.14	0.33	0.42	0.31
0.32	0.21	0.44	0.39	0.38
0.21	0.45	0.45	0.45	0.53
0.21	0.02	0.30	0.42	0.52
0.24	0.35	0.32	0.48	0.52
0.23	0.40	0.44	0.44	0.44
0.25	0.33	0.48	0.42	0.56
0.06	0.38	0.56	0.46	0.57
0.34	0.32	0.52	0.48	0.52
0.23	0.42	0.42	0.26	0.37
0.25	0.36	0.43	0.47	0.35
0.18	0.27	0.41	0.14	0.41
0.34	0.29	0.29	0.31	0.59
0.39	0.10	0.20	0.34	0.44
0.30	0.13	0.45	0.45	0.32
0.44	0.04	0.51	0.50	0.34
0.13	0.10	0.37	0.43	0.45
0.22	0.23	0.32	0.39	0.48
0.30	0.27	0.34	0.31	0.51

(*continued overleaf*)

0.32	0.31	0.16	0.35	0.46
0.23	0.40	0.43	0.34	0.45
0.37	0.43	0.12	0.38	0.47
0.32	0.34	0.20	0.27	0.33
0.29	0.41	0.44	0.33	0.48
0.29	0.44	0.35	0.27	0.45
0.38	0.32	0.41	0.32	0.42
0.28	0.33	0.47	0.29	0.40
0.41	0.39	0.25	0.25	0.33

Plot frequency, relative frequency, cumulative frequency, and cumulative relative frequency diagrams for the data. (Data from Texas Weather Bureau records, 1964–1968, at Dennison Dam Station.)

3.13 In a batch process for the manufacture of ethyl cellulose, ethyl chloride is reacted with alkali cellulose to form the cellulose ether. The degree of reaction is dependent on the cellulose/ethyl chloride ratio, temperature, time, and agitator speed, each of which has been standardized. Product properties are strongly dependent on percent ethoxyl, a measure of degree of reaction. The following data on ethoxyl content (%) were taken on 201 batches of ethyl cellulose, all manufactured when the process was under normal conditions.

44.0	41.5	49.0	45.25	49.0	46.0
43.5	45.25	46.25	47.25	47.25	42.5
41.75	43.5	43.75	46.5	46.0	45.5
45.25	46.5	44.5	47.75	44.5	48.0
43.75	48.0	50.5	48.0	45.0	42.0
42.75	41.5	40.0	45.75	46.0	42.0
45.75	40.75	43.75	42.0	50.25	44.75
48.5	45.0	46.5	44.0	47.0	47.0
43.0	44.0	47.0	43.25	45.25	47.5
45.0	47.5	44.5	48.0	43.5	48.75
43.0	46.25	47.0	45.25	44.0	45.25
43.0	47.0	45.5	42.0	44.5	46.25
45.0	43.75	47.5	45.5	44.75	47.0
48.0	39.0	44.5	47.0	47.75	43.25
43.25	45.5	44.0	47.0	44.5	47.5
45.0	41.0	44.5	43.0	46.5	46.0
45.0	41.25	49.5	44.0	48.0	44.5
46.5	44.25	45.0	42.25	43.5	43.0
47.0	45.25	45.0	45.0	45.25	45.25
43.0	45.5	45.0	42.0	44.0	42.5
44.75	41.5	42.75	48.5	43.5	43.0

44.5	47.0	41.75	46.25	44.75	45.5
42.75	48.5	41.5	43.0	45.5	45.5
43.0	45.5	46.75	49.5	45.25	43.5
46.0	42.25	42.5	41.5	48.0	43.75
46.0	44.0	47.0	47.75	45.25	47.75
48.0	43.5	43.25	43.0	41.75	49.0
45.0	42.5	42.0	44.75	44.0	46.75
44.0	44.5	44.25	42.5	46.5	46.75
44.5	40.0	45.0	46.0	44.5	46.25
45.75	45.5	44.5	42.0	45.25	44.25
43.0	44.0	46.75	45.25	43.75	47.0
41.0	50.0	50.0	45.5	45.25	44.5
47.5	45.75	50.0			

(a) Tabulate the frequency and relative frequency of each value.

(b) Construct a frequency and relative frequency histogram, selecting appropriate class marks and class lengths.

(c) Construct cumulative frequency and relative frequency diagrams of the data using the same class marks and class lengths as in part (b).

3.14 The National Engineers Register contains the results* of a survey made in 1967 as to where chemical engineers are and what they are doing (basis: 3112 AIChE members).

Years of experience	Number of members	Percent
1 year or less	135	4
2–4	425	14
5–9	473	15
10–14	400	13
15–19	566	18
20–24	379	12
25–29	325	10
30–34	182	6
35–39	84	3
40 or more	83	3
No report	60	2

---

* Data from *Chem. Eng. Prog.* 65(4): 18–19 (1969). Copyright 1969 by the American Institute of Chemical Engineers. Reprinted by permission of the copyright owner.

Age	Number of members	Percent
24 or under	150	5
25–29	423	14
30–34	462	15
35–39	452	14
40–44	467	15
45–49	469	15
50–54	350	11
55–59	166	5
60–64	89	3
65–69	53	2
70 and over	29	1
No report	2	—

Plot the relative and cumulative relative distribution functions for ages and years of experience. Comment on the striking similarity of the pairs of graphs so obtained.

3.15 A new air pollution regulation requires that the total particulates (includes aerosols, fly ash, dust, and soot) be kept below $70 \pm 5 \mu g/m^3$. Using the data of Problem 3.1:

(a) What is the probability that the particulate concentration on any day will fall within this allowed range?
(b) What is the probability of exceeding the upper limit?
(c) What is the probability of operating in the region below 65 $\mu g/m^3$?

3.16 The time-weighted average (TWA) values of cotton dust concentration in a textile plant have been measured every semester for 2 years. For 1981:

	Concentration ($\mu g/m^3$) by shift							
	Spring TWAs				Fall TWAs			
Area	A	B	C	D	A	B	C	D
Wastehouse	731.3	1597.0	1222.0	1335.0	1734.0	2143.0	664.7	610.8
Warehouse	114.5	148.5	148.1	56.6	363.3	446.3	406.5	282.3

(continued)

Area	Concentration (μg/m^3) by shift							
	Spring TWAs				Fall TWAs			
	A	B	C	D	A	B	C	D
Opening	389.6	634.7	568.8	422.3	967.6	839.9	594.0	615.9
Carding	515.2	524.3	596.2	508.7	607.8	470.6	699.3	558.4
Drawing	307.2	234.4	293.4	274.9	299.3	295.2	313.0	251.3
Spinning	200.2	178.7	285.5	223.2	278.4	155.9	148.8	219.1
Warping	77.5	76.4	151.2	133.0	144.3	148.1	226.1	172.8
Dyeing	99.9	181.5	102.0	108.9	91.9	87.1	125.9	346.0
Sizing	381.2	345.9	327.6	309.8	341.0	404.4	368.9	374.4
Beaming	114.7	89.5	146.0	131.0	84.2	65.9	83.8	63.5
Weaving	75.0	88.6	88.6	68.9	140.2	217.7	89.3	64.7

For 1982:

Area	Concentration (μg/m^3) by shift							
	Spring TWAs				Fall TWAs			
	A	B	C	D	A	B	C	D
Wastehouse	1979.0	439.8	2184.0	289.4	615.7	1375.0	354.0	316.3
Warehouse	98.8	166.8	229.1	152.3	483.4	368.3	579.7	348.3
Opening	752.4	738.6	871.4	643.3	903.9	962.0	1017.6	735.2
Carding	621.2	594.6	729.7	547.9	539.0	499.2	510.1	357.6
Drawing	320.7	268.8	323.1	319.6	230.3	246.3	398.1	285.5
Spinning	249.0	192.9	212.1	209.0	149.8	139.2	188.3	165.2
Warping	193.2	186.6	188.4	167.2	139.1	211.8	153.7	169.5
Dyeing	168.8	65.7	194.1	253.7	83.9	100.8	74.6	134.2
Sizing	210.8	119.6	261.0	229.6	164.8	196.2	163.9	131.4
Beaming	87.2	130.8	120.6	110.9	90.7	66.1	66.2	74.5
Weaving	130.8	96.0	101.9	130.8	67.8	56.8	92.2	58.6

Prepare a suitable histogram and cumulative relative frequency polygon for these data (a) by season as assigned by your instructor, (b) by designated area over all seasons, or (c) for the entire 2-year period for the whole mill. Using the CRF polygon, what portion of the data is at or below the current OSHA standard of 200 $\mu g/m^3$ in opening through dyeing?

3.17   Repeated measurements of the flow rate of a cotton dust sampler gave the following flow rates (liters/min).

7.53	7.56	7.53	7.55	7.50	7.51
7.45	7.55	7.58	7.51	7.56	7.56
7.39	7.53	7.47	7.44	7.42	7.45
7.41	7.49	7.43	7.46	7.39	7.52
7.55	7.51	7.54	7.54	7.53	7.53
7.49	7.52	7.51	7.52	7.54	7.53
7.55	7.55	7.54	7.47	7.53	7.51
7.58	7.57	7.57	7.56	7.37	7.36

Prepare an appropriate histogram and cumulative relative frequency polygon for these data.

3.18   Experimental work has indicated that the probability for any group of seniors in the unit operations laboratory to obtain the correct value of α, the specific cake resistance in the filtration experiment, is 0.84.

(a)   What variability, as measured by the standard deviation, can be expected in their results?

(b)   If 12 groups of students are involved, what is the expected standard error?

3.19   For Problem 3.18, what are the chances that at least 11 of the groups will obtain the correct value for α?

3.20   If the determination of the correct value of α in Problems 3.18 and 3.19 follows the Poisson distribution, what is the probability of obtaining at least 11 correct α values from the 12 lab groups? As before, the probability for any group to obtain the correct answer is 0.84.

3.21   One of the input gases for an acetylene unit is purchased and piped from another plant. The gas is usually within 5% (relative) of the 98% purity required for the reactions. If the gas composition ever deviates from the required purity by more than 15%, an explosion is very likely to occur. If there have been three explosions in the past 10 years, what is the probability, now that the plant is operating again, of its operating safely for the next 5 years? No new safety or warning devices have been installed and the same gas producer is being used.

3.22   In an analytical instrumentation class, matched cuvettes must be chosen for a colorimetric determination of fluoride concentration. If the chance of one cuvette being the same as a previously chosen standard cuvette is 80%, what is the probability of getting exactly four matched cuvettes from a box containing 12?

3.23 The fluidized reactor cooling coils are guaranteed against rupture for 10 years. The plant is in its ninth year of operation. What is the probability of the coils failing for the first time in this year of operation? The probability of failure in any year is 6%.

3.24 For the production of a certain item by compression molding, it is known that the probability of making a defective unit is 0.06. If five items are inspected at a time:

(a) What is the probability that one of the five items checked is defective?

(b) What is the probability of getting two out of five that have no defects?

3.25 In the production of humidographs, suppose that control limits were set between two and three. That is, if two or fewer defective units in a sample of 10 are found, the quality (20% defective) has not changed. If three or more are found defective, it is assumed that a change in quality has occurred and that corrective action must be taken. What is the probability of finding six, seven, or eight defective units in a sample of 10?

$$\mathscr{P} = 0.2$$

Y	$\mathscr{P}(Y)$
0	0.1074
1	0.2684
2	0.3020
3	0.2013
4	0.0881
5	0.0264
6	0.0055
7	0.0008
8	0.0001
9	0.0000
10	0.0000

3.26 A polyurethane plant in charge of making "flexible foam" for dashboards in automobiles makes one 35,000-lb batch of polymer per 12-h shift. The plant runs continuously with 4 h in each shift being required for downtime to clean the reactors. The polymer is judged on-grade or off-grade by a rate of reaction vs. time test. If the system follows a Poisson distribution with $\lambda = 8.5$ on a basis of 1 week's operation, what is the probability of at least 1.5 million pounds of on-grade output per month?

3.27   In a certain industrial plant, vacancies have occurred in the technical department from 1960 to 1967 in the following manner:

Number of vacancies	Number of months
0	59 (months with 0 vacancies)
1	27 (months with 1 vacancy)
2	9
3	1

Using the equation for the Poisson distribution, calculate the theoretical values for the number of months and compare with the data given.

3.28   In a batch process for producing an oil viscosity additive, the probability of successive batches being on-grade is a variable whose distribution follows a Poisson distribution with $\lambda = 6$. What is the probability of the unit going for a week without being cleaned if one batch can be made every day?

3.29   A manufacturer guarantees that a shipment of Raschig rings has no more than 2% broken rings. A stockman takes a random sample of 100 rings. Assuming that the Poisson distribution applies, what is the highest number of broken rings the stockman should accept to be (a) 95% confident of the manufacturer's guarantee? (b) 99% confident?

3.30   The following table shows the days in a 50-day period during which varying numbers of automobile accidents occurred in a city.

Number of accidents	Number of days
0	21
1	18
2	7
3	3
4	1

Fit a Poisson distribution to the data.

3.31   If the number of graduate students in engineering departments is a random variable depending on student financial status, military status, intellectual drive, and scholastic achievement, and the distribution is closely approximated by the Poisson distribution with $\lambda = 10$, what is the probability of having fewer than 15 graduate students during the spring of any given year?

3.32–3.45 Using the data from Problems 3.1 through 3.14, determine the type of distribution followed in each case. Fit the distribution involved to each data set and estimate the distribution parameters.

3.46 Suppose that on a particular six-engine airplane it has been determined that the probability of an engine failing is 0.05. If the plane can fly on any three engines, what is the probability of a crash?

3.47 Given 12 resistors of which four are defective, what is the probability that a randomly selected sample of four (without replacement) will have at least two defectives?

3.48 Suppose that 10 television sets are available for testing. If five are selected randomly and classified as defective or nondefective, what is the probability that two are classified as defective if it is known that there are three defective sets in the 10 sets available?

3.49 The grades for the first exam in heat transfer follow. Prepare a histogram for these data assuming 10-point intervals for A, B, C, D, and F, with the lowest passing grade being 61. After you have the histogram, prepare the cumulative relative frequency polygon. What is the average grade as estimated from the polygon?

90	76	92	69
70	86	87	90
79	83	81	76
85	79	75	84
78	78	76	65
70	79	88	79
79	70	85	

4

Descriptive Statistics

4.1 INTRODUCTION

In using statistics it is necessary to make a clear distinction between populations and samples taken from the population. A population might for example, be represented by a collection of units such as books in a library or by the products resulting from a continuous chemical process. In many cases the population is quite large. In general, we desire estimates of parameters, such as the population mean and variance. These estimates are calculated from a sample taken from the population. Depending on the size of the population, the sample might represent anywhere from a fraction of 1% to the entire population. To draw conclusions about the population from the sample taken, the method known as *statistical inference* must be used.

Obviously, the sample taken from a population must be representative of the population or no useful inferences can be made. Therefore, the sample used must be random. A *random sample* can be thought of as a collection of values selected from a population in such a way that each value in the population has the same chance of being independently selected as a sample member. This definition of random sample is valid for infinite populations or when sampling from finite populations with replacement. In sampling from a finite population without replacement, a random sample of size n is one selected in such a way that every sample of size n has the same chance of being selected. The idea of a random sample is of utmost importance in statistical inference since most of its theory is based on the underlying assumption that the sample is random.

A value computed from a series of observations, or the values in a random sample, is called a *statistic*. The statistics most commonly encountered in engineering and scientific work are measures of location, such as the mean, median, and mode and measures of variability such as the variance, standard deviation, and range of a sample. We now discuss some statistics that describe a sample, and save the topic of statistical inference until Chapter 6.

4.2 MEASURES OF LOCATION

A measure of location is one that indicates where the "center" of the data is located. The first such measure we consider is the sample mean. The *sample mean*, which is the most important and often used single statistic, is defined as the sum of all the sample values divided by the number of observations in the sample and is denoted by

$$\bar{X} = \frac{\sum_{i=1}^{n} X_i}{n}, \tag{4.1}$$

where $\bar{X}$ is the mean of n values and X_i is any given value in the sample. It is an estimate of the value of the mean of the population from which the sample was drawn.

If a constant, say A, is added to the data (i.e., the data have a fixed offset, $W_i = X_i + A$), the mean can be found from

$$\bar{W} = \frac{\sum_{i=1}^{n} W_i}{n} = \frac{\sum_{i=1}^{n} (X_i + A)}{n} = \bar{X} + A. \tag{4.2}$$

That is, the mean is shifted by an amount A. In like manner, for the case where the data have been multiplied by a constant k, $Z_i = kX_i$, we have

$$\bar{Z} = \frac{\sum_{i=1}^{n} Z_i}{n} = \frac{\sum_{i=1}^{n} kX_i}{n} = k\bar{X}. \tag{4.3}$$

Combining the results in (4.2) and (4.3), the mean of the data transformed according to $U_i = (X_i - A)/B$ is

$$\bar{U} = \frac{\bar{X} - A}{B}. \tag{4.4}$$

The result (4.4) is frequently used to "standardize" a given set of data, as will be seen later.

If the items $X_1, X_2, \ldots, X_n$ in a random sample have weights $f_1, f_2, \ldots, f_n$ associated with them, the *weighted mean* is defined by

$$\bar{X} = \frac{\sum_{i=1}^n f_i X_i}{\sum_{i=1}^n f_i}.$$

For example, in an empirical frequency distribution such as the one in Table 3.5, the sample mean could be approximated by

$$\bar{X} = \frac{f_1 X_1 + f_2 X_2 + \cdots + f_c X_c}{f_1 + f_2 + \cdots + f_c},$$

where $X_1, X_2, \ldots, X_c$ denote the class marks and $f_1, f_2, \ldots, f_c$ the frequencies of the corresponding classes.

Example 4.1   Successive determinations of the opened bottles of HCl were found to be, expressed in normalities, N_i:

N_1	N_2
15.75	15.58
15.64	15.49
15.92	15.72

What was the average concentration of each bottle?
Solution   From Eq. (4.1), $\bar{X} = \sum_{i=1}^n X_i/n$; for bottle 1,

$$\bar{x}_1 = \frac{15.75 + 15.64 + 15.92}{3} = 15.770,$$

and for bottle 2,

$$\bar{x}_2 = \frac{15.58 + 15.49 + 15.72}{3} = 15.597.$$

Another measure of location that is frequently used is the median. The *median* is defined to be a value such that half of the observations in the sample have values smaller than the median and half have values larger than the median. If the sample has an odd number of observations, the median would be the "middle" value in the sample. If the sample has an even number of observations, a median would be any value between the two middle values in the sample. A third measure of location is the *mode*, which is defined as the most frequently occurring value in a sample.

4.3 MEASURES OF VARIABILITY

Having determined the location of our data as expressed by statistics such as the mean, median, and mode, the next thing to be considered is how the data are spread about these central values. Since spread about the median and mode is rarely, if ever, used in engineering work, we will restrict this discussion to the spread of data about the mean. The most popular method of reporting variability is by use of the *sample variance*, defined as

$$S_X^2 = \frac{\sum_{i=1}^{n}(X_i - \bar{X})^2}{n-1},$$ (4.5)

for which useful calculation formulas are

$$S_X^2 = \frac{\sum_{i=1}^{n} X_i^2 - n\bar{X}^2}{n-1} = \frac{\sum_{i=1}^{n} X_i^2 - (\sum X)^2/n}{n-1}.$$ (4.6)

Caution should be exercised, however, if one is writing a computer code to compute S_X^2. The formula in Eq. (4.6) should not be used in that case since it can involve the difference of two similar large numbers. The sample variance is the sum of the squares of the deviations of the data points from the mean value of the sample divided by $(n-1)$. The *standard deviation* of a sample, denoted by S_X, is defined to be the positive square root of the variance, that is,

$$S_X = \sqrt{S_X^2}.$$ (4.7)

Another term frequently encountered is the *coefficient of variation* (CV), defined by

$$CV = \frac{S_X}{\bar{X}}.$$

Another useful statistic is the sample *standard error* of the mean, defined by

$$S_{\bar{X}} = \sqrt{\frac{S_X^2}{n}} = \frac{S_X}{\sqrt{n}}.$$

This says that sample means tend to vary less than do the observations themselves.

It should be noted that the sum of the deviations of the sample values about the sample mean is zero, that is,

$$\sum_{i=1}^{n}(X_i - \bar{X}) = 0.$$

This prompts one to use the squares of the deviations, thus yielding (4.5). One could use the sum of the absolute values of the deviations, that is,

$$\sum_{i=1}^{n} |X_i - \bar{X}|.$$

However, (4.5) is mathematically more tractable and is the one most widely used.

If the sample $X_1, X_2, \ldots, X_n$ is transformed according to $Z_i = kX_i + A$, then

$$S_Z^2 = \frac{\sum_{i=1}^{n} (Z_i - \bar{Z}_i)^2}{n-1}$$

$$= \frac{\sum_{i=1}^{n} (kX_i + A - k\bar{X} - A)^2}{n-1}$$

$$= k^2 S_X^2. \tag{4.8}$$

From (4.8) it is seen that adding a constant value to each data point has no effect on the sample variance.

Equation (4.8) will prove useful in finding the variance of the data standardized according to $U_i = (X_i - A)/B$. According to (4.8) with $k = 1/B$ and A replaced with $-A/B$, we have

$$S_U^2 = \frac{1}{B^2} S_X^2. \tag{4.9}$$

In particular, if $A = \bar{X}$ and $B = \sqrt{S_X^2} = S_X$, we obtain

$$S_U^2 = \frac{1}{S_X^2} S_X^2 = 1.$$

By Eq. (4.4) we also have

$$\bar{U} = \frac{\bar{X} - \bar{X}}{S_X} = 0.$$

Data from any process that fluctuates over a wide range of values will have a large variance. Conversely, a large variance indicates that the data have a wide spread about the mean. In contrast, if all the sample values are very close together, the sample variance will be quite small. The latter case is usually highly desirable.

For example, in sampling a population one is interested in using the data in the sample to make inferences about the population. The inferences will be more precise if the variance of the sampled population is small. This is seen in Chapter 6 in the discussion of the estimation of the mean μ of a population.

The sample variance S_X^2 is merely an estimate of the population variance σ_X^2. The population variance of a population consisting of the values $\{x_1, x_2, \ldots, x_N\}$ is given by

$$\sigma_X^2 = \frac{\sum_{i=1}^{N}(x_i - \mu)^2}{N}, \tag{4.10}$$

where μ, the population mean, is given by

$$\mu = \frac{\sum_{i=1}^{N} x_i}{N}.$$

For a sample of size n, $X_1, X_2, \ldots, X_n$, S_X^2 estimates σ_X^2. Similarly, $\bar{X}$ estimates μ. The estimation of parameters will be taken up in Chapter 6 in more detail.

Another measure of variation that is sometimes used is the *range*. The range of a sample is defined to be the difference between the largest and smallest values in a sample.

Example 4.2 Successive colorimetric determinations of the normality of a $K_2Cr_2O_7$ solution were as follows (expressed as molarities $\times 10^{-4}$): 1.22, 1.23, 1.18, 1.31, 1.25, 1.22, and 1.24. What are the (a) mean, (b) median, (c) mode, (d) standard deviation of a single determination, (e) standard deviation of the mean, (f) range, and (g) coefficient of variation for this sample?

Solution

(a) Mean $= \bar{x} = \dfrac{\sum_{i=1}^{n} x_i}{n} = 1.2357 \times 10^{-4}\,M.$

(b) The median is found to be $1.23 \times 10^{-4}\,M$ from the definition.

(c) The mode is found to be $1.22 \times 10^{-4}\,M.$

(d) $s_X = \sqrt{s_X^2} = \left(\dfrac{\sum_{i=1}^{n} x_i^2 - n\bar{x}^2}{n-1}\right)^{1/2} M$

$\quad = \left(\dfrac{10.6983 \times 10^{-8} - 10.6889 \times 10^{-8}}{6}\right)^{1/2} M$

$\quad = \left(\dfrac{96 \times 10^{-12}}{6}\right)^{1/2} M \qquad = 3.95209 \times 10^{-6}\,M.$

(e) $s_{\bar{x}} = \sqrt{\dfrac{s_X^2}{n}} = \sqrt{\dfrac{96 \times 10^{-12}}{7(6)}}\,M \qquad = 1.49375 \times 10^{-6}\,M.$

(f) Range $= R = (1.31 \times 10^{-4} - 1.18 \times 10^{-4})\,M$
$\quad = 13.0 \times 10^{-6}\,M.$

(g) $CV = \dfrac{s_X}{\bar{x}} = 0.03198.$

Example 4.3 Obtain the average evaporation in inches per day from the evaporation data of Problem 3.12.

(a) What variance can be expected in the daily water evaporation rate?
(b) Find the standard error of the mean.
(c) Find the coefficient of variation.

Solution

(a) $\bar{x} = \dfrac{\sum_{i=1}^{n} x_i}{n} = \dfrac{54.22 \text{ in./day}}{155} = 0.3498 \text{ in./day}.$

(b) $s_x^2 = \dfrac{\sum_{i=1}^{n}(x_i - \bar{x})^2}{n-1} = \dfrac{\sum_{i=1}^{n} x_i^2 - n\bar{x}^2}{n-1}$

$= \dfrac{21.0108 - 155(0.3498)^2}{154} = 0.01327 \text{ in.}^2/\text{day}^2$

$s_x = \sqrt{s_x^2} = 0.1152 \text{ in./day}$

$s_{\bar{x}} = \dfrac{s_x}{\sqrt{n}} = \dfrac{0.1152}{\sqrt{155}} = 0.009253 \text{ in./day}.$

(c) $\text{CV} = \dfrac{s_x}{\bar{x}} = 0.3293.$

Example 4.4 Consider the orifice coefficient data of Example 3.1. What are the mean, range, standard deviation, standard error, variance, and coefficient of variation?
Solution To solve this using **PROC MEANS** (*SAS Introductory Guide*), the simplest approach would have been to insert the two statements

```
PROC MEANS STD MIN MAX RANGE VAR STDERR CV DATA = CO;
TITLE 'DESCRIPTIVE STATISTICS FOR ORIFICE COEFFICIENT DATA';
```

after the **PROC PRINT** statement in the original program in Example 3.1. The results [note that the coefficient of variation (CV) is just the standard deviation expressed as a percentage of the mean ($\bar{X}$)] are

MEAN = 0.63673077	RANGE = 0.4760	CV = 15.329
STANDARD DEVIATION = 0.09760467		VARIANCE = 0.00952667
MINIMUM VALUE = 0.4880		MAXIMUM VALUE = 0.9640
STANDARD ERROR OF MEAN = 0.01353533		

PROBLEMS

4.1 For convenience, it has been decided that the scores of the aptitude test which were presented in Problem 3.2 will be coded for calculational purposes by the following methods: (a) by dividing all values by 250 and (b) by subtracting 25 from all values. Let us say that this has been done and that the means of the resulting coded data have then been determined. How will you determine the mean of the original data from the means of the coded data? Answer for both coding methods.

4.2–4.13 Calculate the means and variances for the variables given in Problems 3.1 to 3.10 and 3.17 and by season, shift, or area for the data for Problem 3.16.

4.14 From the following tables of lignite conversion by a liquefaction reaction, calculate the mean, median, variance, standard error, and coefficient of variation.

69.2	52.7	74.8	60.2	65.4
75.6	57.5	75.2	89.4	51.4
41.8	86.8	76.7	71.4	77.2

4.15 During naval refueling operations, a sample of the fuel is taken, filtered, and then checked for conductivity measured in siemens (S) as an estimate of the degree of contamination. Samples were taken every 10 min during refueling, usually a 1- to 2-hr operation. The mean of the conductivities was used as a guideline for running the main fuel-oil filters on the standby tanks being filled. The standard deviations of the conductivities were used as a guide to how often the fuel-oil filters/coalescers should be checked. If $S_x \leq 0.06$ S, the filter/coalescer had to be checked only once per watch (4 h). Calculate the mean, standard deviation, standard error, and coefficient of variation for the fuel-oil conductivities that follow.

10-min interval	Conductivity (S)
1	0.2
2	0.01
3	0.1
4	0.15
5	0.03
6	0.09
7	0.08
8	0.07
9	0.05
10	0.02

5

Expected Values and Moments

5.1 INTRODUCTION

The expectation or expected value of a random variable is obtained by finding the average value of the variable over all possible values of the variable. Expectation of a random variable was introduced in Section 3.3, in discussing the mean and variance of a population. In this chapter we summarize those results and present a more complete treatment of expectation.

As an example, consider the tossing of two ideal coins. The distribution of X, the number of heads that appear, is given by the binomial probability function

$$\mathscr{P}(X = x) = b(x) = \binom{2}{x}\left(\frac{1}{2}\right)^2, \qquad x = 0,1,2.$$

This indicates that we expect no heads to appear with a relative frequency of $\frac{1}{4}$, one head to appear with a relative frequency of $\frac{1}{2}$, and two heads to appear with the relative frequency of $\frac{1}{4}$ in a large number of trials. Now let us find the average number of heads in 1000 trials. The total number of heads is expected to be

$$250 \times 0 + 500 \times 1 + 250 \times 2 = 1000$$

in the 1000 tosses of two coins; thus the average is expected to be one head per trial. This is the expected value, or mean value, of X, denoted by $\mu = E(X)$. The same result would be obtained if we merely multiplied all possible values of X by their probabilities and added the results; thus

$$E(X) = 0 \times \tfrac{1}{4} + 1 \times \tfrac{1}{2} + 2 \times \tfrac{1}{4} = 1.0.$$

This expected value is a theoretical or ideal average. A random variable does not necessarily take on its expected value in a given trial: however, we might reasonably expect the average value of the random variable in a great number of trials to be somewhere near its expected value.

It is often desirable to obtain the expected value of a function of a random variable. In the example above, one may want the expected value of $f(X) = X^2$. This is obtained by multiplying all possible values of X^2 by their probabilities and adding the results; thus

$$E(X^2) = (0)^2 \times \tfrac{1}{4} + (1)^2 \times \tfrac{1}{2} + (2)^2 \times \tfrac{1}{4} = 1.5.$$

The variance of a population, denoted by σ_X^2 or σ^2, is defined as the expected value of the square of the difference between X and the population mean μ, that is,

$$\sigma^2 = \sigma_X^2 = E[(X - \mu)^2].$$

In the example above the variance of the number of heads is

$$\begin{aligned}
\sigma_X^2 &= E[(X - \mu)^2] \\
&= E[(X - 1)^2] \\
&= (0 - 1)^2 \times \tfrac{1}{4} + (1 - 1)^2 \times \tfrac{1}{2} + (2 - 1)^2 \times \tfrac{1}{4} \\
&= \tfrac{1}{2}.
\end{aligned}$$

In general, if X is a random variable, then a function of X, say $g(X)$, is a random variable also, and its expected value is denoted by $E[g(X)]$.

We now consider the problem of computing certain expected values for discrete and continuous random variables.

5.2 DISCRETE DISTRIBUTIONS

A discrete probability function assigns probabilities to each of all (at most countable) possible outcomes of an experiment. Its distribution function may be looked on as a step function which changes only at those values that the corresponding random variable can take on. For a discrete distribution such that $\mathscr{P}(X = x_i) = p_i = f(x_i)$, for $i = 1, 2, \ldots, n$, the mean of X is

$$E(X) = \sum_{i=1}^{n} x_i p_i = \sum_{i=1}^{n} x_i f(x_i), \tag{5.1}$$

where x_i is the value of an outcome and p_i is the corresponding probability. The variance of X is given by

$$\sigma_X^2 = E[(X - \mu)^2] = \sum_{i=1}^{n} (x_i - \mu)^2 f(x_i). \tag{5.2}$$

Three properties of expectations will now be stated. These are that if k is a constant and X is distributed in accord with $f(x)$, and if $g(x)$ and $h(x)$ are any specific functions of X, then

$$E[k] = k, \tag{5.3}$$
$$E[kg(X)] = kE[g(X)], \tag{5.4}$$

and

$$E[g(X) + h(X)] = E[g(X)] + E[h(X)] \tag{5.5}$$

for discrete distributions. Properties (5.3), (5.4), and (5.5) can be proved very easily. Suppose that the probability function of X is given by $f(x_1)$, $f(x_2), \ldots, f(x_n)$; then

$$E[k] = \sum_{i=1}^{n} kf(x_i) = k \sum_{i=1}^{n} f(x_i) = k, \quad \text{as} \quad \sum_{i=1}^{n} f(x_i) = 1,$$

which proves (5.3). Now

$$E[kg(X)] = \sum_{i=1}^{n} kg(x_i)\mathscr{P}(X = x_i)$$

$$= \sum_{i=1}^{n} kg(x_i)f(x_i)$$

$$= k \sum_{i=1}^{n} g(x_i)f(x_i)$$

$$= kE[g(X)],$$

which proves (5.4). Finally,

$$E[g(X) + h(X)] = \sum_{i=1}^{n} [g(x_i) + h(x_i)] f(x_i)$$

$$= \sum_{i=1}^{n} [g(x_i) f(x_i) + h(x_i) f(x_i)]$$

$$= \sum_{i=1}^{n} g(x_i) f(x_i) + \sum_{i=1}^{n} h(x_i) f(x_i)$$

$$= E[g(X)] + E[h(X)],$$

which proves (5.5).

Properties (5.3), (5.4), and (5.5) may be used to derive a very useful method of evaluating the variance of a random variable. By (5.3)–(5.5) one has

$$\sigma_X^2 = E[(X - \mu)^2]$$
$$= E(X^2 - 2X\mu + \mu^2)$$
$$= E(X^2) - 2\mu E(X) + E(\mu)^2$$
$$= E(X^2) - 2\mu \cdot \mu + \mu^2$$
$$= E(X^2) - \mu^2$$
$$= E(X^2) - [E(X)]^2. \tag{5.6}$$

5.3 CONTINUOUS DISTRIBUTIONS

A continuous distribution is described in terms of a probability density function that assigns probabilities to intervals of values of the corresponding random variable. The probability that the random variable X takes on a value between two specific real values is given by

$$\mathscr{P}(a \le X \le b) = \int_a^b f(x)\, dx, \tag{5.7}$$

where $f(x)$ is the probability density function that models the population. The expected value of a continuous random variable is obtained by integration rather than the summation technique required for discrete distributions. The expected value of a continuous random variable X is defined by

$$E(X) = \int_{-\infty}^{\infty} x f(x)\, dx. \tag{5.8}$$

Properties similar to (5.3), (5.4), and (5.5) also hold for continuous distributions, as can be observed from the properties of integrals:

$$E(k) = \int_{-\infty}^{\infty} kf(x)\,dx = k \int_{-\infty}^{\infty} f(x)\,dx = k, \qquad (5.9)$$

$$E[kg(X)] = \int_{-\infty}^{\infty} kg(x)f(x)\,dx = k \int_{-\infty}^{\infty} g(x)f(x)\,dx$$
$$= kE[g(X)], \qquad (5.10)$$

$$E[g(X) + h(X)] = \int_{-\infty}^{\infty} [g(x) + h(x)]f(x)\,dx$$
$$= \int_{-\infty}^{\infty} g(x)f(x)\,dx + \int_{-\infty}^{\infty} h(x)f(x)\,dx$$
$$= E[g(X)] + E[h(X)]. \qquad (5.11)$$

5.4 JOINT DISTRIBUTIONS AND INDEPENDENCE OF RANDOM VARIABLES

Consider the experiment of tossing two fair dice. Let X and Y denote the outcomes of the first die and second die, respectively. The probability functions for X and Y are

$$f(x_i) = \mathscr{P}(X = x_i) = \tfrac{1}{6}, \qquad x_i = 1, 2, \dots, 6$$

and

$$g(y_i) = \mathscr{P}(Y = y_i) = \tfrac{1}{6}, \qquad y_i = 1, 2, \dots, 6.$$

We assume here that the outcomes of the two dice occur *independent* of each other. Thus we can make statements such as

$$\mathscr{P}(X = 1 \quad \text{and} \quad Y = 3) = \mathscr{P}(X = 1)\mathscr{P}(Y = 3)$$
$$= f(1)g(3)$$
$$= \tfrac{1}{6} \cdot \tfrac{1}{6}$$
$$= \tfrac{1}{36}, \qquad (5.12)$$

since the events $X = 1$ and $Y = 3$ are independent (see Chapter 2 for independence of two events A and B). To make *joint probability* statements involving X and Y, we use the model

$$h(x_i, y_j) = \mathscr{P}(X = x_i \quad \text{and} \quad Y = y_j) = \mathscr{P}(X = x_i)\mathscr{P}(Y = y_j)$$
$$= f(x_i)g(y_j) \qquad (5.13)$$

for all values of x_i and y_j. The function $h(x, y)$ is called the *joint probability function* of X and Y. Equation (5.13) prompts the following definition.

Definition 5.1 The random variables X and Y are said to be statistically independent if their joint probability function $h(x, y)$ can be written as $h(x, y) = f(x)g(y)$ for all values of x and y.

Definition 5.1 holds also if X and Y are continuous random variables. That is, two continuous random variables are statistically independent if their *joint probability density function* $h(x, y)$ factors into $f(x)g(y)$ for all real values x and y, where $f(x)$ and $g(y)$ denote the density functions of X and Y, respectively.

From Definition 5.1 it follows that if X and Y are independent, then for any functions $q(x)$ and $r(y)$,

$$E[q(X)r(Y)] = E[q(X)]E[r(Y)]. \tag{5.14}$$

Equation (5.14) follows from

$$
\begin{aligned}
E[q(X)r(Y)] &= \sum_{i=1}^{n} \sum_{j=1}^{m} q(x_i)r(y_j)h(x_i, y_j) \\
&= \sum_{i=1}^{n} \sum_{j=1}^{m} q(x_i)r(y_j)f(x_i)g(y_j) \\
&= \left[\sum_{i=1}^{n} q(x_i)f(x_i) \right]\left[\sum_{j=1}^{m} r(y_j)g(y_j) \right] \\
&= E[q(X)]E[r(Y)]
\end{aligned}
$$

for discrete random variables. For continuous random variables the similar result is obtained by replacing summations by integrals. A special case of (5.14) is obtained when $g(X) = X$ and $r(Y) = Y$, in which case we have

$$E(XY) = E(X)E(Y)$$

if X and Y are independent random variables.

The idea of a random sample has been discussed in Chapter 3. The outcomes in drawing a random sample from a population occur at random. Thus the random sample of size n is denoted by the random variables $X_1, X_2, \ldots, X_n$, which are taken to be statistically independent. That is, the joint distribution of $X_1, X_2, \ldots, X_n$ is given by

$$h(x_1, x_2, \ldots, x_n) = f_1(x_1)f_2(x_2) \cdots f_n(x_n) \tag{5.15}$$

for all real values $x_1, x_2, \ldots, x_n$, where $f_i(x_i)$ denotes the probability function or density function of X_i. The result in (5.15) is an extension of the independence of two random variables given in Definition 5.1.

Since each of the random variables $X_1, X_2, \ldots, X_n$ in a random sample comes from the same underlying population, each of the functions $f_i(x_i)$ in (5.15) may be replaced by the same function, say $f(x)$. Thus for a random sample $X_1, X_2, \ldots, X_n$ the corresponding joint distribution is described by

$$h(x_1, x_2, \ldots, x_n) = f(x_1) f(x_2) \cdots f(x_n) = \prod_{i=1}^{n} f(x_i). \tag{5.16}$$

The results given by Eqs. (5.5) and (5.11) can easily be extended to the case of a finite number of functions, that is,

$$E[g_1(X) + g_2(X) + \cdots + g_n(X)] = E[g_1(X)] + \cdots + E[g_n(X)]. \tag{5.17}$$

Now suppose that $X_1, X_2, \ldots, X_n$ is a random sample from some population with mean μ and variance $\sigma_X^2 = \sigma^2$. The sample mean has been defined to be

$$\bar{X} = \frac{\sum_{i=1}^{n} X_i}{n}.$$

By (5.17) we have

$$E(\bar{X}) = E\left(\frac{\sum_{i=1}^{n} X_i}{n}\right) = \frac{\sum_{i=1}^{n} E(X_i)}{n} = \frac{\sum_{i=1}^{n} \mu}{n} = \frac{n\mu}{n} = \mu, \tag{5.18}$$

which shows that the expected value or mean value of the sample means is equal to the population mean. The sample variance has been defined to be

$$S^2 = S_X^2 = \frac{\sum_{i=1}^{n} (X_i - \bar{X})^2}{n - 1}.$$

The expected value of S^2 is

$$E(S^2) = E\left[\frac{\sum_{i=1}^{n} (X_i - \bar{X})^2}{n - 1}\right]$$

$$= \frac{1}{n-1} E\left[\sum_{i=1}^{n} X_i^2 - \frac{(\sum_{i=1}^{n} X_i)^2}{n}\right].$$

$$= \frac{1}{n-1} \left\{\sum_{i=1}^{n} E(X_i^2) - \frac{1}{n} E\left[\left(\sum_{i=1}^{n} X_i\right)^2\right]\right\}.$$

However, $E(X_i^2) = \sigma^2 + \mu^2$, by Eq. (5.6). Furthermore,

$$E\left[\left(\sum_{i=1}^{n} X_i\right)^2\right] = E\left[\left(\sum_{i=1}^{n} X_i\right)\left(\sum_{j=1}^{n} X_j\right)\right]$$

$$= E\left(\sum_{i=1}^{n} \sum_{j=1}^{n} X_i X_j\right)$$

$$= E\left(\sum_{i=1}^{n} X_i^2 + \sum_{\substack{i=1 \\ i \neq j}}^{n} \sum_{j=1}^{n} X_i X_j\right)$$

$$= \sum_{i=1}^{n} E(X_i^2) + \sum_{\substack{i=1 \\ i \neq j}}^{n} \sum_{j=1}^{n} E(X_i)E(X_j)$$

$$= \sum_{j=1}^{n} (\sigma^2 + \mu^2) + \sum_{\substack{i=1 \\ i \neq j}}^{n} \sum_{j=1}^{n} \mu^2$$

$$= n(\sigma^2 + \mu^2) + n(n-1)\mu^2$$

$$= n\sigma^2 + n^2\mu^2.$$

Thus

$$E(S_X^2) = \frac{1}{n-1}\left[n(\sigma^2 + \mu^2) - \frac{1}{n}(n\sigma^2 + n^2\mu^2)\right]$$

$$= \frac{1}{n-1} \cdot [(n-1)\sigma^2]$$

$$= \sigma^2. \tag{5.19}$$

The results given by (5.18) and (5.19) say that the average values of $\bar{X}$ and S^2 are given by μ and σ^2, respectively. In estimation theory we say that $\bar{X}$ and S^2 are "unbiased" for μ and σ^2. (This topic is considered further in Chapter 6.) We note in passing that the reason for dividing by $(n-1)$ in the definition of S^2 was so that $E(S^2) = \sigma^2$.

5.5 MOMENTS

In general, the expected value $E[(X - a)^k]$, for k a positive integer and any number a, is referred to as the *k*th *moment about a* of the random variable X. In particular, if $a = 0$, we have $E(X^k)$, which is called the *k*th moment of X about the origin or simply the *k*th moment and is denoted by μ_k'. Letting $k = 1$ yields the first moment of X about the origin, that is, $\mu_1' = \mu$. The general form for moments about the origin for discrete distributions is

$$\mu_k' = E(X^k) = \sum_{i=1}^{N} x_i^k f(x_i), \qquad (5.20)$$

and for continuous distributions it is

$$\mu_k' = E(X^k) = \int_{-\infty}^{\infty} x^k f(x) \, dx. \qquad (5.21)$$

For many statistical analyses, values of the kth moment of X about the mean are most useful. These moments are called central moments and are denoted by μ_k. For discrete distributions, the general form for the kth central moment is

$$\mu_k = E(X - \mu)^k = \sum_{i=1}^{N} (x_i - \mu)^k f(x_i). \qquad (5.22)$$

For continuous distributions, the kth central moment is

$$\mu_k = E(X - \mu)^k = \int_{-\infty}^{\infty} (x - \mu)^k f(x) \, dx. \qquad (5.23)$$

For any distribution, the first moment about the mean is μ_1, where

$$\mu_1 = E(X - \mu) = E(X) - \mu = \mu - \mu = 0.$$

The second moment about the mean, μ_2, is the variance of X, denoted by σ_X^2 (or more often, σ^2). A convenient computational form, according to Eq. (5.6), is

$$\sigma^2 = \mu_2' - (\mu_1')^2 = \mu_2' - \mu^2. \qquad (5.24)$$

5.6 EXAMPLES

We now consider some examples to illustrate the main results in this chapter.
Example 5.1 The grades from the first hour exam in this course were

i	1	2	3	4	5	6	7	8	9	10	11
x_i	100	84	83	75	70	67	63	52	50	35	28
$p(x_i)$	$\frac{3}{14}$	$\frac{1}{14}$	$\frac{1}{14}$	$\frac{2}{14}$	$\frac{1}{14}$	$\frac{1}{14}$	$\frac{1}{14}$	$\frac{1}{14}$	$\frac{1}{14}$	$\frac{1}{14}$	$\frac{1}{14}$

Assuming that these data represent the entire population, calculate the mean and variance of the test grades from the appropriate moments.

Solution   We obtain

$$\mu = \mu_1' = \sum_{i=1}^{11} x_i p(x_i) = 70.14,$$

$$\sigma^2 = \mu_2 = \sum_{i=1}^{11} x_i^2 p(x_i) - (\mu_1')^2 = 492.24.$$

The standard deviation is 22.19.

Example 5.2   Given the function $f(x) = ce^{-x/4}$, $x \ge 0$, what must be the value of c if $f(x)$ is to be a probability density function?

Solution   The domain of the function is $0 < x < \infty$. Thus

$$\int_0^x f(x)\,dx = 1 \Rightarrow c \int_0^x e^{-x/4}\,dx = 1 \Rightarrow c = \tfrac{1}{4}.$$

Example 5.3   Find the mean and variance of X if $f(x) = 3/x^4$ for the range $1 < x < \infty$.

Solution

$$\mu_1' = \int_1^\infty x \cdot \frac{3}{x^4}\,dx = \int_1^x 3\,\frac{dx}{x^3} = \frac{3}{2} = \mu,$$

$$\mu_2' = \int_1^\infty x^2 \cdot \frac{3}{x^4}\,dx = \int_1^x 3\,\frac{dx}{x^2} = 3,$$

$$\sigma^2 = \mu_2 = \mu_2' - \mu^2 = 3 - \tfrac{9}{4} = \tfrac{3}{4}.$$

Example 5.4   Find the mean μ and variance σ^2 of the binomial distribution with parameters n and p.

Solution   The binomial probability function is given by Eq. (3.14) as

$$b(x) = \binom{n}{x} p^x (1-p)^{n-x}, \qquad x = 0,1,2,\ldots,n.$$

The mean and variance could be found directly by using the function $b(x)$ and the definition of mean and variance. However, to illustrate some of the ideas in this chapter, we use a somewhat different (but easier) approach. The binomial random variable is the number of successes in a series of n independent trials of an experiment that can yield one of two outcomes on each trial. Let $Y_1, Y_2, \ldots, Y_n$ denote the random outcomes of the n trials. Thus $\mathscr{P}(Y_i = 1) = p$ and $\mathscr{P}(Y_i = 0) = q = 1 - p$, for each i. The mean and variance of each Y_i are

$$E(Y_i) = 1 \cdot p + 0 \cdot (1 - p) = p$$

and

$$\sigma_Y^2 = E(Y_i^2) - [E(Y_i)]^2$$
$$= (1)^2 \cdot p + (0)^2(1-p) - p^2$$
$$\sigma_Y^2 = p(1-p)$$
$$= pq.$$

Now

$$X = \sum_{i=1}^{n} Y_i$$

and the mean and variance of X are

$$\mu = E(X) = \sum_{i=1}^{n} E(Y_i) = \sum_{i=1}^{n} p = np$$

and

$$\sigma^2 = E[(X-\mu)^2]$$
$$= E(X^2) - \mu^2 \qquad \text{[by (5.6)]}$$
$$= E\left(\sum_{i=1}^{n} Y_i \right)^2 - \mu^2$$
$$= E\left(\sum_{i=1}^{n} Y_i^2 + \sum_{\substack{i=1 \\ i \neq j}}^{n} \sum_{j=1}^{n} Y_i Y_j \right) - \mu^2$$
$$= \sum_{i=1}^{n} p + \sum_{\substack{i=1 \\ i \neq j}}^{n} \sum_{j=1}^{n} p^2 - \mu^2$$
$$= np + n(n-1)p^2 - n^2 p^2$$
$$= np - np^2$$
$$= np(1-p)$$
$$= npq.$$

In this example it should be noted that $Y_1, Y_2, \ldots, Y_n$ is actually a random sample of size n from the population modeled by the point binomial probability function

$$f(x) = p^x(1-p)^{1-x}, \qquad x = 0,1.$$

Example 5.5 Past history indicates that the probability of a given batch of chemical being on-grade is 0.95. Let X denote the number of off-grade batches obtained in 200 runs. What are the mean and variance of X?

Solution   In this example $p = 1 - 0.95 = 0.05$ and $n = 200$. Thus the mean and the variance of X are

$$\mu = E(X) = np = 200(0.05) = 10$$

and

$$\sigma^2 = npq = 200(0.05)(0.95) = 9.5.$$

PROBLEMS

5.1   Given the function $F(x) = x^2/\theta$ for which $\theta > 0$ and $0 < x < k$:

(a)   For what value of k is $F(x)$ a cumulative distribution function?
(b)   What is the associated probability density function $f(x)$?

5.2   If $f(x)$ is the probability density function given by $f(x) = ax^2$ where $0 \leq x \leq 1$, find a.

5.3   A random variable X has values in the range $2 \leq x \leq 4$, with probability density function $f(x) = A(3 + 2x)$.

(a)   What is the value of A?
(b)   What are the mean and variance of X?
(c)   What is the probability that X is between 3 and 4?
(d)   What is the probability that X is less than 3?

5.4   Given that $f(x) = 1/\sqrt{2\pi} \exp(-x^2/2)$ for $-\infty < x < \infty$, what is the corresponding cumulative distribution function?

5.5   (a)   What are the required conditions for $f(x)$ to be a probability density function?

(b)   If $f(x) = c/x^2$ for $x > 600$, find the value of x_0 for which $\mathscr{P}(X \geq x_0) = \frac{1}{2}$.

5.6   Find the mean and variance of X whose probability density function is $f(x) = 2e^{-2x}$ over the range $0 < x < \infty$.

5.7   Find the mean and variance of the Poisson distribution.

5.8   The life for a particular transistor in an oscilloscope can be expressed by the probability density function $f(x) = 100/x^2$, where $100 < x < \infty$. What is the probability that not one of the four transistors in our scope will have to be replaced within the first 250 hours of operation?

5.9   A probability density function is $f(x) = 2x, 0 \leq x \leq 1$. Find:

(a)   The cumulative distribution function $F(x)$
(b) $\mathscr{P}(\frac{1}{4} \leq X \leq \frac{1}{2})$
(c)   The mean and variance of X

5.10 Given that X_1, X_2, and X_3 are statistically independent and each has a mean of 1 and a variance of 2, find $E(Y)$ and $E(Z)$ for

$$Y = 2X_1 - X_2 - X_3,$$
$$Z = X_2 + X_3.$$

5.11 A probability density function may be expressed as $f(x) = C(\theta)(1 - \sin x)^\theta \cos x$ for $0 < x < \pi/2$. Find $C(\theta)$.

5.12 The following values were obtained for a probability density function: $\mu'_1 = \frac{3}{2}$, $\mu'_2 = 3$, $\sigma^2 = \frac{3}{4}$. If the range of X is $1 < x < \infty$, what is the probability density function? You have reason to believe that $f(x)$ may be expressed as an inverse integral power of x.

5.13 Given that the variance of each $\bar{Y}_i$ is 12, what is the variance of

$$X = 3\bar{Y}_1 - 2\bar{Y}_2 + 3\bar{Y}_3 - \bar{Y}_4$$

assuming that the $\bar{Y}_i$ are independent?

5.14 If $X_1, X_2, \ldots, X_n$ denotes a random sample from a continuous distribution, show that the variance of $\bar{X}$ is σ^2/n.

5.15 If $X_1, X_2, \ldots, X_n$ are statistically independent with density functions $f_1, \ldots, f_n$, show that the variance of $\sum_{i=1}^n X_i$ is $\sum_{i=1}^n \sigma_{X_i}^2$, where $\sigma_{X_i}^2$ denotes the variance of X_i.

6

Statistical Inference: Estimation

6.1 INTRODUCTION

The area of statistical inference may be divided into two broad categories: estimation and hypothesis testing. In particular, these two categories pertain to the estimation of parameters or testing hypotheses involving the parameters of a given distribution that models a corresponding underlying population. The parameters that are involved in a distribution are, in general, unknown. Consequently, it is of interest and importance to be able to obtain information regarding them. This information is obtained in part by estimating them and involves the calculation of numerical values from sample data that estimate the true values of the corresponding population parameters. These numerical values are called estimates of the actual population parameters. In this chapter we develop statistical procedures for estimating the parameters of various populations. The corresponding ideas concerned with hypothesis testing are the subject of Chapter 7.

6.2 STATISTICAL ESTIMATION

Engineers are frequently faced with the problem of using data to calculate quantities that they hope will describe the behavior of a process under consideration. This calculated estimate is subject to error because of random fluctuations of the process as well as random measurement errors. The ability to estimate the magnitude of the errors involved is important. This is where the method of statistical estimation can be most useful.

Statistical estimation procedures use sample data to obtain the "best" possible estimates of the corresponding population parameters. In addition, the estimates furnish a quantitative measure of the probable error involved in the estimation. There are two types of estimates: point estimates and interval estimates. *Point estimates* use sample data to calculate a single best value that estimates a population parameter. Point estimates alone give no idea of the magnitude of the error involved in the estimation process. On the other hand, an *interval estimate* gives a range of values that can be expected to include the correct value a certain specified percentage of the time. This gives the engineer a measure of the probable error involved in the estimation.

6.3   POINT ESTIMATES

The usefulness of a point estimate depends on the criteria by which it is judged. The properties of estimators which are used to help us pick that particular estimator that is most useful for our purposes are unbiasedness, efficiency, and consistency. Ideally, an estimator should be unbiased, highly efficient, and consistent.

We will take an estimator to be a statistic (actually, a random variable) and an estimate to be a particular value of the estimator. An estimator (or estimate) of θ is conventionally denoted by $\hat{\theta}$ ($\hat{}$ is a circumflex) in statistical literature. For example, estimates of μ, σ^2, and $\sigma_{\bar{X}}^2$ would be denoted by $\hat{\mu}$, $\hat{\sigma}^2$, and $\hat{\sigma}_{\bar{X}}^2$, respectively. We will use this notation together with the usual notation for random variables (i.e., uppercase Latin letters) to denote estimators.

An estimator is unbiased if on the average it predicts the correct value. More precisely, we have the following definition.

Definition 6.1   An estimator $T = \hat{\theta}$ is an unbiased estimator of the parameter θ if its expected value is equal to that of the parameter itself, that is, $E(T) = \theta$.

For example, the sample mean $\bar{X} = \hat{\mu}$ is an unbiased estimator of the population mean μ since

$$E(\bar{X}) = \mu, \tag{6.1}$$

which was shown in Section 5.4. In the same section it was shown that the expected value of the sample variance

$$\hat{\sigma}^2 = S^2 = \frac{\sum_{i=1}^{n}(X_i - \bar{X})^2}{n-1} \tag{6.2}$$

is σ^2; that is, $E(S^2) = \sigma^2$. Thus S^2 is an unbiased estimator of σ^2.

An estimator T will be said to be efficient for θ if it yields estimates that are "close" to the value of θ. Now one way of measuring the degree of closeness of the values of T to θ is by considering the variation of T about its expected value θ, that is, by considering the variance σ_T^2 of T. The variance of an unbiased estimator T measures the spread of values of the estimator about the parameter θ. The estimator that has the smallest variance is said to be the most efficient estimator. That is, the estimator T is the most efficient estimator for θ if $E(T) = \theta$ and if $\sigma_{T_1}^2 \geq \sigma_T^2$, where T_1 denotes any other unbiased estimator for θ. An efficient estimator is sometimes called the *best* estimator and is an estimator that is unbiased and has the smallest possible variance.

If $X_1, X_2, \ldots, X_n$ denotes a sample of size n from a normal population with mean μ and variance σ^2, the sample mean $\bar{X}$ and the sample variance S^2 are the best estimators for μ and σ^2. The efficiency of any other unbiased estimator T of μ relative to $\bar{X}$ is given by

$$\mathrm{eff}(T, \bar{X}) = \frac{\sigma_{\bar{X}}^2}{\sigma_T^2} \times 100\%.$$

The variance of $\bar{X}$ is given by $\sigma_{\bar{X}}^2 = \sigma^2/n$. This can easily be shown by the methods of Section 5.4 (see also Problem 5.14). Suppose that

$$T = \frac{X_1 + X_2 + \cdots + X_{n-1}}{n - 1};$$

that is, T is the sample mean based on the first $n - 1$ observations. Then $E(T) = \mu$ and $\sigma_T^2 = \sigma^2/(n - 1)$. Thus

$$\mathrm{eff}(T, \bar{X}) = \frac{\sigma^2/n}{\sigma^2/(n - 1)} \times 100\%$$

$$= \frac{n - 1}{n} \times 100\%.$$

If $n = 10$, then $\mathrm{eff}(T, \bar{X}) = 90\%$.

The following definition summarizes the idea of best estimator.

Definition 6.2 An estimator T of θ is said to be best for estimating θ if

$$E(T) = \theta$$

and

$$\sigma_T^2 \leq \sigma_{T_1}^2,$$

where T_1 is any other estimator such that $E(T_1) = \theta$.

The third property of estimators to be considered is that of consistency. An estimator is consistent for θ if it yields values that get closer and closer to the true value of the parameter θ as the size of the sample is increased. As an example, as the size of the sample increases, the sample mean $\bar{X} = \hat{\mu}$ approaches the population mean μ. Therefore, we can say that $\bar{X}$ is a consistent estimator for μ. To say that $\bar{X}$ approaches μ as the sample size increases means that $\mathscr{P}(|\bar{X} - \mu| > \varepsilon)$ tends to zero as the sample size n increases without bound where ε denotes any small positive number. This says that the values of $\bar{X}$ are concentrated very close to μ, with a very high probability (close to 1) if the sample size is large. This also follows from the central limit theorem since $\bar{X}$ has an approximate normal distribution with mean μ and variance σ^2/n. However, σ^2/n tends to zero as n tends to ∞, which also says that the values of $\bar{X}$ are close to μ with a high probability. In the case of a normal population with mean μ and variance σ^2, the sample mean $\bar{X}$ has an exact normal distribution with mean μ and variance σ^2/n. Thus

$$
\begin{aligned}
\mathscr{P}(|\bar{X} - \mu| > \varepsilon) &= 1 - \mathscr{P}(|\bar{X} - \mu| \le \varepsilon) \\
&= 1 - \mathscr{P}(-\varepsilon \le \bar{X} - \mu \le \varepsilon) \\
&= 1 - \mathscr{P}\left(\frac{-\varepsilon}{\sigma/\sqrt{n}} \le \frac{\bar{X} - \mu}{\sigma/\sqrt{n}} \le \frac{\varepsilon}{\sigma/\sqrt{n}} \right) \\
&= 1 - \int_{-\varepsilon/(\sigma/\sqrt{n})}^{\varepsilon/(\sigma/\sqrt{n})} \frac{1}{\sqrt{2\pi}} e^{-x^2/2} \, dx.
\end{aligned}
\tag{6.3}
$$

The integral in Eq. (6.3) tends to 1 as n increases without bound since the limits of integration approach $-\infty$ to $+\infty$. Thus $\mathscr{P}(|\bar{X} - \mu| > \varepsilon)$ tends to 0 as n tends to ∞ and $\bar{X}$ is consistent for μ.

The requirement of consistency is not a demanding one since most reasonable estimators satisfy it. We will not pursue it any further but rather go to the more interesting and useful idea of interval estimation.

6.4 INTERVAL ESTIMATES

From Section 6.3 it has been seen that a point estimate is a single value that gives information about the parameter for which it is an estimate. T is a good estimator for θ if the values of T fall close to θ, where the degree of closeness is measured by σ_T^2. After an estimate has been made, the engineer may be asked about how good it is. It seems that if $E(T) = \theta$ and σ_T^2 is small, one should be able to determine an interval of values about a given estimate t of θ such that the interval is "reasonably short" and one is "reasonably sure" that the true value of θ is in that interval. Such an interval, denoted by (t_L, t_U), is an interval estimate of θ; that is, $t_L < \theta < t_U$. The endpoints of the interval depend on the point estimate t and on the distribution of the estimator T. Since σ_T^2, the variance of T, measures how closely the values of T are concentrated about θ, it follows that the length of an interval estimate will vary according to the magnitude of σ_T^2.

The most important issue involving interval estimates is a quantification of what we mean when we say that we should be "reasonably sure" that the interval estimate contains the true value of the unknown parameter θ. Every sample from the population for which θ is a parameter yields a value of T, the estimator of θ, which, in turn, yields an interval estimate (t_L, t_U) of θ. Now every sample will produce a different estimate of θ and thus a different interval estimate of θ. The distribution of T will allow us to find a t_L and a t_U for all possible samples such that a specified proportion of the intervals will contain θ. For example, t_L and t_U may be computed so that 95% of the intervals contain θ. Thus choosing a sample and computing (t_L, t_U) is comparable to choosing one interval from the entire collection of intervals of which 95% contain θ. The particular interval (t_L, t_U) is called a *95% confidence interval*, and we say we are 95% confident that the interval (t_L, t_U) contains θ. The quantities t_L and t_U are called the confidence limits.

In general the endpoints t_L and t_U are particular values of the random variables T_L and T_U. The interval (T_L, T_U) is a *random interval* of which (t_L, t_U) is a particular outcome. When we say that (t_L, t_U) is a 95% confidence interval for θ we actually mean, prior to taking the sample, that $\mathscr{P}(T_L \leq \theta \leq T_U) = 0.95$. However, once the sample has been observed and (t_L, t_U) has been computed, we say that (t_L, t_U) is a 95% confidence interval; that is, 95% of all possible outcomes of (T_L, T_U) contain θ. In general, we may desire (t_L, t_U) to be such that the proportion $1 - \alpha$, $0 < \alpha < 1$, of all intervals contains θ. The proportion $1 - \alpha$ is called the confidence coefficient and (t_L, t_U) is called a $(1 - \alpha) \times 100\%$ confidence interval for θ.

Ideally, one wants α to be small (between 0.01 and 0.10) and $t_U - t_L$ to be small. In the ensuing sections we present the most efficient estimators for certain parameters, corresponding to certain populations and the corresponding confidence intervals. The next three sections discuss important distributions that arise in constructing confidence intervals on the parameters of normal populations and binomial populations. These distributions were introduced in Section 3.4. However, a more detailed discussion is presented here for ready reference.

The parameters of the normal distribution are the mean μ and the variance σ^2. Sections 6.8 to 6.10 contain statistical inference procedures for the mean(s) of a normal population(s). These procedures are given in the form of confidence intervals, derived under the assumption of normality. These methods work rather well for other types of distributions: the assumption of normality is not critical in the interval estimation of μ.

In Sections 6.11 to 6.13 we discuss techniques for the estimation of the variance(s) under the assumption of normality. The lack of normality in this case is quite serious in that it can affect the results quite adversely. If nonnormality is expected, one should use other techniques, such as those found in Conover [1].

6.5 CHI-SQUARE DISTRIBUTIONS

The chi-square (χ^2) distribution was introduced in Section 3.4.1. It will be used to make inferences concerning population variances and for other inferential purposes in Chapter 7 not directly involving parameters, such as the problem of testing a hypothesis of independence of two attributes.

The random variable χ^2 has a chi-square distribution with v degrees of freedom (d.f.) if it has the density function

$$f(\chi^2) = \frac{(\chi^2)^{v/2-1}e^{-\chi^2/2}}{[(v/2)-1]!\,2^{v/2}} = \frac{(\chi^2)^{v/2-1}e^{-\chi^2/2}}{2^{v/2}\Gamma(v/2)}, \qquad \chi^2 > 0, \quad v > 0, \quad (6.4)$$

which is a continuous function. The parameter v can be any positive real value; however, for our purposes v can be considered to take on integer values, that is, $v = 1,2,\dots$. These values of v arise quite naturally in many situations. For example, if $X_1, X_2, \dots, X_n$ denotes a random sample of size n from a standard normal population, the random variable

$$V = \sum_{i=1}^{n} X_i^2$$

has a χ^2-distribution with $v = n$ d.f. If the sample is from a normal population with mean μ and variance σ^2, the random variable

$$V = \frac{\sum_{i=1}^{n}(X_i - \bar{X})^2}{\sigma^2}$$

has a χ^2-distribution with $v = n - 1$ d.f. All these results can be proved using the notions from advanced statistical theory.

The mean and variance of the χ^2-distribution (6.4) are

$$E(\chi^2) = v$$

and

$$\sigma_{\chi^2}^2 = \text{var}(\chi^2) = 2v.$$

Thus as the mean v increases, so does the variance, and the region of "heavy concentration" of the distribution is shifted to the right. Figure 6.1 gives the graph of a typical χ^2-density function ($v = 8$ d.f.). The quantity $\chi_{v,\alpha}^2$ is that

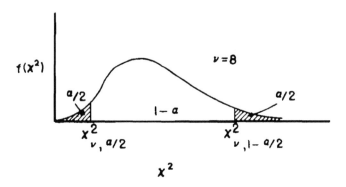

FIGURE 6.1 χ^2-distribution.

number that has the fraction α of the area under the curve to the left. That is,

$$\mathscr{P}(\chi^2 \le \chi^2_{v,\alpha}) = \alpha. \tag{6.5}$$

For example, $\chi^2_{v,\alpha/2}$ and $\chi^2_{v,1-\alpha/2}$ in Fig. 6.1 are such that

$$\mathscr{P}(\chi^2_{v,\alpha/2} \le \chi^2 \le \chi^2_{v,1-\alpha/2}) = 1 - \alpha.$$

Equation (6.5) is actually the distribution function of a χ^2 random variable and is tabulated in Table V of Appendix B for various values of α and v.

For values of v greater than 30, the χ^2-distribution may be approximated from the standard normal distribution as follows:

$$\chi^2_{v,\alpha} = \tfrac{1}{2}(z_\alpha + \sqrt{2v - 1})^2, \tag{6.6}$$

where z_α is the equivalent percentile of the standard normal variable. For example, $z_{0.975} = 1.96$, so that for $v = 30$ we find from Eq. (6.6) that $\chi^2_{30,0.975} = 46.5$ compared to the value $\chi^2_{30,0.975} = 47.0$ found in Table V.

6.6   THE t-DISTRIBUTION

The t-distribution, introduced in Section 3.4.2, will be used in this chapter in regard to the interval estimation of the mean μ when the variance is unknown. A random variable T has the t-distribution with v degrees of freedom if its density function is

$$f(t) = \frac{\Gamma[(v + 1)/2]}{\sqrt{\pi v}\,\Gamma(v/2)[1 + t^2/v]^{(v+1)/2}}, \qquad -\infty < t < \infty.$$

The density function contains the parameter v, which for our purposes can be assumed to take on the values $v = 1, 2, \ldots$.

The density function $f(t)$ resembles the standard normal in that its graph is also bell-shaped with a mean of zero (if $v > 1$). However, its variance exceeds that of the normal, for all v (i.e., it has "heavier" tails). It can be shown that

$$\lim_{v \to \infty} f(t) = \frac{1}{\sqrt{2\pi}} e^{-t^2/2};$$

that is, for large values of v the t-distribution can be approximated by the standard normal. The approximation is actually quite good for $v > 30$.

Let $X_1, X_2, \ldots, X_n$ denote a random sample from a normal population with mean μ and variance σ^2. A well-known result in statistical theory is that the random variable

$$\frac{(\bar{X} - \mu)\sqrt{n}}{\sqrt{\sum_{i=1}^{n}(X_i - \bar{X})^2/(n-1)}} \tag{6.7}$$

has a t-distribution with $v = n - 1$ d.f.

The relation of the t-distribution to the normal and χ^2-distributions is as follows. If Z has a standard normal distribution and U has a χ^2-distribution with v d.f., and if Z and U are independent, then the random variable

$$\frac{Z}{\sqrt{U/v}} \tag{6.8}$$

has a t-distribution with v d.f. The random variable above follows from (6.8) by letting

$$Z = \frac{\bar{X} - \mu}{\sigma/\sqrt{n}},$$

$$U = \sum_{i=1}^{n} \frac{(X_i - \bar{X})^2}{\sigma^2},$$

and

$$v = n - 1.$$

Figure 6.2 illustrates a typical t curve in comparison to the standard normal curve. The quantity $t_{v,\alpha}$ denotes that number such that a fraction α of the area is to the left of $t_{v,\alpha}$; that is, $\mathscr{P}(T \le t_{v,\alpha}) = \alpha$. For example, $t_{v,\alpha/2}$ and $t_{v,1-\alpha/2}$ in Fig 6.2 are such that $\mathscr{P}(t_{v,\alpha/2} \le T \le t_{v,1-\alpha/2}) = 1 - \alpha$. This notation will be used later. The cumulative t-distribution function $\mathscr{P}(T \le t_{v,\alpha})$ is contained in Table IV of Appendix B) for several values of v and α.

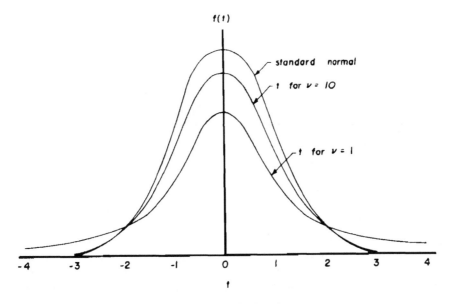

FIGURE 6.2 Graphs of standard normal and *t*-density curves.

THE F-DISTRIBUTION

The *F*-distribution, introduced in Section 3.4.3, will be used in the comparison of two variances and in the analysis of variance and related ideas. A random variable *F* has an *F*-distribution with v_1 and v_2 degrees of freedom if its density function is

$$g(f) = \left(\frac{\Gamma((v_1 + v_2)/2)(v_1/v_2)^{v_1/2}}{\Gamma(v_1/2)\Gamma(v_2/2)}\right)\left(\frac{f^{(v_1 - 2)/2}}{(1 + v_1 f/v_2)^{(v_1 + v_2)/2}}\right), \qquad f > 0.$$

The two parameters v_1 and v_2 can take on noninteger values: however, we will use $v_1 = 1,2,\ldots$ and $v_2 = 1,2,\ldots$.

The *F*-distribution can arise in a variety of ways. For example, if *V* has a χ^2-distribution with v_1 d.f. and *U* has a χ^2-distribution with v_2 d.f., and if *U* and *V* are independent, then

$$F = \frac{V/v_1}{U/v_2} = \frac{\chi_1^2/v_1}{\chi_2^2/v_2}$$

has an *F*-distribution with v_1 and v_2 d.f. Similarly,

$$F = \frac{U/v_2}{V/v_1}$$

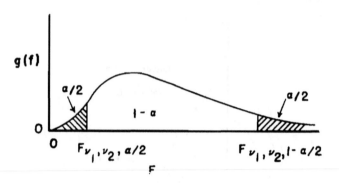

FIGURE 6.3 *F*-distribution.

has an *F*-distribution with v_2 and v_1 d.f. Note that the "first" d.f. are associated with the numerator and the "second" d.f. with the denominator.

If $X_1, X_2, \ldots, X_{n_1}$ and $Y_1, Y_2, \ldots, Y_{n_2}$ are independent random samples from the two normal populations with parameters μ_1, σ_1^2 and μ_2, σ_2^2, respectively, then

$$F = \frac{S_1^2/\sigma_1^2}{S_2^2/\sigma_2^2} = \frac{\sum_{i=1}^{n_1} [(X_i - \bar{X})^2/(n_1 - 1)]/\sigma_1^2}{\sum_{i=1}^{n_2} [(Y_i - \bar{Y})^2/(n_2 - 1)]/\sigma_2^2}$$

has an *F*-distribution with $v_1 = n_1 - 1$ and $v_2 = n_2 - 1$ d.f. If $\sigma_1^2 = \sigma_2^2$, then *F* becomes S_1^2/S_2^2, the ratio of two sample variances.

Figure 6.3 gives the graph of a typical *F*-density function. The value $F_{v_1, v_2, \alpha}$ is such that

$$\mathscr{P}(F \leq F_{v_1, v_2, \alpha}) = \alpha. \tag{6.9}$$

Figure 6.3 illustrates the values $F_{v_1, v_2, \alpha/2}$ and $F_{v_1, v_2, 1 - \alpha/2}$ such that $\mathscr{P}(F_{v_1, v_2, \alpha/2} \leq F \leq F_{v_1, v_2, 1 - \alpha/2}) = 1 - \alpha$. The distribution corresponding to (6.9) is tabulated in Table VI of App. B, for several values of α and v_1, v_2. The parameter v_1 corresponds to the numerator d.f.

6.8 ESTIMATION OF THE MEAN

Suppose that $X_1, X_2, \ldots, X_n$ is a sample from a normal population with mean μ and *known* variance σ^2. The most efficient estimator of the population mean μ is given by the sample mean $\hat{\mu} = \bar{X} = \sum_{i=1}^{n} X_i/n$. Thus the best point estimate of μ is $\bar{x}$. Recall that the expected value and variance of $\bar{X}$ are $E(\bar{X}) = \mu$ and $\sigma_{\bar{X}}^2 = \sigma^2/n$, respectively.

The statistic $\bar{X}$ has a normal distribution with mean μ and variance σ^2/n. Thus the random variable

$$Z = \frac{\bar{X} - \mu}{\sigma/\sqrt{n}}$$

has the standard normal distribution; that is, $\mu_Z = E(Z) = 0$ and $\sigma_Z^2 = 1$. Thus

$$\mathscr{P}(-z_{1-\alpha/2} < Z < z_{1-\alpha/2}) = 1 - \alpha, \tag{6.10}$$

where $z_{1-\alpha/2}$ is that value in the standard normal distribution that has $1 - \alpha/2$ area to the left, that is, $\mathscr{P}(Z < -z_{1-\alpha/2}) = \alpha/2$, and $z_{\alpha/2} = -z_{1-\alpha/2}$.

Substituting for Z in (6.10) gives

$$\mathscr{P}\left(-z_{1-\alpha/2} < \frac{\bar{X} - \mu}{\sigma/\sqrt{n}} < z_{1-\alpha/2}\right) = 1 - \alpha. \tag{6.11}$$

Multiplying each member of the inequality by $\sigma/\sqrt{n}$ and subtracting $\bar{X}$ from each member yields

$$\mathscr{P}\left(-\bar{X} - z_{1-\alpha/2}\frac{\sigma}{\sqrt{n}} < -\mu < -\bar{X} + z_{1-\alpha/2}\frac{\sigma}{\sqrt{n}}\right) = 1 - \alpha.$$

Then multiplying each member by -1, we have

$$\mathscr{P}\left(\bar{X} - z_{1-\alpha/2}\frac{\sigma}{\sqrt{n}} < \mu < \bar{X} + z_{1-\alpha/2}\frac{\sigma}{\sqrt{n}}\right) = 1 - \alpha. \tag{6.12}$$

The interval

$$(T_L, T_U) = \left(\bar{X} - z_{1-\alpha/2}\frac{\sigma}{\sqrt{n}}, \bar{X} + z_{1-\alpha/2}\frac{\sigma}{\sqrt{n}}\right)$$

is a random interval. For a particular sample we have $\bar{x}$ as a point estimate of μ and

$$(t_L, t_U) = \left(\bar{x} - z_{1-\alpha/2}\frac{\sigma}{\sqrt{n}}, \bar{x} + z_{1-\alpha/2}\frac{\sigma}{\sqrt{n}}\right) \tag{6.13}$$

is a $(1 - \alpha)100\%$ confidence interval for μ, and the endpoints of the interval are called $(1 - \alpha)100\%$ confidence limits for μ.

Note that (6.11) is a bona fide probability statement. However, once the sample is taken and the random interval takes on the form in (6.13), we use the term *confidence interval* to mean that $(1 - \alpha)100\%$ of all such intervals cover the mean μ.

   The confidence interval in (6.13) was derived under the assumption of normality. However, the central limit theorem states that, regardless of what the underlying distribution is (as long as σ^2 is finite), the distribution of $\bar{X}$ is approximately normal (Section 3.5.1). Thus (6.13) gives confidence limits for μ which are generally used without regard to the underlying distribution as long as n is sufficiently large. Furthermore, $\bar{X}$ is a rather efficient estimator of μ regardless of the underlying distribution.

Example 6.1   Referring to the data of Example 4.2, suppose that the true variance of an individual observation had been $49 \times 10^{-10} M$. Since seven samples were analyzed,

$$\sigma_{\bar{x}} = \sqrt{\frac{\sigma^2}{n}} = \sqrt{\frac{49 \times 10^{-10}}{7}} = \frac{7 \times 10^{-5}}{\sqrt{7}} = 2.646 \times 10^{-5} M.$$

At the 95% confidence level, $z_{1-\alpha/2} = z_{0.975} = 1.96$.

   Recalling that $\bar{x} = 12.36 \times 10^{-5}$, the 95% confidence interval for the mean μ of the $K_2Cr_2O_7$ determinations can be found from Eq. (6.13) as

$$12.36 \times 10^{-5} M - 2.646 \times 10^{-5} M (1.96)$$
$$< \mu < 12.36 \times 10^{-5} M + 2.646 \times 10^{-5} M (1.96)$$

or

$$7.17 \times 10^{-5} M < \mu < 17.55 \times 10^{-5} M$$

or approximately,

$$0.72 \times 10^{-4} M < \mu < 1.76 \times 10^{-4} M.$$

   The confidence limits for μ given in (6.13) were obtained under the assumption that σ^2, the population variance, is known. If σ^2 is *not known*, the usual procedure is to replace it by its estimate, $s^2 = \sum_{i=1}^{n} (x_i - \bar{x})^2/(n-1)$. However, the distribution of $Z = (\bar{X} - \mu)/(S/\sqrt{n})$ is no longer normal, so that the development which led to (6.13) is no longer valid. We are saved by observing that the distribution of $T = (\bar{X} - \mu)/S_{\bar{x}}$ is the t-distribution with $n-1$ degrees of freedom, which is discussed in Section 6.6. The t-distribution has a curve that is symmetrical, just like the normal curve. Furthermore, if $n > 30$, one could just as well use the normal distribution since the t and normal curves agree very closely for all such sample sizes.

   Proceeding as in (6.10), we have

$$\mathscr{P}(-t_{n-1,1-\alpha/2} < T < t_{n-1,1-\alpha/2}) = 1 - \alpha,$$

where $t_{n-1,1-\alpha/2}$ is that value in the t-distribution with $n-1$ degrees of freedom that has $1 - \alpha/2$ to the left; that is,

$$\mathscr{P}(T < t_{n-1,1-\alpha/2}) = 1 - \frac{\alpha}{2} \quad \text{or} \quad \mathscr{P}(T < -t_{n-1,1-\alpha/2}) = \frac{\alpha}{2}.$$

Replacing T by $(\bar{X} - \mu)/S_{\bar{X}}$, we have

$$\mathcal{P}\left(-t_{n-1,1-\alpha/2} < \frac{\bar{X} - \mu}{S_{\bar{X}}} < t_{n-1,1-\alpha/2}\right) = 1 - \alpha,$$

which yields

$$\mathcal{P}\left(\bar{X} - t_{n-1,1-\alpha/2}\frac{S_X}{\sqrt{n}} < \mu < \bar{X} + t_{n-1,1-\alpha/2}\frac{S_X}{\sqrt{n}}\right) = 1 - \alpha. \qquad (6.14)$$

The result in (6.14) can be obtained from (6.12) by replacing σ and z in (6.12) by S and t, respectively. From (6.14) we obtain

$$\left(\bar{x} - t_{n-1,1-\alpha/2}\frac{S_X}{\sqrt{n}}, \bar{x} + t_{n-1,1-\alpha/2}\frac{S_X}{\sqrt{n}}\right) \qquad (6.15)$$

as the $(1 - \alpha)100\%$ confidence limits for μ when σ^2 is unknown. The result (6.15) is very popular and useful even when the sampled population is only approximately normal.

Example 6.2 The following data were obtained for the calibration of the Ruska dead weight gauge used with our Burnett P-V-T apparatus. The weights corresponding to 100 psi had the following apparent masses[*]:

26.03570	26.03575	26.03599
26.03581	26.03551	26.03533
26.03529	26.03588	26.03570
26.03573	26.03586	

What is the 95% confidence limit for the apparent mass of the 100-psi weight? *Solution* The data are first coded according to $W_i = (X_i - 26.03500)10^5$. For $n = 11$, the following values were obtained: $\sum_i w_i = 755$; $\sum w_i^2 = 56{,}787$; $\bar{w} = \sum w_i/n = 68.6$; $\bar{x} = 26.03500 + \bar{w}(10^{-5}) = 26.03568$ psi; and $s_w = [(\sum (w_i)^2 - \bar{w}\sum w_i)/(n-1)]^{1/2} = 22.36$ psi. s_X is therefore found from s_w by $s_X = s_w(10^{-5}) = 22.36(10^{-5})$ psi. The standard error of the mean, $s_{\bar{x}}$, is then $6.741(10^{-5})$ psi. Since the degrees of freedom are $n - 1 = 10$, we find that $t_{10,0.025} = -2.228$ from Table IV of App. B. The desired 95% confidence limits for the mean μ in psi (remembering that $\bar{x}$ and $s_{\bar{x}}$ are used for particular or calculated values) are, by (6.15),

$$\bar{x} - s_{\bar{x}}t_{10,1-\alpha.2} < \mu < \bar{x} + s_{\bar{x}}t_{10,1-\alpha/2}$$

---

[*] C. E. Miks. *Test Report, Ruska Dead Weight Gauge*, Ruska Instruments Corp., Houston, Texas (1964). Gauge and report owned by Chemical Engineering Department, Texas Tech University, Lubbock, Texas. Data reprinted by permission of the department.

$$26.03568 - 2.228(6.741 \times 10^{-5})$$
$$< \mu < 26.03568 + 2.228(6.741 \times 10^{-5})$$
$$26.03553 < \mu < 26.03583.$$

Example 6.3   Chief design engineer Gant of the Frigid-Flow Corporation has received final test results on the company's new heat exchanger. The following values are overall heat-transfer coefficients (U_c, Btu/hr ft^2 °F) for the exchanger as determined by Gant's testing section: 60, 63, 60, 68, 70, 72, 65, 61, 69, 67. Frigid-Flow employs a 5% *minimizing factor* in their calculations before offering specifications to buyers. At the 99% confidence level, what minimum value for the exchanger's overall heat-transfer coefficient can Gant suggest for Frigid-Flow?

Solution   The sample mean and variance are

$$\bar{x} = \frac{\sum_{i=1}^{n} x_i}{n} = 65.5 \text{ Btu/hr ft}^2 \text{ °F}$$

and

$$s_X = \sqrt{s_X^2} = 4.3525 \text{ Btu/hr ft}^2 \text{ °F}.$$

Furthermore,

$$s_{\bar{x}} = \frac{4.3525}{\sqrt{10}} = 1.3764 \text{ Btu/hr ft}^2 \text{ °F}.$$

At the 99% confidence level, $\alpha/2 = 0.005$. Since the degrees of freedom are 9, the t-table yields $t_{9,0.995} = 3.250$. Then the 99% confidence limits for the mean U_c are

$$\bar{x} - t_{9,1-\alpha/2}s_{\bar{x}} < U_c < \bar{x} + t_{9,1-\alpha/2}s_{\bar{x}}$$
$$65.5 - 3.25(1.3764) < U_c < 65.5 + 3.25(1.3764)$$
$$61.03 \text{ Btu/hr ft}^2 \text{ °F} < U_c < 69.97 \text{ Btu/hr ft}^2 \text{ °F}.$$

Using the minimizing factor, we have

$$(0.05)61.03 = 3.05 \text{ Btu/hr ft}^2 \text{ °F}.$$

Thus the "minimum" value is

$$61.03 - 3.05 = 57.98 \text{ Btu/hr ft}^2 \text{ °F}.$$

Gant will suggest a value of 58 Btu/hr ft^2 °F for Frigid-Flow.

6.9 COMPARISON OF TWO MEANS

The next problem we consider is that of comparing the means of two populations. Consider two normal populations, the first with mean μ_1 and variance σ_1^2 and the second with mean μ_2 and variance σ_2^2. The means μ_1 and μ_2 can be compared by considering their difference, $\mu_1 - \mu_2$. Thus the aim here will be to construct a confidence interval for $\mu_1 - \mu_2$. Our development will be broken down into three cases: (1) σ_1^2 and σ_2^2 *known*, (2) σ_1^2 and σ_2^2 *unknown* but *equal*, and (3) σ_1^2 and σ_2^2 *unknown* and *unequal*.

Case 1 (σ_1^2 and σ_2^2 Known)

Suppose that $X_{11}, X_{12}, \ldots, X_{1n_1}$ is a random sample of size n_1 from a normal population with mean μ_1 and variance σ_1^2, and $X_{21}, X_{22}, \ldots, X_{2n_2}$ is a random sample from a normal population with mean μ_2 and variance σ_2^2. Note that the first subscript in X_{ij} identifies the population (1 or 2) and the second subscript identifies the observation within the sample. The best estimators for μ_1 and μ_2 are given by

$$\bar{X}_1 = \frac{\sum_{i=1}^{n_1} X_{1i}}{n_1}$$

and

$$\bar{X}_2 = \frac{\sum_{i=1}^{n_2} X_{2i}}{n_2},$$

respectively. Since the variances are assumed known there is no need to estimate them. Now

$$E(\bar{X}_1 - \bar{X}_2) = \mu_1 - \mu_2,$$

and the variance of $\bar{X}_1 - \bar{X}_2$ is given by

$$\sigma_{\bar{X}_1 - \bar{X}_2}^2 = \sigma_{\bar{X}_1}^2 + \sigma_{\bar{X}_2}^2 = \frac{\sigma_1^2}{n_1} + \frac{\sigma_2^2}{n_2} \tag{6.16}$$

since $\bar{X}_1$ and $\bar{X}_2$ are independent. Thus

$$Z = \frac{\bar{X}_1 - \bar{X}_2 - (\mu_1 - \mu_2)}{\sigma_{\bar{X}_1 - \bar{X}_2}} \tag{6.17}$$

has mean 0 and variance 1; furthermore, Z has a standard normal distribution. Therefore, we may write

$$\mathscr{P}(-z_{1-\alpha/2} < Z < z_{1-\alpha/2}) = 1 - \alpha,$$

and replacing Z according to Eq. (6.17), we have

$$\mathscr{P}\left(-z_{1-\alpha/2} < \frac{\bar{X}_1 - \bar{X}_2 - (\mu_1 - \mu_2)}{\sigma_{\bar{X}_1 - \bar{X}_2}} < z_{1-\alpha/2}\right) = 1 - \alpha. \tag{6.18}$$

By solving the inequality in Eq. (6.18) for $\mu_1 - \mu_2$, in exactly the same way as Eq. (6.12) was obtained, we have

$$\mathscr{P}(\bar{X}_1 - \bar{X}_2 - z_{1-\alpha/2}\sigma_{\bar{x}_1 - \bar{x}_2} < (\mu_1 - \mu_2) < \bar{X}_1 - \bar{X}_2 + z_{1-\alpha/2}\sigma_{\bar{x}_1 - \bar{x}_2})$$
$$= 1 - \alpha, \tag{6.19}$$

where

$$\sigma_{\bar{x}_1 - \bar{x}_2} = \sqrt{\frac{\sigma_1^2}{n_1} + \frac{\sigma_2^2}{n_2}}.$$

Thus a $(1 - \alpha)100\%$ confidence interval for $\mu_1 - \mu_2$, where σ_1^2 and σ_2^2 are known, is given by

$$\left(\bar{x}_1 - \bar{x}_2 - z_{1-\alpha/2}\sqrt{\frac{\sigma_1^2}{n_1} + \frac{\sigma_2^2}{n_2}}, \bar{x}_1 - \bar{x}_2 + z_{1-\alpha/2}\sqrt{\frac{\sigma_1^2}{n_1} + \frac{\sigma_2^2}{n_2}} \right). \tag{6.20}$$

Example 6.4 Consider the data presented in Example 4.1. Information received from the manufacturer indicates that the average variance in concentration for single analyses for last year's production runs was 0.016. Construct a 95% confidence interval for $\mu_1 - \mu_2$ where μ_1 and μ_2 are the true means of the analyses for the two bottles.

Solution We take the variance of the analysis for each bottle (population) to be 0.016 (i.e., $\sigma_1^2 = \sigma_2^2 = 0.016$). From Example 4.1 we have $\bar{x}_1 = 15.770$ and $\bar{x}_2 = 15.597$. Furthermore, $z_{1-\alpha/2} = z_{0.975} = 1.96$ and

$$\sigma_{\bar{x}_1 - \bar{x}_2} = \sqrt{\frac{\sigma_1^2}{n_1} + \frac{\sigma_2^2}{n_2}}$$

$$= \sqrt{\frac{0.016}{3} + \frac{0.016}{3}}$$

$$= \sqrt{\frac{0.032}{3}}$$

$$= 0.10327.$$

Thus a 95% confidence interval for $\mu_1 - \mu_2$, according to (6.20), is

$$\bar{x}_1 - \bar{x}_2 - 1.96(0.10327) < \mu_1 - \mu_2 < \bar{x}_1 - \bar{x}_2 + 1.96(0.10327)$$

or

$$-0.029 < \mu_1 - \mu_2 < 0.375.$$

This means we are 95% confident that the interval $(-0.029, 0.375)$ contains the unknown difference $\mu_1 - \mu_2$ in it. Since the interval contains 0, we have no real reason to infer that $\mu_1 \neq \mu_2$ (or $\mu_1 - \mu_2 \neq 0$), that is, that the two reagent bottles were not filled from the same bottle. In light of this reasoning we see that the confidence interval may be used to "test" the hypothesis that

$\mu_1 = \mu_2$. These remarks are examined in more detail in Chapter 7, where testing hypotheses is related to confidence intervals.

We now consider confidence intervals on $\mu_1 - \mu_2$ when σ_1^2 and σ_2^2 are unknown but when $\sigma_1^2 = \sigma_2^2 = \sigma^2$ is a reasonable assumption.

Case 2 (σ_1^2 and σ_2^2 Unknown but Equal)

If the variances are unknown but can be presumed equal, each sample can be used to obtain an estimator for the common variance σ^2. These estimators are given by

$$S_1^2 = \frac{\sum_{i=1}^{n_1}(X_{1i} - \bar{X}_1)^2}{n_1 - 1}$$

and

$$S_2^2 = \frac{\sum_{i=1}^{n_2}(X_{2i} - \bar{X}_2)^2}{n_2 - 1}.$$

Since S_1^2 and S_2^2 are both estimators for σ^2, we may "pool" them to obtain a better estimator for σ^2, given by

$$\begin{aligned} S_p^2 &= \frac{(n_1 - 1)S_1^2 + (n_2 - 1)S_2^2}{n_1 + n_2 - 2} \\ &= \frac{\sum_{i=1}^{n_1}(X_{1i} - \bar{X}_1)^2 + \sum_{i=1}^{n_2}(X_{2i} - \bar{X}_2)^2}{n_1 + n_2 - 2}. \end{aligned} \tag{6.21}$$

The denominator $n_1 + n_2 - 2$ makes S_p^2 unbiased for σ^2; that is, $E(S_p^2) = \sigma^2$. Furthermore, the variance of $\bar{X}_1 - \bar{X}_2$ is given by

$$\sigma_{\bar{X}_1 - \bar{X}_2}^2 = \frac{\sigma^2}{n_1} + \frac{\sigma^2}{n_2} \quad \text{(since } \sigma_1^2 = \sigma_2^2 = \sigma^2\text{)}$$

and the best estimator for $\sigma_{\bar{X}_1 - \bar{X}_2}^2$ is given by

$$S_{\bar{X}_1 - \bar{X}_2}^2 = S_p^2 \left(\frac{1}{n_1} + \frac{1}{n_2} \right). \tag{6.22}$$

The random variable

$$T = \frac{\bar{X}_1 - \bar{X}_2 - (\mu_1 - \mu_2)}{S_{\bar{X}_1 - \bar{X}_2}},$$

which is actually (6.17) with $\sigma_{\bar{X}_1 - \bar{X}_2}$ replaced by $S_{\bar{X}_1 - \bar{X}_2}$, has a t-distribution with $v = n_1 + n_2 - 2$ degrees of freedom. Thus we can say that

$$\mathscr{P}(-t_{v,1-\alpha/2} < T < t_{v,1-\alpha/2}) = 1 - \alpha \tag{6.23}$$

or

$$\mathscr{P}\left(-t_{v,1-\alpha/2} < \frac{\bar{X}_1 - \bar{X}_2 - (\mu_1 - \mu_2)}{S_{\bar{X}_1 - \bar{X}_2}} < t_{v,1-\alpha/2}\right) = 1 - \alpha, \qquad (6.24)$$

and consequently,

$$\mathscr{P}(\bar{X}_1 - \bar{X}_2 - t_{v,1-\alpha/2}S_{\bar{X}_1 - \bar{X}_2} \le \mu_1 - \mu_2 \le \bar{X}_1 - \bar{X}_2 + t_{v,1-\alpha/2}S_{\bar{X}_1 - \bar{X}_2})$$
$$= 1 - \alpha. \qquad (6.25)$$

A $(1 - \alpha)100\%$ confidence interval in Case 2 is thus

$$\bar{x}_1 - \bar{x}_2 - t_{v,1-\alpha/2}s_P\sqrt{\frac{1}{n_1} + \frac{1}{n_2}} < \mu_1 - \mu_2 < \bar{x}_1 - \bar{x}_2$$

$$+ t_{v,1-\alpha/2}s_P\sqrt{\frac{1}{n_1} + \frac{1}{n_2}}, \qquad (6.26)$$

where s_p is defined by Eq. (6.21) and $v = n_1 + n_2 - 2$.

The results for Cases 1 and 2 are quite robust; that is, they are valid even when the underlying population is only approximately normal.

Example 6.5 To compare the effects of two solid catalyst component concentrations on NO_2 reductions, six groups of observations were made. Each group consisted of three replicates of five observations each; that is, a total of 15 determinations were made for each concentration. The concentrations (in mass percent) were 0.5 and 1.0%. The reduction data are summarized below.

Group	Replicate means (5 observations each)			Group mean (based on 15 observations)
	A	B	C	
1	5.18	5.52	5.42	$5.37333 = \bar{x}_1$
2	5.58	5.62	5.82	$5.67333 = \bar{x}_2$

Construct a 95% confidence interval on $\mu_1 - \mu_2$, the difference of the means corresponding to concentrations of 0.5% and 1.0%.

Solution The replicate means are treated as individual observations. Thus $\bar{x}_1 = (5.18 + 5.52 + 5.42)/3 = 5.37$ and $\bar{x}_2 = 5.67$. Furthermore, the sample variance of 5.18, 5.52, and 5.42 is $s_1^2 = 0.0305334$ and similarly, $s_2^2 = 0.0165334$.

It is assumed that $\sigma_1^{2'} = \sigma_2^2$, so that

$$s_p^2 = \frac{2s_1^2 + 2s_2^2}{3 + 3 - 2} = \frac{0.0610668 + 0.0330667}{4} = 0.0235334.$$

Since the confidence level is 95% and $n_1 = 3$, $n_2 = 3$, we have from the t-distribution table with 4 d.f., $t_{4,0.975} = 2.776$. Thus by using the result given in (6.26), we have

$$\bar{x}_1 - \bar{x}_2 - 2.776(0.153406\sqrt{\tfrac{1}{3} + \tfrac{1}{3}}) < \mu_1 - \mu_2 < \bar{x}_1 - \bar{x}_2$$
$$+ 2.776(0.153406\sqrt{\tfrac{2}{3}}), \qquad \bar{x}_1 - \bar{x}_2 \pm 0.347709$$

or

$$-0.647709 < \mu_1 - \mu_2 < 0.047709.$$

The 95% confidence interval contains 0, and on this basis the results of the experiment do not warrant inferring that $\mu_1 \neq \mu_2$. Thus we conclude that one catalyst component is not more (or less) effective than the other. From this example we again see that we can use a confidence interval to "test" the hypothesis that $\mu_1 = \mu_2$.

We now consider the third case in constructing confidence intervals for $\mu_1 - \mu_2$.

Case 3 (σ_1^2 and σ_2^2 Unknown and Unequal)

Suppose that $X_{11}, X_{12}, \ldots, X_{1n_1}$ and $X_{21}, X_{22}, \ldots, X_{2n_2}$ are random samples from normal populations with mean and variance (μ_1, σ_1^2) and (μ_2, σ_2^2). σ_1^2 and σ_2^2 are *not* known and are *not* assumed to be equal.

If the two unknown population variances cannot be presumed equal, the individual sample variances, s_1^2 and s_2^2, cannot be pooled, and construction of the $(1 - \alpha)100\%$ confidence interval is based on the random variable

$$T = \frac{\bar{X}_1 - \bar{X}_2 - (\mu_1 - \mu_2)}{(S_1^2/n_1 + S_2^2/n_2)^{1/2}}, \tag{6.27}$$

where T has an approximate t-distribution with f degrees of freedom, where f is defined by

$$f = \frac{(s_1^2/n_1 + s_2^2/n_2)^2}{(s_1^2/n_1)^2/(n_1 - 1) + (s_2^2/n_2)^2/(n_2 - 1)}. \tag{6.28}$$

We have

$$\mathscr{P}(-t_{f,1-\alpha/2} < T < t_{f,1-\alpha/2}) = 1 - \alpha, \tag{6.29}$$

where $t_{f,1-\alpha/2}$ is obtained from the t-table with degrees of freedom f given by (6.28). (If $\sigma_1^2 = \sigma_2^2 = \sigma^2$, recall that the d.f. were $n_1 + n_2 - 2$.) By substituting T defined in (6.27) into (6.29), we obtain the $(1 - \alpha)100\%$ confidence limits

$$\bar{x}_1 - \bar{x}_2 \pm t_{f,1-\alpha/2} \sqrt{\frac{s_1^2}{n_1} + \frac{s_2^2}{n_2}}. \qquad (6.30)$$

The fact that T defined by Eq. (6.27) has an approximate t-distribution causes (6.30) to approximate a $(1 - \alpha)100\%$ confidence interval.

We postpone an example for Case 3 until Chapter 7, where we look at the analogous problem concerning the hypothesis that $\mu_1 = \mu_2$.

6.10   ESTIMATION INVOLVING PAIRED OBSERVATIONS

In some studies involving data that consist of two samples, one cannot make the assumption that the two samples are independent of each other as was done in Section 6.5. This happens when, by the very nature of the study, the observations occur in pairs. For example, in studying the weather resistance of two types of paints, each of five shingles is painted with both paints, one-half with one paint and the other half with the other paint. An observation from one shingle is an ordered pair. Another example involves the question of whether two methods of measuring the strength of materials are numerically equivalent (except for statistical imprecision). Each specimen is scored by each method, giving rise to a pair of values.

Suppose that the samples $X_1, X_2, \ldots, X_n$ and $Y_1, Y_2, \ldots, Y_n$ occur in a way that results in the natural pairings $(X_1, Y_1), (X_2, Y_2), \ldots, (X_n, Y_n)$, where X_i and Y_i are not independent, but (X_i, Y_i) is independent of (X_j, Y_j) for each $i \neq j$. Assume that X and Y have normal distributions with means μ_X and μ_Y, respectively. The fundamental problem here is that of obtaining an inference concerning $\mu_X - \mu_Y$. (For example, can we say that μ_X and μ_Y are different?) The appropriate procedure is to calculate the difference $D_i = X_i - Y_i, i = 1, 2, \ldots, n$, and observe that $\mu_D = \mu_X - \mu_Y$. We have thus reduced a two-sample problem to a one-sample problem which is exactly like the one in Section 6.4. To construct a confidence interval on $\mu_X - \mu_Y$, we need only to construct one on μ_D. Let σ_D^2 denote the population variance of $D = X - Y$. According to the confidence interval (6.15), a $(1 - \alpha)100\%$ confidence interval on $\mu_D = \mu_X - \mu_Y$ is given by

$$\bar{d} - t_{n-1,1-\alpha/2} \frac{s_D}{\sqrt{n}} < \mu_D < \bar{d} + t_{n-1,1-\alpha/2} \frac{s_D}{\sqrt{n}}, \qquad (6.31)$$

where

$$\bar{d} = \frac{\sum_{i=1}^{n} d_i}{n} = \frac{\sum_{i=1}^{n} (x_i - y_i)}{n}$$

$$= \bar{x} - \bar{y} \qquad (6.32)$$

$$s_D^2 = \frac{\sum_{i=1}^{n}(d_i - \bar{d})^2}{n-1},\qquad(6.33)$$

and $t_{v,1-\alpha/2}$ is the usual tabular t value using $n-1$ degrees of freedom.

Example 6.6 In studying weather resistance of two types of paint, each of five shingles was painted with both paints, one-half of the shingle with one of the paints and the other half of the shingle with the other paint. After a month's exposure, wear, measured in suitable units, was found to be (1.57,1.45), (1.46,1.59), (1.53,1.27), (1.30,1.48), (1.37,1.40). Is it plausible that μ_D is greater than zero? less than zero? Between what two limits may μ_D reasonably be expected to lie?

Solution   We have

$$d_i = x_i - y_i,$$
$$d_1 = +0.12,$$
$$d_2 = -0.13,$$
$$d_3 = +0.26,$$
$$d_4 = -0.18,$$
$$d_5 = -0.03,$$

and

$$\bar{d} = 0.008.$$

By (6.33) we have

$$s_D = 0.1815764$$

and

$$s_{\bar{D}} = \frac{s_D}{\sqrt{n}} = 0.0812034.$$

Using $n-1 = 4$ d.f., we have $t_{4,0.975} = 2.776$ and the 95% confidence interval is

$$\bar{d} - t_{4,0.975}s_{\bar{D}} \le \mu_D \le \bar{d} + t_{4,0.975}s_{\bar{D}}$$
$$-0.2174 \le \mu_D \le 0.2334.$$

From the 95% confidence interval on μ_D, we come to the conclusion that μ_D is "very nearly" 0. If we had conjectured $\mu_D = 0$ prior to collecting the data, the resulting confidence interval does not seem to contradict that conjecture. It should be noted that by using both types of paint on each shingle we have eliminated any possible errors due to differences in the individual shingles.

Example 6.7   The question is whether two methods of measuring the strength of materials are numerically equivalent, except for statistical imprecision (i.e., we recognize that the two techniques cannot be expected to yield identical results even for identical specimens, but we are wondering whether "on the average" nearly identical strength values tend to be scored for nearly identical specimens). Fifteen matched test specimens have been scored by the two methods, with the following results: (338,327), (156,232), (243,248), (267,246), (195,192), (203,222), (262,261), (225,223), (214,216), (292,285), (218,230), (151,142), (168,181), (223,234), (260,236). One of the 30 measurements above is in error. Attempt to locate this measurement on sight. Delete the maverick pair, and on the basis of the remaining 14 d's, compute $\bar{d}$ and $s_{\bar{D}}$. What about the question of the equivalence of the two methods?
Solution   Pair 2 is obviously the maverick pair. If we throw out the maverick pair, we find that $t_{13,0.975} = 2.160$, from Table IV of App. B. Also, we find that $\bar{d} = 1.14285$,

$$s_D = 12.672545, \qquad s_{\bar{D}} = 3.38688,$$

or

$$\bar{d} - t_{n-1,1-\alpha/2} s_{\bar{D}} \le \mu_D \le \bar{d} + t_{n-1,1-\alpha/2} s_{\bar{D}}$$
$$-6.173 \le \mu_D \le 8.458.$$

Note that the degrees of freedom used for s_D and t are 13 now that the maverick pair has been discarded. The resulting confidence interval on μ_D appears to indicate that there is no reason to doubt that the two methods are numerically equivalent.

6.11   VARIANCE

The next thing to be considered is the variability of the data that are obtained and used. In other words, how far do the data scatter about a central point? If the scatter is too large, a given observation is less reliable than for the case where the data points are very close together. A measure of the scatter or variability of data is the *variance*. Statistical techniques that are derived concerning interval estimation of the variance are based on the χ^2-distribution and the F-distribution. More specifically, interval estimation on a single variance σ^2 is based on the χ^2-distribution, and estimation concerning two variances is based on the F-distribution. In the next two sections we consider these two ideas.

6.12   ESTIMATION OF A VARIANCE

Let $X_1, X_2, \ldots, X_n$ denote a random sample from a normal population with mean μ and variance σ^2. According to Section 6.5, we know that the random variable

$$V = (n-1)\frac{S^2}{\sigma^2} = \frac{\sum_{i=1}^{n}(X_i - \bar{X})^2}{\sigma^2}, \tag{6.34}$$

where S^2 is the sample variance of $X_1, X_2, \ldots, X_n$ and has a χ^2 distribution with $(n-1)$ degrees of freedom. Hence

$$\mathscr{P}(V < \chi^2_{n-1,\alpha/2}) = \frac{\alpha}{2}$$

and

$$\mathscr{P}(V > \chi^2_{n-1,1-\alpha/2}) = \frac{\alpha}{2}.$$

Thus

$$\mathscr{P}(\chi^2_{n-1,\alpha/2} < V < \chi^2_{n-1,1-\alpha/2}) = 1 - \alpha. \tag{6.35}$$

Replace V in (6.35) according to (6.34) to obtain

$$\mathscr{P}\left(\chi^2_{n-1,\alpha/2} < \frac{(n-1)S^2}{\sigma^2} < \chi^2_{n-1,1-\alpha/2}\right) = 1 - \alpha. \tag{6.36}$$

Now

$$\chi^2_{n-1,\alpha/2} < \frac{(n-1)S^2}{\sigma^2} < \chi^2_{n-1,1-\alpha/2}$$

is equivalent to $\chi^2_{n-1,\alpha/2} < (n-1)S^2/\sigma^2$ and $(n-1)S^2/\sigma^2 < \chi^2_{n-1,1-\alpha/2}$ or to $\sigma^2 < (n-1)(S^2/\chi^2_{n-1,\alpha/2})$ and $(n-1)(S^2/\chi^2_{n-1,1-\alpha/2}) < \sigma^2$. Therefore, (6.36) may be written as

$$\mathscr{P}\left(\frac{(n-1)S^2}{\chi^2_{n-1,1-\alpha/2}} < \sigma^2 < \frac{(n-1)S^2}{\chi^2_{n-1,\alpha/2}}\right) = 1 - \alpha. \tag{6.37}$$

Equation (6.37) yields a $(1-\alpha)100\%$ confidence interval for σ^2, which is

$$\frac{(n-1)s^2}{\chi^2_{n-1,1-\alpha/2}} \le \sigma^2 \le \frac{(n-1)s^2}{\chi^2_{n-1,\alpha/2}}. \tag{6.38}$$

Example 6.8   Five similar determinations of the cold water flow rate to a heat exchanger were, in gal/min, 5.84, 5.76, 6.03, 5.90, and 5.87. Compute a 95% confidence interval for the imprecision (on a per measurement basis) affecting this measuring operation.

Solution   We have

$$\bar{x} = \frac{\sum_{i=1}^{n} x_i}{n} = 5.88$$

and

$$s_{\bar{X}}^2 = \frac{\sum x_i^2 - n\bar{x}^2}{n-1} = 0.00975.$$

Since $n = 5$, $v = n - 1 = 4$. The confidence interval is obtained from Eq. (6.38). The values of $\chi_{4,0.025}^2$ and $\chi_{4,0.975}^2$ are 0.484 and 11.1, respectively, according to Table V of App. B. Therefore, the calculated 95% confidence interval on σ^2 is

$$\frac{4(0.00975)}{11.1} \le \sigma^2 \le \frac{4(0.00975)}{0.484}$$

or

$$0.00351 \le \sigma^2 \le 0.08057.$$

If the mean μ is known, then $\bar{x}$ is replaced by μ in (6.34) to yield

$$V = \frac{\sum_{i=1}^{n}(x_i - \mu)^2}{\sigma^2},$$

where V has a χ^2-distribution with n d.f. Thus in this case the confidence interval in (6.36), on replacing $(n-1)s^2$ with $\sum_{i=1}^{n}(x_i - \mu)^2$, becomes

$$\sum_{i=1}^{n}\frac{(x_i - \mu)^2}{\chi_{n,1-\alpha/2}^2} \le \sigma^2 \le \sum_{i=1}^{n}\frac{(x_i - \mu)^2}{\chi_{n,\alpha/2}^2}, \tag{6.39}$$

where $\chi_{n,1-\alpha/2}^2$ and $\chi_{n,\alpha/2}^2$ are tabular values corresponding to n d.f. The result (6.39) is not very useful, however, since μ is generally unknown.

6.13 COMPARISON OF TWO VARIANCES

The F-distribution is used to compare the variances of two populations. Suppose that we let S_1^2 and S_2^2 denote the sample variances corresponding to the independent samples $X_{11}, X_{12}, \ldots, X_{1n_1}$ and $X_{21}, X_{22}, \ldots, X_{2n_2}$, respectively, from two normal populations with means μ_1 and μ_2 and variances σ_1^2 and σ_2^2. By Section 6.7 the random variable

$$F = \frac{S_1^2/\sigma_1^2}{S_2^2/\sigma_2^2} = \frac{\chi_1^2/(n_1 - 1)}{\chi_2^2/(n_2 - 1)}, \tag{6.40}$$

where

$$(n_1 - 1)S_1^2 = \sum_{i=1}^{n_1}(X_{1i} - \bar{X}_1)^2 = \chi_1^2\sigma_1^2$$

and

$$(n_2 - 1)S_2^2 = \sum_{i=1}^{n_2}(X_{2i} - \bar{X}_2)^2 = \chi_2^2\sigma_2^2$$

has an F-distribution with $v_1 = (n_1 - 1)$ and $v_2 = (n_2 - 1)$ d.f., respectively. Recall that $(n_1 - 1)$ is the d.f. corresponding to the numerator in (6.40). Let $F_{v_1, v_2, \alpha/2}$ and $F_{v_1, v_2, 1 - \alpha/2}$ denote the tabular F values with $\alpha/2$ and $1 - \alpha/2$ area to the left, respectively. That is,

$$\mathscr{P}(F_{v_1, v_2, \alpha/2} < F < F_{v_1, v_2, 1 - \alpha/2}) = 1 - \alpha. \qquad (6.41)$$

Replacing F in (6.41) according to (6.40) gives

$$\mathscr{P}\left(F_{v_1, v_2, \alpha/2} < \frac{\sigma_2^2}{\sigma_1^2} \frac{S_1^2}{S_2^2} < F_{v_1, v_2, 1 - \alpha/2} \right) = 1 - \alpha$$

and

$$\mathscr{P}\left(\frac{S_2^2}{S_1^2} F_{v_1, v_2, \alpha/2} < \frac{\sigma_2^2}{\sigma_1^2} < \frac{S_2^2}{S_1^2} F_{v_1, v_2, 1 - \alpha/2} \right) = 1 - \alpha.$$

Therefore, a $(1 - \alpha)100\%$ confidence interval for σ_2^2/σ_1^2 is given by

$$\frac{S_2^2}{S_1^2} F_{n_1 - 1, n_2 - 1, \alpha/2} < \frac{\sigma_2^2}{\sigma_1^2} < \frac{S_2^2}{S_1^2} F_{n_1 - 1, n_2 - 1, 1 - \alpha/2}. \qquad (6.42)$$

Similarly, a $(1 - \alpha)100\%$ confidence interval for σ_1^2/σ_2^2 is given by

$$\frac{S_1^2}{S_2^2} F_{n_2 - 1, n_1 - 1, \alpha/2} < \frac{\sigma_1^2}{\sigma_2^2} < \frac{S_1^2}{S_2^2} F_{n_2 - 1, n_1 - 1, 1 - \alpha/2}. \qquad (6.43)$$

Confidence intervals corresponding to (6.42) and (6.43) could be derived in the case where μ_1 or μ_2 (or both) are known; however, the result would not be very useful, so we shall not pursue it.

Example 6.9 Reaction temperatures in degrees Celsius (measured on two different days) for two catalyst concentrations were:

x_1	x_2
310.95	308.94
308.86	308.23
312.80	309.98
309.74	311.59
311.03	309.46
311.89	311.15
310.93	311.29
310.39	309.16
310.24	310.68
311.89	311.86
309.65	310.98
311.85	312.29
310.73	311.21

Find a 98% confidence interval for σ_1/σ_2.
Solution   We have

$$\bar{x}_1 = 310.8423°\text{C}, \qquad \bar{x}_2 = 310.5246°\text{C},$$
$$s_1^2 = 1.1867, \qquad\qquad s_2^2 = 1.5757.$$

Now, since the d.f. are $n_1 - 1 = 12$ and $n_2 - 1 = 12$, we have, from Table VI of App. B, $F_{12,12,0.01} = 0.241$ and $F_{12,12,0.99} = 4.16$. Thus a 98% confidence interval on σ_1^2/σ_2^2 is [from Eq. (6.43)]

$$\frac{1.1867}{1.5757}(0.241) \le \frac{\sigma_1^2}{\sigma_2^2} \le \frac{1.1867}{1.5757}\ (4.16) \qquad (4.16)$$

or

$$0.18104 \le \frac{\sigma_1^2}{\sigma_2^2} \le 3.13300.$$

The corresponding 98% confidence interval on σ_1/σ_2 is

$$0.42548 \le \frac{\sigma_1}{\sigma_2} \le 1.7700.$$

Note that since the confidence interval contains the value 1, we cannot "justifiably" say that $\sigma_1/\sigma_2 \ne 1$ or that $\sigma_1 \ne \sigma_2$.

6.14   ESTIMATION OF A PROPORTION *P*

Consider a population that contains two types of objects. This type of population was discussed in Section 3.4.1. Each of the objects can be represented by one of the values 0 or 1. Such a population is labeled a binomial population and we associate success and failure with the values 1 and 0. The probability function corresponding to the population is

$$\mathscr{P}(X = x) = f(x) = P^x(1 - P)^{1-x}, \qquad x = 0,1, \quad 0 < P < 1.$$

The parameter P is the proportion of 1's in the population. The only values that X can take on are 0 and 1.

   If a random sample $X_1, X_2, \ldots, X_n$ is chosen from the population, $S = X_1 + X_2 + \cdots + X_n$ has the binomial probability function

$$\mathscr{P}(S = x) = b(x) = \binom{n}{x} P^x(1 - P)^{n-x}, \qquad x = 0,1,2,\ldots,n. \qquad (6.44)$$

The random variable S is actually the number of successes (1's) in the sample. In this section we consider the problem of estimating the parameter P, that is, the probability of obtaining a success in one draw from the population.

It was shown in Example 5.4 that

$$E(S) = E\left(\sum_{i=1}^{n} X_i\right) = nP$$

and

$$\sigma_S^2 = nP(1 - P) = nPQ.$$

The best estimator for P is given by

$$\hat{P} = \frac{S}{n} = \frac{\sum_{i=1}^{n} X_i}{n} = \bar{X},$$

that is, the sample mean. We have

$$E(\hat{P}) = E\left(\frac{S}{n}\right) = \frac{1}{n} E(S) = P$$

and

$$\sigma_{\hat{P}}^2 = \sigma_{\bar{X}}^2 = \frac{\sigma_{\bar{X}}^2}{n} = \frac{PQ}{n}. \tag{6.45}$$

Thus no other estimator of P has a smaller variance than PQ/n.

We turn now to the problem of constructing an interval estimate for P. To find confidence intervals for P one must solve each of the equations

$$\sum_{x=s}^{n} \binom{n}{x} P^x (1 - P)^{n-x} = \frac{\alpha}{2} \tag{6.46}$$

and

$$\sum_{x=0}^{s} \binom{n}{x} P^x (1 - P)^{n-x} = \frac{\alpha}{2} \tag{6.47}$$

for P, where s is the number of successes in the sample. The solution, say P_1, of (6.46) is the lower confidence limit and the solution, say P_2, of (6.47) is the upper confidence limit. To solve (6.46) and (6.47) one must resort to extensive binomial tables which cover a wide range of values for P.

Example 6.10   Let P denote the unknown proportion of batches of chemical that are on-grade. Suppose that 100 batches of chemical are obtained and it is determined that 90 are on-grade. The best estimate of P is $\hat{P} = 90/100 = 0.9$. To find a 95% confidence interval for P it is necessary to solve (6.46) and (6.47) for P with $s = 90$, $n = 100$, and $\alpha = 0.05$. The solution of (6.46) is the lower confidence limit, and the solution of (6.47) is the upper confidence limit. This example illustrates that determining exact confidence limits is a very tedious chore. The exact confidence limits are not given for this example but will be determined approximately later by a very popular technique which will now be discussed.

According to the central limit theorem, the random variable $\bar{X}$ has an approximate normal distribution with mean μ and variance σ^2/n even if the distribution of X is discrete. Thus the random variable S/n has an approximate normal distribution with mean P and variance PQ/n. Therefore,

$$Z = \frac{S/n - P}{\sqrt{PQ/n}} = \frac{\hat{P} - P}{\sigma_{\hat{p}}} \tag{6.48}$$

has an approximate standard normal distribution. We modify the random variable (6.48) further by replacing PQ in the denominator with $\hat{P}(1 - \hat{P})$. Thus we have

$$Z = \frac{\hat{P} - P}{\sqrt{\hat{P}\hat{Q}/n}} = \frac{\hat{P} - P}{S_{\hat{p}}}.$$

Proceeding in the usual fashion, we set

$$\mathscr{P}(z_{\alpha/2} < Z < z_{1-\alpha/2}) = 1 - \alpha,$$

which is the same as

$$\mathscr{P}(-z_{1-\alpha/2} < Z < z_{1-\alpha/2}) = 1 - \alpha$$

and can be converted directly to

$$\mathscr{P}\left(-z_{1-\alpha/2} < \frac{\hat{P} - P}{S_{\hat{p}}} < z_{1-\alpha/2}\right) = 1 - \alpha. \tag{6.49}$$

This last equation is of exactly the same form as that of some obtained previously [see, e.g., Eqs. (6.11) and (6.18)]. Equation (6.49) can be written as

$$\mathscr{P}\left(\hat{P} - z_{1-\alpha/2}\sqrt{\frac{\hat{P}(1-\hat{P})}{n}} < P < \hat{P} + z_{1-\alpha/2}\sqrt{\frac{\hat{P}(1-\hat{P})}{n}}\right) = 1 - \alpha,$$

from which we obtain an approximate $(1 - \alpha)100\%$ confidence interval for P:

$$\hat{P} - z_{1-\alpha/2}\sqrt{\frac{\hat{P}(1-\hat{P})}{n}} < P < \hat{P} + z_{1-\alpha/2}\sqrt{\frac{\hat{P}(1-\hat{P})}{n}}. \tag{6.50}$$

Example 6.11   In Example 6.10 we had a sample of 100 batches of chemical of which 90 were on-grade. Find a 95% confidence interval for P, the true unknown probability that a batch is on-grade.

Solution The best estimate of $\hat{P}$ is 0.90. Furthermore, $n = 100$ and $z_{0.975} = -z_{0.025} = 1.96$. Thus a 95% confidence interval for P is

$$0.90 - 1.96 \sqrt{\frac{0.9(0.1)}{100}} < P < 0.90 + 1.96 \sqrt{\frac{0.9(0.1)}{100}},$$

$$0.90 - 1.96(0.03) < P < 0.90 + 1.96(0.03),$$

$$0.90 - 0.0588 < P < 0.90 + 0.588,$$

$$0.841 < P < 0.959.$$

Since the interval contains the value 0.90, we cannot safely say that $P \neq 0.90$. (The exact confidence limits in this example are 0.82 and 0.95.)

Example 6.12 Each compression ring in a population (lot) of compression rings either fits properly or it does not. Let P denote the true proportion of those rings that fit properly. In a sample of 400 rings, 10% were observed to be faulty; that is, 10% did not seat properly. Find the 99% confidence interval for the proportion faulty in the lot from which the sample was taken.
Solution The best estimate of P is $\hat{P} = 0.10$. Since $n = 400$ and $z_{0.005} = -2.58$, the 99% confidence limits for P are, by (6.50),

$$0.10 \pm 2.58 \sqrt{\frac{0.1(0.9)}{400}},$$

which simplifies to

$$0.10 \pm 2.58 \left(\frac{0.3}{20} \right)$$

and reduces to

$$(0.061, 0.139).$$

6.15 COMPARISON OF TWO PROPORTIONS

Consider the problem of comparing the parameters P_1 and P_2 of two binomial populations. Let $X_1, X_2, \ldots, X_{n_1}$ and $Y_1, Y_2, \ldots, Y_{n_2}$ denote random samples from the two binomial populations with parameters P_1 and P_2, respectively. The best estimates of P_1 and P_2 are

$$\hat{P}_1 = \frac{\sum_{i=1}^{n_1} X_i}{n_1} \quad \text{and} \quad \hat{P}_2 = \frac{\sum_{i=1}^{n_2} Y_i}{n_2}.$$

Furthermore,

$$S_{\hat{p}_1}^2 = \frac{\hat{P}_1(1 - \hat{P}_1)}{n_1} \quad \text{and} \quad S_{\hat{p}_2}^2 = \frac{\hat{P}_2(1 - \hat{P}_2)}{n_2}.$$

To compare P_1 and P_2 we consider the difference $P_1 - P_2$. The best estimate of $P_1 - P_2$ is $\hat{P}_1 - \hat{P}_2$. Since $\hat{P}_1$ and $\hat{P}_2$ are independent, an estimate of the variance of $\hat{P}_1 - \hat{P}_2$ is

$$S_{\hat{P}_1 - \hat{P}_2}^2 = S_{\hat{P}_1}^2 + S_{\hat{P}_2}^2 = \frac{\hat{P}_1(1 - \hat{P}_1)}{n_1} + \frac{\hat{P}_2(1 - \hat{P}_2)}{n_2}.$$

The random variables $\hat{P}_1$ and $\hat{P}_2$ are each approximately normally distributed, and consequently, $\hat{P}_1 - \hat{P}_2$ has an approximate normal distribution with mean $P_1 - P_2$ and variance

$$\sigma_{\hat{P}_1 - \hat{P}_2}^2 = \frac{P_1(1 - P_1)}{n_1} + \frac{P_2(1 - P_2)}{n_2}.$$

Thus

$$\frac{\hat{P}_1 - \hat{P}_2 - (P_1 - P_2)}{\sqrt{P_1(1 - P_1)/n_1 + P_2(1 - P_2)/n_2}} \tag{6.51}$$

has approximately the standard normal distribution. Replacing P_1 and P_2 by $\hat{P}_1$ and $\hat{P}_2$ in the denominator of (6.51), it follows that

$$Z = \frac{\hat{P}_1 - \hat{P}_2 - (P_1 - P_2)}{\sqrt{\hat{P}_1(1 - \hat{P}_1)/n_1 + \hat{P}_2(1 - \hat{P}_2)n_2}}$$

has approximately the standard normal distribution. Thus

$$\mathscr{P}(-z_{1-\alpha/2} < Z < z_{1-\alpha/2}) = 1 - \alpha,$$

from which we obtain the $(1 - \alpha)100\%$ confidence limits for $P_1 - P_2$:

$$\hat{P}_1 - \hat{P}_2 \pm z_{1-\alpha/2} \sqrt{\frac{\hat{P}_1(1 - \hat{P}_1)}{n_1} + \frac{\hat{P}_2(1 - \hat{P}_2)}{n_2}}. \tag{6.52}$$

Example 6.13 Two different methods (say, 1 and 2) of manufacturing compression rings are being used. One hundred rings were sampled for each method. It was found that 10 rings using method 1 and 3 rings from method 2 did not seat properly. Construct a 99% confidence interval for $P_1 - P_2$.

Solution The best estimate for $P_1 - P_2$ is $\hat{P}_1 - \hat{P}_2 = 0.10 - 0.03 = 0.07$. We have $n_1 = 100 = n_2$ and $z_{0.995} = 2.58$. Thus

$$s_{\hat{P}_1 - \hat{P}_2}^2 = s_{\hat{P}_1}^2 + s_{\hat{P}_2}^2 = \frac{0.10(0.90)}{100} + \frac{0.03(0.97)}{100}$$
$$= 0.001191$$

and

$$s_{\hat{P}_1 - \hat{P}_2} = \sqrt{0.001191} = 0.0345.$$

The 99% confidence limits are, by (6.44),

$$\hat{P}_1 - \hat{P}_2 \pm z_{1-\alpha/2} s_{\hat{P}_1 - \hat{P}_2},$$
$$0.07 \pm 2.58(0.0345),$$
$$0.07 \pm 0.080,$$
$$-0.01, 0.15.$$

Since the interval contains 0, we cannot safely infer that $P_1 - P_2 \neq 0$ or that $P_1 \neq P_2$.

PROBLEMS

6.1 Data taken from the plate and frame filter press located in the unit operations laboratory are used to determine α, the specific cake resistance, of a calcium carbonate slurry. Several values of α, expressed in ft/lb, have been calculated from data taken during the fall semester.

2.49×10^{11}	2.67×10^{11}
2.40×10^{11}	2.60×10^{11}
2.43×10^{11}	2.50×10^{11}
2.30×10^{11}	2.54×10^{11}
2.53×10^{11}	2.55×10^{11}

Based on these values, predict the interval within which 90% of all such values calculated in the future must fall.

6.2 The endurance limit of a material is a function of its surface roughness. If this department asked you to design a new elevator, specifying an endurance limit for a bearing material of 105,000 psi, what specification for surface roughness would you submit to the contractor? The firm limits surface finish of all bearings to 9 µin. roughness as an economic factor. The following data are available for 12 different bearings:

Surface roughness (μin.)	Endurance limit (psi)
18	104,800
16	107,000
6	115,000
17	106,000
15	107,300
12	109,000
7	114,000
14	108,000
13	108,900
8	112,500
11	110,000
20	104,000

6.3 Due to the burning of cotton plant wastes (hulls, leaves, etc.), sulfate content in the air over Lubbock is highest during the month of November. The following data are the mean values of sulfate content ($\mu g/m^3$ of air) during the month of November (analyses of air performed daily) over the past 10 years: 10.83, 8.90, 14.71, 12.35, 11.86, 13.80, 11.75, 9.68, 9.33, 10.9. What value for the sulfate content in Lubbock air can the Lubbock City–County Department of Health predict at a 95% confidence level during next November?

6.4 Twenty companies have submitted bids for a batch reactor that the Department of Chemical Engineering is planning to install in the unit operations lab.

Company	Capacity (gal)	Cost ($)
A	7.5	425
B	8.0	500
C	9.0	525
D	10.0	550
E	6.0	450
F	16.5	675
G	12.0	575
H	7.0	475
I	8.5	510

(*continued*)

Company	Capacity (gal)	Cost
J	8.75	525
K	8.25	510
L	9.25	540
M	10.25	625
N	5.75	350
O	6.00	375
P	9.00	500
Q	8.50	525
R	8.00	450
S	9.25	550
T	10.00	600

As part of the financial review, you are asked to submit a price range based on these bids within which the price of a suitable reactor will fall. The reactor must have a minimum capacity of 8 gal and a maximum cost of \$550. Your answer should be based on the 95% confidence interval of reactor cost.

6.5 The following values have been obtained for the critical moisture content of sawdust (lb moisture/lb bone-dry solid): 1.1000, 1.0500, 1.0800, 1.1200, 1.0900, 1.1300, 1.1100, 1.0700. Based on these data, your lab partner submits a confidence interval of $1.0779 < \mu < 1.1194$.

(a) What must be the value of z to obtain such an interval?
(b) What percent confidence interval is represented?

6.6 The Gulp-a-Cup Coffee Company utilizes spray drying in their coffee production process. In the past, Gulp-a-Cup has utilized banks of single atomizer nozzles with external mixing of the gas and liquid phases in all their drying chambers. Two-fluid nozzles employing internal mixing were recently installed in chamber 12 of the plant and trial runs were made to determine the optimum drying conditions. The following data are for entrance gas pressure (psig): 52.00, 51.00, 51.80, 51.75, 51.30, 50.85, 50.25, 49.00, 48.65, 48.00. Above 52 psig the coffee particle size was too fine. Below 48 psig the coffee was not dried sufficiently. What new value for gas pressure should be specified for chamber 12? Use the 99% confidence limit.

6.7 If the interval estimate for the optimum reference gas flow rate in mL/min during a series of chromatographic determinations of ketones in a sample mixture was $11.36 < \mu < 14.14$, what was the confidence level of the estimate? Data are (reference gas flow, mL/min): 10.0, 10.5, 12.0, 15.0, 13.0, 16.0.

6.8   The lab instructor of Ch.E. 3111 made a binary liquid "unknown" for his students and told them to determine the weight percent of the more volatile component. At the end of the experiment, all the students became upset when the instructor informed them that all their results were outside the limits of analysis:

concentration $= 91\% \pm 0.1\%$

Any concentration outside this range made an "F". Values within the range received grades of 100. The instructor's conclusions were based on the following data (% more volatile component).

91.0	91.0
91.2	90.8
91.1	90.9
91.5	90.5

What value of x corresponds to a grade of 100? Was the instructor fair?

6.9   In a process for the manufacture of ester gum (chemically, glyceryl abietate), rosin is reacted with glycerine and water is taken off as the reaction proceeds. The mean color of the product using the standard source of rosin is 65. A new source of rosin is being considered on the basis of its giving an improved (lower) color product. Test runs were made using rosin from this new source, with the following color results: 55, 62, 54, 57, 65, 64, 60, 63, 58, 67, 63, 61. Each of these values may be considered independent. Find a 95% confidence interval for color of the product made from the new rosin source.

6.10   A random sample of 100 students who smoke cigarettes on this campus showed that the average student smokes 17.8 cigarettes per day. The variance in cigarettes smoked per day for any one person was 480.

(a)   What is the 95% confidence limit for the actual average number of cigarettes smoked per day by these students?

(b)   Discuss the application of these statistics to the population of university smokers.

6.11   A company engaged in the manufacture of cast iron has employed a system of raw material and processing procedures that has produced a product whose overall population average silicon content was 0.85%. A new contract was put into effect in which a new supplier of raw material supplanted the old one. During the first month of operation using the new material, random samples of the product silicon content were found to be

1.13	0.87
0.80	0.92
0.85	0.81
0.60	0.97
0.97	0.48
0.92	1.00
0.94	0.92
0.72	0.61
1.17	0.81
0.87	0.71
0.36	0.97
0.68	0.89
0.73	1.16
0.82	0.68
0.79	1.00
$\sum x = 25.1499$	$\sum x^2 = 22.0863$

What are your 99% confidence limits on the silicon content of the iron using the new raw material?

6.12 Five analyses of the methane content of a natural gas showed: 92.4, 92.8, 92.3, 93.0, 92.5%.

(a) What is the 95% CI for the true methane content?
(b) Can it be safely said that the true methane content averages at least 91.8%? (Evaluate this part by the appropriate t-test.) For these data, $s_{\bar{X}}^2 = 0.085$.

6.13 Use the data in Problem 3.17 to calculate the 95% confidence limit for the flow rate of the elutriator.

6.14 Using the data of Problem 3.16 and the cumulative relative frequency polygon you obtained, what is the 95% confidence interval for the cotton dust concentration in each area? Obtain this value by calculation using any computerized method and compare the results. Are they different? If so, why?

6.15 The following cotton dust concentrations (in $\mu g/m^3$) were measured in accordance with the provisions of the OSHA standard for two shifts in the spinning room of a cotton mill.

C	D
96.7	58.7
149.7	75.9
107.6	101.3
73.5	117.8
213.1	141.9
195.5	143.3

Is there any significant difference ($\alpha = 0.05$) in the average concentration for the two shifts?

6.16 The hydrogen gas content available from two sources of supply have been analyzed with the results

$$\bar{x}_1 = 69.1, \qquad \bar{x}_2 = 66.2,$$
$$n_1 = 7, \qquad n_2 = 5,$$
$$s_{X_1}^2 = 26.1, \qquad s_{X_2}^2 = 11.$$

What is the 95% CI for the difference in hydrogen content for these sources?

6.17 An experiment conducted to compare the tensile strengths of two types of synthetic fibers gave the following breaking loads in thousands of pounds force per square inch.

Fiber A	14	4	10	6	3	11	12		
Fiber B	16	17	13	12	7	16	11	8	7

Calculate the 99% confidence interval for the difference between the means.

6.18 Two gaskets were cut from each of eight different sheet stocks. One gasket of each pair was randomly selected for use in dilute HCl service. The other gasket of each pair was for concentrated HCl service. All gaskets were subjected to accelerated life tests for their respective intended uses. From the data so obtained, the estimated average service life in weeks is as follows:

Material	Dilute HCl	Concentrated HCl
1	35	30
2	40	32
3	27	28
4	25	27
5	36	33
6	48	38
7	53	41
8	48	39

(a) Calculate the mean difference in expected service life in the two environments and a 95% confidence interval for this mean difference.

(b) Explain the advantages obtained from pairing the test samples in the manner described.

6.19 Seven samples of a catalyst have been analyzed for carbon content by two technicians (i.e., each technician made one analysis of each sample). The results were:

Sample	$X_1 = X_2 = D$
1	−0.5
2	−0.5
3	−0.2
4	0
5	−0.9
6	−0.1
7	−0.6

Is there any difference between the two technicians? Make whatever comments are appropriate regarding whether one technician is better than the other.

6.20 Using the data in Problem 6.17, calculate a 95% confidence interval for the ratio of variances σ_A^2/σ_B^2.

6.21 The following data give the yields of a product that resulted from trying catalysts from two different suppliers in a process:

Catalyst I	36	33	35	34	32	34	
Catalyst II	35	39	37	38	39	38	40

Calculate (a) the 90% confidence interval for σ_I^2/σ_{II}^2 and (b) the 95% confidence interval. What do you infer from the differences in the estimated ranges of the variance ratio?

6.22   A sulfuric acid plant produces acid with a long-term mean concentration of 60%. What is the maximum value of the standard error allowable to have a 99% confidence interval that the acid concentration is between 56 and 65%?

6.23   The weight of a particular medication capsule has a mean of 40 mg and a standard deviation of 2 mg. If the capsule weight is NID, what is the probability that a capsule chosen at random will weigh between 39 and 40 mg?

6.24   A computer manufacturer is considering using one of two types of components in their home computers. Ninety components of type 1 are tested and five fail. One hundred of type 2 are tested and seven fail. Based on these data, compute a 95% confidence interval for the difference $P_1 - P_2$, where P_1 and P_2 are the true proportions of failures for the type 1 and type 2 components, respectively.

6.25   In a sample of 140 batches of chemical A, 120 were on-grade. In a sample of 110 batches of chemical B, only 90 were on-grade. Find a 99% confidence interval for $P_A - P_B$, where P_A and P_B are the true proportions of A and B that are on-grade.

6.26   The heats of ionic reactions are to be observed by reacting 10 mL of 2.5 M NaOH with two different acid solutions in a solution calorimeter. The two solutions were 100 mL of 0.25 M HCl and 100 mL of 0.25 M NaHR. Baseline temperature readings (°C) were taken for each reaction. Four runs were performed on each acid solution. The data are as follows:

HCl	NaHR
20.68	21.10
21.09	21.08
20.81	21.00
20.84	20.91

Is there a difference in the two heats of reaction? Use a 95% confidence level.

6.27   Two loading methods were simulated with a sample of fresh catalyst. The loading was simulated using a 200-mL graduated cylinder. Seventy-six grams of catalyst was weighed and used for the simulation. The volumes (in mL) recorded from four trials for each loading method are as follows:

Method 1	Method 2
22	15
23	17
22	16
24	17

Is there a significant difference in the variance between the two types of loading?

6.28 A typical batch of used catalyst removed from a microwax hydrofinishing unit is tested for coke content. The long-term average variance of coke deposits (wt %) is approximately 0.5. Tests for coke content were done on samples of the catalyst removed from four different sections of the reactor. The test results were reported in weight %.

4.45   (top)
4.445
4.215
4.211   (bottom)

Has the variance of coke deposit across the catalytic bed changed significantly ($\alpha = 0.05$) from the long-term average value of 0.5?

6.29 A premixed acid solution is marketed as containing at least 22 wt % nitric acid. Some of your experiments have failed for no apparent reason. Finally, you decide to check the acid concentration by a standard $NaHCO_3$ titration to the phenolphthalein endpoint. The results are as follows (in wt %):

22.01	20.54	22.20
22.48	23.06	21.92
21.73	22.20	21.55
22.11	21.00	21.09

Does this acid meet the minimum concentration specification of $\alpha = 0.05$?

6.30 Pilot-plant studies were conducted for the measurement of pressure drop across a fixed-bed catalytic reactor containing a fouled catalyst sample removed from the lead reactor of a hydrofinishing train. A feedstock with a 175°F melting point was used. The feedstock and the hydrogen gas were run through the reactor at 360 g/h and 1.71 scf/hr, respectively, to obtain a reference pressure drop. Once a pressure drop of 83 psig was established, an antifoulant agent was injected at 800 ppm(v) for 12 h. The procedure was repeated, but the antifoulant agent was injected at a dosage of 300 ppm(v). The temperature was held constant throughout the process. The pressure drop data are listed below.

ΔP at 800 ppm(v) (psig)	ΔP at 300 ppm(v) (psig)
83	81
82	80
83	82
83	83
84	81
81	82
81	81
80	82
82	83
83	83
82	82
82	81

Is a greater change in pressure drop experienced at the lower dosage than at the higher dosage of 800 ppm(v)? Use a 98% CI.

REFERENCE

1. Conover, W. J., *Practical Nonparametric Statistics*, 2nd ed., Wiley, New York (1980) Ch. 5.

7

Statistical Inference: Hypothesis Testing

7.1 INTRODUCTION

The second area of statistical inference is that of hypothesis testing. A statistical *hypothesis* is simply a statement concerning the probability distribution of a random variable. Before the hypothesis is formulated, it is necessary to choose a probability model for the population. There are two types of general hypotheses: simple and composite. A *simple hypothesis* is one which states that the data in question are represented by a distribution that is a specific member of a particular family of distributions. For example, the hypothesis that $\mu = 6$, $\sigma^2 = 1.7$ for a normal population uniquely specifies a particular normal distribution. If the value of the variance is unknown and we stated the hypothesis as $\mu = 6$, we would have a *composite hypothesis*. This is because the hypothesis states that the distribution involved can be any member of the family determined by the parameters $\mu = 6$ and $\sigma^2 > 0$.

Once the hypothesis has been stated, appropriate statistical procedures are used to determine whether it is an acceptable conjecture or an unacceptable one. In general, we cannot prove that a hypothesis is absolutely true or false. If the information furnished by the data supports the hypothesis, we do not reject it. If the data do not support the hypothesis, we reject it.

The hypothesis being tested is referred to as the *null hypothesis* and is denoted by H_0. Another hypothesis is generally stipulated that complements the null hypothesis. It is referred to as the *alternative hypothesis* and is denoted by H_A. If we accept H_0, we automatically reject H_A, and vice versa. Regardless of which we do, we are susceptible to the chance of having committed an error if the wrong choice is made. The possible errors are discussed in Section 7.2.

7.2   TYPES OF ERRORS

Many statistical hypotheses are of the type that specify a value, or range of values, of one or more parameters of the distribution in question. Such hypotheses can be tested by using the sample characteristics obtained from the data. The values of these characteristics vary from sample to sample, and thus they may by chance be quite different from those of the population. Since this is true, it is possible to make an error in accepting or rejecting any given hypothesis. A *type 1 error* occurs when a true hypothesis H_0 is rejected. The Greek letter α will denote the probability of rejecting H_0 when it is true. The term α is also referred to as the *level of significance* of the test. A *type 2 error* consists of accepting H_0 when it is false. The Greek letter β is used to represent the probability of making a type 2 error. Ideally, we would prefer to use a test that minimizes both types of errors. Unfortunately, as α decreases, β tends to increase. The reverse is also true. The *power* of a test is defined as the probability of rejecting H_0 when it is false (i.e., power $= 1 - \beta$).

The choice of values for α and β is not always easy to make. No hard and fast rules can be made for the choice of α and β. This decision must be made for each problem as it arises based on economics and quality-control considerations.

Consider the routine inspection of thermostats for baby incubators. If from a suitably sized sample we conclude that the lot is satisfactory based on some predetermined significance level, say $\alpha = 0.05$, we still run the risk of being wrong some of the time in accepting the thermostats. If, however, we reject the hypothesis that the lot is satisfactory at the 5% level, the probability of producing an incubator with unacceptable temperature control has been virtually eliminated whether or not the thermostats are defective. If the thermostats are satisfactory and the lot has been rejected, a type 1 error has occurred. Consider the alternative: The lot in question contains a significant number of defective thermostats. If we accept the lot under these circumstances, we will certainly produce faulty incubators unless this defect is caught later in the quality-control process. As the production of improperly regulatable incubators is unacceptable, the power of the test used should be quite high; that is, the hypothesis that the termostats are acceptable should be rejected with a high probability if they are in fact faulty. We must, in other words, minimize the probability of committing a type 2 error.

Much more extreme examples are available from spacecraft design work, where the reliability must be at least 0.9999. To achieve this, β must be almost vanishingly small so that no defective parts will be used.

7.3 TESTING OF HYPOTHESES

To test a hypothesis, sample data are collected and used to calculate a test statistic. Depending on the value of this statistic, the null hypothesis H_0 is accepted or rejected. The *critical region* for H_0 is defined as the range of values of the test statistic that corresponds to a rejection of the hypothesis at some fixed probability of committing a type 1 error. The test statistic itself is determined by the specific probability distribution sampled and by the parameters selected for testing.

The general procedure used for testing a hypothesis is as follows:

1. Assume an appropriate probability model to describe the behavior of the random variable under investigation. This choice should be based on previous experience or intuition.
2. Formulate a null hypothesis and an alternative hypothesis. This must be done carefully to permit meaningful conclusions to be drawn from the test.
3. Specify the test statistic.
4. Choose a level of significance α for the test.
5. Determine the distribution of the test statistic and the critical region for the test statistic.
6. Calculate the value of the test statistic from a random sample of data.
7. Accept or reject H_0 by comparing the calculated value of the test statistic with the values defining the critical region.

7.4 ONE-TAILED AND TWO-TAILED TESTS

The statistical tests we are going to consider will be either *one-* or *two-tailed* (sometimes referred to as *one-* or *two-sided*), depending on the exact nature of the null hypothesis H_0 and the alternative hypothesis H_A. Consider a test of the null hypothesis $H_0: \mu = \mu_0$ when sampling from a normal distribution with known variance σ^2, against $H_A: \mu \neq \mu_0$. We would expect our test statistic to be based on the sample mean $\bar{X}$, since it is the best estimator for

μ. Also, intuitively, if $\bar{X}$ has a value quite different from the value μ_0 specified by the null hypothesis, we would tend not to believe H_0. Hence we use the test statistic

$$Z = \frac{\bar{X} - \mu_0}{\sigma/\sqrt{n}},$$

where n is the sample size. Since $\bar{X}$ has a normal distribution with mean μ_0 (if H_0 is true) and variance σ^2/n, Z has the standard normal distribution. Thus, whenever the value of Z falls far enough out under either tail of the normal distribution, the hypothesis $H_0: \mu = \mu_0$ is rejected.

The distribution of Z when H_0 is true is the standard normal. This information can be used to find α, the probability of committing a type 1 error. Since we want α to be fixed, we must find the values $z_{\alpha/2}$ and $z_{1-\alpha/2}$ in the standard normal table such that

$$\mathscr{P}(Z > z_{1-\alpha/2} \text{ or } Z < z_{\alpha/2}) = \alpha.$$

Since the standard normal distribution is symmetric about zero, the "critical" values for rejecting H_0 can be chosen symmetrically about zero. Recall that

$$\mathscr{P}(Z > z_{1-\alpha/2}) = \frac{\alpha}{2} = \mathscr{P}(Z < z_{\alpha/2}).$$

The two-sided rejection region (hatched) for testing $H_0: \mu = \mu_0$ against $H_A: \mu \neq \mu_0$ is illustrated in Fig. 7.1. Note that the inequality symbols in the alternative hypotheses point toward the rejection region(s) for this or any other null hypothesis.

Suppose, however, that we wanted to test $H_0: \mu \geq \mu_0$ against the alternative $H_A: \mu < \mu_0$, again sampling from a normal distribution with known variance σ^2. In this case, if $\bar{X}$ is larger than μ_0, we would tend to believe H_0 rather than H_A. Thus we want to reject H_0 only when $\bar{X}$ is sufficiently smaller than μ_0. After normalizing, this corresponds to having $Z < z_\alpha$, where z_α is such that $\mathscr{P}(Z < z_\alpha) = \alpha$ with α fixed. Thus our rejection region is under the left tail of the standard normal distribution, and we have a one-tailed test. This critical region (hatched) is illustrated in Fig. 7.2. Similarly, for $H_0: \mu \leq \mu_0$ versus $H_A: \mu > \mu_0$, we would have a rejection region of size α under the right tail of the standard normal distribution.

If we desire to test $H_0: \mu = \mu_0$ against either one of the three alternatives $H_A: \mu \neq \mu_0$, $H_A: \mu > \mu_0$, or $H_A: \mu < \mu_0$, when sampling from a normal distribution with *unknown* variance σ^2 we can no longer use the statistic $Z = (\bar{X} - \mu_0)/(\sigma/\sqrt{n})$. The reason for this is that we could not carry out the test since we would not be able to evaluate $Z = (\bar{X} - \mu_0)/(\sigma/\sqrt{n})$ given the

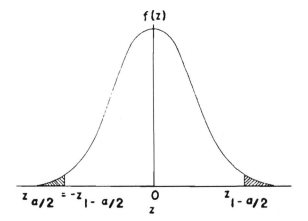

FIGURE 7.1   Standard normal distribution showing rejection region for a two-tailed test.

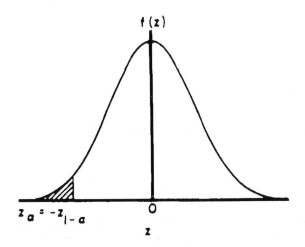

FIGURE 7.2   Standard normal distribution showing rejection region for a one-tailed test.

sample values, since σ would be unknown. If the variance is unknown, as will be seen later, Z is replaced with $T = (\bar{X} - \mu_0)/(S/\sqrt{n})$, where S is the sample standard deviation. The distribution of T is the t-distribution with $(n - 1)$ degrees of freedom.

Example 7.1   Suppose that X has a normal distribution with mean μ and variance $\sigma^2 = 4$. Consider the simple null hypothesis $H_0: \mu = 1$ against the simple alternative $H_A: \mu = 3$. If a sample of size 1 is chosen from the population, find the critical region for testing the hypothesis at the $\alpha = 0.05$ level of significance and find the power of the test.

Solution   The choice of sample size 1 is not realistic; however, the purpose here is to illustrate the basic ideas. Intuitively, if the value in the random sample is sufficiently larger than 1, we would reject H_0. Thus the procedure would be to reject $H_0: \mu = 1$ if $X_1 - 1 > c$, where c is to be determined. Since $\alpha = 0.05$, c must be determined so that

$$\mathcal{P}(X_1 - 1 > c) = 0.05, \tag{7.1}$$

where the probability is computed under the assumption that $H_0: \mu = 1$ is true. $\mathcal{P}(X_1 - 1 > c)$ is the probability of rejecting H_0 when H_0 is in fact true. If $H_0: \mu = 1$ is true, then since $\sigma^2 = 4$ and $n = 1$,

$$Z = \frac{X_1 - \mu_0}{\sigma/\sqrt{n}} = \frac{X_1 - 1}{2}$$

has a standard normal distribution. Thus, by (7.1),

$$\mathcal{P}(X_1 - 1 > c) = \mathcal{P}\left(\frac{X_1 - 1}{2} > \frac{c}{2}\right)$$

$$= \mathcal{P}\left(Z > \frac{c}{2}\right) = 0.05.$$

The last equality implies that $c/2 = 1.645$, from the standard normal table. Hence $c = 3.290$ and the procedure would be to reject $H_0: \mu = 1$ if $X_1 - 1 > 3.290$ or $X_1 > 4.290$. The critical region is the set of all real values x_1 such that $x_1 > 4.290$. To carry out the test a sample of size 1 is taken and $H_0: \mu = 1$ is rejected or not rejected according as $X_1 > 4.290$ or $X_1 \leq 4.29$ [note that $\mathcal{P}(X_1 = 4.29) = 0$].

To find the power of the test, we must find the probability of rejecting $H_0: \mu = 1$ when it is false or when $H_A: \mu = 3$ is true. If $\mu = 3$, we must find $\mathcal{P}(X_1 > 4.29)$, where X_1 has a normal distribution with mean 3 and variance 4. This probability is given by

$$\mathcal{P}(X_1 > 4.29) = \mathcal{P}\left(\frac{X_1 - 3}{2} > \frac{4.29 - 3}{2}\right)$$

$$= \mathcal{P}(Z > 0.645)$$

$$= 0.740.$$

The value 0.740 was obtained from Table III by linear interpolation. Thus the probability of rejecting $H_0: \mu = 1$ when $H_A: \mu = 3$ is true is 0.740.

We now consider some important and popular hypothesis-testing problems. These will include all the problems that were discussed in Chapter 6 in terms of interval estimation.

7.5 TESTS CONCERNING THE MEAN

The tests pertaining to the mean of a population and those comparing two means will be developed under the assumption of normality. The tests are also valid in cases where only approximate normality exists.

7.5.1 Parametric Methods for the Mean (Variance Is Known)

To test the hypothesis that a random sample of size n comes from a population with mean $\mu = \mu_0$, we use the statistic

$$Z = \frac{\bar{X} - \mu_0}{\sigma/\sqrt{n}},$$

where we have assumed that the population variance σ^2 is known. The problem is that of testing the hypothesis $H_0: \mu = \mu_0$ against the two-sided alternative $H_A: \mu \neq \mu_0$. The hypothesis H_0 is rejected if the observed value of Z is "too large" or "too small." Since Z has a standard normal distribution, $H_0: \mu = \mu_0$ is rejected if $Z = (\bar{X} - \mu)/(\sigma/\sqrt{n}) > z_{1-\alpha/2}$ or $Z < z_{\alpha/2} = -z_{1-\alpha/2}$. The level of significance α is such that

$$\mathscr{P}\left(\frac{\bar{X} - \mu_0}{\sigma/\sqrt{n}} > z_{1-\alpha/2} \quad \text{or} \quad \frac{\bar{X} - \mu_0}{\sigma/\sqrt{n}} < -z_{1-\alpha/2}\right) = \alpha. \tag{7.2}$$

The probability implied in Eq. (7.2) is that of rejecting $H_0: \mu = \mu_0$ when it is true. The critical region consists of all those values of $\bar{X}$ such that $Z > z_{1-\alpha/2}$ or $Z < -z_{1-\alpha/2}$. This means that

$$\frac{\bar{X} - \mu_0}{\sigma/\sqrt{n}} > z_{1-\alpha/2} \quad \text{or} \quad \frac{\bar{X} - \mu_0}{\sigma/\sqrt{n}} < -z_{1-\alpha/2},$$

which reduces to

$$\bar{X} - z_{1-\alpha/2}\frac{\sigma}{\sqrt{n}} > \mu_0 \quad \text{or} \quad \bar{X} + z_{1-\alpha/2}\frac{\sigma}{\sqrt{n}} < \mu_0. \tag{7.3}$$

We will reject $H_0: \mu = \mu_0$ if either inequality in (7.3) is satisfied. Consequently, we will not reject (i.e., we will accept) H_0 if μ_0 satisfies neither inequality in (7.3), that is, if

$$\bar{X} - z_{1-\alpha/2}\frac{\sigma}{\sqrt{n}} < \mu_0 < \bar{X} + z_{1-\alpha/2}\frac{\sigma}{\sqrt{n}},$$

Table 7.1   Testing μ When σ^2 Is Known

$$\text{Statistic: } Z = \frac{\bar{X} - \mu_0}{\sigma/\sqrt{n}}$$

Null hypothesis	Alternative	Rejection region
$H_0: \mu = \mu_0$	$H_A: \mu \neq \mu_0$	$Z > z_{1-\alpha/2}$ or $Z < -z_{1-\alpha/2}$
$H_0: \mu \leq \mu_0$	$H_A: \mu > \mu_0$	$Z > z_{1-\alpha}$
$H_0: \mu \geq \mu_0$	$H_A: \mu < \mu_0$	$Z < -z_{1-\alpha}$

which is just another way of saying that $H_0: \mu = \mu_0$ is not rejected if μ_0 is contained in the $(1 - \alpha)100\%$ confidence interval for μ. Thus every confidence interval in Chapter 6 can be used to test a two-sided hypothesis.

To test $H_0: \mu \leq \mu_0$ against $H_A: \mu > \mu_0$, we reject H_0 if $Z = (\bar{X} - \mu_0)/(\sigma/\sqrt{n}) > z_{1-\alpha}$; that is,

$$\mathscr{P}\left(\frac{\bar{X} - \mu_0}{\sigma/\sqrt{n}} > z_{1-\alpha}\right) = \alpha$$

for the one-sided hypothesis. But $(\bar{X} - \mu_0)/(\sigma/\sqrt{n}) > z_{1-\alpha}$ reduces to $\bar{X} - z_{1-\alpha}(\sigma/\sqrt{n}) > \mu_0$. Thus corresponding to this situation is the *one-sided* $(1 - \alpha)100\%$ confidence interval

$$\left(\bar{X} - z_{1-\alpha}\frac{\sigma}{\sqrt{n}}, \infty\right).$$

If the hypothesized μ_0 falls in $(\bar{X} - z_{1-\alpha}(\sigma/\sqrt{n}), \infty)$, then $(\bar{X} - \mu_0)/(\sigma/\sqrt{n}) < z_{1-\alpha}$ and $H_0: \mu \leq \mu_0$ is accepted.

Since all the parameters for which confidence intervals were constructed in Chapter 6 have hypothesis-testing counterparts, we shall simply summarize the results and give some examples. They can all be developed in exactly the same way as was done in the first part of this section.

The procedures for testing the mean μ when the variance is *known* are summarized in Table 7.1.

Example 7.2   Referring to the data of Example 4.2, suppose that the true variance of an individual observation had been $49 \times 10^{-10}\,M$. Seven samples were analyzed for which $\bar{x} = 1.236 \times 10^{-4}\,M$. Furthermore,

$$\frac{\sigma}{\sqrt{n}} = \sqrt{\frac{49 \times 10^{-10}}{7}} = 2.646 \times 10^{-5}\,M.$$

Can we assume that the mean μ of the $K_2Cr_2O_7$ is $1.0 \times 10^{-4}\,M$?

Solution   We will follow the procedure set forth in Section 7.3 for testing a hypothesis:

1. We assume that the data came from a normal population (or approximately so) with known variance.
2. The hypothesis and alternative are $H_0: \mu = 1.0 \times 10^{-4}\ M$ and $H_A: \mu \neq 1.0 \times 10^{-4}\ M$.
3. The test statistic is

$$Z = \frac{\bar{X} - \mu_0}{\sigma/\sqrt{n}}.$$

4. Let $\alpha = 0.05$.
5. Z has a standard normal distribution and the critical region is given in Table 7.1 as $|Z| > z_{1-\alpha/2} = z_{0.975} = 1.96$.
6. The value of the test statistic is

$$z = \frac{1.236 \times 10^{-4} - 1.0 \times 10^{-4}}{2.646 \times 10^{-5}} = 0.8919.$$

7. Since the value 0.8919 does not exceed $z_{0.975} = 1.96$, we accept $H_0: \mu = 1.0 \times 10^{-4}\ M$ (we do not have sufficient evidence to reject H_0). This example was considered in Example 6.1 in terms of confidence intervals.

7.5.2   Parametric Methods for the Mean (Variance Is Unknown)

To test hypotheses on μ when σ^2 is *unknown*, we use the *t*-statistic $T = (\bar{X} - \mu_0)/(S/\sqrt{n})$ and replace z with t and σ^2 with the sample variance S^2 in Table 7.1 to obtain Table 7.2.

TABLE 7.2   Testing μ When σ^2 Is Unknown

$$\text{Statistic: } T = \frac{\bar{X} - \mu_0}{S/\sqrt{n}}, \quad n - 1 \text{ d.f.}$$

Null hypothesis	Alternative	Rejection region
$H_0: \mu = \mu_0$	$H_A: \mu \neq \mu_0$	$T > t_{n-1,1-\alpha/2}$ or $T < -t_{n-1,1-\alpha/2}$
$H_0: \mu \leq \mu_0$	$H_A: \mu > \mu_0$	$T > t_{n-1,1-\alpha}$
$H_0: \mu \geq \mu_0$	$H_A: \mu < \mu_0$	$T < -t_{n-1,1-\alpha}$

Example 7.3   The following data were obtained for the calibration of the Ruska dead weight gauge used with our Burnett P-V-T apparatus (see Example 6.2). The weights corresponding to the 1000-psi loading had the following apparent masses:

26.03570	26.03575	26.03599
26.03581	26.03551	26.03533
26.03529	26.03588	26.03570
26.03573	26.03586	

Can we say that the mean apparent mass μ falls below 26.5?
Solution

1.   Assume approximate normality.
2. $H_0: \mu \geq 26.5$ against $H_A: \mu < 26.5$ (actually, the hypothesis was set up before any data were observed).
3.   Since σ^2 is unknown, use $T = (\bar{X} - 26.5)/S_{\bar{x}}$.
4.   Let $\alpha = 0.05$.
5. T has the t-distribution with $v = 10$ degrees of freedom.
6. $\bar{x} = 26.03568$ and $s/\sqrt{11} = 6.741 \times 10^{-5}$ by Example 6.2. Thus the calculated t value is

$$\frac{26.03568 - 26.5}{6.741 \times 10^{-5}} = -\frac{0.46432}{6.741} \times 10^5.$$

7.   The critical region is, by Table 7.2,

$$\frac{\bar{X} - 26.5}{S/\sqrt{n}} < -t_{10,0.95} = -1.812.$$

As the calculated t value is $-0.46432/6.741 \times 10^5 = -6888 < -1.812$, H_0 is rejected. Thus we conclude that there is sufficient evidence from the data to say that μ falls below 26.5. The chances that we are making an error in so saying are 5 out of 100, which corresponds to $\alpha = 0.05$.

Example 7.4   Fifty determinations of a certain concentration yielded the following values:

54.20	53.82	55.13	55.77	56.52
51.73	54.15	51.12	52.22	56.91
52.56	53.10	53.73	54.55	52.35
53.55	51.56	55.01	56.78	52.02
56.15	53.43	55.57	56.00	58.16
57.50	53.77	53.95	57.27	57.73
52.94	55.88	53.39	54.89	55.33
54.25	54.96	54.30	57.05	54.13
54.46	58.51	52.89	56.25	56.60
53.08	54.65	57.35	56.35	55.21

From the data we obtain

$$\bar{x} = \frac{\sum_{i=1}^{n} x_i}{n} = 54.76,$$

$$s^2 = \frac{\sum_{i=1}^{n} (x_i - \bar{x})^2}{n - 1} = 4.216,$$

and

$$s = \sqrt{s^2} = 2.05329.$$

To test $H_0: \mu \geq 55$, the test statistic is, according to Table 7.2, $T = (\bar{X} - \mu_0)/(S/\sqrt{n})$. Thus

$$T = \frac{54.76 - 55.0}{2.05329/\sqrt{50}} = \frac{-0.24}{0.29038} = -0.8265.$$

Since $t_{49,0.01} \cong -2.4053$, we do not reject H_0 at the 1% significance level.

7.5.3 Nonparametric Methods for the Mean

In Section 7.5.2 we saw how the standard normal statistic $Z = (\bar{X} - \mu_0)/(\sigma/\sqrt{n})$ and the t-statistic $T = (\bar{X} - \mu_0)/(S/\sqrt{n})$ are used to test hypotheses concerning the mean of a normal distribution. These tests are valid when sampling is from a normal population or when the sample sizes are sufficiently large so that deviations from normality do not vitiate the results. Sometimes it is not reasonable to assume that the sampled population is normal. In some cases the sample sizes may be small. When these events happen, the experimenter is advised to use a distribution-free or nonparametric test for the mean (location) rather than the usual f- or t-test.

The terms *nonparametric* and *distribution-free* are used interchangeably in statistics. The terms mean that in conducting a test, no assumption is made about the specific form of the probability density function that models the population from which the sample has been taken. Generally, the only assumption is that the distribution is continuous, so that the assumption of normality need not be made.

If the form of the probability density function is known, the best procedure for testing the mean of that distribution should be employed. For example, if the population is normal, the test of Section 7.5.2 should be used. A feature that characterizes many nonparametric tests and makes them very popular is that they maintain a prespecified significance level, α, and that they have reasonably high power across several forms of the probability density function. That is, they compare well with the Z- and T-tests under normality and outperform them when the sampled distribution is not normal.

Sign Test for the Mean

We assume here that the random variable is continuous and that the sampled distribution is symmetric. The result is that the mean and the median are the same. The random variable X is continuous with mean (and median) μ. Let $X_1,\ldots,X_n$ denote a random sample of size n. The null hypotheses is $H_0: \mu = \mu_0$ and the alternatives are $H_A: \mu \neq \mu_0$, $H_A: \mu < \mu_0$, or $H_A: \mu > \mu_0$, where μ_0 is the hypothesized mean.

If $\mu = \mu_0$, then under the assumption of a continuous distribution, each of the differences $X_i - \mu_0$ has a probability of $\frac{1}{2}$ of being positive, a probability of $\frac{1}{2}$ of being negative, and a probability 0 of being zero. Let B denote the number of $X_i - \mu_0$ that are positive. If $H_0: \mu = \mu_0$ is true, B has a binomial distribution with parameters n and $\frac{1}{2}$. Thus, if H_0 is true, one would expect half of the differences to be positive and half to be negative. In testing H_0 against $H_A: \mu > \mu_0$, we want to detect a situation in which the true median μ lies above the hypothesized mean μ_0. If H_0 is not true, we would expect more than half of the differences to be positive (or less than half to be negative). A reasonable procedure is thus to reject H_0 in favor of $H_A: \mu > \mu_0$ if the observed value of B is too large to have occurred by chance. Similarly, in testing H_0 against $H_A: \mu < \mu_0$, one would reject H_0 if B is too small. In testing H_0 against $H_A: \mu \neq \mu_0$, one would reject H_0 if B is too small or too large. Use of the sign test is illustrated by applying it to the data of Example 7.3.

Example 7.5 In Example 7.3 we tested $H_0: \mu \geq 26.5$ against $H_A: \mu < 26.5$. From the $n = 11$ weights in Example 7.3, we can see that they all fall below $\mu_0 = 26.5$, so that all the differences $X_i - 26.5$ are negative. Thus the value of the sign-test statistic $B = 0$ and we would reject $H_0: \mu \geq 26.5$ in favor of $H_A: \mu < 26.5$. From the binomial distribution table we can see, to three decimal places, that for $n = 11$ and $p = \frac{1}{2}$, $P_2(B = 0) = 0$. The value $B = 0$ is very unlikely to have occurred if $\mu \geq 26.5$, so we conclude that $\mu < 26.5$.

If the sample size n is larger than 20, the normal approximation for the distribution of B may be used. That is, for $n > 20$,

$$Z = \frac{B - np}{\sqrt{np(1-p)}} = \frac{B - n/2}{\sqrt{n/4}} \tag{7.4}$$

has an approximate standard normal distribution.

Example 7.6 For the $n = 50$ determinations of a certain concentration in Example 7.4, we are interested in testing $H_0: \mu \geq 55$ against $H_A: \mu < 55$. Of the 50 observations, 22 are larger than 55, so that the value of the sign test statistic is $B = 22$. The standardized value of B is

$$Z = \frac{B - n/2}{\sqrt{n/4}} = \frac{22 - 25}{\sqrt{25/2}} = -0.8485.$$

The test is to reject $H_0: \mu \geq 55$ in favor of $H_A: \mu < 55$ if Z is too small. At the $\alpha = 0.05$ significance level, we reject H_0 if $Z < -1.645$. Since the computed Z-value is -0.849, we do not reject $H_0: \mu \geq 55$.

The assumption of the underlying distribution being continuous means that zero differences occur with probability zero; however, in practice, zero differences can occur. The zero differences can be handled by assigning to the zero difference the sign that is compatible with the null hypothesis. Thus, for $H_A: \mu > \mu_0$, use negative signs for zero differences, for $H_A: \mu < \mu_0$, use positive signs, and for $H_A: \mu \neq \mu_0$, we use the sign of the less frequently occurring sign. If the number of zeros is small relative to the sample size, reduce the sample size by not using the differences in the analysis.

The sign test uses only the signs of the differences but not their magnitude. We now present the Wilcoxon signed-ranks test, one that uses the sign and magnitude of each difference.

Wilcoxon Signed-Ranks Test

Assume a random sample $X_i, \ldots, X_n$ from a continuous distribution that is symmetric. The hypotheses are the same as for the sign test, that is, we are testing $H_0: \mu = \mu_0$ against $H_A: \mu \neq \mu_0$, $H_A: \mu < \mu_0$, or $H_A: \mu > \mu_0$. The differences $X_i - \mu_0$, $i = 1, 2, \ldots, n$, are a random sample from a distribution that is symmetric about zero. The Wilcoxon signed-ranks test is based on the sign and magnitude of the differences $X_i - \mu_0$, $i = 1, 2, \ldots, n$. To carry out the test, first form the absolute differences $|x_i - \mu_0|$, $i = 1, 2, \ldots, n$. Then rank the absolute differences in increasing order and assign the ranks $1, 2, \ldots, n$ to these differences with the smallest absolute difference reciving the rank 1. Each rank is assigned the sign of the corresponding difference $X_i - \mu_0$. The Wilcoxon signed-ranks test statistic is the sum of the positive ranks chosen from the n signed ranks $R_1, R_2, \ldots, R_n$. The Wilcoxon signed ranks test statistic is thus

$$W_n^+ = \sum_{\substack{\text{positive} \\ \text{ranks}}} R_i. \qquad (7.5)$$

In testing $H_0: \mu = \mu_0$ against $H_A: \mu > \mu_0$, the null hypothesis is rejected for large values of W_n^+. When testing H_0 against $H_A: \mu < \mu_0$, H_0 is rejected for small values of W_n^+. When testing H_0 against $H_A: \mu \neq \mu_0$, H_0 is rejected when W_n^+ is too small or too large. Selected percentiles of the statistic W_n^+ are provided in Table VII of App. B.

If ties occur in the difference values $X_i - \mu_0$, the values for each tied group should be given the average rank of the corresponding ranks.

Example 7.7 To illustrate the procedure, consider the data from Example 7.3. The null hypothesis and its alternative in that example are $H_0: \mu \geq 26.5$ and $H_A: \mu < 26.5$. The $n = 11$ differences $X_i - 26.5$ are all negative, so that $W_{11}^+ = 0$. The null hypothesis is rejected since the value of W_{11}^+ is too small. The critical value in this case is $W_{0.05} = 14$ at a significance level $\alpha = 0.05$.

If n is greater than 20, the normal approximation for the distribution of W_n^+ may be used. The mean and variance of W_n^+ may be used. The mean and variance of W_n^+, under $H_0: \mu = \mu_0$, are

$$E(W_n^+) = \frac{n(n+1)}{4} \quad \text{and} \quad \text{var}(W_n^+) = \frac{n(n+1)(2n+1)}{24}.$$

Thus the random variable

$$Z = \frac{W_n^+ - E(W_n^+)}{\sqrt{\text{var}(W_n^+)}} = \frac{W_n^+ - n(n+1)/4}{\sqrt{n(n+1)(2n+1)/24}}$$

has an approximate standard normal distribution under the hypothesis $H_0: \mu = \mu_0$. The hypothesis $H_0: \mu = \mu_0$ is rejected in favor of $H_A: \mu > \mu_0$ if z is large, in favor of $H_A: \mu < \mu_0$ if Z is small, and in favor of $H_A: \mu \neq \mu_0$ if z is small or large.

Example 7.8 Thirty determinations of a certain concentration yielded the following values (including the ranks in parentheses):

(17) 54.20	(6) 52.94	(8) 52.10	(30) 58.51	(25) 55.57
(3) 51.73	(18) 54.25	(2) 51.56	(21) 54.65	(15) 53.95
(4) 52.56	(20) 54.46	(10) 53.43	(24) 55.13	(9) 53.39
(11) 53.55	(7) 53.08	(13) 53.77	(1) 51.12	(19) 54.30
(27) 56.15	(14) 53.82	(26) 55.88	(12) 53.73	(5) 52.89
(29) 57.50	(16) 54.15	(22) 54.96	(23) 55.01	(28) 57.35

We desire to test $H_0: \mu \geq 55$ against $H_A: \mu < 55$. To accomplish this using the Wilcoxon signed-ranks test we first need to determine the signed ranks of the differences $|X_i - 55|$. The differences $D_i = X_i - 55$, in decreasing order, and the corresponding ranks of the $|D_i|$ are given below.

$D_i = X_i - 55$	Rank of $|D_i|$	$D_i = X_i - 55$	Rank of $|D_i|$
3.51	29	−1.05	12
2.50	26	−1.18	14
2.35	24	−1.23	15
1.15	13	−1.27	16
0.88	11	−1.45	17
0.57	6	−1.57	18
0.13	3	−1.61	19
0.01	1	−1.90	20
−0.04	2	−1.93	21
−0.35	4	−2.06	22
−0.54	5	−2.11	23
−0.70	7	−2.44	25
−0.75	8	−3.27	27
−0.80	9	−3.44	28
−0.85	10	−3.88	30

Since $n > 20$ we will use the normal approximation for the distribution of W_n^+. The mean and variance of W_n^+, under H_0, are $E(W_n^+) = n(n + 1)/4 = 232.5$, and $\mathrm{var}(W_n^+) = n(n + 1)(2n + 1)/24 = 2363.75$. There are eight X-values larger than 55 in the data, and the sum of the signed ranks corresponding to them, from the table, is $w_n^+ = 29 + 26 + \cdots + 3 + 1 = 113$. The standardized value of W_n^+ is

$$Z = \frac{W_n^+ - E(W_n^+)}{\sqrt{\mathrm{var}(W_n^+)}} = \frac{113 - 232.5}{\sqrt{2362.75}} = -2.458.$$

At $\alpha = 0.01$, the decision is to reject $H_0: \mu \geq 55$ if $z < -2.326$. Since the computed z-value is $-2.458 < -2.326$, we reject $H_0: \mu \geq 55$ and conclude that $\mu < 55$ (at $\alpha = 0.01$).

7.6   PARAMETRIC TESTS ON THE DIFFERENCE OF TWO MEANS

There are three cases to be considered in comparing the means of two populations: namely, the same ones that were considered in Section 6.9.

Case 1 (σ_1^2 and σ_2^2 Known)

Suppose that we have two random samples from two different populations with parameters μ_1, σ_1^2 and μ_2, σ_2^2, where σ_1^2 and σ_2^2 are known. The test statistic used to test $H_0: \mu_1 = \mu_2 (\mu_1 - \mu_2 = 0)$ against $H_A: \mu_1 \neq \mu_2$ is the standard normal random variable

$$Z = \frac{\bar{X}_1 - \bar{X}_2 - (\mu_1 - \mu_2)}{\sqrt{\sigma_1^2/n_1 + \sigma_2^2/n_2}}, \tag{7.6}$$

where n_1 and n_2 are the sample sizes. The statistic in (7.4) can actually be used to test the more general hypothesis $H_0: \mu_1 - \mu_2 = d$, where d is a specified number. However, the case $\mu_1 - \mu_2 = 0$ is of special interest. The test rejects $H_0: \mu_1 = \mu_2$ if $Z > z_{1-\alpha/2}$ or $Z < z_{\alpha/2} = -z_{1-\alpha/2}$. This result and other possibilities are included in Table 7.3.

A confidence interval corresponding to each H_0 in Table 7.3 can be derived. We leave those derivations to the interested reader and turn our attention to Case 2.

TABLE 7.3   Testing $\mu_1 - \mu_2$ When σ_1^2 and σ_2^2 Are Known

$$\text{Statistic: } Z = \frac{\bar{X}_1 - \bar{X}_2}{\sqrt{\sigma_1^2/n_1 + \sigma_2^2/n_2}}$$

Null hypothesis	Alternative	Rejection region
$H_0: \mu_1 - \mu_2 = 0$	$H_A: \mu_1 - \mu_2 \neq 0$	$Z > z_{1-\alpha/2}$ or $Z < -z_{1-\alpha/2}$
$H_0: \mu_1 - \mu_2 \leq 0$	$H_A: \mu_1 - \mu_2 > 0$	$Z > z_{1-\alpha}$
$H_0: \mu_1 - \mu_2 \geq 0$	$H_A: \mu_1 - \mu_2 < 0$	$Z < -z_{1-\alpha}$

Example 7.9   Consider the data presented in Example 4.1. From those data, is it reasonable to believe that the two bottles of reagent were filled from the same production run? Information received from the manufacturer indicates that the average variance in concentration for single analyses for the last year's production runs was 0.016.

Solution   We set up the hypothesis that the two bottles were indeed filled out of the same batch. This signifies that the means of the analyses for the two bottles should be equal and we write $H_0: \mu_1 = \mu_2$, against $H_A: \mu_1 \neq \mu_2$.

We take the variance for both samples, σ_1^2 and σ_2^2, to be 0.016 and we have a Case 1 situation. From Table 7.3 we choose the appropriate statistic, which when evaluated yields

$$z = \frac{15.770 - 15.597 - (\mu_1 - \mu_2)}{\sqrt{0.016(\sqrt{\frac{1}{3}} + \frac{1}{3})}},$$

since $\bar{x}_1 = 15.770$, $\bar{x}_2 = 15.597$, $n_1 = n_2 = 3$, and $\sigma_1^2 = \sigma_2^2 = 0.016$. If H_0: $\mu_1 = \mu_2$ is true, z reduces to

$$z = \frac{15.770 - 15.597}{0.12649(0.8165)} = \frac{0.173}{0.10327} = 1.6751.$$

If we choose $\alpha = 0.05$, then $z_{1-\alpha/2} = z_{0.975} = 1.96$, and since $-1.96 < 1.675 = z < 1.96$, we cannot reject H_0: $\mu_1 = \mu_2$ and so conclude that $\mu_1 = \mu_2$. From Example 6.4, a 95% confidence interval for $\mu_1 - \mu_2$ is $-0.029 < \mu_1 - \mu_2 < 0.375$. Since the hypothesized difference, $\mu_1 - \mu_2 = 0$, falls in the 95% confidence interval, we reach the same conclusion regarding H_0 by that route.

Case 2 (σ_1^2 and σ_2^2 Unknown but Assumed Equal)

In this case the population variances are unknown but assumed equal. The common variance σ^2 is estimated by the pooled variance

$$S_p^2 = \frac{\sum_{i=1}^{n_1} (X_{1i} - \bar{X}_1)^2 + \sum_{i=1}^{n_2} (X_{2i} - \bar{X}_2)^2}{n_1 + n_2 - 2},$$

as given in Eq. (6.21). The test statistic is

$$T = \frac{\bar{X}_1 - \bar{X}_2 - (\mu_1 - \mu_2)}{S_p\sqrt{1/n_1 + 1/n_2}}, \tag{7.7}$$

which follows the t-distribution with $n_1 + n_2 - 2$ degrees of freedom. A summary of test criteria for Case 2 is presented in Table 7.4.

TABLE 7.4 Testing $\mu_1 - \mu_2$ When σ_1^2 and σ_2^2 Are Unknown but Equal

Statistic: $T = \dfrac{\bar{X}_1 - \bar{X}_2}{S_p\sqrt{1/n_1 + 1/n_2}}$, $\quad v = n_1 + n_2 - 2$ d.f.

Null hypothesis	Alternative	Rejection region
H_0: $\mu_1 - \mu_2 = 0$	H_A: $\mu_1 - \mu_2 \neq 0$	$T > t_{v, 1-\alpha 2}$ or $T < -t_{v, 1-\alpha 2}$
H_0: $\mu_1 - \mu_2 \leq 0$	H_A: $\mu_1 - \mu_2 > 0$	$T > t_{v, 1-\alpha}$
H_0: $\mu_1 - \mu_2 \geq 0$	H_A: $\mu_1 - \mu_2 < 0$	$T < -t_{v, 1-\alpha}$

Example 7.10 To compare the effects of two solid catalyst component concentrations on NO_2 reduction, six groups of observations were made.

Each group consisted of three replicates of five observations each; that is, a total of 15 determinations were made for each concentration. The concentrations (in %) were 0.5% and 1.0%. The reduction data are summarized below.

Group	Replicate mean, $\bar{X}_i$ (5 observations each)			Group mean, $\bar{\bar{X}}_i$ (based on 15 observations)
	A	B	C	
1	5.18	5.52	5.42	5.37
2	5.58	5.62	5.82	5.67

Can we conclude that there is no difference between the means μ_1 and μ_2? The null hypothesis is thus $H_0: \mu_1 - \mu_2 = 0$.

Solution In this case we treat the group means A, B, C as individual observations (see Example 6.5). Thus $\bar{x}_1 = 5.37\bar{3}$, $\bar{x}_2 = 5.67\bar{3}$, and $n_1 = n_2 = 3$. We assume that $\sigma_1^2 = \sigma_2^2$, so we have a Case 2 situation. As σ^2 is unknown, we cannot use Z for our test statistic but must use T, which we choose from Table 7.4.

From Example 6.5 we already have $s_p^2 = 0.02353\bar{3}$.

$$T = \frac{\bar{x}_1 - \bar{x}_2 - (\mu_1 - \mu_2)}{s_p\sqrt{\frac{1}{3} + \frac{1}{3}}}$$

$$= \frac{5.37 - 5.67 - 0}{\sqrt{0.02353\bar{3}}\sqrt{\frac{2}{3}}} = -2.3951.$$

If we choose $\alpha = 0.05$, then $t_{4,0.025} = -2.7764$, and since $t = -2.3951$ does not fall below -2.7764, we accept $H_0: \mu_1 - \mu_2 = 0$. On calculating the 95% confidence interval for $\mu_1 - \mu_2$, we have $-0.647709 < \mu_1 - \mu_2 < 0.047709$, which leads us to the same conclusion: that is, there is no difference between μ_1 and μ_2 at the 5% level of significance.

Case 3 (σ_1^2 and σ_2^2 Unknown and Cannot Be Presumed Equal)

To test hypotheses involving $\mu_1 - \mu_2$, we use the statistic

$$T_f = \frac{\bar{X}_1 - \bar{X}_2 - (\mu_1 - \mu_2)}{\sqrt{S_1^2/n_1 + S_2^2/n_2}}, \qquad (7.8)$$

where n_1 and n_2 are the sample sizes and S_1^2 and S_2^2 are the sample variances. The degrees of freedom associated with the random variable T is f, where

$$f = \frac{(s_1^2/n_1 + s_2^2/n_2)^2}{(s_1^2/n_1)^2/(n_1 - 1) + (s_2^2/n_2)^2/(n_2 - 1)}. \qquad (7.9)$$

Thus T_f in (7.8) has a t-distribution with f d.f. Since this result is not used as frequently as that in Cases 1 and 2, we do not give the summary results in tabular form but rather, consider an example.

For all values of f, truncate the calculated value of f and use t for the corresponding number of degrees of freedom. This procedure will give reasonable values of t_f for low values of f and conservative values of t_f at higher values of f, as shown by Satterthwaite [5].

Example 7.11   To test the effect of a solid catalyst component on NO_2 reduction, six groups of observations were made. Each group consisted of three replicates of five observations each. The component concentrations by groups were:

Group	Concentration (wt %)
1	0.0
2	0.05
3	0.1
4	0.5
5	1.0
6	2.5

The following reduction data were obtained on standardized activity tests.

Group	Replicate means, $\bar{X}_i$			Group mean, $\bar{\bar{X}}_i$
	A	B	C	
1	4.96	4.98	4.82	4.92
2	4.86	4.84	4.88	4.86
3	4.94	5.14	5.10	5.03
4	5.18	5.52	5.42	5.37
5	5.58	5.62	5.82	5.67
6	6.08	5.82	5.72	5.87

From these data, the variances, sample standard deviations, and standard errors of the group means are calculated. For these data, $s_{\bar{x}}$ has the connotation associated with s_x, the sample standard deviation, because the calculations were made using the replicate means as individual data points. By the same token, $s_{\bar{x}}$ is the standard error of the group means.

Group	$\sum\limits_{i=1}^{3} \bar{x}_i^2$	$s_{\bar{x}}^2$	$s_{\bar{x}}$	$s_{\bar{x}}$
1	72.6344	0.0076	0.0872	0.05036
2	70.8596	0.0004	0.0200	0.01155
3	75.8232	0.03975	0.1994	0.11513
4	86.6792	0.08425	0.2903	0.16761
5	96.5932	0.07325	0.2707	0.15676
6	103.5572	0.09325	0.3054	0.17630

The null hypothesis $H_0: \bar{\mu} = \bar{\mu}_1$ (group 1 was the control group) is used to determine whether or not the addition of the catalyst component had a significant effect on NO_2 reduction. By visual inspection, we conclude that all the group variances are not equal. Certainly, the variances of groups 1, 2, and 3 are different from each other and from any of the variances of the other three groups. Many users of statistics might stretch a point and say that the variances of groups 4, 5, and 6 are close enough to be considered equal. Yet as all the variances are not equal or even nearly so, we will use Eq. (7.8) to calculate the test statistic T_f needed to test the hypothesis.

We must also determine the number of degrees of freedom f with which T_f is distributed and the corresponding tabular value of $t_{f,0.975}$, which will be used in testing the hypothesis. The 95% confidence level in a two-tailed test was chosen because one group mean $\bar{x}_2$ falls below $\bar{x}_1$ while all the other $\bar{x}_i$ fall above $\bar{x}_1$. Using group 1 as the control group and the data of group 5 for illustration,

$$
\begin{aligned}
f &= \frac{(s_5^2/n_5 + s_1^2/n_1)^2}{(s_5^2/n_5)^2/(n_5 - 1) + (s_1^2/n_1)^2/(n_1 - 1)} \\
&= \frac{(0.07325/3 + 0.0076/3)^2}{(0.07325/3)^2/2 + (0.0076/3)^2/2} \\
&= \frac{(0.02695)^2}{0.0030129} = 2.4106 \simeq 2.411,
\end{aligned}
$$

$$T = \frac{x_5 - x_1}{\sqrt{s_5^2/n_5 + s_1^2/n_1}} = \frac{5.67 - 4.92}{\sqrt{0.02695}} = 4.5686.$$

Truncating the value of f to 2, we have $t_{2,0.975} = 4.303$. As the rejection region for H_0 is

$$T \le -t_{2,0.975} \quad \text{and} \quad T \ge t_{2,0.975},$$

the hypothesis is rejected. Results for all groups are shown below.

Group	f	$t_{f,0.975}$	H_0
2	2	± 4.303	Accept
3	2	± 4.303	Accept
4	2	± 4.303	Accept
5	2	± 4.303	Reject
6	2	± 4.303	Reject

From examination of the original data and by comparing corresponding values of $t_{f,0.975}$, we suspect that the inclusion of the catalyst component could have a beneficial effect on NO_2 reduction if present in concentrations at or above 1.0 wt %. Attempts to draw further conclusions from these results should not be made.

    The use of PROC TTEST (*SAS for Linear Models*, SAS Institute, Cary, NC, 1981, 231 pp.) to evaluate the difference of two means is quite simple. This procedure does not depend on your knowing in advance whether the variances can be presumed equal or not, as it calculates the value of T both ways, as illustrated in the following example.

Example 7.12   Two of the orifice holes in a Saybolt viscometer are used to make simultaneous readings of time for flow of 60-mL samples. The times (in seconds) for each orifice for a sample of CCl_4 are as shown below in the SAS program. The students running the experiment ask the instructor if the orifices should be cleaned and the trials rerun. If the instructor operates on a 95% confidence level, what should her answer be?

(put initial JCL cards/statements here)

DATA;

INPUT HOLE 2 TIME 5-8;

CARDS;

1	27.8
1	27.9
1	28.0
1	27.8
2	29.5
2	29.7
2	29.4
2	29.8

PROC TTEST;

CLASS HOLE;

TITLE 'EVALUATION OF VISCOMETER';

(put final JCL cards/statements here)

The results are as shown below.

EVALUATION OF VISCOMETER

TTEST PROCEDURE

VARIABLE: TIME

HOLE	N	MEAN	STD DEV	STD ERROR	MINIMUM	MAXIMUM
1	4	27.87500000	0.09574271	0.04787136	27.80000000	28.00000000
2	4	29.60000000	0.18257419	0.09128709	29.40000000	29.80000000

VARIANCES	T	DF	PROB > |T|
UNEQUAL	-16.7350	4.5	0.0001
EQUAL	-16.7350	6.0	0.0001

Solution The variances for Case 2 and Case 3 *t*-tests have been used to calculate the degrees of freedom and the corresponding T values for comparing the means. For the hypothesis that the variances are equal, the program calculates $F = 3.64$ with a probability of 0.3172, indicating no evidence that the variances are unequal. Thus the calculated T for Case 2 (equal but unknown population variances) would be used in evaluating the null hypothesis that there is no difference in the mean times of the two viscometer holes. For that test, $t_{6,0.975} = 2.4469 \nless T = -16.7350$. The last column shows the probability that the absolute value of T will exceed the corresponding computed value (e.g., $|-16.735|$).

7.6.1 Wilcoxon Rank-Sum Test

We next consider a nonparametric test for testing $H_0: \mu_1 = \mu_2$ or $H_0: \mu_1 - \mu_2 = 0$, a hypothesis on the difference of two means. We assume that the distributions are continuous. In this case we combine the samples $X_{11}, X_{12}, \ldots, X_{1n_1}$, and $X_{21}, X_{22}, \ldots, X_{2n_2}$ and order the $n_1 + n_2$ observations from the smallest to the largest. The ranks $1, 2, \ldots, n_1 + n_2$ are then assigned to the ordered observations. The Wilcoxon rank-sum test statistic is, for $N = n_1 + n_2$,

$$W_N = \sum_{i=1}^{n_1} R_{1,i}, \tag{7.10}$$

where $\sum_{i=1}^{n_1} R_{1,i}$ is the sum of the ranks of the observations from the first sample, in the combined ordered sample.

In testing $H_0: \mu_1 = \mu_2$ against $H_A: \mu_1 > \mu_2$, the hypothesis is rejected if W_N is too large. In testing against $H_A: \mu_1 < \mu_2$, H_0 is rejected if W_N is too small, and in testing against $H_A: \mu_1 \neq \mu_2$, H_0 is rejected if W_N is too small or too large. Table VIII of App. B contains selected percentiles of the statistic W_N.

Example 7.13 In Example 7.9 we have $n_1 = 3$ and $n_2 = 3$. The hypothesis is $H_0: \mu_1 = \mu_2$ and the alternative is $H_A: \mu_1 \neq \mu_2$. The ordered observations are $x_{1,2} = 15.49$, $x_{2,2} = 15.58$, $x_{2,1} = 15.64$, $x_{3,2} = 15.72$, $x_{1,1} = 15.75$, and $x_{3,1} = 15.92$. The ranks of the observations from the first sample are $R_{1,1} = 5$, $R_{2,1} = 3$, and $R_{3,1} = 6$, so that

$$w_6 = 5 + 3 + 6 = 14.$$

The critical values at $\alpha = 0.05$ are $w_{0.025} = 6$ and $w_{0.975} = 21 - w_{0.025} = 15$. Since $w_6 = 14$ falls between the two critical values, we do not reject $H_0: \mu_1 = \mu_2$, the same conclusion as that reached in Example 7.9.

If n_1 and n_2 are both large, the normal approximation for the distribution of W_N may be used. Under $H_0: \mu_1 = \mu_2$, the mean and variance of W_N are

$$E(W_N) = E\left[\sum_{i=1}^{n_1} R_{1,i}\right] = \frac{n_1(n_1 + n_2 + 1)}{2} \tag{7.11}$$

and

$$\mathrm{var}(W_N) = \mathrm{var}\left(\sum_{i=1}^{n_1} R_{1,i}\right) = \frac{n_1 n_2(n_1 + n_2 + 1)}{12}. \tag{7.12}$$

Thus

$$Z = \frac{W_N - n_1(n_1 + n_2 + 1)/2}{\sqrt{n_1 n_2 (n_1 + n_2 + 1)/12}}$$

has approximately a standard normal distribution. The statistic Z can be used to carry out tests of $H_0: \mu_1 = \mu_2$, using percentiles from the standard normal distribution. The percentiles of W_N are approximated by

$$w_\alpha \approx n_1(n_1 + n_2 + 1)/2 + z_\alpha \sqrt{\frac{n_1 n_2 (n_1 + n_2 + 1)}{12}}, \tag{7.13}$$

where Z_α is a standard normal percentile.

7.7 PAIRED *t*-TEST

7.7.1 Parametric Test for Paired Observations

In the case of paired observations $(X_1, Y_1), \ldots, (X_n, Y_n)$ we wish to test $H_0: \mu_1 = \mu_2$ or $H_0: \mu_D = 0$, where $\mu_D = \mu_1 - \mu_2$. The procedure is to consider the differences $D_i = X_i - Y_i$, $i = 1, 2, \ldots, n$, and observe that $\mu_D = \mu_X - \mu_Y$. The $H_0: \mu_D = 0$ can be tested by using the statistic

$$T = \frac{\bar{D} - \mu_D}{S_D/\sqrt{n}},$$

which has a t-distribution with $n - 1$ d.f. Recall from Section 6.10 that

$$\bar{D} = \sum_{i=1}^{n} \frac{X_i - Y_i}{n} = \bar{X} - \bar{Y}$$

and

$$S_D^2 = \sum_{i=1}^{n} \frac{(D_i - \bar{D})^2}{n - 1}.$$

The test procedures are summarized in Table 7.5.

Example 7.14 Let us reexamine the data in Example 6.6. In studying the weather resistance of two types of paint, each of five shingles was painted with both paints, one-half with one paint and the other half with the other paint. After a month's exposure, the wear, measured in suitable units, was found to be (1.57,1.45), (1.46,1.59), (1.53,1.27), (1.30,1.48), (1.37,1.40). Is it plausible that $\mu_D = \mu_1 - \mu_2$ is greater than zero?

Solution From Example 6.6, if $D_i = X_i - Y_i$, we have $d_1 = 0.12$, $d_2 = -0.13$, $d_3 = 0.26$, $d_4 = -0.18$, and $d_5 = -0.03$. Also, $\bar{d} = 0.008$ and $S_D = 0.181576$.

TABLE 7.5 Testing $\mu_D = 0$ When Observations Are Paired

$$\text{Statistic: } T = \frac{\bar{X} - \bar{Y}}{S_D/\sqrt{n}}, \quad n-1 \text{ d.f.}$$

Null hypothesis	Alternative	Rejection region
$H_0: \mu_D = 0$	$H_A: \mu_D \neq 0$	$T > t_{n-1,1-\alpha/2}$ or $T < -t_{n-1,1-\alpha/2}$
$H_0: \mu_D \leq 0$	$H_A: \mu_D > 0$	$T > t_{n-1,1-\alpha}$
$H_0: \mu_D \geq 0$	$H_A: \mu_D < 0$	$T < -t_{n-1,1-\alpha}$

We want to test $H_0: \mu_D \leq 0$ against $H_A: \mu_D > 0$. The statistic to be used is

$$T = \frac{\bar{D} - \mu_D}{S_D/\sqrt{n}} = \frac{\bar{D}}{S_D/\sqrt{n}}.$$

From Table 7.5 we must reject $H_0: \mu_D \leq 0$ if $T = \bar{D}/(S_D/\sqrt{n}) > t_{4,1-\alpha}$. Let $\alpha = 0.05$, so that $t_{4,0.95} = 2.1318$. Since $T = 0.008/0.081203 = 0.0985 < t_{4,0.95} = 2.1318$, we do not have sufficient evidence to reject $H_0: \mu_D \leq 0$. We therefore conclude that it is not plausible that $\mu_D > 0$.

Example 7.15 The use of computerized data reduction for handling paired comparisons is illustrated below for Example 6.6. The SAS program in this case is as follows.

```
(put initial JCL cards/statements here)

DATA;
INPUT TOP BOTTOM;
D = TOP - BOTTOM;
CARDS;
```

(data are entered here, one pair per line with the values separated by one space)

```
PROC PRINT;
PROC MEANS MEAN STD STDERR T PRT;
VAR D;
TITLE 'PAIRED COMPARISON FOR WEATHERING DATA';
```

(put final JCL cards/statements here)

The output from this program consists of two parts: the results of the PRINT procedure and those from the MEANS procedure.

SAS

OBS	TOP	BOTTOM	D
1	1.46	1.59	-0.13
2	1.53	1.27	0.26
3	1.30	1.48	-0.18
4	1.37	1.40	-0.03
5	1.57	1.45	0.12

PAIRED COMPARISON OF WEATHERING DATA

VARIABLE	MEAN	STANDARD DEVIATION	STD ERROR OF MEAN	T	PR>|T|
D	0.00800000	0.18157643	0.08120345	0.10	0.9263

From PROC MEANS, we see that the calculated value of T is 0.10. The probability 0.9263 means that $T = 0.10$ is not significantly different from 0, so that $H_0: \mu_D = 0$ is not rejected.

7.7.2 Wilcoxon Signed-Ranks Test for Paired Data

The Wilcoxon signed-ranks test may be used in the case of paired data $(X_1, Y_1), (X_2, Y_2), \ldots, (X_n, Y_n)$. The hypothesis is $H_0: \mu_1 = \mu_2$ or $H_0: \mu_D = 0$, where $\mu_D = \mu_2 - \mu_1$. The hypotheses in question are summarized in Table 7.5. In this case, the absolute differences $|D_i| = |Y_i - X_i|$ are ranked in increasing order, and ranks are assigned to these absolute differences, with 1 being assigned to the smallest absolute difference. Each rank is assigned the sign of the corresponding difference $Y_i - X_i$. The Wilcoxon signed-ranks test statistic is defined by

$$W_n^+ = \sum_{\substack{\text{positive} \\ \text{ranks}}} R_i. \tag{7.14}$$

The test procedure is to reject $H_D: \mu_D = 0$ (a) in favor of $H_A: \mu_D > 0$ if W_n^+ is large, (b) in favor of $H_A: \mu_D < 0$ if W_n^+ is small, and (c) in favor of $H_A: \mu_D \neq 0$ if W_n^+ is small or large. In case of zero differences, the method of average ranks is used in assigning ranks.

Example 7.16   In Example 6.6 we have a sample of size $n = 5$: namely, (1.57,1.45), (1.46,1.59), (1.53,1.27), (1.30,1.48), and (1.37,1.40). Suppose that we are interested in testing $H_D: \mu_D = 0$ against $H_D: \mu_D > 0$, where $\mu_D = \mu_2 - \mu_1$. The differences are

$$D_i = Y_i - X_i: \quad d_i = 0.12, \quad d_2 = -0.13,$$
$$d_3 = 0.26, \quad d_4 = -0.18, \quad d_5 = -0.03.$$

The absolute differences are

$$|d_1| = 0.12, \quad |d_2| = 0.13, \quad |d_3| = 0.26, \quad |d_4| = 0.18, \quad |d_5| = 0.03$$

so

$$|d_5| = 0.03, \quad |d_1| = 0.12, \quad |d_2| = 0.13, \quad |d_4| = 0.18, \quad |d_3| = 0.26.$$

The associated signed ranks are $R_1 = -1$, $R_2 = +2$, $R_3 = -3$, $R_4 = -4$, and $R_5 = +5$. Thus $w_s^+ = R_2 + R_5 = 7$. At the 5% significance level, we reject $H_0: \mu_D = 0$ in favor of $H_A: \mu_D = \mu_2 - \mu_1 < 0$ if $w_s^+ \leq 1$. Since $w_s^+ = 7 > 1$, we do not reject $H_0: \mu_D = 0$.

7.8 TESTING A PROPORTION *P*

To test the hypothesis $H_0: P = P_0$ involving a proportion P which is the parameter of the binomial distribution, we use the statistic

$$Z = \frac{\hat{P} - P_0}{\sqrt{P_0 Q_0/n}}, \tag{7.15}$$

which has an approximate standard normal distribution. This statistic was discussed in Section 6.12, where confidence intervals on P were presented. The hypothesis $H_0: P = P_0$ is rejected in favor of $H_A: P \neq P_0$ if either

$$Z > z_{1-\alpha/2} \quad \text{or} \quad Z < -z_{1-\alpha/2},$$

where α is the level of significance. This and other possible situations are summarized in Table 7.6.

Example 7.17 It is desired to know if the proportion of all batches of a particular chemical formulation which are on-grade is larger than 90%. That is, if a batch is chosen, is the probability P that it is on-grade larger than 0.9? Choose $\alpha = 0.05$.

TABLE 7.6 Testing a Proportion *P*

$$\text{Statistic: } Z = \frac{\hat{P} - P_0}{\sqrt{P_0(1 - P_0)/n}}$$

Null hypothesis	Alternative	Rejection region
$H_0: P = P_0$	$H_A: P \neq P_0$	$Z > z_{1-\alpha/2}$ or $Z < -z_{1-\alpha/2}$
$H_0: P \leq P_0$	$H_A: P > P_0$	$Z > z_{1-\alpha}$
$H_0: P \geq P_0$	$H_A: P < P_0$	$Z < -z_{1-\alpha}$

Solution   A sample of 100 batches of chemical are chosen and 95 are on grade: $\hat{P} = 0.95$. Furthermore,

$$Z = \frac{\hat{P} - P_0}{\sqrt{P_0(1 - P_0)/n}} = \frac{0.95 - 0.90}{\sqrt{[0.90(0.10)]/100}} = \frac{0.05}{0.03} = 1.66\bar{6}.$$

The hypothesis is $H_0: P \leq 0.9$ and the alternative is $H_A: P > 0.9$. From Table 7.6 the rejection region is

$$Z > z_{1-\alpha} = z_{0.95} = 1.645.$$

Since $1.667 > 1.645$, we reject H_0 and conclude that $P > 0.90$. The probability that we have reached a wrong conclusion is $\alpha = 0.05$.

7.9   TESTING THE DIFFERENCE OF TWO PROPORTIONS

To test the hypothesis $H_0: P_1 = P_2$ of equality of two proportions corresponding to two binomial populations, we use the statistic

$$Z = \frac{\hat{P}_1 - \hat{P}_2}{\sqrt{\hat{P}(1 - \hat{P})/n_1 + \hat{P}(1 - \hat{P})/n_2}} \tag{7.16}$$

where n_1 and n_2 are the sample sizes, $\hat{P}_1$ and $\hat{P}_2$ are the estimators of P_1 and P_2, and $\hat{P} = (x_1 + x_2)/(n_1 + n_2)$, the pooled estimate of the common value of $P_1 = P_2$. Note that x_1 and x_2 are the numbers of successes in the two random samples. The statistic Z was discussed in Section 6.15 and has an approximate standard normal distribution. The results pertaining to the use of Eq. (7.16) are presented in Table 7.7.

Example 7.18   Two different methods of manufacturing compression rings for $\frac{1}{4}$-in. copper tubing are being used. It is desired to know if the proportions P_1 and P_2 of rings that do not seat properly are the same for the two methods. Can we say that $P_1 - P_2 \neq 0$? Use $\alpha = 0.01$.

TABLE 7.7   Testing $P_1 = P_2$

Statistic: $Z = \dfrac{\hat{P}_1 - \hat{P}_2}{\sqrt{\hat{P}(1 - \hat{P})/n_1 + \hat{P}(1 - \hat{P})/n_2}}$

Null hypothesis	Alternative	Rejection region
$H_0: P_1 = P_2$	$H_A: P_1 \neq P_2$	$Z > z_{1-\alpha/2}$ or $Z < -z_{1-\alpha/2}$
$H_0: P_1 \leq P_2$	$H_A: P_1 > P_2$	$Z > z_{1-\alpha}$
$H_0: P_1 \geq P_2$	$H_A: P_1 < P_2$	$Z < -z_{1-\alpha}$

Solution   One hundred rings are sampled from each method and it is found that 10 rings using method 1 and 3 rings using method 2 do not seat properly. The null hypothesis and alternative are $H_0: P_1 = P_2$ and $H_A: P_1 \neq P_2$. From Table 7.8, the critical region is

$$\frac{\hat{P}_1 - \hat{P}_2}{\sqrt{\hat{P}(1 - \hat{P})/n_1 + \hat{P}(1 - \hat{P})/n_2}} > z_{0.995} \quad \text{or} \quad < -z_{0.995}.$$

Now $n_1 = n_2 = 100$, $\hat{P} = 0.065$, and the computed value for Z (under H_0) is

$$\frac{0.10 - 0.03}{\sqrt{0.065(0.935)/100 + 0.065(0.935)/100}} = \frac{0.07}{0.034864} = 2.0078 \simeq 2.008.$$

Since $z_{0.995} \simeq 2.575$, we have $-2.575 < 2.008 < 2.575$, so we cannot reject $H_0: P_1 = P_2$. From Example 6.13, a 99% confidence interval is $(-0.01, 0.15)$, which contains $P_1 - P_2 = 0$ and leads us to the same conclusion.

7.10   TESTS CONCERNING THE VARIANCE

Several hypotheses about a single variance are possible. If we wish, for example, to test whether a random sample is drawn from a population having a specific known variance σ_0^2, we have the hypothesis

$$H_0: \sigma^2 = \sigma_0^2$$

against the alternative

$$H_A: \sigma^2 \neq \sigma_0^2.$$

The alternative could be of the form $H_A: \sigma^2 > \sigma_0^2$ or $H_A: \sigma^2 < \sigma_0^2$. Assuming that H_0 is correct, the test statistic used is that given by the random variable

$$\chi^2 = \frac{(n - 1)S^2}{\sigma_0^2}, \tag{7.17}$$

which has a χ^2-distribution with $n - 1$ degrees of freedom. The critical region consists of the values in the two tails of the chi-square distribution when $H_0: \sigma^2 = \sigma_0^2$ is true. The hypothesis $H_0: \sigma^2 = \sigma_0^2$ will therefore be rejected if either $\chi^2 = (n - 1)S^2/\sigma_0^2 \leq \chi^2_{n-1,\alpha/2}$ or $\chi^2 \geq \chi^2_{n-1,1-\alpha/2}$, which represents the conditions where the calculated χ^2 value $(n - 1)s^2/\sigma_0^2$ is too large or too small under the assumption that $\sigma^2 = \sigma_0^2$. The significance level is determined from

$$\mathscr{P}(\chi^2_{n-1,\alpha/2} \leq \chi^2 \leq \chi^2_{n-1,1-\alpha/2}) = 1 - \alpha.$$

If we wish to test whether the variance of the population exceeds a given value, we have a one-sided hypothesis $H_0: \sigma^2 \leq \sigma_0^2$ against $H_A: \sigma^2 > \sigma_0^2$.

The corresponding one-sided critical region for which H_0 will be rejected is $\chi^2 = (n-1)S^2/\sigma_0^2 > \chi^2_{n-1,1-\alpha}$.

For the other possible one-tailed test, we have $H_0: \sigma^2 \geq \sigma_0^2$ as opposed to $H_A: \sigma^2 < \sigma_0^2$. The corresponding rejection region for this H_0 is $\chi^2 = (n-1)S^2/\sigma_0^2 < \chi^2_{n-1,\alpha}$. Table 7.8 summarizes the critical regions needed for each of the possible alternatives.

TABLE 7.8   Testing the Variance σ^2

$$\text{Statistic}: \chi^2 = \frac{(n-1)S^2}{\sigma_0^2}, \quad n-1 \text{ d.f.}$$

Hypothesis	Alternative	Rejection region
$H_0: \sigma^2 = \sigma_0^2$	$H_A: \sigma^2 \neq \sigma_0^2$	$\chi^2 > \chi^2_{n-1,1-\alpha/2}$ or $\chi^2 < \chi^2_{n-1,\alpha/2}$
$H_0: \sigma^2 \leq \sigma_0^2$	$H_A: \sigma^2 > \sigma_0^2$	$\chi^2 > \chi^2_{n-1,1-\alpha}$
$H_0: \sigma^2 \geq \sigma_0^2$	$H_A: \sigma^2 < \sigma_0^2$	$\chi^2 < \chi^2_{n-1,\alpha}$

Example 7.19   Five similar determinations of the cold water flow rate to a heat exchanger were, in gal/min, 5.84, 5.76, 6.03, 5.90, and 5.87. It is of importance that the imprecision (on a per measurement basis) does not greatly affect this measuring operation. Can we be reasonably sure that the variance per measurement is less than 0.01?
Solution

1. Assume a normal population.
2. The hypothesis is $H_0: \sigma^2 \geq 0.01$ against $H_A: \sigma^2 < 0.01$.
3. The test statistic from Table 7.8 is

$$\chi^2 = \frac{(n-1)S^2}{\sigma_0^2}.$$

4. Let $\alpha = 0.025$.
5. The test statistic in (3) has a χ^2-distribution with 4 d.f. The critical region from Table 7.8 is $\chi^2 < \chi^2_{4,0.025} = 0.4844$.
6. From the data we obtain

$$\chi^2 = \frac{4(0.00975)}{0.01} = 3.900.$$

   The sample variance $s^2 = 0.00975$, as was given in Example 6.8.
7. Since $\chi^2 = 3.9 > 0.484$ we do not reject $H_0: \sigma^2 \geq 0.01$ and thus conclude that it is not reasonable to believe that $\sigma^2 < 0.01$.

7.11 GOODNESS-OF-FIT TESTS

The chi-square test was developed to facilitate testing the significance of data in light of their observed scatter. Indeed, this is the fundamental principle underlying all tests of significance. The differences studied may be those between comparable sets of data or between theoretically proposed and experimentally observed distributions and their attributes. One of the most frequent uses of the χ^2-test is in the comparison of observed frequencies (percentages cannot be used unless the sample size is exactly 100) and the frequencies we might expect from a given theoretical explanation of the phenomenon under investigation.

The χ^2 statistic used in this case is defined by

$$\chi^2 = \sum_{i=1}^{k} \frac{(O_i - E_i)^2}{E_i},\tag{7.18}$$

where O_i is the observed frequency, E_i the expected frequency, and k the number of classes. The random variable defined by Eq. (7.18) has an approximate χ^2-distribution with $k - 1$ degrees of freedom.

Example 7.20 In the manufacture of compression rings it is desired to test whether the proportion P of rings that do not seat properly is significantly different from some specified number P_0. If n rings are chosen, then under the hypothesis $H_0: P = P_0$ against $H_A: P \neq P_0$ we would *expect* $E_1 = nP_0$ not to seat properly and $E_2 = n(1 - P_0)$ to seat properly. However, given the sample, we observe $O_1 = s$ that do not seat properly and $O_2 = n - s$ that do seat properly. In this case we have $k = 2$ classes and the statistic

$$\chi^2 = \frac{(O_1 - E_1)^2}{E_1} + \frac{(O_2 - E_2)^2}{E_2}$$

will be used to determine whether what we observe is significantly different from what we expect. Suppose that $n = 100$ with $s = 10$. Under $H_0: P = 0.08$ we obtain

$$E_1 = 100(0.08) = 8, \qquad O_1 = 10,$$
$$E_2 = 100(0.92) = 92, \qquad O_2 = 90,$$

Thus the computed χ^2 value is

$$\chi^2 = \frac{(10 - 8)^2}{8} + \frac{(90 - 92)^2}{92}$$
$$= \tfrac{4}{8} + \tfrac{4}{92}$$
$$= \tfrac{1}{2} + \tfrac{1}{23}$$
$$= \tfrac{25}{46}$$
$$= 0.5435.$$

We will reject $H_0: P = 0.08$, at the 10% level, if $\chi^2 > \chi^2_{1,0.95}$ or $\chi^2 < \chi^2_{1,0.05}$. Since $\chi^2_{1,0.95} = 3.8415$ and $\chi^2_{1,0.05} = 0.0039$, we accept $H_0: P = 0.08$ since $0.0039 < 0.5435 < 3.8415$.

It is of interest to note that in this example

$$
\begin{aligned}
\chi^2 &= \frac{(s - nP_0)^2}{nP_0} + \frac{[(n - s) - n(1 - P_0)]^2}{n(1 - P_0)} \\[2mm]
&= \frac{(s - nP_0)^2}{n - P_0} + \frac{(s - nP_0)^2}{n(1 - P_0)} \\[2mm]
&= (s - nP_0)^2 \cdot \frac{n}{n^2 P_0 (1 - P_0)} \\[2mm]
&= \frac{(s - nP_0)^2}{nP_0(1 - P_0)} = \frac{(\hat{P} - P_0)^2}{P_0(1 - P_0)/n}.
\end{aligned}
\tag{7.19}
$$

Thus in this example the statistic χ^2 is just the square of the statistic

$$
\frac{\hat{P} - P_0}{\sqrt{P_0(1 - P_0)/n}},
$$

which has an approximate standard normal distribution. Using the technique of Section 7.8 to test $H_0: P = 0.08$ would yield the same conclusion.

The value of χ^2, as calculated by Eq. (7.18) can be used to characterize experimental distributions. In this case, the null hypothesis is that the experimental distribution follows some theoretical (normal, binomial, exponential, etc.) distribution. In this case the corresponding confidence limit on χ^2 is

$$
\mathscr{P}(\chi^2_{v,\alpha/2} \le \chi^2 \le \chi^2_{v,1-\alpha/2}) = 1 - \alpha,
\tag{7.20}
$$

where $v = k - 1 - c$, k is the number of classes, and c is the number of estimates of population parameters calculated from the sample data. In this goodness-of-fit test, the O_i are the observed experimental frequencies and the E_i are the theoretically expected frequencies in each class.

To evaluate the null hypothesis that the experimental distribution can be represented by the assumed theoretical distribution, a one-tailed test is used as described by

$$
\mathscr{P}(\chi^2 \le \chi^2_{v,1-\alpha}) = 1 - \alpha.
\tag{7.21}
$$

If $\chi^2 > \chi^2_{v,1-\alpha}$, the hypothesis should be rejected.

Example 7.21   Test the melting point data in Table 3.2 to determine whether they came from a normal distribution. Use the empirical distribution in Table 3.5. The sample mean and standard deviation for these data are $\bar{x} = 108.05$

and $s = 0.960$. The class boundaries have to be standardized according to $x = \bar{x} + zs = 108.05 + 0.960z$ or $z = (x - \bar{x})/s$. The following table contains the needed information.

Class (z)	Observed (O_i)	Expected (E_i)	Contribution to χ^2
$-3.646 < z \le -2.604$	1	0.644	0.19679
$-2.604 < z \le -1.563$	12	7.837	2.21138
$-1.563 < z \le -0.521$	33	35.929	0.23878
$-0.521 < z \le 0.521$	57	56.200	0.01139
$0.521 < z \le 1.563$	35	35.929	0.02402
$1.563 < z \le 2.604$	4	7.837	1.87860
$2.604 < z \le 3.646$	2	0.644	2.85518

Solution The expected numbers are computed by first finding the probabilities of the seven classes using the standard normal distribution table. For example, the first class has the probability

$$\mathscr{P}(-3.646 \le z \le -2.604) = \mathscr{P}(z \le -2.604) - \mathscr{P}(z \le -3.646)$$
$$= 0.00460 - 0.00013 = 0.00447.$$

The expected number is then $E_1 = 144(0.00447) = 0.644$. The computed χ^2 value for all classes is

$$\chi^2 = \sum_{i=1}^{n} \frac{(O_i - E_i)^2}{E_i} = 7.4161.$$

Since $\chi^2_{4,0.95} = 9.4877$, we do not reject the hypothesis that the underlying distribution of the melting point data is normal.

In Example 7.21, the degrees of freedom for the tabular χ^2 value were $(7 - 1) - 2 = 4$. The reduction from 6 degrees of freedom to 4 degrees of freedom was used because of having to estimate the two parameters μ and σ^2 in the hypothesized distribution.

If some of the expected values are "small," the results of the goodness-of-fit test may be incorrect. To be conservative, accept the test if all the $E_i \ge 5$. If any $E_i < 5$, that ith class should be combined with an adjacent class so that the resulting expected value is at least 5. A less conservative restriction [2,5] is that each E_i must be greater than 1 and that no more than 20% of the E_i may be less than 5. In addition, for those cases where $v = k - 1 - c = 1$, the absolute value of each $(O_i - E_i)$ pair must be reduced by 0.5 before calculating the contribution of each class to χ^2.

In Example 7.21, the first two categories should be combined, as should the last two categories. This yields five categories, with all categories having expected values greater than 5. The calculated value of $\chi^2 = 3.4079$, which is less than $\chi^2_{2,0.95} = 5.9915$, so there is insufficient evidence to reject the hypothesis that the melting point data are normally distributed.

7.12   CONTINGENCY TESTS

Another use of the χ^2 statistic is in contingency testing, where n randomly selected items are classified according to two different criteria. The most common example is in the testing of materials subjected to a specific physical test. Here it is desired to determine whether or not some protective measure or sample preparation technique has been effective. Other examples are in quality control, where it is desired to determine whether all inspectors are equally stringent. For testing the hypothesis that the outcomes are significant from such a series of binomial (yes or no) populations, the χ^2 test is a very popular procedure. The comments made in Section 7.11, on the approximate nature of the χ^2 test and how the performance of the test is affected if some of the expected numbers are small, are applicable here also.

Example 7.22   Consider the case where pressure gauges are being hydraulically tested by three inspectors prior to shipment. It has been noted that their acceptances and rejections for some period of time have been as follows:

	Inspector			
	A	B	C	Total
Passed	300	100	200	600
Failed	40	20	40	100
Total	340	120	240	700

To test the hypothesis H_0: all inspectors are equally stringent, the test statistic used is

$$\chi^2 = \sum_{i=1}^{2} \sum_{j=1}^{3} \frac{(O_{ij} - E_{ij})^2}{E_{ij}}, \tag{7.22}$$

where O_{ij} are the number observed to be passed by the jth inspector and E_{ij} are the number expected to be passed by the jth inspector. From the data in the table we see that

$$O_{11} = 300, \qquad O_{12} = 100, \qquad O_{13} = 200,$$
$$O_{21} = 40, \qquad O_{22} = 20, \qquad O_{23} = 40,$$
$$\sum_{j=1}^{3} O_{1j} = 600, \qquad \sum_{j=1}^{3} O_{2j} = 100, \qquad \sum_{i=1}^{2} O_{i1} = 340,$$
$$\sum_{i=1}^{2} O_{i2} = 120, \qquad \sum_{i=1}^{2} O_{i3} = 240, \qquad \sum_{i=1}^{3}\sum_{j=1}^{3} O_{ij} = n = 700.$$

In general, a contingency table will have r rows and c columns. The statistic is then

$$\chi^2 = \sum_{i=1}^{r} \sum_{j=1}^{c} \frac{(O_{ij} - E_{ij})^2}{E_{ij}}, \tag{7.23}$$

which has an approximate χ^2-distribution with $(r-1)(c-1)$ d.f. The statistic is used to test whether one attribute (corresponding to row) is contingent on the other attribute (corresponding to column) or whether the two attributes are independent. Under H_0: the attributes are independent and the probability p_{ij} that an item selected at random falls in the class corresponding to the ith row and jth column is $p_{i.}p_{.j}$, where $p_{i.}$ is the probability that the item falls in the ith row and $p_{.j}$ is the probability that the item falls in the jth column. The hypothesis can then be written as H_0: $P_{ij} = p_{i.}p_{.j}$ for $i = 1,2,\ldots,r$, $j = 1,2,\ldots,c$. The expected numbers E_{ij} under H_0 are $E_{ij} = np_{ij} = np_{i.}p_{.j}$. Now the probabilities $p_{i.}$ and $p_{.j}$ are not specified in H_0. Therefore, we estimate them with

$$E_{ij} = np_{i.}p_{.j} = n \frac{\sum_{j=1}^{c} O_{ij}}{n} \frac{\sum_{i=1}^{r} O_{ij}}{n}$$
$$= \frac{(\sum_{j=1}^{c} O_{ij})(\sum_{i=1}^{r} O_{ij})}{n},$$

and these are the E's that go into Eq. (7.23). In other words, to find the expected number for a given cell, simply multiply the corresponding row and column totals together and divide by the total number of items selected. Thus the expected number in each case for this example is found from

$$E_{ij} = \frac{\sum_i (\text{passed})_i \sum_j (\text{inspector})_j}{\sum_i \sum_j (\text{total})}$$

or

$$E_{ij} = \frac{(\sum_{j=1}^{3} O_{ij})(\sum_{i=1}^{2} O_{ij})}{\sum_{i=1}^{2} \sum_{j=1}^{3} O_{ij}}.$$

We have

$$E_{11} = \frac{600(340)}{700} = 291.428,$$

$$E_{12} = \frac{600(120)}{700} = 102.857,$$

$$E_{13} = \frac{600(240)}{700} = 205.714,$$

$$E_{21} = \frac{100(340)}{700} = 48.571,$$

$$E_{22} = \frac{100(120)}{700} = 17.143,$$

$$E_{23} = \frac{100(240)}{700} = 34.286.$$

Thus the computed χ^2 value is

$$\chi^2 = \frac{(O_{11} - E_{11})^2}{E_{11}} + \frac{(O_{12} - E_{12})^2}{E_{12}} + \frac{(O_{13} - E_{13})^2}{E_{13}}$$

$$+ \frac{(O_{21} - E_{21})^2}{E_{21}} + \frac{(O_{22} - E_{22})^2}{E_{22}} + \frac{(O_{23} - E_{23})^2}{E_{23}}$$

$$= \frac{(300 - 291.428)^2}{291.428} + \frac{(100 - 102.875)^2}{102.875} + \frac{(200 - 205.714)^2}{205.714}$$

$$+ \frac{(40 - 48.571)^2}{48.571} + \frac{(20 - 17.143)^2}{17.143} + \frac{(40 - 34.286)^2}{34.286}$$

$$= 0.4916.$$

This value of χ^2 is compared with the tabular value $\chi^2_{(r-1)(c-1),1-\alpha}$, which, as r = number of inspectors = 3 and c = number of outcomes = 2, is, by Table V, $\chi^2_{2,0.95} = 5.9915$. Since $\chi^2 < \chi^2_{2,0.95}$, the hypothesis is accepted and we may say that no one inspector is more demanding than the other two.

7.13   BARTLETT'S TEST FOR EQUALITY OF VARIANCES

The analysis-of-variance technique for testing equality of means is a rather robust procedure. That is, when the assumption of normality and homogeneity of variances is "slightly" violated, the F-test remains a good procedure to use. In the one-way model, for example, with an equal number of observations per column it has been exhibited that the F-test is not affected significantly. However, if the sample size varies across columns, the validity of the F-test can be greatly affected. There are various techniques for testing the equality of k variances $\sigma_1^2, \sigma_2^2, \ldots, \sigma_k^2$. We discuss here a very widely used technique, Bartlett's χ^2-test [1] for examining homogeneity of variances, which is valid when the sample populations are *normal*.

Let $S_1^2, S_2^2, \ldots, S_k^2$ be k independent sample variances corresponding to k normal populations with means μ_i and variance σ_i^2, $i = 1, 2, \ldots, k$. Suppose that $n_1 - 1, n_2 - 1, \ldots, n_k - 1$, are the respective degrees of freedom. Bartlett [1] proposed the statistic

$$\chi^2 = \frac{(\ln V) \sum_{i=1}^{k} (n_i - 1) - \sum_{i=1}^{k} (n_i - 1) \ln S_i^2}{l},$$ (7.24)

where

$$V = \frac{\sum_{i=1}^{k} (n_i - 1) S_i^2}{\sum_{i=1}^{k} (n_i - 1)}$$

and ln denotes the natural logarithm. The denominator in Eq. (7.24) is defined by

$$l = 1 + \frac{1}{3(k-1)} \left[\sum_{i=1}^{k} \frac{1}{n_i - 1} - \frac{1}{\sum_{i=1}^{k} (n_i - 1)} \right].$$

It can be shown that the statistic in Eq. (7.24) has an approximate χ^2-distribution with $k - 1$ degrees of freedom when used as a test statistic for $H_0: \sigma_1^2 = \cdots = \sigma_k^2$. Given k random samples of sizes $n_1, n_2, \ldots n_k$, from k independent normal populations, the statistic χ^2 in Eq. (7.24) can be used to test H_0. Recall that a sample variance S^2 is

$$S^2 = \frac{\sum_{i=1}^{n} Y_i^2 - n \bar{Y}^2}{n - 1}.$$

If all the samples are of the same size, say n, then Bartlett's statistic in Eq. (7.24) becomes

$$\chi^2 = \frac{(n-1)(k \ln V - \sum_{i=1}^{k} \ln S_i^2)}{l},$$ (7.26)

where $l = 1 + [(k + 1)/3k(n - 1)]$. This statistic can be written in terms of common logarithms by observing that $\ln a = 2.3026 \log a$, where a is any positive real number. The statistic in Eq. (7.26) becomes

$$\chi^2 = \frac{2.3026(n-1)(k \log V - \sum_{i=1}^{k} \log S_i^2)}{l}.$$ (7.27)

If $n_i = n$, $i = 1,2,\ldots,k$, then

$$V = \frac{\sum_{i=1}^{k} S_i^2}{k}.$$

The rejection region for testing $H_0: \sigma_1^2 = \sigma_2^2 = \cdots = \sigma_k^2$ is $\chi^2 > \chi^2_{(k-1),1-\alpha}$.

Example 7.23 In an experiment to determine the effects of sample size and amount of liquid phase on the height equivalent to a theoretical plate in gas chromatography, it was necessary to utilize solid support material from different batches. It was therefore imperative that the resulting data be checked for homogeneity prior to attempting to develop any quantitative expressions regarding the effects of these variables on HETP. Several sets of data points were selected at random and examined using Bartlett's test.

In particular, a set of four HETP values obtained for cyclohexane for a 4-μL sample injected into a 40 wt % β,β'-oxydipropionitrile column were

$$
\begin{array}{lll}
y_1 = 0.44 & y_1^2 = 0.1936 & \bar{y} = \sum y_i/n = 1.71/4 = 0.4275 \\
y_2 = 0.44 & y_2^2 = 0.1936 & \bar{y}^2 = (0.4275)^2 = 0.18275625 \\
y_3 = 0.40 & y_3^2 = 0.1600 & n\bar{y}^2 = 0.731025 \\
y_4 = 0.43 & y_4^2 = 0.1849 & \\
\hline
\sum y_i = 1.71 & \sum y_i^2 = 0.7321 &
\end{array}
$$

$$s^2 = \frac{\sum_{i=1}^{n} y_i^2 - n\bar{y}^2}{n-1} = \frac{0.7321 - 0.731025}{3} = \frac{0.001075}{3} = 3.583 \times 10^{-4}.$$

The variance of 10 cyclohexane data sets, each consisting of four observations, are thus calculated and presented below. In this case $n_i = 4$, $i = 1,2,\ldots,10$, and $k = 10$.

Liquid-phase β,β'-oxydipropionitrile	s_i^2	log s_i^2
40%; 4-μL sample	0.0003583	−3.44575
30%; 8-μL sample	0.0002250	−3.64782
20%; 10-μL sample	0.0002250	−3.64782
20%; 4-μL sample	0.0000916	−4.03810
10%; 4-μL sample	0.0000916	−4.03810
5%; 10-μL sample	0.0003000	−3.52288
5%; 2-μL sample	0.0002250	−3.64782
3%; 8-μL sample	0.0002250	−3.64782
3%; 6-μL sample	0.0003000	−3.52288
10%; 2-μL sample	0.0002250	−3.64782
	0.0022665	−36.80681

The computations needed to calculate χ^2 according to Eq. (7.27) are

$$v = \frac{0.0022665}{10} = 0.00022665,$$

$$\log v = -3.64464,$$

$$k \log v = 10(-3.64464) = -36.4464,$$

$$l = 1 + \frac{11}{30(3)} = 1.1222,$$

$$\chi^2 = \frac{2.3026(n-1)(k \log v - \sum \log s_i^2)}{l}$$

$$= \frac{2.3026(3)[-36.4464 - (-36.8068)]}{1.1222}$$

$$= \frac{6.9078(0.3604)}{1.1222} = 2.218.$$

Reference to a χ^2 table (Table V) shows that $\chi^2_{9,0.975} = 19.0228$ and $\chi^2_{9,0.99} = 21.6660$, which says that the computed χ^2 value is not significantly large and therefore we do not reject the hypothesis of homogeneous variances. These data were therefore used subsequently for quantitative effects determination by the method of least squares (discussed in Chapter 9).

The F-test for testing the equality of two variances is quite sensitive to nonnormality. Bartlett's test suffers from the same undesirable characteristic. If nonnormality is suspected, one should resort to other methods, such as those discussed in Conover [2]. Nonparametric procedures for testing equality of variances are surveyed in Duran [4].

7.14 TESTING THE EQUALITY OF VARIANCES

Suppose that σ_1^2 and σ_2^2 are the variances of two normal populations. The problem now is that of testing the hypothesis $H_0: \sigma_1^2 = \sigma_2^2$ against one of the three alternatives $H_A: \sigma_1^2 \neq \sigma_2^2$, $H_A: \sigma_1^2 > \sigma_2^2$, or $H_A: \sigma_1^2 < \sigma_2^2$. The hypotheses can be stated $H_0: \sigma_1^2/\sigma_2^2 = 1$ against either $H_A: \sigma_1^2/\sigma_2^2 \neq 1$, $H_A: \sigma_1^2/\sigma_2^2 > 1$, or $H_A: \sigma_1^2/\sigma_2^2 < 1$.

If $X_1, X_2, \ldots, X_{n_1}$ and $Y_1, Y_2, \ldots, Y_{n_2}$ denote random samples from two normal populations, the statistic to be used in testing the hypotheses concerning variances is

$$F = \frac{S_1^2/\sigma_1^2}{S_2^2/\sigma_2^2}. \tag{7.28}$$

TABLE 7.9   Testing $\sigma_1^2 = \sigma_2^2$

$$\text{Statistic: } F = \frac{S_1^2}{S_2^2}, \quad \nu_1 = n_1 - 1, \, \nu_2 = n_2 - 1 \text{ d.f.}$$

Null hypothesis	Alternative	Rejection region
$H_0: \sigma_1^2 = \sigma_2^2$	$H_A: \sigma_1^2 \neq \sigma_2^2$	$F > F_{\nu_1, \nu_2, 1 - \alpha/2}$ or $F < F_{\nu_1, \nu_2, \alpha/2}$
$H_0: \sigma_1^2 \leq \sigma_2^2$	$H_A: \sigma_1^2 > \sigma_2^2$	$F > F_{\nu_1, \nu_2, 1 - \alpha}$
$H_0: \sigma_1^2 \geq \sigma_2^2$	$H_A: \sigma_1^2 < \sigma_2^2$	$F < F_{\nu_1, \nu_2, \alpha}$

Under $H_0: \sigma_1^2/\sigma_2^2 = 1$, the statistic in (7.28) becomes

$$F = \frac{S_1^2}{S_2^2},$$

which has an F-distribution with $n_1 - 1$ and $n_2 - 1$ degrees of freedom if $H_0: \sigma_1^2/\sigma_2^2 = 1$ is true. Thus H_0 is rejected if S_1^2/S_2^2 is sufficiently different from 1. That is, α, the significance level, is such that

$$\mathscr{P}(F_{n_1 - 1, n_2 - 1, \alpha/2} \leq F \leq F_{n_1 - 1, n_2 - 1, 1 - \alpha/2}) = 1 - \alpha,$$

and H_0 is rejected in favor of $H_A: \sigma_1^2 \neq \sigma_2^2$ if the calculated value $f = s_1^2/s_2^2$ falls in either tail of the F-distribution. These hypotheses and their critical regions are presented in Table 7.9.

Example 7.24   Reaction temperature in degrees Celsius (measured on two different days) for two catalyst concentrations is given in Example 6.9. There are 13 observations corresponding to each catalyst (i.e., $n_1 = n_2 = 13$). Is there any significant difference in the temperature variations corresponding to the two catalysts? Choose $\alpha = 0.02$.

Solution   From Example 6.9 we obtain the following statistics:

$$\bar{x}_1 = 310.8423°C, \qquad \bar{x}_2 = 310.5246°C,$$

$$s_1^2 = 1.1867, \qquad s_2^2 = 1.5757.$$

The null hypothesis is $H_0: \sigma_1^2 = \sigma_2^2$ against $H_A: \sigma_1^2 \neq \sigma_2^2$. From Table 7.9 we see that the statistic to test H_0 is $F = S_1^2/S_2^2$ with critical region $F > F_{12,12,0.99} = 4.16$ and $F < F_{12,12,0.01} = 0.241$. Now $f = s_1^2/s_2^2 = 1.1867/1.5757 = 0.753$. Since $f = 0.753$ is between 0.241 and 4.16, the null hypothesis cannot be rejected. We conclude that there is no significant difference at the 2% level of significance.

This type of analysis might precede using the t-test to test $\mu_1 - \mu_2 = 0$. If it is concluded that $\sigma_1^2 = \sigma_2^2$, then s_1^2 and s_2^2 can be pooled to estimate

$\sigma_1^2 = \sigma_2^2$ and Case 2 in Section 7.6 is in effect. In fact, for the data of this example,

$$T = \frac{\bar{x}_1 - \bar{x}_2}{s_p\sqrt{\frac{1}{13} + \frac{1}{13}}} = \frac{0.3177}{0.46095} = 0.689,$$

where

$$s_p^2 = \frac{12s_1^2 + 12s_2^2}{24} = \frac{12(1.1867) + 12(1.5757)}{24} = 1.3812.$$

Since $t_{24,0.01} = -2.4922$ and $t_{24,0.99} = 2.4922$, we see that $-2.4922 < 0.689 < 2.4922$ and we accept H_0: $\mu_1 = \mu_2$.

There is no truly nonparametric test for testing the hypothesis of equality of variances. Nonparametric tests based on the ranks of the observations require knowledge about the location parameters of the population from which the samples are taken. This information will not be available, especially in analysis-of-variance situations, where the equality of variances is required to carry out a test on location parameters. There have been some "quick" tests proposed for testing the hypothesis of equality of variances, but there does not appear to be any one test that maintains a stable type I error while yielding a sufficiently high power. Conover et al. [3] have done a comparative study of tests of homogeneity of variances and make recommendations on three tests that appear to be superior in terms of stable type I error and power. The user who suspects the appropriateness of the normality assumption should refer to the paper by Conover et al. [3] to choose an appropriate test.

PROBLEMS

7.1 Random samples from two batches of material were taken and analyzed in the same laboratory. Twenty analyses of samples from batch I averaged 46.0% component A with a variance s^2 of 120. Eighteen analyses from batch II had a mean value of 39.1% compound A and a variance of 180. There is no reason to expect the variances to be significantly different for the two batches. Is batch I significantly (at the 5% significance level) higher in component A than batch II?

7.2 Using the data presented in Problem 6.9, is there adequate evidence (you choose a satisfactory risk level) that the new rosin source is better than the old one from a product color standpoint?

7.3 The chief engineer wants you to find out if the company's current ratio (now 3.0) for the year 1966 is equal to, greater, or less than the mean value of the other 129 leading companies in the chemical process industries.

Current ratio is the ratio obtained by dividing current assets by current liabilities. The 95% confidence interval is required. Available current-ratio data are*:

2.3	2.4	3.6	1.9	2.6	2.4	1.6
1.4	2.7	2.0	2.4	2.7	2.3	2.8
3.5	4.1	2.3	2.2	2.4	2.5	2.8
2.4	1.7	1.9	2.3	6.2	2.0	2.8
2.8	4.6	2.3	3.0	2.2	2.3	4.3
2.3	2.6	4.7	1.9	4.0	2.4	2.5
2.7	2.4	1.6	2.4	5.1	2.5	2.5
3.6	2.8	1.9	2.2	1.9	2.6	4.0
3.1	1.9	2.7	2.5	3.7	2.2	2.3
1.7	2.4	2.3	1.9	2.4	2.7	2.6
1.8	1.6	2.4	3.0	2.5	3.2	2.8
1.4	2.1	2.2	4.6	1.8	2.4	1.8
2.9	3.2	2.3	3.3	2.0	2.7	2.2
4.6	1.4	2.0	3.4	1.4	2.4	1.7
2.9	2.5	2.7	2.1	2.2	2.8	2.4
2.6	2.7	2.3	3.6	2.7	3.7	2.2
1.6	1.9	3.0	1.9	2.9	3.1	3.9
2.9	2.1	2.1	1.8	3.0	2.8	3.0
2.6	2.5	2.6	2.6	2.1	3.3	2.5
2.1	2.3	2.1	1.2	2.6	3.9	

7.4   A manufacturer of light bulbs claims that on the average less than 1% of all light bulbs manufactured by his firm are defective. A random sample of 400 light bulbs contained 12 defective. On the evidence of this sample, do you believe the manufacturer's claims? Assume that the maximum risk you wish to run of falsely rejecting the manufacturer's claim has been set at 2%.

7.5   A lot of rolls of paper is acceptable for making bags for grocery stores if its mean breaking strength on a standard sample is not less than 40 lb. A random sample has a mean breaking strength of 39 lb and a standard deviation s of 2.4 lb. Twenty samples were used.

(a)   Should the lot be rejected with a predetermined acceptable chance of rejecting good lots of 0.010?

(b)   What is the chance of nonrejection of a lot that has a true mean breaking strength of 39 lb?

---

* Data reprinted with permission from *Chem. Eng. News* 45(34): 33A–47A (Sept. 4, 1967). Copyright 1967 by the American Chemical Society.

7.6 Using the silicon analyses in Problem 6.11, use $\alpha = 0.02$ to test the hypothesis that the silicon content is at least as good as that when using the old supplier's raw material.

7.7 Among the design criteria for a machine producing molding powder are that it will produce cylindrical pellets by extrusion through $\frac{1}{8}$-in. spinneret holes. These pellets are produced by cutting the strings of polymer so produced into pieces $\frac{3}{16}$ in. long. To determine whether the unit was performing properly, a sample of 100 pellets was selected and examined. The mean length determined was 0.197 in. The standard deviation was 0.005 in. Test the hypothesis that the unit was in proper working order using a two-tailed t-test at the 0.05 and 0.01 significance levels.

7.8 For the fiber data of Problem 6.17, test the hypothesis (at the $\alpha = 0.02$ level) that the mean breaking loads are the same for the two types of fibers.

7.9 A manufacturer of electrical cable ties claims to have developed a new material with a mean breaking strength of 25 lb$_f$. From a sample of 50 ties tested, the mean breaking strength was found to be 24 lb$_f$ with a standard deviation of 0.82 lb$_f$. State and test the appropriate hypotheses at the 2% significance level.

7.10 In the determination of carbon in benzoic acid, 10 technicians each analyzed a pair of samples, with the following results*:

69.03	68.96
69.18	69.22
69.58	69.43
68.79	68.98
69.23	69.17
69.14	69.42
68.86	68.73
68.86	68.81
68.80	68.83
69.14	69.24

The theoretical amount of carbon is 68.84%.

(a) What is the highest significance level at which you can say that there is no statistical difference between the theoretical value and the average value determined in this test?

(b) What is the standard error of the analysis for all technicians?

---

* Data reprinted with permission from W. J. Youden, *Anal. Chem.* 20: 1136–1140 (1948). Copyright 1948 by the American Chemical Society.

7.11 A large textile mill recently installed improved air-handling and air-cleaning equipment in an attempt to meet the OSHA standard of $200 \ \mu g/m^3$. The results of 13 before and after tests are as follows:

Original	Modified
738.6	129.9
839.9	173.9
729.7	159.6
230.3	173.9
308.5	188.5
356.7	167.1
339.1	147.4
398.1	177.6
524.0	168.0
513.1	175.7
469.7	182.2
275.9	146.9
359.5	158.9

Was the expenditure for the new dust control system justified by the results?

7.12 The unit for producing the protective atmosphere for a steel heat-treating process consists of a combustion chamber that is fed with a fuel gas and a limited amount of air. The fuel gas may be assumed to be nitrogen-free. In making material balances about the unit, the air rate is calculated from a nitrogen balance and a measurement of the product gas rate and composition. For this particular situation the nitrogen balance equation reduces to

$$Ax_a = Px_p,$$

where A and P are air and product rates in mol/h and x_a and x_p are mole fractions nitrogen in the two streams. The mole fraction of nitrogen in the product gas is determined from an Orsat analysis by difference. For a given test period the following data were obtained on the product gas (volumetric analysis, dry basis):

	x_i	σ_i
CO	0.122	0.002
CO_2	0.011	0.002
O_2	0.022	0.001
N_2	0.865	(by difference)

$P = 84.2$ lb mol/h, $\sigma_P = 2.2$ lb mol/h

The air composition is assumed to be 79.0% nitrogen, but there is some variability with σ_{N_2} in the air of 0.0025 mole fraction. Estimate the air rate and 95% confidence limits of that flow rate. Twenty samples were used.

7.13 Twelve observations were made for the flow rate (in L/min) for two cotton dust samplers.

Sampler 1			Sampler 2		
7.49	7.49	7.48	7.55	7.51	7.54
7.46	7.47	7.48	7.49	7.52	7.51
7.51	7.46	7.47	7.52	7.52	7.52
7.50	7.48	7.49	7.56	7.48	7.57

Are these two samplers equivalent? Use the 98% confidence interval.

7.14 Fluid flow through a packed bed was examined on a pilot unit using 2-mm glass beads in a plastic column (ID = 1.51 in.). A test was made to find the pressure drop to be expected from a bed length of $60\frac{5}{8}$ in. and a flow rate of approximately 13.5 mL water/s. Two pressure readings were made for each flow rate measurement.

$-\Delta p$ (cm oil, sp. gr. = 1.75)		Flow (mL/s)
58.0	60.5	13.60
57.5	59.5	13.45
57.0	59.0	13.59
58.0	60.0	13.52
57.0	60.5	13.45
58.0	61.0	13.52

Determine the mean value, sample variance, and 95% confidence interval (using Student's t-test) for both the pressure drop and the flow rate. Does the experimental $-\Delta p$ value differ significantly from the theoretical value of 60.0 cm oil?

7.15 In a routine evaluation of laboratory procedures, a revised method (number 2) was checked against the current method (number 1). The results follow.

Revised method	Current method
0.789	0.784
0.790	0.792
0.785	0.743
0.786	0.734
0.789	0.732
0.790	0.738
0.800	0.743
0.775	0.772
	0.748
	0.736

Is the accuracy the same for both methods? Is the precision, as measured by the standard deviation, of the revised method as good as that of the current method?

7.16 A group of seniors did a series of experiments to determine the overall heat-transfer coefficient for the vacuum drying of sand. The group worked two successive afternoons, making five trials per day. Their results (Btu/hr ft^2 $^\circ$F) are presented below.

First day	Second day
4.68	4.70
4.73	4.67
4.65	4.63
4.69	4.58
4.70	4.57

During analysis of the data, however, the group suddenly makes the startling discovery that nobody remembered to check the pressure gauge periodically during the experiment. Hence the steam may have varied somewhat from the desired value of 20 psig. Must the group discard their second day's work? They must be 95% sure that the steam pressure did not drift during their work. Reference to old unit operations reports shows that the average variance for all prior trials is 0.20 (Btu/hr ft^2 $^\circ$F)^2.

7.17 The average wastage per day from a particular synthetic fiber process is 2650 lb. The standard deviation for the amount wasted on a daily basis as measured over the past 9 months is 140 lb. A new mixer has been installed in the solvent–monomer blending system in an attempt to reduce the amount of scrap material produced. It is fairly safe to assume that the daily variation in material wasted will not be significantly affected by this change. The following data were obtained from a 2-week trial of a new mixer (values are in pounds per day):

2540	2390	2530	2250	2170
2380	2400	2160	2250	2430
2310	2490	2200	2610	

(a) Was the modification successful? Use the 5% significance level to support your answer statistically.

(b) Calculate the 98% confidence interval for the daily amount of waste to be expected when using the new mixer.

(*Hint*: Code the data by first subtracting 2400 and then dividing by 10 for ease in calculations.)

7.18 Uniformity of abrasive coverage* within and between samples of coated abrasives can be measured by a uniformity index K. Consider the following data:

	Number of grits	
Test	Sample 1	Sample 2
1	22	33
2	27	21
3	31	31
4	21	27
5	18	38
6	24	25
7	18	24
8	35	35
9	24	26
10	18	32

---

* C. A. Bickering, New angles on old problems of measurement and data analysis, *Ind. Qual. Control 22*: 510–512 (1966). Copyright American Society for Quality Control. Reprinted by permission.

Assuming the samples to have been randomly selected, what is the 98% confidence interval on the difference in grit count?

7.19 A certain heat exchanger that had been performing poorly was taken out of service and cleaned thoroughly. To test the effectiveness of the cleaning, measurements were made before and after so as to determine the heat-transfer coefficient. These were the results:

Run	Before	After
1	90.5	93.4
2	87.6	90.4
3	91.3	99.6
4	93.2	93.7
5	85.7	89.6
6	89.3	88.1
7	92.4	96.7
8	95.3	94.2
9	90.1	98.6
10	83.2	91.1

Did the cleaning of the heat exchanger significantly affect the heat-transfer coefficient?

7.20 A random sample of 2640 students who live in the dormitories were asked about their daily eating habits by means of questionnaire X. Similar information was obtained from another sample of 2640 students living in the dormitories by means of questionnaire Y. The results were as follows:

	Number of students	
Class interval	X	Y
Miss no meals per week	385	316
Miss 1–4 meals per week	1490	1581
Miss 5–8 meals per week	728	719
Miss 9 or more meals per week	37	24

Can the difference in these distributions be due purely to chance? Support your answer from a statistical point of view.

7.21 In the manufacture of a synthetic fiber, the material, still in the form of a continuous flow of monofilaments, is subjected to high temperatures to improve its shrinkage properties. The shrinkage test results (percent shrinkage) of fibers from the same source treated at two different temperatures are given below.

140°C	120°C
3.45	3.72
3.64	4.03
3.57	3.60
3.62	4.01
3.56	3.40
3.44	3.76
3.60	3.54
3.56	3.96
3.49	3.91
3.53	3.67
3.43	

Is the shrinkage at 140°C significantly less (at $\alpha = 0.01$) than that for material treated at 120°C? For data coded by the transformation $U = 100$ $(x - 3.40)$,

$$\bar{U}_1 = 13.54, \qquad \bar{U}_2 = 36.0,$$
$$S^2_{U_1} = 90.3, \qquad S^2_{U_2} = 456.89.$$

7.22 Pilot-plant runs were made on two variations of a process to produce crude naphthalene. Product purities (% naphthalene) for the several runs are given below.

Conditions A	76.0	77.5	77.0	75.5	75.0		
Conditions B	80.0	76.0	80.5	75.5	78.5	79.0	78.5

In each series, all conditions were controlled in the normal manner and there is no evidence from the log sheets of bad runs.

(a) The development engineer reports that on the basis of these data, conditions B give better purities but that control is poorer at these conditions. Do you agree with him? What are the chances that he is wrong in each of these conclusions?

(b) For a single run under conditions B, what is the probability that the product purity will be greater than 76.0%?

(c) What are the 95% confidence limits of the variance of the condition B population based on these data?

7.23 The following percent conversion data were obtained with two different catalysts used for the oxidation of organic materials containing N, S, H, O, and C atoms in spacecraft simulators to CO_2, H_2O, NO_2, and SO_2.

MnO_2	55	62	64	63	58	61	60	62	64
CuO	50	57	52	55	57	54	56	51	55

As MnO_2 is more expensive than CuO, it will be selected only if its efficiency is clearly superior to that of CuO. It has been decided that superiority can be adequately demonstrated if the conversion when using MnO_2 is at least 4% higher than that attainable with CuO. A significance level of 0.01 is required.

(a) Are the data groups above subject to equal variability?

(b) Should MnO_2 be specified for the catalytic oxidizers?

(c) What is the 99% confidence interval for the catalysts' mean percent conversion?

$$\bar{x}_1 = 61, \qquad \sum x_1^2 = 33,559, \qquad \sum x_2^2 = 26,405,$$

$$\bar{x}_2 = 54.111, \qquad (\sum x_1)^2 = 301,401, \qquad (\sum x_2)^2 = 237,168.9.$$

7.24 The silica contents for two different cement samples were determined in triplicate by numerous laboratories in a cooperative testing program.[*] Selected results for sample 1 are:

---

* Data reprinted with permission from P. J. Elving and M. G. Mellon, *Anal. Chem. 20*: 1140–1143 (1948). Copyright 1948 by the American Chemical Society.

Test	X_1	X_2	X_3
5	9.96	10.12	10.32
6	10.26	10.28	10.36
7	10.22	10.11	—
8	10.31	10.32	10.32
9	9.88	9.97	10.32
10	10.03	10.06	10.12
11	9.83	10.08	10.10
12	9.56	9.58	9.59
13	9.74	10.01	10.02
14	10.20	10.01	10.09

For sample 2:

Test	X_1	X_2	X_3
5	9.34	10.17	10.26
6	9.45	10.03	10.23
7	9.98	9.99	10.02
8	10.06	10.06	10.08
9	10.37	10.13	10.10
10	9.76	10.02	10.11
11	10.04	10.20	10.00
12	10.05	10.19	10.28
13	10.05	10.09	10.14
14	10.08	10.10	10.26

(a) From these data, was there any difference in the silica content of the samples?

(b) What are the 98% confidence intervals for mean silica content for the samples?

7.25 Two different methods for predicting the pressure drop through gate valves are to be evaluated with regard to relative variability. The X data were calculated from the classic method; the Y data were calculated from a supposedly improved method.

X	Y
0.363	0.240
0.277	0.208
0.536	0.201
0.634	0.278
0.549	0.371
0.714	
0.326	

$$\sum X = 3.399 \qquad \sum Y = 1.298$$
$$\sum X^2 = 1.815223 \qquad \sum Y^2 = 0.35619$$

The classic method uses only one flow system parameter; the improved method uses two. Does the improved method provide results that are less variable than those from the classic method? Note that the design equations for using the X and Y data are different, so that you cannot calculate and compare the mean pressure drops. Use an F-test to evaluate the null hypothesis.

7.26   Random number generators are essential in simulation studies. A generator that is to be used yielded $n = 100$ values, which are tabulated as follows:

Interval	Frequency
$0 \le x < 0.25$	18
$0.25 \le x < 0.5$	20
$0.5 \le x < 0.75$	40
$0.75 \le x < 1$	22

Test the hypothesis that these data conform to a uniform distribution in the interval (0,1). Use $\alpha = 0.05$.

7.27   An electronic component is under study. A sample of 100 components yields the following empirical distribution of life lengths (in 100 h), where x denotes a typical life length:

Class	Frequency, O_i
$0 < x \le 2$	45
$2 < x \le 4$	16
$4 < x \le 6$	10
$6 < x \le 8$	12
$8 < x \le 10$	7
$10 < x \le 12$	4
$x > 12$	6

It is desired to test the hypothesis that these data came from an exponential distribution with mean $\mu = 4$; that is, $f(x) = \frac{1}{4}e^{-x/4}$, $x > 0$. Carry out a goodness-of-fit χ^2-test for this hypothesis.

7.28 Perform a χ^2 goodness-of-fit test on the data in Example 7.16, after combining the first two categories and the last two categories, thus yielding five categories. Use $\alpha = 0.05$.

7.29 Samples of a synthetic lipid were applied to soil samples to determine their effectiveness in reducing evaporative water losses in small-scale tests. All the soil samples were made up from the same batch of well-mixed materials. Twelve soil samples were available. Half were sprayed with a wetting agent prior to application of the lipid; the other six samples were not. The results obtained in grams of water lost per square decimeter per unit time for a particular set of temperature and humidity conditions were as follows:

	Sample					
	1	2	3	4	5	6
Lipid	11.5	13.4	14.0	13.6	11.6	14.6
Wetting agent + lipid	10.8	10.8	12.5	12.1	12.1	13.5

Did the inclusion of the wetting agent significantly affect the water loss rate? Use the 5% significance level and work this problem by the t-test.

7.30 Regional differences in the resistance of mosquitoes to DDT are believed to be one cause of the variation in the severity of encephalitis outbreaks in the southwest during the summer of 1968. Samples of 200 mosquitoes were randomly collected in four regions. These samples were exposed to a standard DDT level, with the following results:

Area	Died	Lived
1	132	68
2	110	90
3	124	76
4	114	86

(a) State and test the null hypothesis.

(b) What inferences are possible from these data and the outcome in part (a)?

7.31 An engineer has developed a correlation for the prediction of property *A* of a certain product line from information that is gathered routinely. He proposes that measurement of the property be discontinued since it can be predicted accurately by his correlation method and the factors on which his correlation is based are determined with high precision. The quality control engineer is not convinced and has tests made on 18 samples of material in order to compare measured and predicted values. The results are as follows:

Sample	Measured	Predicted
1	78	74
2	59	71
3	56	52
4	94	68
5	84	68
6	81	85
7	66	79
8	78	70
9	59	64
10	56	39
11	88	77

(*continued*)

Sample	Measured	Predicted
12	88	83
13	75	62
14	75	74
15	72	74
16	81	83
17	84	78
18	73	70

Do you recommend discontinuance of the test? Set up a test procedure in detail showing exactly what will be calculated and how the decision is made. Do not perform calculations.

7.32 The observed values and those predicted by the Euler formula for the Prandtl number, N_{Pr},

$$N_{Pr} = \frac{C_p \mu}{k}$$

of gases at 1 atm and 273.2 K are given below.*

Gas	Predicted	Observed
Ne	0.67	0.66
Ar	0.67	0.67
H_2	0.73	0.70
N_2	0.74	0.73
O_2	0.74	0.74
SO_2	0.79	0.86
CO_2	0.78	0.78

Are there any differences at the 95% confidence level between the predicted and observed values for the gases listed?

---

* R. B. Bird, W. E. Stewart, and E. N. Lightfoot, *Transport Phenomena*, pp. 256–257. Copyright 1960 by John Wiley & Sons, Inc. Data reprinted by permission of John Wiley & Sons, Inc.

7.33   White River Lake was recently stocked with bass. As one of the most avid fishermen in this department, you wanted to know how many bass are caught each week and what proportion of these bass are at or over the legal "keep" limit for length before you knock off working on the design project to go fishing. The boat patrol has the following information available from a sample of five fishermen out of the 100 at the lake week before last.

Fisherman (arbitrary numbering)	Bass caught	"Keepers"
1	0	0
2	5	4
3	2	2
4	3	2
5	5	2

From this information, what are the answers to your questions? You actually went fishing at the lake yesterday and caught three bass, two of which were of legal size. Are you an average fisherman? Better than average? Worse than average? Support your answers to these last three questions by testing the appropriate hypotheses.

7.34   A quarter is flipped 100 times. Heads turn up 65 times. Can we justifiably state that the coin is biased at the 1% significance level?

7.35   In a chemical process, the amount of a particular impurity is directly dependent on temperature as given by the relation $y = 0.0319x - 2.44$, where y is the % impurity and x is the temperature (°C). The impurity concentration can be determined uniformly with a standard deviation of 0.002%. What variation in impurity concentration can be expected at 86°C?

7.36   In the development of a fluorescent tracer material for cotton ginning, two samples of leaf were crushed. Both samples came from the same field. One sample was used as the control for the experiment. It was subjected to the full dyeing regimen but omitted the dye. The other sample was dyed by the standard procedure. The results, in terms of friability (ease of crushing) in each particle size range, are given below.

Size	Control	Dyed
12	123	132
20	509	477
40	312	323
100	44	61
140	3	4
270	3	2
Pan	6	1

Did the dye itself affect the friability at the 5% level?

7.37 In the experimental evaluation of a manufacturing facility, the observed and predicted noise levels are shown below in decibels (A scale).

Area	Predicted	Measured
1	70	82
2	54	62
3	80	90
4	60	75
5	60	82
6	60	68

Was the noise suppression system installed properly?

7.38 The following data* were obtained for the actual and predicted weight percent of the total stream for C_3 and lighter components in a hydrocarbon cracking unit.

Run	Actual	Predicted
1	6.5	5.6
2	6.3	5.7
3	5.0	4.6
4	4.9	4.7
5	6.5	5.8
6	6.5	6.0

---

* P. J. White, How cracker feed influences yield, *Hydrocarb. Proc.* 47(5): 103–107 (1968). Reproduced with permission from Hyrocarbon Processing, May 1968.

Are the predicted values acceptable estimates of system performance? Work this problem by the *t*-tests.

7.39 A chemist is trying to extend the shelf life of a certain rubber product. Ten batches have been prepared. Five of those batches have an increased amount of retardant compared to the five batches made using the standard mix. The curing times in minutes are shown for these two formulations.

Standard mix	Increased retardant
8.31	8.25
8.25	8.32
8.20	8.30
8.27	8.35
8.21	8.22

Based on these results, did the increased amount of retardant increase the shelf life of the product? Use $\alpha = 0.05$.

7.40 Two companies manufacture the same type of diskette. The disks are 1.44-MB 3.5-in. floppies. Based on a sample of 10 diskettes from each of two sources, is there a difference in the mean storage capacity (MB) of the disks manufactured by these sources?

Source 1	Source 2
1.34	1.41
1.41	1.42
1.36	1.40
1.36	1.39
1.42	1.37
1.43	1.39
1.43	1.40
1.36	1.41
1.44	1.41
1.41	1.42

7.41 The following cotton dust concentrations (in $\mu g/m^3$) were measured in accordance with the provisions of the OSHA standard for 10 shifts in the opening and carding areas of a cotton mill.

Opening	Carding
389.6	515.2
634.7	524.3
568.8	596.2
422.3	508.7
967.6	607.8
839.9	470.6
594.0	699.3
588.4	615.9
621.2	729.7
357.6	510.1

Is there any significant difference ($\alpha = 0.05$) in the average concentration for the two areas?

7.42 The following values for the orifice discharge coefficient C_0 were obtained as a function of Reynolds number, N_{Re}.

N_{Re}	C_0
15,430	0.602
17,170	0.608
18,830	0.613
19,620	0.616
20,730	0.600
21,380	0.608
23,420	0.622
22,630	0.605
23,590	0.609
24,670	0.614
25,800	0.620
26,350	0.615
27,110	0.608
27,940	0.611

For flange taps, do these results support the accepted value of $C_0 = 0.61$ for $N_{Re} > 10^4$ and $D_0/D_p \geq 0.6$? For this experiment, $D_0/D_p = 0.61$. As part of your solution, examine the seventh datum point. Is it an outlier? If so, proceed as usual.

7.43 A sales representative claims that his company's flowmeters (which sell for one-half the price of the current meters) *outperform* (in both precision and accuracy) the liquid flowmeters that your plant uses currently. However, the sales representative seems unwilling to discuss actual statistical values with you and has presented no data to back her claims. You have decided to order one of her company's meters and conduct your own tests. You instruct your lab technican to perform the tests and report the mean and variance to you. The following flow rates (ft^3/min), taken over a 3-day period with one observation per 8-h shift, were listed in the technician's report for the two meters:

X_{old}	48.00	49.00	50.85	51.75	51.00	52.00	51.80	51.30	50.25
x_{new}	50.85	50.05	51.15	51.39	50.10	49.90	51.10	50.85	51.50

Your lab technician comes back with his report stating that he has calculated with 95% confidence that the meters are probably equivalent. However, as you read through his report, you realized that the technician assumed equality of variances when he determined that the means of the samples from both meters are probably equivalent. You realize that you must first compare the variances of the two sets of values before comparing the means. Do so and then comment on the sales rep's claims.

REFERENCES

1. Bartlett, M. S., *Proc. Roy. Soc. (London) A901:* 273–275 (1946).
2. Conover, W. J., *Practical Nonparametric Statistics*, 2nd ed., Wiley, New York (1980), Ch. 5.
3. Conover, W. J., M. E. Johnson, and M. M. Johnson, A comparative study of tests for homogeneity of variances, with applications to the outer continental shelf bidding data, *Technometrics 23*: 351–361 (1981).
4. Duran, B. S., A survey of nonparametric tests for scale, *Commun. Statist. Theory Methods A5:* 1287–1312 (1976).
5. Satterthwaite, F. An approximate distribution of estimates of variance components, *Biometrics Bull.* 2: 110–114 (1946).
6. Snedecor, G. W., and W. G. Cochran, *Statistical Methods*, 7th ed., Iowa State University Press, Ames, Iowa (1967), pp. 230–256.

8

Analysis of Variance

8.1 INTRODUCTION

The technique known as analysis of variance (AOV) employs tests based on variance ratios to determine whether or not significant differences exist among the means of several groups of observations, where each group follows a normal distribution. The analysis-of-variance technique extends the t-test (Section 7.6) used to determine whether or not two means differ to the case where there are three or more means.

Analysis of variance is particularly useful when the basic differences between the groups cannot be stated quantitatively. A *one-way analysis of variance* is used to determine the effect of one independent variable on the dependent variable. A two-way analysis of variance is used to determine the effects of two independent variables on a dependent variable, and so on. As the number of independent variables increases, the calculations become much more complex and are best carried out on a high-speed digital computer. The term *independent variable* used here is what is also referred to as a *factor* or *treatment*. We now develop some useful and common notation.

Suppose that we have the array of data in Table 8.1, where I and J denote the number of rows and columns, respectively. Each column of observations can be thought of as a random sample of size I from a normal population. Each data point is subscribed first to identify its row location and second to identify its column location. Thus $_{ij}$ is the ith entry in the jth

TABLE 8.1 Data Array

				Column		
Row	1	2	· · ·	j	· · ·	J
1	Y_{11}	Y_{12}	· · ·	Y_{1j}	· · ·	Y_{1J}
2	Y_{21}	Y_{22}	· · ·	Y_{2j}	· · ·	Y_{2J}
i	Y_{i1}	Y_{i2}	· · ·	Y_{ij}	· · ·	Y_{iJ}
I	Y_{I1}	Y_{I2}	· · ·	Y_{Ij}	· · ·	Y_{IJ}

column. The sum of the observations in the ith row is designated by $Y_{i\cdot}$, defined by

$$Y_{i\cdot} = \sum_{j=1}^{J} Y_{ij}, \tag{8.1}$$

where the dot refers to the variable that has been summed. The mean of the values in the ith row is then

$$\bar{Y}_{i\cdot} = \frac{Y_{i\cdot}}{J}. \tag{8.2}$$

Similarly, the sum of the jth column is

$$Y_{\cdot j} = \sum_{i=1}^{I} Y_{ij} \tag{8.3}$$

and the mean is

$$\bar{Y}_{\cdot j} = \frac{Y_{\cdot j}}{I}. \tag{8.4}$$

The sum of all the values in the array (matrix) is designated by $Y_{\cdot\cdot}$, where

$$Y_{\cdot\cdot} = \sum_{i=1}^{I} \sum_{j=1}^{J} Y_{ij} = \sum_{i=1}^{I} Y_{i\cdot} = \sum_{j=1}^{J} Y_{\cdot j}. \tag{8.5}$$

The mean of all the values in the array is called the *grand mean* $\bar{Y}_{\cdot\cdot}$, where

$$\bar{Y}_{\cdot\cdot} = \frac{Y_{\cdot\cdot}}{IJ}. \tag{8.6}$$

From here on, to simplify the equations we designate

$$\sum_{i=1}^{I} \quad \text{by} \quad \sum_{i} \quad \text{and} \quad \sum_{j=1}^{J} \quad \text{by} \quad \sum_{j}. \tag{8.7}$$

Furthermore, as before, capital letters denote random variables and lowercase letters denote particular values of the corresponding random variables.

8.2 GENERAL LINEAR MODEL

The general linear model is described by means of an equation that gives the relationship between a set of independent variables and parameters and a dependent (or response) variable Y. The term *linear* signifies that the expression containing the independent variables and the parameters is linear in the parameters. Symbolically, the model may be written as

$$Y = f(X_1, X_2, \ldots, X_p; \beta_1, \beta_2, \ldots, \beta_k) + \varepsilon,$$

where $X_1, X_2, \ldots, X_p$ denote the independent variables; $\beta_1, \beta_2, \ldots, \beta_k$ denote the parameters; ε denotes a random error; and Y denotes the dependent variable. The function f is linear in $\beta_1, \beta_2, \ldots, \beta_k$. As an example, the regression model

$$Y = \beta_0 + \beta_1 X_1 + \beta_2 X_2 + \frac{\beta_3}{\sqrt{X_3}} + \varepsilon \tag{8.8}$$

is linear because the parameters β_i all appear to the first power. The fact that some of the independent variables have exponents other than 1 is immaterial. On the other hand, the relation

$$Y = \beta_0 \beta_1^X + \varepsilon \tag{8.9}$$

is nonlinear, as one of the parameters, β_1, does not appear to the first power. In both cases, ε is the experimental error term.

In some situations the independent variables are not continuous but may take on the value 0 or 1. In such cases the general linear model is termed an *analysis-of-variance* (or *experimental design*) *model*. This type of model is discussed in this chapter and in Chapter 11, and may include terms for the effects of different "factors" and their interactions (similar to synergism) on the dependent variable Y. Terms for different sources of error (variation) are included as appropriate. The number of independent variables (or treatments) and the design used to perform the experiment (more of this in Chapter 11) are used to label specific models. If only one independent variable τ_j is involved, the result is a one-way analysis-of-variance model:

$$Y_{ij} = \mu + \tau_j + \varepsilon_{ij}. \tag{8.10}$$

If two parameters α_i and β_j are involved, the model is

$$Y_{ij} = \mu + \alpha_i + \beta_j + \varepsilon_{ij}. \qquad (8.11)$$

When more than one sample is taken for some experimental conditions, the model is unaffected: the experimental results should become more precise. If, however, the experiment results in samples that can be divided (subsampled), a sampling error η_{ijk} exists and must be evaluated separately from the experimental error. An example of the model for one-way analysis of variance when subsampling is involved is

$$Y_{ijk} = \mu + \tau_j + \varepsilon_{ij} + \eta_{ijk}. \qquad (8.12)$$

Suitable limits on the subscripts are given where each of these models is discussed in detail.

8.3   ONE-WAY ANALYSIS OF VARIANCE

One-way analysis of variance is used when we wish to test the equality of J population means. The procedure is based on the assumptions that each of J groups of observation is a random sample from a normal distribution and that the population variance $\sigma_Y^2 = \sigma^2$ is constant among the groups. There are two possible ways to estimate the population variance: pooled estimates and calculation of the variance of group means around the grand mean. We shall see that these two ways of estimating σ^2 will lead to a test of the equality of the J population means.

8.3.1   Pooled Variance Estimates

In the pooled estimate the sample variance for each group (each column of data in a one-way analysis) is calculated. These estimates are then weighted by the corresponding degrees of freedom to obtain a pooled sample variance. For any column j the sample variance is

$$S_j^2 = \frac{\sum_i (Y_{ij} - \bar{Y}_{.j})^2}{I - 1}. \qquad (8.13)$$

From Section 5.4 we know that $E(S_j^2) = \sigma^2$ for every j. We are assuming, for the sake of simplicity, that each column contains the same number of values. We then have J estimates in the form of Eq. (8.13). To pool them we weight each by its degrees of freedom $(I - 1)$, add them up, and divide by the total degrees of freedom, which is $J(I - 1)$. The pooled sample variance is then

$$S_p^2 = \frac{\sum_j [(I - 1)S_j^2]}{J(I - 1)}, \qquad (8.14)$$

which may be combined with Eq. (8.13) to give

$$S_p^2 = \frac{\sum_j [\sum_i (Y_{ij} - \bar{Y}_{\cdot j})^2]}{J(I-1)} = \frac{SS_W}{J(I-1)} = MS_W. \tag{8.15}$$

In this equation, the term SS_W is referred to as the *sum of squares within groups* or the *error sum of squares*. The quantity SS_W when divided by the appropriate degrees of freedom $J(I-1)$ is referred to as the mean square for within groups (or error mean square) and is denoted by MS_W.

As Eq. (8.15) is not particularly convenient for calculation purposes, it can be presented in the more usable form

$$S_p^2 = \frac{\sum_i \sum_j (Y_{ij})^2 - (\sum_j Y_{\cdot j}^2)/I}{J(I-1)} = \frac{SS_W}{J(I-1)} = MS_W. \tag{8.16}$$

The pooled estimator of the population variance, S_p^2, is an unbiased estimator for σ^2 regardless of whether the population means $\mu_1, \mu_2, \ldots, \mu_J$ are equal or not, because it takes into account deviations from each group mean $\bar{Y}_{\cdot j}$, $j = 1, 2, \ldots, J$. Unbiasedness follows from Eq. (8.15) since

$$E(S_p^2) = \frac{\sum_j \{E[\sum_i (Y_{ij} - \bar{Y}_{\cdot j})^2/(I-1)]\}}{J}$$

$$= \frac{\sum_j E(S_j^2)}{J}$$

$$= \sigma^2. \tag{8.17}$$

8.3.2   Variance of Group Means

A second method of estimating the population variance σ^2 is to calculate the sample variance of the group means around the grand mean by use of

$$S_{\bar{Y}}^2 = \frac{\sum_j (\bar{Y}_{\cdot j} - \bar{Y}_{\cdot \cdot})^2}{J-1}. \tag{8.18}$$

If the group population means are *all* equal, $S_{\bar{Y}}^2$ is an unbiased estimate of the variance of the population mean $\sigma_{\bar{Y}}^2$. To obtain an estimator of the population variance σ^2, recall that

$$S_{\bar{Y}}^2 = \frac{S_{\bar{Y}}^2}{I}. \tag{8.19}$$

Combining Eqs. (8.18) and (8.19), we have our second estimator of the population variance σ^2:

$$S_{\bar{Y}}^2 = \frac{I \sum_j (\bar{Y}_{\cdot j} - \bar{Y}_{\cdot \cdot})^2}{J-1} = \frac{SS_B}{J-1} = MS_B. \tag{8.20a}$$

If there are unequal numbers of observations for the groups, Eq. (8.20a) may be written as

$$S_Y^2 = \frac{\sum_j I_j(\bar{Y}_{\cdot j} - \bar{Y}_{\cdot \cdot})^2}{J-1} = \frac{SS_B}{J-1} = MS_B. \tag{8.20b}$$

The quantities SS_B and MS_B are usually referred to as the *between-groups sum of squares* and *mean square between groups*, respectively. The estimator S_Y^2 can be written as

$$S_Y^2 = \frac{\sum_j Y_{\cdot j}^2/I - Y_{\cdot \cdot}^2/IJ}{J-1} = \frac{SS_B}{J-1} = MS_B, \tag{8.21}$$

which is a more useful form for calculation purposes.

The estimator S_Y^2 given in (8.20a), (8.20b), or (8.21) is an unbiased estimator of σ^2 only when the group population means are equal. If the population means $\mu_1, \mu_2, \ldots, \mu_J$ are not all equal, the estimator S_Y^2 over-estimates σ^2; that is, $E(S_Y^2) > \sigma^2$. The estimators S_Y^2 and S_p^2 are linked by a very important identity given by

$$\sum_j \sum_i (Y_{ij} - \bar{Y}_{\cdot \cdot})^2 = I \sum_j (\bar{Y}_{\cdot j} - \bar{Y}_{\cdot \cdot})^2 + \sum_j \sum_i (Y_{ij} - \bar{Y}_{\cdot j})^2. \tag{8.22}$$

The left-hand side of Eq. (8.22) is usually referred to as the *total sum of squares corrected for the mean* and is denoted by SS_{TC}. To verify (8.22), we write

$$\sum_j \sum_i (Y_{ij} - \bar{Y}_{\cdot \cdot})^2 = \sum_j \sum_i [(Y_{ij} - \bar{Y}_{\cdot j}) + (\bar{Y}_{\cdot j} - \bar{Y}_{\cdot \cdot})]^2$$

$$= \sum_j \sum_i [(Y_{ij} - \bar{Y}_{\cdot j})^2 + (\bar{Y}_{\cdot j} - \bar{Y}_{\cdot \cdot})^2$$

$$+ 2(Y_{ij} - \bar{Y}_{\cdot j})(\bar{Y}_{\cdot j} - \bar{Y}_{\cdot \cdot})]$$

$$= \sum_j \sum_i (Y_{ij} - \bar{Y}_{\cdot j})^2 + \sum_j \sum_i (\bar{Y}_{\cdot j} - \bar{Y}_{\cdot \cdot})^2. \tag{8.23}$$

The cross-product term vanishes since

$$\sum_j \sum_i 2(Y_{ij} - \bar{Y}_{\cdot j})(\bar{Y}_{\cdot j} - \bar{Y}_{\cdot \cdot}) = \sum_j 2(\bar{Y}_{\cdot j} - \bar{Y}_{\cdot \cdot}) \sum_i (Y_{ij} - \bar{Y}_{\cdot j})$$

and $\sum_i (Y_{ij} - \bar{Y}_{\cdot j}) = 0$. Equation (8.22) thus follows from (8.23). The identity given by Eq. (8.22) can be written as $SS_{TC} = SS_B + SS_W$. In other words, the total variation is partitioned into two components, a component SS_B, which reflects variation among groups, and a component SS_W, which reflects experimental error or sampling variation. The degrees of freedom associated with SS_{TC} are also partitioned into the degrees of freedom associated with SS_B and SS_W [i.e., $IJ - 1 = (J - 1) + J(I - 1)$].

If the means $\mu_1, \mu_2, \ldots, \mu_J$ are all equal, then S_p^2 and S_Y^2 are independent, so that the random variable

$$F = \frac{S_Y^2}{S_p^2} = \frac{SS_B/(J-1)}{SS_W/J(I-1)} = \frac{MS_B}{MS_W} \qquad (8.24)$$

has an F-distribution with $J-1$ and $J(I-1)$ degrees of freedom. Thus under H_0: $\mu_1 = \mu_2 = \cdots = \mu_J$, we would expect the value of F to be close to 1. If H_0 is not true, the value of S_Y^2 would tend to be larger than S_p^2, which would force F to be larger than 1. Consequently, based on the data, the hypothesis H_0 would be rejected if the computed F value, $f = s_Y^2/s_p^2$, is too large. That is, the rejection region is of the form $F > F_{v_1, v_2, 1-\alpha}$, where $v_1 = J - 1$ and $v_2 = J(I-1)$.

8.3.3 Model for One-Way Analysis of Variance

Let us consider the model for a one-way analysis of variance. Here it is assumed that the columns of data are J random samples from J independent normal populations with means $\mu_1, \mu_2, \ldots, \mu_J$ and common variance σ^2. The one-way analysis-of-variance technique will give us a procedure for testing the hypothesis H_0: $\mu_1 = \mu_2 = \cdots = \mu_J$ against the alternative H_A: at least two μ_j not equal. The statistical model gives us the structure of each observation in the $I \times J$ array in Table 8.1. The model is

$$Y_{ij} = \mu_j + \varepsilon_{ij}. \qquad (8.25)$$

This model says that the dependent variable Y_{ij} is made up of two parts: the first part, μ_j, which is the mean of the population corresponding to the jth column (population), and the second part, ε_{ij}, the random experimental error, which is taken to have mean 0 [i.e., $E(\varepsilon_{ij}) = 0$]. This must be the case since $E(Y_{ij}) = \mu_j$. The model in Eq. (8.25) can be reduced to Eq. (8.10):

$$\begin{aligned} Y_{ij} &= \mu_j + \mu - \mu + \varepsilon_{ij} \\ &= \mu + (\mu_j - \mu) + \varepsilon_{ij} \\ &= \mu + \tau_j + \varepsilon_{ij}, \end{aligned} \qquad (8.10)$$

where $\mu = \bar{\mu}. = \sum_j \mu_j/J$ is called the *grand population mean* and $\tau_j = \mu_j - \bar{\mu}$ is called the *effect* of the jth population. Equation (8.10) states that any experimental value is the sum of a term representing the general location of the grand population mean plus a term τ_j showing the displacement of a given population from the general location, plus a term giving the random experimental error ε_{ij} of the particular observation. The ε_{ij} are independent and normally distributed with mean 0 and variance σ^2.

If all the column means are equal, $\mu = \mu_j$ and $\tau_j = 0$. Therefore, the hypothesis $H_0: \mu_1 = \mu_2 = \cdots = \mu_J$ is the same as the hypothesis $H_0: \tau_j = 0$ for all j.

One is generally interested in the hypothesis $H_0: \tau_1 = \tau_2 = \cdots = \tau_J = 0$, which states that there is no population or column effect. This means that the variation in $\bar{y}_{.1}, \bar{y}_{.2}, \ldots, \bar{y}_{.J}$ is due to experimental error and not to any difference in population means. To test the hypothesis H_0 we use the F-statistic given in Eq. (8.24),

$$F = \frac{\mathrm{SS_B}/(J-1)}{\mathrm{SS_W}/J(I-1)} = \frac{\mathrm{MS_B}}{\mathrm{MS_W}}.$$

The rejection region at significance level α is $F > F_{J-1, J(I-1), 1-\alpha}$.

Recall from Eq. (8.17) that $E(S_p^2) = E(\mathrm{MS_W}) = \sigma^2$. Furthermore, if H_0: $\tau_1 = \tau_2 = \cdots = \tau_J = 0$ is true, then $E(S_Y^2) = E(\mathrm{MS_B}) = \sigma^2$. However, if H_0 is not true, then $E(\mathrm{MS_B}) \neq \sigma^2$. To show this, consider

$$E(\mathrm{SS_B}) = E\left[\sum_i \sum_j (\bar{Y}_{.j} - \bar{Y}_{..})^2\right]$$

$$= I \sum_j E(\bar{Y}_{.j} - \bar{Y}_{..})^2.$$

Now $Y_{ij} = \mu + \tau_j + \varepsilon_{ij}$. Therefore,

$$\bar{Y}_{.j} = \frac{\sum_i Y_{ij}}{I} = \frac{\sum_i (\mu + \tau_j + \varepsilon_{ij})}{I} = \mu + \tau_j + \bar{\varepsilon}_{.j}$$

and

$$\bar{Y}_{..} = \frac{\sum_j \bar{Y}_{.j}}{J} = \frac{\sum_j (\mu + \tau_j + \bar{\varepsilon}_{.j})}{J} = \mu + \bar{\varepsilon}_{..},$$

since

$$\sum_j \tau_j = \sum_j (\mu_j - \mu) = 0.$$

Thus

$$E(\mathrm{SS_B}) = I \cdot E\left\{\sum_j [\tau_j + (\bar{\varepsilon}_{.j} - \bar{\varepsilon}_{..})]^2\right\}$$

$$= I \cdot E\left\{\sum_j [\tau_j^2 + (\bar{\varepsilon}_{.j} - \bar{\varepsilon}_{..})^2]\right\}$$

since

$$E[2\tau_j(\bar{\varepsilon}_{.j} - \bar{\varepsilon}_{..})] = 2\tau_j E(\bar{\varepsilon}_{.j} - \bar{\varepsilon}_{..}) = 0.$$

Consequently,

$$E(SS_B) = I \sum_j \tau_j^2 + I \cdot E\left[\sum_j (\bar{\varepsilon}_{.j} - \bar{\varepsilon}_{..})^2\right],$$

and since $\bar{\varepsilon}_{.1}, \bar{\varepsilon}_{.2}, \ldots, \bar{\varepsilon}_{.J}$ can be considered to be a random sample of size J from a normal population with mean 0 and variance σ^2/I, it follows that $E[\sum_j (\bar{\varepsilon}_{.j} - \bar{\varepsilon}_{..})^2/(J-1)] = \sigma^2/I$. Thus

$$E(SS_B) = I \sum_j \tau_j^2 + I(J-1)\frac{\sigma^2}{I}$$

$$= I \sum_j \tau_j^2 + (J-1)\sigma^2$$

and

$$E(S_Y^2) = E(MS_B)$$

$$= E\left(\frac{SS_B}{J-1}\right)$$

$$= \sigma^2 + \frac{I \sum_j \tau_j^2}{J-1}.$$

The quantities $E(MS_W)$ and $E(MS_B)$ are called *expected mean squares*. The expected mean squares suggest what ratio to use in testing H_0. That is, since $E(MS_B) = \sigma^2$ when $H_0: \tau_1 = \tau_2 = \cdots = \tau_J = 0$ is true and $E(MS_B) > \sigma^2$ otherwise, and since MS_B/MS_W has an F-distribution when H_0 is true, H_0 will have the rejection region $F > F_{J-1,J(I-1),1-\alpha}$. For the F-test we use the unbiased estimate of σ^2 in the denominator of the F-ratio and the biased estimate in the numerator.

The results needed to test $H_0: \tau_1 = \tau_2 = \cdots = \tau_J = 0$ are usually summarized in the form of Table 8.2. It should be noted that $SS_T = SS_M + SS_B + SS_W$ and that the degrees of freedom add up similarly, that is, $IJ = 1 + (J-1) + J(I-1)$.

Example 8.1 The amount of fluoride in the local water supply was determined by the strontium chloranilate (A), ferric thiocyanate (B), alizarin lanthanum complexan (C), and eriochrome cyanine RC (D) colorimetric methods in a comparative study. Five replications were made for each test. To preclude bias from variations in the sample over the time required for the analyses, all samples were taken from a single 10-gal carboy of water. The results in ppm are as follows:

TABLE 8.2   One-Way Analysis of Variance with an Equal Number of Observations per Treatment

Source	d.f.	SS	MS
Mean	1	$SS_M = y^2_{..}/IJ$	
Between columns	$J-1$	$SS_B = \sum_j y^2_{.j}/I - y^2_{..}/IJ$	$MS_B = \dfrac{SS_B}{J-1}$
Within columns (error)	$J(I-1)$	$SS_W = \sum_i \sum_j y^2_{ij} - \sum_j y^2_{.j}/I$	$MS_W = \dfrac{SS_W}{J(I-1)}$
Total	IJ	$SS_T = \sum_i \sum_j y^2_{ij}$	

A	B	C	D
2	5	1	2
3	4	3	1
6	4	2	1
5	2	4	2
4	3	4	1

(a)   Are the methods equivalent? Use the 5% significance level.
(b)   What are the 95% confidence limits on the values obtained from each method?

Solution   (a) We first calculate the required sums and squares:

$$y_{.A} = 20, \quad \sum_i y^2_{iA} = 90, \quad \bar{y}_{.A} = 4.0, \quad \sum_j y^2_{.j} = 969.$$

$$y_{.B} = 18, \quad \sum_i y^2_{iB} = 70, \quad \bar{y}_{.B} = 3.6,$$

$$y_{.C} = 14, \quad \sum_i y^2_{iC} = 46, \quad \bar{y}_{.C} = 2.8,$$

$$y_{.D} = 7, \quad \sum_i y^2_{iD} = 11, \quad \bar{y}_{.D} = 1.4,$$

From Eq. (8.16) we calculate the pooled sample variance:

$$s_p^2 = \frac{\sum_j \sum_i y_{ij}^2 - (\sum_j y_{\cdot j}^2)/I}{J(I-1)} = \frac{217 - 193.8}{16} = 1.45 = MS_W.$$

We then calculate the variance of the group means around the overall mean using Eq. (8.21):

$$s_{\bar{Y}}^2 = \frac{\sum_j y_{\cdot j}^2/I - y_{\cdot \cdot}^2/IJ}{J-1} = \frac{969/5 - (59)^2/20}{3} = 6.58\overline{3} = MS_B.$$

We next compare the equality of group means by calculating $F = s_{\bar{Y}}^2/s_p^2 = 6.58\overline{3}/1.45 = 4.54$ according to Eq. (8.24). To test the null hypothesis H_0: $\tau_1 = \tau_2 = \tau_3 = \tau_4 = 0$, the calculated value of F is compared to $F_{3,16,0.95} \cong 3.252$. Since $F = 4.54 > 3.252$, we reject H_0 and conclude that there is a significant difference among the four methods. The AOV results are summarized below

Source	d.f.	SS	MS	F
Mean	1	174.05	—	—
Between	3	19.75	6.58$\overline{3}$	4.540
Error	16	23.20	1.450	—
Total	20	217.00		

(b) For 16 degrees of freedom, at the 95% confidence level, $t_{16,0.975} = 2.1199$ and $t_{16,0.025} = -2.1199$. The standard error is found as before from $s_{\bar{p}} = s_p/\sqrt{I} = \sqrt{1.45/5} = 0.5385$. The confidence limits for the means corresponding to the four colorimetric methods are (according to Section 6.8)

μ_A: 4.0 ± 2.1199(0.5385)   or   2.86   to   5.14
μ_B: 3.6 ± 2.1199(0.5385)   or   2.46   to   4.74
μ_C: 2.8 ± 2.1199(0.5385)   or   1.66   to   3.94
μ_D: 1.4 ± 2.1199(0.5385)   or   0.26   to   2.54.

Example 8.2 Evaluate the coke yield data in volume percent given in Problem 8.2.
Solution Using the following SAS program (*SAS for Linear Models*) for the analysis of variance when equal numbers of observations are present for all variables, we prepare the following program.

(put initial JCL cards/statements here)

```
DATA;
INPUT FEED $ 1 YIELD 3-6;
CARDS;
```

(yield data inserted here according to INPUT format)

```
PROC PRINT;
PROC ANOVA;
CLASS FEED;
MODEL YIELD = FEED;
TITLE 'COMPARISON OF HYDROGENATION LEVEL IN FEED ON COKE YIELD';
TITLE 'NOTE: A = 0, B = 300, C = 500';
```

(put final JCL cards/statements here)

The following results were obtained from this program.

SAS

COMPARISON OF HYDROGENATION LEVEL IN FEED ON COKE YIELD

ANALYSIS OF VARIANCE PROCEDURE

CLASS LEVEL INFORMATION

CLASS	LEVELS	VALUES
FEED	3	A B C

NUMBER OF OBSERVATIONS IN DATA SET = 18

NOTE: A = 0, B = 300, C = 500

OBS	FEED	YIELD
1	A	4.2
2	A	6.8
3	A	5.9
4	A	11.0
5	A	16.0
6	A	8.0
7	B	4.1
8	B	4.8
9	B	6.2
10	B	8.0
11	B	14.0
12	B	4.0
13	C	4.0
14	C	4.4
15	C	8.1
16	C	8.3
17	C	11.9
18	C	6.1

COMPARISON OF HYDROGENATION LEVEL IN FEED ON COKE YIELD

ANALYSIS OF VARIANCE PROCEDURE

DEPENDENT VARIABLE: YIELD

SOURCE	DF	SUM OF SQUARES	MEAN SQUARE	F VALUE
MODEL	2	11.24111111	5.62055556	0.41
ERROR	15	206.88333333	13.79222222	PR > F
CORRECTED TOTAL	17	218.12444444		0.6724

SOURCE	DF	ANOVA SS	F VALUE	PR · F
FEED	2	11.24111111	0.41	0.6724

The null hypothesis, as in all one-way analyses of variance, was that the τ_j (treatment = hydrogenation level) do not affect the coke yield. From the calculated value of $F = 0.41$ and the 67.24% probability of finding a larger value, there is no reason to reject the hypothesis: it is thus accepted as true at the $\alpha = 0.05$ level. Note that $F_{2,15,0.95} = 3.68$.

8.3.4 Unequal Observations in One-Way Analysis of Variance

The analysis of variance for a situation involving unequal numbers of observations per treatment is given in Table 8.3, where n_0 is defined as

$$n_0 = \frac{\sum_{j=1}^{t} n_j - \sum_{j=1}^{t} n_j^2 / \sum_{j=1}^{t} n_j}{t - 1}. \tag{8.26}$$

The statistic $F = MS_{Tr}/MS_E$, which has an F-distribution with $(t - 1)$ and $\sum (n_j - 1)$ degrees of freedom, is used to test $H_0: \tau_1 = \tau_2 = \cdots = \tau_t = 0$ and $H_0: \sigma_\tau^2 = 0$, depending on which model is assumed. Model I assumes fixed effects or fixed treatments applied to randomly selected experimental units. An example would be a one-way AOV designed to study the effect of different gasoline additives (treatments) on the performance of your car (the experimental unit). Model II assumes the assignment of randomly selected treatments to randomly selected experimental units [i.e., the effect of flow rate (a random variable) on arbitrarily selected dike sections (the experimental units) in a hydraulic model of a river system]. Such investigations have been used by the U.S. Army Corps of Engineers in levee designs. The difference between the two types of models is thus apparent. The model I case can only involve fixed effects, while the model II case involves random variables even though specific values of the treatments may have been chosen for purposes of the study. The gasoline additives are of fixed composition as purchased at your local service station. However, flow rates can be any value in the range. The fact that you selected 5, 10, 20, 40, and 75 gal/min for use in your study is immaterial. You could have selected *any* number of different values.

TABLE 8.3   One-Way Analysis of Variance with Unequal Number of Observations per Treatment

Source	d.f.	SS	EMS
Mean	1	$\left(\sum_i \sum_j Y_{ij}\right)^2 \Big/ \sum_j n_j = SS_M$	
Among treatments	$t-1$	$SS_{Tr} = \sum_i \left[\left(\sum_i Y_{ij}\right)^2 \Big/ n_j\right] - SS_M$	Model I: $\sigma^2 + \dfrac{\sum_{j=1}^{t} n_j \tau_j^2}{t-1}$ Model II: $\sigma^2 + n_0 \sigma_\tau^2$
Experimental error	$\sum_j (n_j - 1)$	$\sum_i \sum_j Y_{ij}^2 - SS_M - SS_{Tr} = SS_E$	σ^2
Total	$\sum_j n_j$	$\sum_i \sum_j Y_{ij}^2 = SS_T$	

The procedures used to test the validity of equality of treatment effects are shown in Table 8.3.

   If an unequal number of observations is available for the independent variables, the SAS program must be modified. PROC GLM (for "general linear model") must be used instead of PROC ANOVA. This approach (*SAS for Linear Models*) is illustrated in the following example.

Example 8.3   A sample of the coeds taking English 231 in one of our classrooms last summer was classified according to hair color (blonde, brunette, redhead). Their heights in inches were then measured.

Blondes	Brunettes	Redheads
60	70	65
62	61	63
68	64	61
65	67	
63	64	
	66	
	63	

What statistical model would be appropriate for analyzing these data? Indicate what hypotheses and tests would be useful in analyzing these data. Perform the tests of the hypotheses.

Solution The SAS program is:

(put JCL cards/statements here)

DATA;

INPUT COLOR $ 1-2 HEIGHT 4-5;

CARDS;

(type in the data here in format specified by the INPUT statement using BL, BR, and RH for blonde, brunette, and redhead)

PROC GLM;

CLASS COLOR;

MODEL HEIGHT = COLOR;

TITLE 'EFFECT OF HAIR COLOR ON HEIGHT OF FEMALE STUDENTS';

(put final JCL cards/statements here)

The SAS output obtained is shown below.

SAS

EFFECT OF HAIR COLOR ON HEIGHT OF FEMALE STUDENTS

GENERAL LINEAR MODELS PROCEDURE

CLASS LEVEL INFORMATION

CLASS	LEVELS	VALUES
COLOR	3	BL BR RH

NUMBER OF OBSERVATIONS IN DATA SET = 15

DEPENDENT VARIABLE: HEIGHT

SOURCE	DF	SUM OF SQUARES	MEAN SQUARE	F VALUE
MODEL	2	10.53333333	5.26666667	0.65
ERROR	12	97.20000000	8.10000000	PR > F
CORRECTED TOTAL	14	107.73333333		0.5394

SOURCE	DF	TYPE I SS	F VALUE	PR > F
COLOR	2	10.53333333	0.65	0.5394

SOURCE	DF	TYPE III SS	F VALUE	PR > F
COLOR	2	10.53333333	0.65	0.5394

The model sum of squares appears three times in the output: in the AOV table and as separate entries below that table, where it is labeled both as type I and as type III. The type I sum of squares is a result of adding each source of variation sequentially to the model in the order listed in the MODEL statement. Type III sums of squares are used when it is necessary to adjust the effect of each independent variable for the presence of the others in order to eliminate interaction terms. As we have a one-way AOV model in this example, all of the sums of squares should be identical and are.

The interpretation of the results is straightforward. As the calculated value of $F = 0.65$ and there is a 53.94% probability of obtaining a greater value, the null hypothesis that hair color does not affect height is accepted as true, as confirmed by the F-test as $F_{2,12,0.95} = 3.74$.

8.3.5 Subsampling in One-Way Analysis of Variance

For the case of subsampling in a completely randomized design we use the model

$$Y_{ijk} = \mu + \tau_j + \varepsilon_{ij} + \eta_{ijk}, \quad i = 1,2,\ldots,n, \ j = 1,2,\ldots,t, \ k = 1,2,\ldots,m,$$
$$(8.27)$$

where we assume that the ε_{ij}, the experimental errors, are $NID(0,\sigma^2)$ and that the η_{ijk}, the sampling errors, are $NID(0,\sigma_\eta^2)$. ε_{ij} is the true effect of the ith unit subjected to the jth treatment and η_{ijk} is the effect of the kth sample taken from the ith unit, which is subjected to the jth treatment. The one-way analysis of variance for subsampling m times with n observations is as shown in Table 8.4. The experimental error variance σ^2 is estimated by

$$S^2 = \frac{\text{experimental error MS} - \text{sampling error MS}}{m}$$

$$= \frac{MS_{EE} - MS_{SE}}{m}.$$
$$(8.28)$$

The variance of a treatment mean is given by

$$var(\bar{Y}_{j.}) = S_{\bar{Y}_{.}}^2 = \frac{MS_{EE}}{mn}.$$
$$(8.29)$$

Note that in Table 8.4 two expected mean squares, EMS_I and EMS_{II}, are designated for the treatments. These correspond to models I and II discussed previously. The mean squares for treatments for models I and II are

$$E(MS_I) = \sigma_\eta^2 + m\sigma^2 + nm\frac{\sum_{j=1}^t \tau_j^2}{t-1}$$

TABLE 8.4 One-Way Analysis of Variance with Subsampling (Equal Subclass Numbers)

Source	d.f.	SS	EMS
Mean	1	$\left(\sum_i \sum_j \sum_k Y_{ijk}\right)^2 \Big/ tnm = SS_M$	—
Treatments	$t-1$	$\sum_i \left(\sum_i \sum_k Y_{ijk}\right)^2 \Big/ mn - SS_M = SS_{Tr}$	$\begin{cases} E(MS_I) \\ E(MS_{II}) \end{cases}$
Experimental error	$t(n-1)$	$\sum_i \sum_j \left(\sum_k Y_{ijk}\right)^2 \Big/ m - SS_M - SS_{Tr} = SS_{EE}$	$\sigma_\eta^2 + m\sigma^2$
Sampling error	$tn(m-1)$	$\sum_i \sum_j \sum_k Y_{ijk}^2 - SS_M - SS_{Tr} - SS_{EE} = SS_{SE}$	σ_η^2
Total	tnm	$\sum_i \sum_j \sum_k Y_{ijk}^2 = SS_T$	

and

$$E(MS_{II}) = \sigma_\eta^2 + m\sigma^2 + nm\sigma_\tau^2.$$

It should also be noted that the sum of squares for the sampling error is a pooled sum of squares among samples on the same experimental unit. The F-ratios for testing $H_0: \tau_j = 0$, $j = 1,2,\ldots,t$ and $H_0: \sigma_\tau^2 = 0$ can be obtained easily by considering the expected mean squares.

Note that in a subsampling situation, n independent *samples* are taken and then each is divided into m subsamples. The subsamples do not constitute independent observations but merely provide an estimate of the sampling error as shown in Table 8.4. Repeated observations (temperature, flow rate, residence time, attendance in large lecture classes, etc.) by their very nature cannot be subdivided.

Subsampling involves multiple testing, measuring, or examination of the same original sample which has been divided prior to such evaluation. As a result, any variability in the multiple values so obtained for a sample is due to a new error source, that within samples, or sampling error. Because subsampling always involves division of original samples before evaluation or treatments, if there is no division, sampling error is not present. Consider the following example.

Example 8.4   Refractive indices of samples of the liquid on three trays in the methanol–water distillation column were measured in duplicate. The samples were taken simultaneously and divided prior to analysis.

		Tray	
Sample	1	2	3
1	1.3280	1.3375	1.3390
	1.3283	1.3375	1.3391
2	1.3300	1.3380	1.3389
	1.3329	1.3385	1.3391
3	1.3290	1.3380	1.3388
	1.3293	1.3379	1.3389

Is there a difference in the liquid composition (as indicated by the refractive index) on the trays, or is any apparent difference due to sampling error? *Solution*   The SAS program (*SAS User's Guide: Statistics*, 1982 edition) for this situation is given below.

(put initial JCL cards/statements here)

```
DATA;
INPUT TRAY 1 SAMPLE 3 INDEX 5-10;
CARDS;
```

(data here according to INPUT format)

```
PROC PRINT;
PROC GLM;
CLASSES TRAY SAMPLE;
MODEL INDEX = TRAY SAMPLE(TRAY)/SS3;
TITLE 'SUBSAMPLING IN ONE-WAY AOV';
```

(put final JCL cards/statements here)

The results follow.
                          SAS

        SUBSAMPLING IN ONE-WAY AOV

      GENERAL LINEAR MODELS PROCEDURE

          CLASS LEVEL INFORMATION

        CLASS   LEVELS   VALUES

        TRAY   3   1 2 3

        SAMPLE   3   1 2 3

NUMBER OF OBSERVATIONS IN DATA SET = 18

                          SAS

        SUBSAMPLING IN ONE-WAY AOV

      GENERAL LINEAR MODELS PROCEDURE

DEPENDENT VARIABLE: SUB

SOURCE	DF	SUM OF SQUARES	MEAN SQUARE	F VALUE
MODEL	8	0.00032877	0.00004110	83.02
ERROR	9	0.00000446	0.00000050	PR > F
CORRECTED TOTAL	17	0.00033323		0.0001

SOURCE	DF	TYPE III SS	F VALUE	PR > F
TRAY	2	0.00031670	319.54	0.0001
SAMPLE(TRAY)	6	0.00001207	4.06	0.0299

From the f values, we conclude that for this model I situation (trays correspond to fixed effects), experimental error [SAMPLE(TRAY)] is first compared to sampling error (ERROR) by

$$F_{EE} = \frac{MS_{EE}}{MS_{SE}} = \frac{0.00001207/6}{0.0000005} = 4.0594 \cong 4.06,$$

which, compared to $F_{6,9,0.95} = 3.37$, indicates that the experimental error is significant ($H_0: \sigma^2 = 0$ is rejected) compared to sampling error. The treatment (TRAY) effects are tested by calculating

$$F_{Tr} = \frac{MS_{Tr}}{MS_{EE}} = \frac{0.0003167/2}{0.00001207} = 78.716,$$

which, when compared to $F_{2,6,0.95} = 5.14$, shows that the treatments are significant. In the SAS output statement, the source identified as ERROR is the sampling error and that identified as SAMPLE(TRAY) in this example is the experimental error. This error nomenclature is followed for all PROC GLM outputs when subsampling is involved. You can always tell which entry represents the corresponding error term by looking at the degrees of freedom (DF) in the output and comparing those values to the input data.

8.4 TWO-WAY AND THREE-WAY ANALYSIS OF VARIANCE

If we desire to study the effects of two independent variables (factors) on one dependent variable, we will have to use a *two-way analysis of variance*. For this case the columns represent various values or levels of one independent variable and the rows represent levels or values of the other independent variable. Each entry in the array in Table 8.1 of data points then represents one of the possible combinations of the two independent variables and how it affects the dependent variable. Here we consider the case of only one observation per data point.

8.4.1 Model for Two-Way Analysis of Variance

For a two-way analysis of variance the assumed model is

$$Y_{ij} = \mu_{ij} + \varepsilon_{ij}.$$

In this case Y_{ij} is assumed to come from a normal population with mean μ_{ij} and variance σ^2. Furthermore, the Y_{ij} are independently distributed. The two-way model can be written as

$$Y_{ij} = \mu + \alpha_i + \beta_j + \varepsilon_{ij}, \tag{8.30}$$

where $\mu = \bar{\mu}.. = \sum_i \sum_j \mu_{ij}/IJ$, $\alpha_i = \bar{\mu}_{i.} - \bar{\mu}..$, $\beta_j = \bar{\mu}._j - \bar{\mu}..$, $\bar{\mu}_{i.} = \sum_j \mu_{ij}/J$, and $\bar{\mu}._j = \sum_i \mu_{ij}/I$. The parameter μ is the contribution of the grand mean, α_i is the contribution of the ith level of the row variable, β_j is the contribution of the jth level of the column variable, and ε_{ij} is the random experimental error. The model in Eq. (8.30) does not contain what is usually referred to as a row–column interaction; that is, the row and column effects are additive. The idea of interaction is discussed briefly in the next two subsections.

Two restrictions (or assumptions) related to the model in Eq. (8.30) are $\sum_i \alpha_i = 0$ and $\sum_j \beta_j = 0$. These follow from $\alpha_i = \bar{\mu}_{i.} - \bar{\mu}..$ and $\beta_j = \bar{\mu}._j - \bar{\mu}..$.

Two hypotheses related to the model in (8.30) are

$$H_0: \mu_1. = \mu_2. = \cdots = \mu_I.$$

and

$$H_0: \mu._1 = \mu._2 = \cdots = \mu._J.$$

The first hypothesis says that there is no row effect; that is, the means across rows have the same value. Similarly, the second hypothesis says that there is no column effect. The two hypotheses above can be written equivalently as

$$H_0: \alpha_1 = \alpha_2 = \cdots = \alpha_I = 0 \tag{8.31}$$

and

$$H_0: \beta_1 = \beta_2 = \cdots = \beta_J = 0. \tag{8.32}$$

The hypotheses relevant to no row or column effects are usually written in the form of (8.31) and (8.32). If both hypotheses are true, then $Y_{ij} = \mu + \varepsilon_{ij}$, which says that all the observations came from one fixed normal population.

It can be shown that the total variation can be partitioned according to the identity

$$\sum_i \sum_j (Y_{ij} - \bar{Y}_{..})^2 = \sum_i \sum_j (\bar{Y}_{.j} - \bar{Y}_{..})^2 + \sum_i \sum_j (\bar{Y}_{i.} - \bar{Y}_{..})^2$$
$$+ \sum_i \sum_j (Y_{ij} - \bar{Y}_{.j} - \bar{Y}_{i.} + \bar{Y}_{..})^2.$$

This identity can be written as

$$SS_{TC} = SS_C + SS_R + SS_E.$$

That is, the total variation is partitioned into variation due to columns (SS_C), variation due to rows (SS_R), and variation due to experimental error (SS_E). The degrees of freedom are likewise partitioned as

$$IJ - 1 = (J - 1) + (I - 1) + (I - 1)(J - 1).$$

The mean squares are defined by $MS_C = SS_C/(J - 1)$, $MS_R = SS_R/(I - 1)$, and $MS_E = SS_E/(I - 1)(J - 1)$. Furthermore, it can be shown (just as for the one-way model) that the expected mean squares are

$$E[MS_R] = \sigma^2 + J \frac{\sum_i \alpha_i^2}{I - 1} \quad \text{or} \quad E[MS_R] = \sigma^2 + \beta\sigma_\alpha^2$$

$$E[MS_C] = \sigma^2 + I \frac{\sum_j \beta_j^2}{J - 1} \quad \text{or} \quad E[MS_C] = \sigma^2 + \alpha\sigma_\beta^2$$

$$E[MS_E] = \sigma^2 \qquad\qquad\qquad \text{or} \quad E[MS_E] = \sigma^2,$$

where the left group represents the model I situation, and the right group represents the model II situation. Mixed models are possible. For only one observation per treatment combination, the calculations are identical. Thus,

TABLE 8.5 Two-Way Analysis of Variance

Source	d.f.	SS	MS
Mean	1	$SS_M = y_{..}^2/IJ$	
Between columns	$J-1$	$SS_C = \sum_j y_{.j}^2/I - y_{..}^2/IJ$	$MS_C = \dfrac{SS_C}{J-1}$
Between rows	$I-1$	$SS_R = \sum_i y_{i.}^2/J - y_{..}^2/IJ$	$MS_R = \dfrac{SS_R}{I-1}$
Error	$(I-1)(J-1)$	$SS_E = SS_T - SS_C - SS_R - SS_M$	$MS_E = \dfrac{SS_E}{(I-1)(J-1)}$
Total	IJ	$SS_T = \sum_i \sum_j y_{ij}^2$	

to test H_0: all $\alpha_i = 0$, we use the F-statistic with $I - 1$ and $(I - 1)(J - 1)$ degrees of freedom,

$$F = \frac{MS_R}{ME_E}.$$

The rejection region is $F > F_{(I-1),(I-1)(J-1),1-\alpha}$.

Similarly, to test H_0: all $\beta_j = 0$, we use $F = MS_C/MS_E$ with $J - 1$ and $(I - 1)(J - 1)$ degrees of freedom. The rejection region is $F > F_{(J-1),(I-1)(J-1),1-\alpha}$.

The results pertinent to the two-way model are summarized in Table 8.5. We note from Table 8.5 that the sum of squares due to error, SS_E, is $SS_T - SS_C - SS_R - SS_M$. The procedure is to compute SS_T, SS_M, SS_C, and SS_R and then determine SS_E according to the difference in Table 8.5. Statistical theory has shown that MS_E is an unbiased estimate of the population variance σ^2 regardless of whether the hypotheses are true or not. However, MS_R and MS_C are unbiased for σ^2 only if the corresponding hypotheses are true.

Example 8.5 In an experiment to determine the effects of varying the reflux ratio on the number of required stages Y_{ij} used in the separation of benzene and toluene, four different lab groups used the same four reflux ratios with the following results:

Lab group	Reflux ratio (mol liquid/mol vapor)			
	1	2	3	4
1	11.4	9.2	7.5	6.2
2	10.7	8.6	8.3	5.9
3	11.9	8.7	9.3	5.4
4	9.9	9.0	7.1	5.6

You should realize that any differences that occur due to lab groups will affect the total for each column. However, as each group checked each reflux ratio, all column totals should be equally affected by the differences in groups. We now have identified the two major sources of variation: groups and reflux ratio. Prepare an analysis-of-variance table for these results. What can you conclude from the entries in this table? Note that this problem represents a mixed model, as the lab groups are fixed, whereas the reflux ratio is a random variable, as any value in the column operating range could have been chosen.

Solution We obtain

$$y_{.1} = 43.9, \qquad y_{1.} = 34.3, \qquad y_{..} = 134.7,$$

$$y_{.2} = 35.5, \qquad y_{2.} = 33.5,$$

$$y_{.3} = 32.3, \qquad y_{3.} = 35.3,$$

$$y_{.4} = 23.1, \qquad y_{4.} = 31.6,$$

$$SS_T = \sum_i \sum_j y_{ij}^2 = 1195.17,$$

$$SS_M = \frac{y_{..}^2}{IJ} = \frac{18{,}144.09}{16} = 1134.0056,$$

$$SS_C = \frac{\sum_j y_{.j}^2}{I} - SS_M = \frac{4757.91}{4} - SS_M = 55.4719,$$

$$SS_R = \frac{\sum_i y_{i.}^2}{J} - SS_M = \frac{4543.39}{4} - SS_M = 1.8419,$$

$$SS_E = SS_T - SS_C - SS_R - SS_M = 3.8506.$$

The analysis-of-variance table is:

Source	d.f.	Sum of squares	Mean square	F
Mean	1	1134.0056	—	—
Reflux ratio	3	55.4719	18.4906	43.222
Lab group	3	1.8419	0.6139	1.435
Error	9	3.8506	0.4278	—
Total	16	1195.17	—	—

When the calculated values of F are compared to $F_{3,9,0.95} = 3.86$ from Table VI App. B, it is seen at once that the differences in the number of required stages is affected significantly by the reflux ratio. No significant differences between lab groups were found.

The use of PROC ANOVA for two-way AOV is illustrated by the following example.

Example 8.6 Astarita et al. [1] give the following data* for non-Newtonian gravity flow along an inclined plane:

β	δ (mm) at $Q = 2\,cm^3/min$	δ (mm) at $Q = 10\,cm^3/min$
20	1.63	2.89
10	2.06	3.80
6.5	2.42	4.41
3.5	3.06	5.46
2.0	3.84	6.90

where β is the angle of inclination (deg), δ is the liquid layer depth (mm), and Q is the total flow rate (cm³/s). Determine if the angle β and/or the flow rate Q make a significant difference in the liquid depth δ at a 95% significance level.

* Data reprinted by permission of the copyright owner.

Solution   The SAS program is as follows.

(put initial JCL cards/statements here)

```
DATA;
INPUT DELTA 1-4 BETA 6-9 FLOW 11-12;
CARDS;
```

(enter data from problem statement here using format given by INPUT statement)

```
PROC PRINT;
PROC ANOVA;
CLASSES BETA FLOW;
MODEL DELTA = BETA FLOW;
TITLE 'EFFECT OF ANGLE AND THICKNESS ON GRAVITY FLOW';
```

(put (final JCL cards/statements here)

The SAS output for this experiment is shown below. (As the data have already been given, we have omitted that portion of the output.)

SAS

EFFECT OF ANGLE AND FLOW RATE ON LAYER THICKNESS

ANALYSIS OF VARIANCE PROCEDURE

CLASS LEVEL INFORMATION

CLASS	LEVELS	VALUES
BETA	5	2.0 3.5 6.5 10 20
FLOW	2	2 10

NUMBER OF OBSERVATIONS IN DATA SET = 10

EFFECT OF ANGLE AND FLOW RATE ON LAYER THICKNESS

ANALYSIS OF VARIANCE PROCEDURE

DEPENDENT VARIABLE: DELTA

SOURCE	DF	SUM OF SQUARES	MEAN SQUARE	F VALUE
MODEL	5	22.59261000	4.51852200	19.45
ERROR	4	0.92920000	0.23230000	
CORRECTED TOTAL	9	23.52181000		

SOURCE	DF	ANOVA SS	F VALUE	PR > F
BETA	4	11.67236000	12.56	0.0155
FLOW	1	10.92025000	47.01	0.0024

$$F_{4,4,0.95} = 6.39, \qquad F_{1,4,0.95} = 7.71$$

From the calculated $F = 19.45$ for the model as a whole, and the very slight (0.66%) probability of obtaining a higher value of F, we reject the null hypothesis that the model is invalid at $\alpha = 0.05$. Examining the individual model components, $F_\beta = 12.56$ with only a 1.55% probability of being exceeded. We reject as false the null hypothesis that angle of inclination does not affect flow rate. Similarly, $F_Q = 47.01$ with only a 0.24% probability of being exceeded. We thus reject as false the hypothesis that flow rate does not influence the liquid layer depth. The same results are obtained by comparing F_β to $F_{4,4,0.95} = 6.39$ and F_Q to $F_{1,4,0.95} = 7.71$. In both cases, the hypothesis of no treatment effect is rejected as false.

8.4.2   Interaction

Two-way analysis of variance (and higher classifications) lead to the presence of interactions. If, for example, an additive, A, is added to a lube oil stock to improve its resistance to oxidation, and another additive, B, is added to inhibit corrosion by the stock under load or stress, it is entirely possible that the performance of the lube oil in a standard ball-and-socket wear test will be different from that expected if only one additive were present. In other words, the presence of one additive may adversely or helpfully affect the action of the other additive in modifying the properties of the lube oil. These are termed *antagonistic* and *synergistic effects*, respectively. It is important to consider the presence of such interactions in any treatment of multiply classified data. To do this, the two-way analysis-of-variance table is set up as shown in Tables 8.6 and 8.7.

TABLE 8.6 Two-Way Analysis of Variance with Interaction (Model I)

Source	d.f.	SS	EMS
Mean	1	$\left(\sum_i \sum_j \sum_k Y_{ijk}\right)^2 \bigg/ abn = SS_M$	
A	$a-1$	$\sum_i \left(\sum_j \sum_k Y_{ijk}\right)^2 \bigg/ bn - SS_M = SS_A$	$\sigma^2 + nb \sum_i \alpha_i^2/(a-1)$
B	$b-1$	$\sum_j \left(\sum_i \sum_k Y_{ijk}\right)^2 \bigg/ an - SS_M = SS_B$	$\sigma^2 + na \sum_j \beta_j^2/(b-1)$
AB	$(a-1)(b-1)$	$\sum_i \sum_j \left(\sum_k Y_{ijk}\right)^2 \bigg/ n - SS_M - SS_A$ $- SS_B = SS_{AB}$	$\sigma^2 + \dfrac{n \sum_i \sum_j (\alpha\beta)_{ij}^2}{(a-1)(b-1)}$
Error	$ab(n-1)$	$\sum_i \sum_j \sum_k Y_{ijk}^2 - SS_A - SS_B - SS_{AB}$ $- SS_M = SS_E$	σ^2
Total	abn	$\sum_i \sum_j \sum_k Y_{ijk}^2 = SS_T$	

TABLE 8.7 Two-Way Analysis of Variance with Interaction (Model II)

Source	d.f.	SS	EMS
Mean	1	$\left(\sum_i \sum_j \sum_k Y_{ijk}\right)^2 \bigg/ abn = SS_M$	
A	$a-1$	$\sum_i \left(\sum_j \sum_k Y_{ijk}\right)^2 \bigg/ bn - SS_M = SS_A$	$\sigma^2 + n\sigma_{\alpha\beta}^2 + nb\sigma_\alpha^2$
B	$b-1$	$\sum_j \left(\sum_i \sum_k Y_{ijk}\right)^2 \bigg/ an - SS_M = SS_B$	$\sigma^2 + n\sigma_{\alpha\beta}^2 + na\sigma_\beta^2$
AB	$(a-1)(b-1)$	$\sum_i \sum_j \left(\sum_k Y_{ijk}\right)^2 \bigg/ n - SS_M - SS_A$ $- SS_B = SS_{AB}$	$\sigma^2 + n\sigma_{\alpha\beta}^2$
Error	$ab(n-1)$	$\sum_i \sum_j \sum_k Y_{ijk}^2 - SS_A - SS_B - SS_{AB}$ $- SS_M = SS_E$	σ^2
Total	abn	$\sum_i \sum_j \sum_k Y_{ijk}^2 = SS_T$	

8.4.3 Assumptions in Two-Way Analysis of Variance

In some cases, it is highly desirable to use *factorial* treatment combinations in completely randomized designs. Consider the case where T treatments are actually combinations of a levels of factor A and b levels of factor B. Then the statistical model is given as

$$Y_{ijk} = \mu + \alpha_i + \beta_j + (\alpha\beta)_{ij} + \varepsilon_{ijk}, \\ i = 1,2,\ldots,a, \quad j = 1,2,\ldots,b, \quad k = 1,2,\ldots,n, \Bigg\} \quad (8.33)$$

where μ is the true mean effect, α_i the true effect of the ith level of factor A, β_j the true effect of the jth level of factor B, $(\alpha\beta)_{ij}$ the true effect of the interaction of the ith level of factor A with the jth level of factor B, and ε_{ijk} the true effect of the kth experimental unit subjected to the (ij)th treatment combination. The usual assumption is made that the ε_{ijk} are NID$(0, \sigma^2)$.

There are four possible sets of assumptions that can be made concerning the true treatment effects. The first set of assumptions is that we are concerned with only the fixed effects, α and β, of factors A and B. That is, we are concerned only with the a levels of factor A and b levels of factor B. This is our model I (fixed effects). These assumptions can be summarized by

$$\sum_i \alpha_i = \sum_j \beta_j = \sum_i (\alpha\beta)_{ij} = \sum_j (\alpha\beta)_{ij} = 0. \quad (8.34)$$

The analysis of variance for α and β fixed is in Table 8.6. Compare Table 8.6 with Table 8.3.

The second set of assumptions to be considered is that where α and β are random effects. This is model II. Here we are concerned with the population of levels of factor A, of which only a random sample is present in the experiment, and the population of levels of factor B, again of which only a random sample is present in the experiment. These assumptions are summarized as follows:

$$\alpha_i \quad \text{are} \quad \text{NID}(0,\sigma_\alpha^2), \quad\quad\quad\quad\quad\quad\quad\quad (8.35a)$$

$$\beta_j \quad \text{are} \quad \text{NID}(0,\sigma_\beta^2), \quad\quad\quad\quad\quad\quad\quad\quad (8.35b)$$

$$(\alpha\beta)_{ij} \quad \text{are} \quad \text{NID}(0,\sigma_{\alpha\beta}^2). \quad\quad\quad\quad\quad\quad\quad (8.35c)$$

The corresponding analysis of variance for this model is in Table 8.7.

The other two sets of assumptions yield what are called *mixed models*. In the third model (III) α is considered fixed and β is random, and in the fourth model (IV) α is considered random and β fixed. Assumptions underlying these two models are the following:

Model III:
$$\sum_i \alpha_i = \sum_i (\alpha\beta)_{ij} = 0, \qquad \beta_j \quad \text{are} \quad \text{NID}(0,\sigma_\beta^2). \tag{8.36a}$$

Model IV:
$$\sum_j \beta_j = \sum_j (\alpha\beta)_{ij} = 0, \qquad \alpha_i \quad \text{are} \quad \text{NID}(0,\sigma_\beta^2). \tag{8.36b}$$

Examination of the expected mean squares in Tables 8.6 and 8.7 will indicate the proper F-test for hypotheses concerning the various treatments and treatment combinations. These are summarized in Table 8.8. For an example of a two-treatment (two-factor) model, see Example 8.7.

A few comments are in order regarding testing the significance of main effects in models that contain interactions. If $H_0: (\alpha\beta)_{ij} = 0$, $i = 1,2,\ldots,a$, $j = 1,2,\ldots,b$, is rejected, then $H_0: \alpha_i = 0$, $i = 1,2,\ldots,a$, can be tested, but the results are usually of no interest. When interactions are present, the best treatment combinations are of prime interest, rather than the best levels of A or B. If there is a significant interaction present, the acceptance of $H_0: \alpha_i = 0$ should be interpreted as meaning that there are no differences in the various levels of A when averaged over the levels of B. Similarly, if $H_0: \beta_j = 0$, $j = 1,2,\ldots,b$, is accepted, the interpretation is that there is no difference in the levels of B when they are averaged over the levels of A.

TABLE 8.8　*F*-Ratios for Hypothesis Testing in Two-Way Analysis of Variance

Source	Model I	Model II	Model III	Model IV
Effects	α,β: fixed	α,β: random	α fixed β random	a random β fixed
Mean	—	—	—	—
A	MS_A/MS_E	MS_A/MS_{AB}	MS_A/MS_{AB}	MS_A/MS_E
B	MS_B/MS_E	MS_B/MS_{AB}	MS_B/MS_E	MS_B/MS_{AB}
AB	MS_{AB}/MS_E	MS_{AB}/MS_E	MS_{AB}/MS_E	MS_{AB}/MS_E
Error	—	—	—	—

The use of the SAS System (*SAS User's Guide: Statistics, Version 5*) to handle any two-way AOV situation is illustrated below. When multiple observations are available for each treatment combination (or "cell"), you can get the interaction term provided that those extra observations came from replications of the basic experiment and not from subsamples. To get the interaction term, add var1 * var2 to the MODEL statement as shown in this example.

Example 8.7   The effects of 5-bromodeoxyuridine (*A*) and cytochlasin B (*B*) on normal development of sucrase specific activity in the embryonic chicken were studied. Thirty fertilized eggs were selected for study. Ten of the eggs remained uninjected (*N*) during incubation. Ten eggs were injected with 5-bromodioxyuridine prior to day 10 of incubation and 10 were injected with cytochlasin B prior to day 10 of incubation. The embryos were removed at embryonic age 13 days to 17 days and duodenal loop tissue was analyzed for specific invertase activity. Data representing g glucose released/mg protein per hour were obtained. What can be said concerning the age effect, the treatment effect, or any interaction?

	Embryonic age (days)				
Treatment	13	14	15	16	17
N	56.3	60.5	64.3	86.1	97.5
	55.4	61.2	63.9	86.0	98.0
A	51.2	54.4	62.3	78.7	95.5
	50.8	55.6	61.7	78.3	96.5
B	55.4	61.5	62.0	72.8	81.9
	54.8	60.5	63.0	71.2	80.1

Solution   The program is shown below.

(put initial JCL cards/statements here)

```
DATA;
INPUT AGE 1-2 TRT $ 4 SPACT 6-9;
CARDS;
```

(data here according to INPUT format)

```
PROC PRINT;
PROC ANOVA;
CLASSES AGE TRT;
MODEL SPACT = AGE TRT AGE * TRT;
TITLE 'EFFECT OF AGE & TREATMENT ON SUCRASE SPECIFIC ACTIVITY';
```

(put final JCL cards/statements here)

SAS

ANALYSIS OF VARIANCE PROCEDURE

DEPENDENT VARIABLE: SPACT

SOURCE	DF	SUM OF SQUARES	MEAN SQUARE	F VALUE
MODEL	14	6440.83466667	460.05961905	1061.68
ERROR	15	6.50000000	0.43333333	PR > F
CORRECTED TOTAL	29	6447.33466667		0.0001

SOURCE	DF	ANOVA SS	F VALUE	PR > F
TRT	2	226.16266667	260.96	0.0001
AGE	4	5825.07466667	3360.62	0.0001
TRT*AGE	8	389.59733333	112.38	0.0001

The resulting analysis of variance shows by comparing the calculated F-values for this model I situation with the tabular values that all the hypotheses expressed in Eq. (8.34) should be rejected.

Note that an interaction term (AGE*TRT) could be obtained because repeated observations for the same treatment combinations are available. The SAS system does not know the difference between model I and model II hypotheses, so the appropriate F values must be calculated from the mean squares provided in the output.

For the situation when an unequal number or no observation exists for a cell, the PROC ANOVA command must be replaced by PROC GLM. The MODEL statement must also be modified by adding

/SS1 SS2 SS3 SS4;

after the interaction term.

Type I functions (SS1 output) add each source of variation sequentially to the model in the order listed in the MODEL statement. If the observations are balanced, type I results are the same as those from PROC ANOVA. Type II functions (SS2 output) are adjusted sums of squares. For example, for effect U, which may be a main effect or an interaction, SS2 is adjusted for the effect of V (another main effect) if and only if V does *not* contain U. As an example, the main effects A and B are not adjusted for $A*B$ because $A*B$ contains both A and B. A *is* adjusted for the presence of B because B does not contain A. Similarly, B is adjusted for A and $A*B$ is adjusted for A and B. These sums of squares are appropriate for the case where there are no interactions.

Type III functions (SS3 output) are weighted mean-squares analyses and are used when it is necessary to compare main effects in the presence of suspected significant interactions. Each effect is adjusted for all others. The main effects A and B will be adjusted for $A*B$. For unequal observations in the cells, SS3 should also be specified as a type III function is needed for the AOV. Type IV functions (SS4 output) are used for missing data (empty cells). If data are entirely missing for a cell, specify SS4 for the type IV function. For empty cells, omit the entire data line. Otherwise, the SAS software inserts 0.0 in the cell for the missing value.

Example 8.8 Results of a laboratory analysis for the specific rate constant for the saponification of ethyl acetate by NaOH at 0°C are given below. One of the runs was performed with an electric stirrer, while the others were hand stirred. The results were analyzed by the differential and integral methods.

	With stirrer		Without stirrer	
Integral	0.02058	0.02121	0.01849	0.01816
method	0.02214	0.02073	0.01951	0.01884
Differential	0.01995	0.02003	0.01725	0.01752
method	0.01968	0.01982	0.01696	0.01726

Does the electric stirrer make a significant difference in the results? Does the method of analysis of data make a significant difference?

Solution The use of the SAS program for two-way analysis of variance with interaction gave the results shown below.

SAS

EFFECT OF ANALYSIS TECHNIQUE AND STIRRING METHOD ON RATE CONSTANT

ANALYSIS OF VARIANCE PROCEDURE

DEPENDENT VARIABLE: RATE

SOURCE	DF	SUM OF SQUARES	MEAN SQUARE	F VALUE
MODEL	3	0.00003452	0.00001151	71.03
ERROR	12	0.00000194	0.00000016	PR > F
CORRECTED TOTAL	15	0.00003646		0.0001

SOURCE	DF	ANOVA SS	F VALUE	PR > F
METHOD	1	0.00000643	39.67	0.0001
STIR	1	0.00002809	173.39	0.0001
METHOD*STIR	1	0.00000000	0.02	0.8936

We interpret the F-tests as follows for this model I situation. As $F_{1,12,0.95} = 4.75$, we reject the hypotheses of no method effect and no stirring effect and accept the hypothesis of no interaction effect. Thus we say that the rate constants are probably dependent on the stirring and analysis method but that there is no apparent interaction between the two variables.

8.4.4 Model for Three-Way Analysis of Variance

In some instances, the data obtained are the result of three factors and their interactions. The general model for this situation is

$$Y_{ijkl} = \mu + \alpha_i + \beta_j + \gamma_k + (\alpha\beta)_{ij} + (\alpha\gamma)_{ik} + (\beta\gamma)_{jk} + (\alpha\beta\gamma)_{ijk} + \varepsilon_{ijkl},$$

$$i = 1,2,\ldots,a, \quad j = 1,2,\ldots,b, \quad k = 1,2,\ldots,c, \quad l = 1,2,\ldots,n, \quad (8.37)$$

where the three main effects α, β, and γ can be referred to as groups, blocks, and treatments. The random effects ε_{ijkl} are assumed to be independently and normally distributed. The other terms denote the corresponding interactions among the three factors.

The assumptions for the three-way AOV involving the main effects and interactions for a model I situation are

$$\sum_i \alpha_i = \sum_j \beta_j = \sum_k \gamma_k = 0, \tag{8.38a}$$

$$\sum_i (\alpha\beta)_{ij} = \sum_j (\alpha\beta)_{ij} = 0 \quad [\text{similarly for } (\alpha\gamma) \text{ and } (\beta\gamma)], \tag{8.38b}$$

$$\sum_i (\alpha\beta\gamma)_{ijk} = \sum_j (\alpha\beta\gamma)_{ijk} = \sum_k (\alpha\beta\gamma)_{ijk} = 0. \tag{8.38c}$$

The presentation of the data in tabular form is similar to the previous situations and will only be illustrated in the example to follow. The three-way AOV is summarized in Table 8.9. In this case the total number of observations is $N = abcn$ and the computational expressions for the sums of squares are given directly in the AOV table. An example to illustrate these computations follows.

Example 8.9 The tables below give the results of an experiment on the amounts of niacin found in peas after three treatments (R_1, R_2, R_3) on blanched and processed peas of different sieve sizes (A, B, C). In these calculations, the subscripts are assigned as follows: G_i, preparation method; B_j, sieve sizes; and T_k, treatments. Each combination has $l = 10$ replications.

TABLE 8.9 Three-Way Analysis of Variance

Source	d.f.	SS	MS	EMS
Between group means	$a-1$	$SS_G = \sum_i (y_{i\cdots}^2/bcn) - (y_{\cdots}^2/N)$	$SS_G/(a-1)$	$\sigma^2 + bc\sum_{i=1}^{a} \alpha_i^2/(a-1)$
Between block means	$b-1$	$SS_B = \sum_j (y_{\cdot j\cdot}^2/acn) - (y_{\cdots}^2/N)$	$SS_B/(b-1)$	$\sigma^2 + ac\sum_{j=1}^{b} \beta_j^2/(b-1)$
Between treatment means	$c-1$	$SS_T = \sum_k (y_{\cdot\cdot k}^2/abn) - (y_{\cdots}^2/N)$	$SS_T/(c-1)$	$\sigma^2 + ab\sum_{k=1}^{a} \gamma_{\cdot k}^2/(c-1)$
Interactions G × B	$(a-1)(b-1)$	$SS_{GB} = \sum_i\sum_j (y_{ij\cdot}^2/cn) - \sum_i (y_{i\cdots}^2/bcn)$ $- \sum_j (y_{\cdot j\cdot}^2/acn) + (y_{\cdots}^2/N)$	$SS_{GB}/(a-1)(b-1)$	$\sigma^2 + c\sum_{i=1}^{a}\sum_{j=1}^{b} (\alpha\beta)_{ij}^2/(a-1)(b-1)$
G × T	$(a-1)(c-1)$	$SS_{GT} = \sum_i\sum_k (y_{i\cdot k}^2/bn) - \sum_i (y_{i\cdots}^2/bcn)$ $- \sum_k (y_{\cdot\cdot k}^2/abn) + (y_{\cdots}^2/N)$	$SS_{GT}/(a-1)(c-1)$	$\sigma^2 + b\sum_{i=1}^{a}\sum_{k=1}^{c} (\alpha\gamma)_{ik}^2/(a-1)(c-1)$

Source	df	SS	MS	E(MS)
B × T	$(b-1)(c-1)$	$SS_{BT} = \sum_j \sum_k (y^2_{\cdot jk\cdot}/an) - \sum_j (y^2_{\cdot j\cdot\cdot}/acn)$ $- \sum_k (y^2_{\cdot\cdot k\cdot}/abn) + (y^2_{\cdots}/N)$	$SS_{BT}/(b-1)(c-1)$	$\sigma^2 + a\sum_{j=1}^{b}\sum_{k=1}^{c}(\beta\gamma)^2_{jk}/(b-1)(c-1)$
G × B × T	$(a-1)$ $\times$ $(b-1)$ $\times$ $(c-1)$	$SS_{GBT} = \sum_i\sum_j\sum_k(y^2_{ijk\cdot}/n) - \sum_i\sum_j(y^2_{ij\cdot\cdot}/cn)$ $-\sum_i\sum_k(y^2_{i\cdot k\cdot}/bn) - \sum_j\sum_k(y^2_{\cdot jk\cdot}/an)$ $+\sum_i(y^2_{i\cdots}/bcn) + \sum_j(y^2_{\cdot j\cdot\cdot}/acn)$ $+\sum_k(y^2_{\cdot\cdot k\cdot}/abc) - (y^2_{\cdots}/N)$	$\dfrac{SS_{GBT}}{(a-1)(b-1)(c-1)}$	$\sigma^2 = \dfrac{\sum_{i=1}^a\sum_{j=1}^b\sum_{k=1}^c(\alpha\beta\gamma)^2_{ijk}}{(a-1)(b-1)(c-1)}$
Error	$N-abc$	$SS_E = \sum_i\sum_j\sum_k\sum_l y^2_{ijkl} - \sum_i\sum_j\sum_k(y^2_{ijk\cdot}/n)$	$SS_R/(N-abc)$	σ^2
Total corrected	$N-1$	$SS_{TC} = \sum_i\sum_j\sum_k\sum_l y^2_{ijkl} - (y^2_{\cdots}/N)$	$SS_{TC}/(N-1)$	

Blanched peas:

	A			B			C		
	R_1	R_2	R_3	R_1	R_2	R_3	R_1	R_2	R_3
	65	90	44	59	70	83	88	80	123
	87	94	92	63	65	95	60	81	95
	48	86	88	81	78	88	96	105	100
	28	70	75	80	85	99	87	130	131
	20	78	80	76	74	81	68	122	121
	22	65	70	85	73	98	75	130	115
	24	75	73	64	61	95	96	121	127
	47	98	88	96	71	95	98	125	99
	28	95	77	91	63	90	84	172	101
	42	66	72	65	54	76	82	133	111
Total	411	817	759	760	694	900	834	1199	1123

Processed peas:

	A			B			C		
	R_1	R_2	R_3	R_1	R_2	R_3	R_1	R_2	R_3
	62	106	126	150	138	150	146	52	100
	113	107	193	112	120	112	172	97	133
	171	79	122	136	135	126	138	112	125
	135	122	115	120	126	123	124	116	124
	123	125	126	118	120	125	113	121	115
	132	96	110	134	132	110	121	99	122
	120	111	98	125	135	125	125	120	112
	117	116	115	114	124	120	116	121	116
	153	124	112	112	137	110	165	122	99
	132	126	109	120	125	125	137	134	105
Total	1258	1112	1226	1241	1292	1226	1357	1094	1151

From the tables we get $a = 2$, $b = 3$, $c = 3$, $n = 10$, $N = 180$; $\sum\sum\sum\sum y_{ijkl}^2 =$ $(65)^2 + \cdots + (105)^2 = 2{,}054{,}828$ and $y_{\cdots}^2 = 340{,}550{,}116$. Thus SS_{TC} is found to be 162,883.

The necessary statistics can now be calculated:

$$\sum y_{i\cdots}^2 = (7497)^2 + (10{,}957)^2 = 176{,}260{,}858,$$
$$\sum y_{\cdot j\cdot}^2 = (5583)^2 + (6113)^2 + (6758)^2 = 114{,}209{,}222,$$
$$\sum y_{\cdot\cdot k}^2 = (5861)^2 + (6208)^2 + (6385)^2 = 113{,}658{,}810,$$
$$\sum\sum y_{ij\cdot}^2 = (1987)^2 + \cdots + (3602)^2 = 59{,}485{,}522,$$
$$\sum\sum y_{i\cdot k}^2 = (2005)^2 + \cdots + (3603)^2 = 59{,}189{,}998,$$
$$\sum\sum y_{\cdot jk}^2 = (1669)^2 + \cdots + (2274)^2 = 38{,}144{,}306,$$
$$\sum\sum\sum y_{ijk}^2 = (411)^2 + \cdots + (1151)^2 = 20{,}073{,}704.$$

Dividing by the appropriate degrees of freedom and combining the results according to Table 8.9, we get the following:

$$SS_{TC} = 2{,}054{,}828 - 1{,}891{,}945 = 162{,}883,$$
$$SS_G = 1{,}958{,}454 - 1{,}891{,}945 = 66{,}509$$
$$SS_B = 1{,}903{,}487 - 1{,}891{,}945 = 11{,}542,$$
$$SS_T = 1{,}894{,}314 - 1{,}891{,}945 = 2{,}369,$$
$$SS_{GB} = 1{,}982{,}851 - 1{,}958{,}454 - 1{,}903{,}487 + 1{,}891{,}945 = 12{,}855,$$
$$SS_{GT} = 1{,}973{,}000 - 1{,}958{,}454 - 1{,}894{,}314 + 1{,}891{,}945 = 12{,}177,$$
$$SS_{BT} = 1{,}907{,}215 - 1{,}903{,}487 - 1{,}894{,}314 + 1{,}891{,}945 = 1{,}359,$$
$$SS_{GBT} = 2{,}007{,}370 - 1{,}982{,}851 - 1{,}973{,}000 - 1{,}907{,}215 + 1{,}958{,}454$$
$$+ 1{,}903{,}487 + 1{,}894{,}314 - 1{,}891{,}945 = 8614,$$
$$SS_E = 2{,}054{,}828 - 2{,}007{,}370 = 47{,}458.$$

Summarizing these results in an AOV table, we have

Source	d.f.	SS	MS
Between-group means	1	66,509	66,509.00
Between-block means	2	11,542	5,771.00
Between-treatments means	2	2,369	1,184.50
Interactions			
G × B	2	12,855	6,427.50
G × T	2	12,177	6,088.50
B × T	4	1,359	339.75
G × B × T	4	8,614	2,153.5
Error	162	47,458	292.95
Total corrected	179	162,883	909.96

It is clear from the EMS column of Table 8.9 which quotients need to be formed to test hypotheses about various parameters in the model. For example, to test the hypothesis H_0: $(\alpha\beta\gamma)_{ijk} = 0$ for all i, j, and k, we use the statistic $F = MS_{GBT}/MS_E$ since this statistic has an F-distribution if and only if H_0 is true. The rejection region is $F_{GBT} > F_{v_1, v_2, 1-\alpha}$, where $v_1 = (a-1)(b-1)(c-1)$ and $v_2 = N - abc$.

In our example, $MS_{GBT} = 2153.5$, $MS_E = 292.95$, and $F_{GBT} = 2153.5/292.95 = 7.351$. As $F_{4,162,0.95} = 2.37$, we reject H_0 as false since $7.351 > 2.37$.

The complete analysis of variance is given below.

EFFECT OF TREATMENT/SIZE/TYPE ON PEA NIACIN CONC.

General Linear Models Procedure

Dependent Variable: NIACIN

Source	DF	Sum of Squares	Mean Square	F Value	Pr > F
Model	17	115425.311	6789.724	23.18	0.0001
Error	162	47457.600	292.948		
Corrected Total	179	162882.911			

R-Square	C.V.	Root MSE	NIACIN Mean
0.708640	16.69465	17.1157	102.52222

Source	DF	Type I SS	Mean Square	F Value	Pr > F
TYPE	1	66508.8889	66508.8889	227.03	0.0001
SIZE	2	11541.9444	5770.9722	19.70	0.0001
TREAT	2	2368.4111	1184.2056	4.04	0.0194
TYPE*SIZE	2	12854.8111	6427.4056	21.94	0.0001
TYPE*TREAT	2	12177.5444	6088.7722	20.78	0.0001
SIZE*TREAT	4	1359.8556	339.9639	1.16	0.3303
TYPE*SIZE*TREAT	4	8613.8556	2153.4639	7.35	0.0001

Source	DF	Type III SS	Mean Square	F Value	Pr > F
TYPE	1	66508.8889	66508.8889	227.03	0.0001
SIZE	2	11541.9444	5770.9722	19.70	0.0001
TREAT	2	2368.4111	1184.2056	4.04	0.0194
TYPE*SIZE	2	12854.8111	6427.4056	21.94	0.0001
TYPE*TREAT	2	12177.5444	6088.7722	20.78	0.0001
SIZE*TREAT	4	1359.8556	339.9639	1.16	0.3303
TYPE*SIZE*TREAT	4	8613.8556	2153.4639	7.35	0.0001

Extensions of two-, three-, and four-way analysis of variance are continued in Chapter 11. We now turn our attention to confidence intervals associated with differences in treatment means.

8.5 CONFIDENCE INTERVALS AND TESTS OF HYPOTHESES

In the two-way model $Y_{ij} = \mu + \alpha_i + \beta_j + \varepsilon_{ij}$, it may be of interest to compare the effects due to two rows, say α_1 and α_2, if H_0: all $\alpha_i = 0$ has been rejected. The mean for the ith row is $\bar{Y}_{i\cdot} = \mu + \alpha_i + \bar{\beta}_\cdot + \bar{\varepsilon}_{i\cdot}$, where $\bar{\beta}_\cdot = \sum_j \beta_j/J = 0$ and $\bar{\varepsilon}_{i\cdot} = \sum_j \varepsilon_{ij}/J$. Thus $\bar{Y}_{1\cdot} - \bar{Y}_{2\cdot} = \alpha_1 - \alpha_2 + \bar{\varepsilon}_{1\cdot} - \bar{\varepsilon}_{2\cdot}$, and $E(\bar{Y}_{1\cdot} - \bar{Y}_{2\cdot}) = \alpha_1 - \alpha_2$ since $E(\varepsilon_{ij}) = 0$ for all i and j. The MS_E is an unbiased estimator for σ^2 and the random variable

$$T = \frac{\bar{Y}_{1\cdot} - \bar{Y}_{2\cdot} - (\alpha_1 - \alpha_2)}{S_{\bar{Y}_1 - \bar{Y}_2}} = \frac{\bar{Y}_{1\cdot} - \bar{Y}_{2\cdot} - (\alpha_1 - \alpha_2)}{\sqrt{\hat{\sigma}^2(1/J + 1/J)}}$$

$$= \frac{\bar{Y}_{1\cdot} - \bar{Y}_{2\cdot} - (\alpha_1 - \alpha_2)}{\sqrt{MS_E(2/J)}}$$

has a t-distribution with $(I - 1)(J - 1)$ degrees of freedom. The development regarding the two-sample t-test in Section 7.6 yields $(1 - \alpha)100\%$ confidence limits for $\alpha_1 - \alpha_2$:

$$\bar{Y}_{1\cdot} - \bar{Y}_{2\cdot} \pm t_{v,1-\alpha/2}\left(\frac{2MS_E}{J}\right)^{1/2}, \tag{8.39}$$

where $v = (I - 1)(J - 1)$. If these confidence limits cover the value zero, we accept H_0: $\alpha_1 = \alpha_2 = 0$; that is, there is no significant difference between the effects due to rows 1 and 2. The confidence limits (8.39) can be used on any pair of row effects α_i and α_i'. They can also be used for any pair of column effects β_j and β_j' if the difference of the sample means is replaced by $\bar{y}_{\cdot j} - \bar{y}_{\cdot j}'$ and J is replaced with I.

Consider the random variable $\sum_i c_i \bar{Y}_{i\cdot}$ where $c_1, c_2, \ldots, c_I$ are any constants. It is not difficult to show that $\sum_i c_i \bar{Y}_{i\cdot}$ has mean $\sum_i c_i(\mu + \alpha_i)$ and variance $\sum_i c_i^2 \sigma^2/J$. Now if we assume that $\sum_i c_i = 0$, the mean becomes $\sum_i c_i \alpha_i$, since $\sum_i c_i \mu = \mu \sum_i c_i = 0$. The linear combination $\sum_i c_i \alpha_i$ is called a *contrast* if $\sum_i c_i = 0$. Let $L = \sum_i c_i \bar{Y}_{i\cdot}$ be a contrast. Then L has a normal distribution with mean $\sum_i c_i \alpha_i$ and variance $\sigma_L^2 = \sum_i c_i^2 \sigma^2/J$. Thus the random variable

$$T = \frac{L - \mu_L}{\hat{\sigma}_L} = \frac{\sum_i c_i \bar{Y}_{i\cdot} - \sum_i c_i \alpha_i}{\sqrt{(MS_E/J)\sum_i c_i^2}} \tag{8.40}$$

has a t-distribution with $(I - 1)(J - 1)$ degrees of freedom. From this we obtain a $(1 - \alpha)100\%$ confidence interval for the contrast $\sum_i c_i \alpha_i$:

$$\sum_i c_i \bar{Y}_{i\cdot} - t_{v,1-\alpha/2}\left(\frac{MS_E}{J}\sum_i c_i^2\right)^{1/2} < \sum_i c_i \alpha_i < \sum_i c_i \bar{Y}_{i\cdot}$$

$$+ t_{v,1-\alpha/2}\left(\frac{MS_E}{J}\sum_i c_i^2\right)^{1/2}, \tag{8.41}$$

where $v = (I - 1))(J - 1)$. For these contrasts, the c_i and a_j should be small numbers arranged as mirror-image pairs centered about 0. If $I = 8$, the c_i could be $-4, -3, -2, -1, 1, 2, 3$, and 4. Similarly, for the contrast $\sum a_j \beta_j$,

$$\sum_j a_j \bar{Y}_{.j} - t_{v,1-\alpha/2}\left(\frac{MS_E}{I}\sum_j a_j^2\right)^{1/2} < \sum_j a_j \beta_j < \sum_j a_j \bar{Y}_{.j}$$

$$+ t_{v,1-\alpha/2}\left(\frac{MS_E}{I}\sum_j a_j^2\right)^{1/2}. \quad (8.42)$$

We have considered only simple contrasts. However, multiple contrasts arise quite naturally. In Example 8.5, for instance, since $H_0: \beta_1 = \beta_2 = \beta_3 = \beta_4 = 0$ was rejected, we conclude that there is a significant difference due to reflux ratio. In trying to determine where the difference is occurring, we may want to consider $\beta_1 - \beta_2$ or $\beta_1 - (\beta_2 + \beta_3 + \beta_4)/3$, which are contrasts. We can use the confidence intervals above to test $H_0: \beta_1 = \beta_2$ or $H_0: \beta_1 = (\beta_2 + \beta_3 + \beta_4)/3$.

Example 8.10   In Example 8.5 there was a significant difference among the four levels of reflux ratio at the 5% level of significance. Find 95% confidence intervals for the contrasts (a) $\beta_1 - \beta_2$ and (b) $\beta_1 - (\beta_2 + \beta_3 + \beta_4)/3$.

Solution   (a) For the contrast $\beta_1 - \beta_2$ we need to use the interval given in (8.42) with $a_1 = 1$, $a_2 = -1$, and $a_3 = a_4 = 0$. From the analysis of variance of Example 8.5 we have $MS_E = 0.4278$. Also, $I = 4$ and $v = (I - 1)(J - 1) = 9$. From Table IV we find that $t_{9,0.975} = 2.262$. Since $\bar{y}_{.1} = 10.975$ and $\bar{y}_{.2} = 8.875$, the 95% confidence limits are

$$\bar{y}_{.1} - \bar{y}_{.2} \pm 2.262\left[\frac{0.4278}{4}(1 + 1)\right]^{1/2},$$

$$10.975 - 8.875 \pm 2.262(0.2139)^{1/2},$$

$$2.1 \pm 1.046,$$

$$1.054, 3.146,$$

Since the confidence interval does not contain zero, we can say that $\beta_1 - \beta_2 > 0$ or $\beta_1 > \beta_2$ or that the effect from the first reflux ratio is greater than that from the second reflux ratio.

(b) For the contrast $\beta_1 - (\beta_2 + \beta_3 + \beta_4)/3$ we have $a_1 = 1$ and $a_2 = a_3 = a_4 = -\frac{1}{3}$. The 95% confidence limits are, by (8.42),

$$10.975 - \frac{8.875 + 8.05 + 5.775}{3} \pm 2.262\left[\frac{0.4278}{4}(1 + \tfrac{3}{9})\right]^{1/2},$$

$$10.475 - 7.717 \pm 2.262(0.1426)^{1/2},$$

$$3.258 \pm 0.854,$$

$$2.404, 4.112.$$

Since both confidence limits are positive we conclude that

$$\beta_1 > (\beta_2 + \beta_3 + \beta_4)/3;$$

that is, the effect from the first reflux ratio is greater than the average effect from the last three reflux ratios.

Confidence intervals can also be constructed for contrasts in the one-way model $Y_{ij} = \mu + \beta_j + \varepsilon_{ij}$. Confidence limits on

$$\sum_j c_j \beta_j = \sum_j c_j(\mu_j - \mu) = \sum_j c_j \mu_j$$

are

$$\sum_j c_j \bar{y}_{.j} \pm t_{J(I-1),1-\alpha/2} \left[\frac{SS_E}{J(I-1)} \sum c_j^2/I \right]^{1/2}. \tag{8.43}$$

Recall that MS_E from Table 8.2 has $J(I - 1)$ degrees of freedom associated with it.

Example 8.11 For the fluoride data of Example 8.1 determine a 95% confidence interval on $\beta_A - \beta_B$.

Solution The contrast $\beta_A - \beta_B = \mu_A - \mu - (\mu_B - \mu) = \mu_A - \mu_B$. We thus desire a 95% confidence interval on $\mu_A - \mu_B$, the difference of mean responses of the ferric thiocyanate and alizarin lanthanum complexan methods. From the results of Example 8.1 we have $\bar{y}_{.A} = 4.0$, $\bar{y}_{.B} = 3.6$, $J(I - 1) = 4(4) = 16$, and $MS_E = 1.45$. Also, $c_A = 1$, $c_B = -1$, and $t_{16,0.975} = 2.120$. The 95% confidence limits on $\mu_A - \mu_B$ are, from (8.43),

$$4.0 - 3.6 \pm 2.120 \left[\frac{23.2}{16} \left(\frac{1+1}{5} \right) \right]^{1/2},$$

$$0.4 \pm 2.120(0.58)^{1/2},$$

$$0.4 \pm 2.120(0.7616),$$

$$-1.2146, 2.0146.$$

Since the interval contains zero, we conclude that there is no difference between methods A and B.

The information from Tables 8.2 and 8.3 for the one- and two-way models, respectively, can be used to construct a confidence interval on σ^2 or to test hypotheses concerning σ^2. This can be done by observing that MS_E/σ^2 has a χ^2-distribution with the same degrees of freedom as MS_E. The details then follow along the lines of Sections 7.10 and 7.11 and will be omitted.

One can also test any hypothesis concerning contrasts $\sum_i c_i \alpha_i$ or $\sum a_j \beta_j$ for both the one- and two-way models. The confidence intervals presented in this section can, of course, be used to test a corresponding hypothesis. If the hypothesis is one-sided, one can find the critical regions just as was done in Chapter 7.

8.6 MULTIPLE COMPARISONS AMONG TREATMENT MEANS

In comparing treatment means (e.g., in a one-way AOV), the comparisons to be considered should be selected prior to any analysis of the data. Otherwise, the analysis of contrasts such as that discussed in Section 8.5 would not in general be valid. However, there are situations in which the experimenter wishes to gain some insight into the data that have been collected. No clues may be available regarding which contrasts should be examined at the time the experiment is planned. That is, the experiment may be of an exploratory nature and designed to gain more information than a mere statement that the treatment means are (or are not) significantly different.

The method discussed in Section 8.5 of comparing treatment means, or combinations of treatment means, by way of contrasts can be applied to individual contrasts, one at a time. However, if one is interested in studying two or more contrasts *simultaneously*, the method of Section 8.5 is not recommended, especially if the number of treatments is high. If there are I treatments, there are $I(I-1)/2$ treatment differences that might be of interest.

One method for making all possible comparisons among treatment means is due to Scheffé [11]. In a one-way AOV, the probability is $1 - \alpha$ that all possible contrasts will be contained in the intervals given by

$$\hat{L} - S\hat{\sigma}_L \le L \le \hat{L} + S\hat{\sigma}_L, \tag{8.44}$$

where

$$L = \sum_{i=1}^{I} c_i \alpha_i = \sum_{i=1}^{I} c_i \mu_i,$$

$$\hat{L} = \sum_{i=1}^{I} c_i \bar{Y}_{i\cdot},$$

$$S^2 = (I-1)F_{I-1, J(I-1), 1-\alpha},$$

and

$$\hat{\sigma}_L^2 = \frac{MS_E}{J} \sum_{i=1}^{I} c_i^2.$$

In a two-way analysis of variance, the intervals needed to construct Scheffé multiple confidence intervals are given as follows: Comparisons among $\alpha_1, \ldots, \alpha_I$ require that

$$\hat{L} = \sum_{i=1}^{I} c_i \bar{Y}_{i\cdot},$$

$$\hat{\sigma}_L^2 = \frac{MS_E}{J} \sum_{i=1}^{I} c_i^2,$$

and
$$S^2 = (I - 1)F_{I-1,(I-1)(J-1),1-\alpha}.$$

Comparisons among $\beta_1, \ldots, \beta_J$ require that

$$\hat{L} = \sum_{j=1}^{J} a_j \bar{Y}_{.j},$$

$$\hat{\sigma}_L^2 = \frac{MS_E}{I} \sum_{j=1}^{J} a_j^2,$$

and
$$S^2 = (J - 1)F_{J-1,(I-1)(J-1),1-\alpha}.$$

It should be noted that if the AOV F-test rejects the hypothesis of equal means, at least one of the contrasts will be judged different from zero by the Scheffé method. Thus this method allows more conclusions than merely stating that the treatments are not all the same.

Example 8.12 For the fluoride data of Example 8.1, determine 95% multiple confidence intervals on $\alpha_A - \alpha_B$, $\alpha_A - \alpha_C$, and $\alpha_A - \alpha_D$. The contrasts are:

$$L_1 = \alpha_A - \alpha_B = \mu_A - \mu_B,$$
$$L_2 = \alpha_A - \alpha_C = \mu_A - \mu_C,$$
$$L_3 = \alpha_A - \alpha_D = \mu_A - \mu_D.$$

Solution From Example 8.1 we have

$$\hat{L}_1 = \bar{Y}_{.A} - \bar{Y}_{.B} = 4.0 - 3.6 = 0.4,$$
$$\hat{L}_2 = \bar{Y}_{.A} - \bar{Y}_{.C} = 4.0 - 2.8 = 1.2,$$
$$\hat{L}_3 = \bar{Y}_{.A} - \bar{Y}_{.D} = 4.0 - 1.4 = 2.6.$$

Also, S^2 and $\hat{\sigma}_L^2$ for each contrast are

$$S^2 = 3F_{3,16,0.95} = 3(3.24) = 9.72,$$

$$\hat{\sigma}_L^2 = \frac{MS_E}{5} \sum_{i=1}^{5} c_i^2 = \frac{1.45}{5}(2) = 0.58,$$

and
$$S\hat{\sigma}_L = [9.72(0.58)]^{1/2} = 2.3744.$$

Thus the multiple confidence intervals are

$$0.4 - 2.3744 \leq L_1 \leq 0.4 + 2.3744$$
$$1.2 - 2.3744 \leq L_2 \leq 1.2 + 2.3744$$
$$2.6 - 2.3744 \leq L_3 \leq 2.6 + 2.3744$$

or

$$-1.97 \leq \alpha_A - \alpha_B \leq 2.77$$
$$-1.17 \leq \alpha_A - \alpha_C \leq 3.57$$
$$0.23 \leq \alpha_A - \alpha_D \leq 4.97.$$

Based on these confidence intervals, there is probably a difference between methods A and D but not between A and B or between A and C, as those last two confidence intervals contain zero.

In Example 8.5, the 95% confidence interval on $\alpha_A - \alpha_B$ is $(-1.2146, 2.0146)$, based on use of the t-distribution and Eq. (8.43). This interval is shorter than the interval $(-1.97, 2.77)$ obtained in Example 8.11 by using the Scheffé method. The result from Eq. (8.43) can be applied to only one contrast using a 95% confidence level. The Scheffé method can be applied to all possible contrasts, with an overall confidence level of 95%. The price we pay in obtaining multiple confidence intervals is that of having longer intervals. This is to be expected from the nature of the two methods.

There are several other multiple comparison methods that are used. For a discussion and comparison of such methods the reader is referred to Bancroft [3].

Some existing computer statistical packages which provide procedures to perform analyses of variance also have the feature of computing multiple confidence intervals as an option. SAS is one of these packages (*SAS User's Guide: Statistics*, Version 5 edition, SAS Institute, Cary, NC, 1985, 956 pp.). We now consider the analysis of the data in Example 8.1 by means of PROC ANOVA or PROC GLM. This procedure has several multiple comparisons methods as options. For references and documentation on the various multiple comparisons methods used, the reader is referred to the SAS manual referenced above.

Example 8.13   The data of Example 8.1 were processed using the ANOVA procedure in the SAS system to obtain another multiple comparison between the treatment means by the methods of Duncan's multiple range test and Scheffé's method. The program is shown below.

(put initial JCL cards/statements here)

```
DATA FLUORIDE;
INPUT METHOD $ 1 PPM 3;
CARDS;
```

(data here according to INPUT format)

```
PROC ANOVA;
CLASS METHOD;
MODEL PPM = METHOD;
TITLE 'ONE-WAY ANALYSIS OF VARIANCE';
MEANS METHOD/DUNCAN;

MEANS METHOD/SCHEFFE;
```

(put final JCL cards/statements here)

The output contains the AOV table and the results of the multiple comparisons requested, namely, those using the Duncan method and the Scheffé method. The output is summarized below.

SAS

ONE-WAY ANALYSIS OF VARIANCE

ANALYSIS OF VARIANCE PROCEDURE

DEPENDENT VARIABLE: PPM

SOURCE	DF	SUM OF SQUARES	MEAN SQUARE	F VALUE
MODEL	3	19.75000000	6.58333333	4.54
ERROR	16	23.20000000	1.45000000	PR > F
CORRECTED TOTAL	19	42.95000000		0.0174

SOURCE	DF	ANOVA SS	F VALUE	PR > F
METHOD	3	19.75000000	4.54	0.0174

The SAS output was condensed to yield the table below. It gives the corrected total rather than the uncorrected total. The number 0.0174 under PR > F is $\mathscr{P}(F \geq 4.54)$, where F has an F-distribution with 3 and 16 d.f. That is, 0.0174 is the probability of observing an F ratio larger than or equal to 4.54, given that $H_0: \beta_1 = \beta_2 = \beta_3 = \beta_4 = 0$ is true. The SAS output can be compared with that in Example 8.1.

The multiple comparisons results are shown below. Using Duncan's multiple range test, the results for the four means are as follows:

SAS

ONE-WAY ANALYSIS OF VARIANCE

ANALYSIS OF VARIANCE PROCEDURE

DUNCAN'S MULTIPLE RANGE TEST FOR VARIABLE: PPM
NOTE: THIS TEST CONTROLS THE TYPE I COMPARISONWISE ERROR RATE,
      NOT THE EXPERIMENTWISE ERROR RATE.

ALPHA=0.05   DF=16   MSE=1.45

MEANS WITH THE SAME LETTER ARE NOT SIGNIFICANTLY DIFFERENT.

DUNCAN	GROUPING		MEAN	N	METHOD
		A	4.0000	5	A
		A			
		A	3.6000	5	B
		A			
B		A	2.8000	5	C
B					
B			1.4000	5	D

These results indicate that the means $\bar{y}_{.A} = 4.0$, $\bar{y}_{.B} = 3.6$, and $\bar{y}_{.C} = 2.8$ are not significantly different. Also, the means $\bar{y}_{.C} = 2.8$ and $\bar{y}_{.D} = 1.4$ are not significantly different. There are two "homogeneous" groups of means, (A,B,C) and (C,D). The letters A and B in the SAS output are not to be confused with the letters A and B in Example 8.1.

The results of the Scheffé option are as follows:

SAS

ONE-WAY ANALYSIS OF VARIANCE

ANALYSIS OF VARIANCE PROCEDURE

SCHEFFE'S TEST FOR VARIABLE: PPM
NOTE: THIS TEST CONTROLS THE TYPE I EXPERIMENTWISE ERROR RATE
      BUT GENERALLY HAS A HIGHER TYPE II ERROR RATE THAN REGWF
      FOR ALL PAIRWISE COMPARISONS.

ALPHA=0.05   DF=16   MSE=1.45
CRITICAL VALUE OF T=1.79969
MINIMUM SIGNIFICANT DIFFERENCE=2.37395

MEANS WITH THE SAME LETTER ARE NOT SIGNIFICANTLY DIFFERENT.

SCHEFFE	GROUPING		MEAN	N	METHOD
		A	4.0000	5	A
		A			
B		A	3.6000	5	B
B		A			
B		A	2.8000	5	C
B					
B			1.4000	5	D

There are two groups deemed homogeneous by the Scheffé method: (A,B,C) and (B,C,D). The minimum significant difference in the previous example is $\hat{s}\sigma_{\hat{L}} = [9.75(0.58)]^{1/2} = 2.378$, compared with 2.37395 from the AOV results. From the previous example, we can see that $\bar{y}_{.B} - \bar{y}_{.D} = 2.2$ and $\bar{y}_{.C} - \bar{y}_{.D} = 1.4$ even though the Scheffé method indicates that neither B and D nor C and D are different. The Duncan method does not place B in the same group with C and D. One can infer from the Duncan method that D is different from A and B, whereas from the Scheffé method one can only say that D is different from A.

It has been established that the Scheffé method is quite conservative in the number of comparisons declared significant. It is more conservative than the Duncan method, as can be seen from this example. For further details on multiple comparisons, see Bancroft [3] or the references on the ANOVA and GLM procedures in the SAS publication referred to previously in this section.

8.7 NONPARAMETRIC METHODS IN ANALYSIS OF VARIANCE

As seen in Chapter 7, there were alternative nonparametric methods in testing the equality of means of two normal distributions in the case of two independent samples and in the case of paired samples. In this chapter the normal theory two-sample techniques have been extended to the one-way analysis of variance and the two-way analysis of variance. Both of these analysis-of-variance techniques assume underlying normal distributions. There are well-known nonparametric alternatives for both of these procedures when the normal theory test assumptions are not valid. The nonparametric test for the one-way analysis of variance is the Kruskal–Wallis test. The Friedman test is the nonparametric alternative for the randomized complete block model (two-way analysis of variance with no interaction where one factor is a fixed effect and the other factor is a block or random effect, not of primary importance). The Kruskal–Wallis test is discussed here and the Friedman test is deferred until Chapter 11.

8.7.1   Kruskal–Wallis Test

We assume that J independent random samples of sizes $n_1, n_2, \ldots, n_J$ are drawn from continuous distributions. The Kruskal–Wallis procedure is used to test the hypothesis that the J samples came from J identical distributions. The test, however, is sensitive to location differences, the null hypothesis being stated in terms of the equality of J population means (the same as the one-way analysis-of-variance hypothesis in Section 8.3.3). The null hypothesis is thus

$$H_0: \mu_1 = \mu_2 = \cdots = \mu_J,$$

and the alternative hypothesis is

$$H_A: \text{at least two } \mu_j \text{ are not equal.}$$

To define the Kruskal–Wallis test statistic for testing H_0, the $N = n_1 + n_2 + \cdots + n_J$ observations are combined and ranked from smallest to largest, while retaining group identity (as was done for the Wilcoxon test statistic with $J = 2$). Any ties among observations are assigned the average rank for the set of tied observations. Let $R_j, j = 1, 2, \ldots, J$, denote the sum of the ranks corresponding to the observations in the jth sample. The Kruskal–Wallis test statistic is defined by

$$H = \frac{12}{N(N+1)} \sum_{j=1}^{J} n_j \left(\bar{R}_j - \frac{N+1}{2} \right)^2, \tag{8.45}$$

where $\bar{R}_j = R_j / n_j$ denotes the mean of the ranks assigned to the observations in the jth sample. If H_0 is true, it can be shown that

$$E[\bar{R}_j] = \frac{N+1}{2}.$$

If H_0 is true, one can expect the value of H to be small, since under H_0, all the R_j would be expected to be close to $(N+1)/2$, resulting in a small value of H. Consequently, H_0 is rejected if the value of H is large. The exact distribution of H has been tabulated for small values of J and n_j. However, if all the $n_j \geq 5$, H has been shown to have approximately a chi-squared distribution with $J - 1$ degrees of freedom. An equivalent form of the statistic H which is computationally convenient is

$$H = \frac{12}{N(N+1)} \sum_{j=1}^{J} \frac{R_j^2}{n_j} - 3(N+1). \tag{8.46}$$

Example 8.14 Consider the data of Example 8.2, which were taken from Problem 8.2. In this case, $J = 3$ and $n_1 = n_2 = n_3 = 6$, giving a value of $N = n_1 + n_2 + n_3 = 18$. The observations with the corresponding ranks (in parentheses) are given below.

Feed A	Feed B	Feed C
4.2 (4)	4.1 (3)	4.0 (1.5)
6.8 (10)	4.8 (6)	4.4 (5)
5.9 (7)	6.2 (9)	8.1 (13)
11.0 (15)	8.0 (11.5)	8.3 (14)
16.0 (18)	14.0 (17)	11.9 (16)
8.0 (11.5)	4.0 (1.5)	6.1 (8)

The rank sums are

$$R_1 = 4 + 10 + 7 + 15 + 18 + 11.5 = 65.5,$$
$$R_2 = 3 + 6 + 9 + 11.5 + 17 + 1.5 = 48,$$
$$R_3 = 1.5 + 5 + 13 + 14 + 16 + 8 + 57.5.$$

The corresponding value of the Kruskal–Wallis statistic is

$$H = \frac{12}{N(N+1)} \sum_{j=1}^{3} \frac{R_j^2}{n_j} - 3(N+1)$$
$$= \frac{12}{18(19)} \left[\frac{(65.5)^2}{6} + \frac{(48)^2}{6} + \frac{(57.5)^2}{6} \right] - 3(19)$$
$$= 0.898.$$

From the chi-square distribution table, the value $\chi^2_{2, 0.995}$ is 10.5966. Thus there is strong evidence that H_0: $\mu_1 = \mu_2 = \mu_3$ should not be rejected since the value of H is 0.898.

The Kruskal–Wallis test is very robust relative to the usual normal theory test. For further discussion and other topics, including multiple comparison, related to the Kruskal–Wallis test, the reader is referred to Conover [4].

PROBLEMS

8.1 The following data give the yields of a product that resulted from trying catalysts I to IV from four different suppliers in a process.

I	II	III	IV
36	35	35	34
33	37	39	31
35	36	37	35
34	35	38	32
32	37	39	34
34	36	38	33

(a)   Are yields influenced by catalysts?
(b)   What are your recommendations in the selection of a catalyst? Assume that economics dictate 95% probability of being right on a decision. (*Hint*: Reexamine the catalyst group means to find the corresponding confidence intervals.)

8.2   Pilot-plant runs were made to determine the effect of percent conversion (defined as volume converted per 100 volumes fed) on coke yields in catalytic cracking of cracked cycle gas oil, hydrogenated to varying degrees. It was also desired to determine whether hydrogenation has any effect on the coke-yield relationship and if so, to estimate its effect. Three samples of gas oil were subjected to cracking runs, each made at a different level of severity to vary conversion. Coke yields were determined by measuring the CO_2 in fuel gas from burning off coke from the cracking catalyst following each run. Feed A is untreated; feed B is A hydrogenated to a hydrogen consumption of 300 standard cubic feet per barrel of feed; feed C is A hydrogenated to 500 standard cubic feet per barrel. The following data were obtained.

Feed A		Feed B		Feed C	
Conv. (vol %)	Coke yield (wt %)	Conv. (vol %)	Coke yield (wt %)	Conv. (vol %)	Coke yield (wt %)
20.1	4.2	20.3	4.1	25.8	4.0
32.2	6.8	27.6	4.8	31.6	4.4
24.7	5.9	35.4	6.2	38.8	8.1
41.2	11.0	38.0	8.0	45.0	8.3
47.0	16.0	49.0	14.0	50.2	11.9
32.7	8.0	24.2	4.0	36.1	6.1

Correlate the data to yield as much information as possible.

8.3 Random samples of size 10 were drawn from normal populations $A, B, C,$ and D. The measurements of property x are given in the following table along with totals and sums of squares in columns.

	y_A	y_B	y_C	y_D	
	3.355	0.273	3.539	3.074	
	1.086	2.155	2.929	3.103	
	2.367	1.725	3.025	2.389	
	0.248	0.949	4.097	4.766	
	1.694	0.458	2.236	2.553	
	1.546	1.455	3.256	3.821	
	1.266	2.289	3.374	1.905	
	0.713	2.673	1.781	2.350	
	0.000	1.800	2.566	1.161	
	3.406	2.407	2.510	2.122	
$\sum y_{i \cdot}$	15.681	16.184	29.313	27.244	$y_{\cdot\cdot} = 88.422$
$\bar{y}$	1.5681	1.6184	2.9313	2.7244	
$\sum\limits_{i} y_{ij}^2$	37.071327	32.339668	90.081121	83.620342	$\sum\limits_{i}\sum\limits_{j} y_{ij}^2 =$
					243.112458

(a) Are there significant differences between population means? If so, which are different?

(b) An independent estimate of a population variance σ^2 is 1.0 (property units)^2. Does the variation in these data differ from what would be expected on the basis of the independent estimate?

(c) Compare each group with every other group by appropriate t-tests. Discuss the advantages of using the F-test for group comparisons as a composite t-test.

8.4 Six vertical elutriators were calibrated after 100-h service intervals. The flow rates in l/min are given below.

Time (h)	Flow rate (L/min) for vertical elutriator:					
	VE_1	VE_2	VE_3	VE_4	VE_5	VE_6
0	7.51	7.71	7.45	7.44	7.35	7.59
100	7.58	7.72	7.48	7.48	7.38	7.63
200	7.57	7.71	7.43	7.44	7.34	7.60
300	7.60	7.70	7.44	7.48	7.32	7.61
400	7.60	7.71	7.50	7.53	7.37	7.65
500	7.67	7.67	7.55	7.58	7.40	7.64

Is there a difference in flow rates? If so, does the change depend on cumulative time in service?

8.5 You work for one of two companies that manufactures thermocouples for caustic service. The rival company advertises that its new thermocouple lasts longer than any other on the market. From a great many tests you know that the estimated mean length of life for your company's unit is 400 days. The rival's advertising compaign is damaging to your company, but if the claims are publicly disputed and the rival's thermocouples are shown to last as long or longer than yours, the resulting publicity will be even more damaging to your company than the rival's advertising campaign.

(a) Explain how you would test the rival's claim and give the null and alternative hypotheses for your experiment.
(b) In view of the above, explain if you would dispute the claim or not and explain why in each of the following situations:
   (1) The mean of your sample of the rival's thermocouples is 4250 h and the estimated variance of the estimated mean is 6250.
   (2) The mean of the sample is 4250 h and the estimated variance of the estimated mean is 25,000.
   (3) The sample mean is 3500 h and the estimated variance of the estimated mean is 640.

8.6 The additives shown below were introduced into a trioxane system [8]* in hopes of strengthening its mechanical properties without altering its physical properties, as estimated by the following intrinsic viscosities:

---

* Data reproduced with permission from *Hydrocarbon Processing*.

None	Formic acid	H_2O	Methanol	HCHO	Acetic anhydride
2.1	2.3	2.2	2.2	1.9	1.9
1.9	1.6	1.9	2.6	1.9	2.1
2.1	1.5	1.6	2.0	1.9	2.4

Do any of the additives affect intrinsic viscosity significantly? Use the $\alpha = 0.01$ significance level.

8.7 Samples of steel from four different batches were analyzed for carbon content. The results are shown below for quadruplicate determinations by the same analyst.

1	2	3	4
0.39	0.36	0.32	0.43
0.41	0.35	0.36	0.39
0.36	0.35	0.42	0.38
0.38	0.37	0.40	0.41

Are the carbon contents (in wt %) of these batches the same? What are the 99% confidence limits on the average carbon content of each?

The following additional information is available.

$y_{.1} = 1.54,$ $\quad \sum_i y_{i1}^2 = 0.5942,$ $\quad \bar{y}_{..} = 0.3850,$ $\quad \bar{y}_{.1} = 0.3850,$

$y_{.2} = 1.43,$ $\quad \sum_i y_{i2}^2 = 0.5115,$ $\quad \sum_i \sum_j y_{ij}^2 = 2.3236,$ $\quad \bar{y}_{.2} = 0.3575,$

$y_{.3} = 1.50,$ $\quad \sum_i y_{i3}^2 = 0.5684,$ $\quad y_{..}^2 = 36.9664,$ $\quad \bar{y}_{.3} = 0.3750,$

$y_{.4} = 1.61,$ $\quad \sum_i y_{i4}^2 = 0.6495,$ $\quad \sum_j y_{.j}^2 = 9.2586,$ $\quad \bar{y}_{.4} = 0.4025.$

8.8 Auerbach et al. [2]* have performed an exhaustive check of wavelength fidelity in the ultraviolet- and visible-wavelength portion of spectrophotometers. A holmium glass filter was used for this purpose. The wavelengths corresponding to the maximum absorption peaks for this filter

* Data reprinted by permission; copyright American Society for Quality Control, Inc.

were recorded at 241, 361, 446, 454, and 460.5 nm by two different operators on each of 27 instruments tested. This study was performed to see if spectral findings in different laboratories on different instruments could be used interchangeably. The results are given below for 361 and 454 nm.

$\lambda = 361$		$\lambda = 454$		$\lambda = 361$		$\lambda = 454$	
1	2	1	2	1	2	1	2
361	361	453	453	361.3	361	453.5	452.8
361	361	453	453	—	—	454	454
361	361	454	453.5	—	—	454	454
361	361	454	453.5	362.5	362	455	454
361	360	454	454	362	361	455	454
360.8	360	454.2	455	361.5	361.5	455.5	455.5
360.7	361	453.5	454	361.5	361.5	456	455
360.7	361	453.5	454	361	361.5	455.5	455
367	367	458	458	361	361.5	455	455
367	367	458	458	360	361	454	454
360.8	361	453	453	360	361	454	454
360.8	360.5	453	453	361.6	362	455.8	455.5
361	361	454	454	361.6	361.8	456	456
361	360.8	454	454	361	360.5	454	454
361	361	453	453	360.5	360.5	454	454
361	361	453	453	360	360	453	454
360.5	—	453	—	360	360	452.5	454
360.4	—	452.2	—	363	—	460	—
362	362	456	456	360.5	—	453.5	—
362	362	456	456	360.5	—	453.5	—
360	360	—	—	361	361	454	—
360	360	—	—	361	361	454	—
359.5	359.7	451.5	454.2	361	361	454	454
359.5	359.1	451	450.8	361	361	454	454
361	361.1	—	455	360.3	—	452.8	—
361	351.1	—	454.7	360.4	—	453.0	—

With what confidence can you state that the absorption maxima at these two wavelengths will deviate no more than ± 0.3 nm from the true value?

8.9 Harper and El Sahrigi [6] give the following data* for the

---

* Data reprinted by permission of the copyright owner; copyright 1964 by the American Chemical Society.

measured thermal conductivity of a quick-frozen pear containing the following gases in the pores.

$$k_e = \frac{\text{Btu}}{\text{hr ft}^2 \, ^\circ\text{F/ft}} \times 10^3$$

F-12	CO$_2$	N$_2$	Ne	He
12.6	12.7	12.6	12.6	12.6
12.6	12.7	12.6	12.9	12.6
12.8	13.0	12.6	13.7	12.6
13.5	13.4	12.6	14.2	12.8

Do the various gases make a significant difference in the thermal conductivity of the pear at the 95% confidence level?

8.10 A solution of HPC (hot potassium carbonate) is used to scrub CO$_2$ from gas feedstock in the production of nylon intermediate materials. Pilot-plant data for the effect of HPC flow rate on CO$_2$ leakage are given in the following table.

Leakage (weight percent) for HPC flow rate (gal/hr):			
0.5	0.7	0.9	1.1
2.00	1.90	1.65	1.31
2.15	1.87	1.54	1.29
2.21	1.75	1.49	1.18
2.12	1.82	1.60	1.20
2.35	1.78	1.52	1.09

Prepare an analysis of variance for these results. Does flow rate affect CO$_2$ leakage?

8.11 Four groups in the physical chemistry lab are told to find the molecular weight of dichloroethane by the method of Dumas using the same equipment over a period of 4 weeks. Test the following data to see if there is a significant difference in the lab techniques of groups 1 to 4.

1	2	3	4
120.0	105.8	130.8	116.2
127.0	116.7	123.6	108.8
118.0	104.8	104.2	111.3
121.6	109.7	100.3	115.3
110.8	110.4	102.8	107.4

8.12 Suppose that the data given in Problem 8.7 had been obtained by four different technicians, each working with one sample from each batch of steel. Are there significant differences between analysts, or is the major source of variation in wt % C due to the differences between steel batches? The following calculations have been made:

$$y_{1.} = 1.50, \quad \sum_j y_{1j}^2 = 0.5690, \quad \sum_i y_i^2 = 9.2348.$$

$$y_{2.} = 1.51, \quad \sum_j y_{2j}^2 = 0.5723,$$

$$y_{3.} = 1.51, \quad \sum_j y_{3j}^2 = 0.5729,$$

$$y_{4.} = 1.56, \quad \sum_j y_{4j}^2 = 0.6094,$$

8.13 The 6 ft × 3 in. packed gas absorber in the unit operations laboratory was used by four different groups, each of which found the concentration (in ppm) of ammonia in outlet air at three different inlet NH_3 concentrations. A constant value of moles of liquid per mole of total gas was used. The following data were obtained:

Group	Concentration (%) at inlet:		
	1	3	5
1	49.8	92.1	138.4
2	52.6	89.3	135.9
3	47.9	93.4	136.2
4	48.7	90.6	139.9

Do either of the variables affect the outlet ammonia concentration?

8.14 In an experiment to determine the effects of varying inlet water flow rate on the HTU (in.) for the cross-flow, pilot-scale cooling tower, different lab groups used the same water flow rates to yield the following results:

Group	flow				

	Flow rate (gal/min) for test:				
Group	1	2	3	4	5
1	12.6	6.2	4.8	3.5	2.9
2	13.1	7.0	5.1	3.8	3.0
3	12.7	6.7	4.8	3.6	2.7

Is the difference in HTU due to group differences or flow rates?

8.15 The torque output for several pneumatic actuators at several different supply pressures are given in the following table.

	Torque (in. lb) at pressure:		
Model	60 psi	80 psi	100 psi
A	205	270	340
B	515	700	880
C	1775	2450	3100
D	7200	9600	12100

(a) Do the different supply pressures significantly affect the output?
(b) Does the output vary for the different models?

8.16 In the colorimetric analysis of chloride ion in aqueous solution, the following data were obtained using the modified Iwasaki method (wavelength = 465 nm):

| | % Transmittance for: | |
Sample	Trial 1	Trial 2
1	100	99.8
2	98.5	99.0
3	94.0	98.0
4	84.0	97.0
5	88.0	96.0
6	90.0	95.0
7	91.0	92.2
8	89.0	92.0
9	85.0	88.5
10	84.0	70.5

By the mercuric chloranilate method (wavelength = 485 nm):

| | % Transmittance for: | |
Sample	Trial 1	Trial 2
1	96.4	99.8
2	99.0	97.2
3	94.5	92.5
4	91.3	96.8
5	99.2	97.0
6	92.8	92.8
7	94.5	93.2
8	90.2	92.8
9	92.4	91.2
10	95.5	94.3

(a) Show statistically that the procedures yield different results.

(b) When downgraded for the inconsistency of the results in the pairs of trials, the students who obtained these data said that the results of these trials were obtained with reagents prepared from two different water sources: trial 1 using laboratory distilled water and trial 2 using commercial distilled water. They contended that the difference in water sources is the major source of error. Perform suitable calculations to make a definite comment on their contention.

8.17 In the unit operations lab, three different people read the outlet water temperature for four different inlet hot water flow rates. The data are shown below.

Observer	Flow rate (gal/min) for inlet:			
	1	2	3	4
1	67	72	77	80
2	71	75	80	81
3	69	71	76	80

Was there a difference in temperature due to flow rate, or are the results due to the observer, or both?

8.18 It is proposed that the addition of an amine to the carbonate solution in Problem 8.10 will reduce CO_2 leakage in the purified or scrubbed gas stream. Pilot-plant data (HPC flow rate, gal/hr) are given in the following table.

Amine conc. (wt %)	Leakage (weight percent) for HPC flow rate (gal/hr):			
	0.5	0.7	0.9	1.1
0.5	1.64	1.32	0.76	0.62
1.0	1.31	1.09	0.47	0.38
3.0	0.99	0.81	0.15	0.08
6.0	0.61	0.45	0.04	0.01

Does the additive have the desired effect?

8.19 The following data show the relationship between the temperature of molasses to its velocity through several pipe diameters. The velocity is due to gravity at sea level.

Smooth pipe	Velocity (ft/min) at:		
diameter (in.)	70°F	100°F	212°F
1	2.0	15	20
3	4.4	17	21
5	5.1	21	21
6	5.2	21	24

(a)   Are the flows influenced by the temperature?
(b)   Are the flows influenced by the pipe diameter?

8.20   The following data show the total conversion for the catalytic cracking of acetic acid to acetic anhydride at 750°C with different concentrations of triethyl phosphate (TEP) as the catalyst. The feed rate was also changed for each run.

Feed rate	% Conversion at:		
(gal/hr)	0.5% TEP	0.3% TEP	0.1% TEP
200.0	77.85	76.03	73.87
147.0	89.12	88.94	87.65
98.5	99.09	97.14	91.78
61.0	99.55	99.51	97.60

(a)   Are yields influenced by the amount of catalyst?
(b)   Are yields influenced by feed rate?

8.21   In an air–water contact utilizing a packed tower, the following [7]* gas film heat-transfer coefficient data were recorded in Btu/hr ft^2 °F, where L is the liquid rate, lb/hr ft^2, G the gas rate, lb/hr ft^2 and H the heat-transfer coefficient, Btu/hr ft^2 °F.

---

* Data reprinted by permission of the Copyright owner; copyright 1952 by the American Institute of Chemical Engineers.

	Liquid rate, L			
Gas rate, G	190	250	300	400
200	200	226	240	261
400	278	312	330	381
700	369	416	462	517
1100	500	575	645	733

(a) Perform the indicated two-way AOV for these data. Use the 0.01 significance level.

(b) Which variable has the greater effect on the heat transfer coefficient?

8.22 Dana et al. [5] have presented data* on the lightness (L), greenness (a), and yellowness (b) of a semimatte opaque glaze for use on ceramic tile as a function of the amounts of blue and yellow colorants added in formulation. Their data are given below.

Percent yellow colorant	Properties for percent blue colorant:							
		1.4		2.0		2.6		3.2
1.1	L	78.8	L	77.6	L	76.5	L	75.3
	a	11.5	a	12.4	a	13.4	a	14.3
	b	5.9	b	10.0	b	2.5	b	0.8
1.5	L	78.6	L	77.3	L	75.9	L	74.7
	a	12.0	a	12.9	a	13.9	a	14.8
	b	8.8	b	4.2	b	5.3	b	3.5
1.9	L	78.3	L	76.8	L	75.4	L	74.1
	a	12.5	a	13.4	a	14.3	a	15.2
	b	11.8	b	7.1	b	8.1	b	6.3

* Data reprinted by permission; copyright American Society for Quality Control, Inc.

For the data presented above, determine whether the blue or yellow colorant has the more pronounced effect on the lightness of the glaze. Present the complete AOV table as part of your answer.

8.23 Repeat Problem 8.22 for the effect of both colorants on greenness (*a*). Give the appropriate AOV table. State and test the appropriate null hypothesis. Show the model involved and fully define all its terms.

8.24 Repeat Problem 8.22 for the effect of both colorants on yellowness (*b*). Give the model involved and state the null hypothesis. Test it, interpret the results at the 5% significance level, and present a complete AOV table.

8.25 Lemus [9] presented the force required to cause an electrical connection to separate as a function of angle of pull. The data are given below.* Each entry is in pounds-force.

Angle (deg)	Force required (lb-force) for connector:				
	1	2	3	4	5
0	45.3	42.2	39.6	36.8	45.8
2	44.1	44.1	38.4	38.0	47.2
4	42.7	42.7	42.6	42.2	48.9
6	43.5	45.8	47.9	37.9	56.4

What conclusions are possible from these data? Support your answer by stating and testing the appropriate hypothesis. Show a complete analysis of variance.

8.26 Mouradian [10] has presented the results of a study to determine the pull strength of adhesive systems on primed and unprimed surfaces. The data for test specimens of two thicknesses are given below.† For adhesive systems with primer:

---

* Data reprinted by permission; copyright American Society for Quality Control, Inc.
† Data reprinted by permission; copyright American Society for Quality Control, Inc.

I	II	III	IV
60.0*	57.0*	19.8*	52.0*
73.0	52.0	32.0	77.0
63.0*	52.0*	19.5*	53.0*
79.0	56.0	33.0	78.0
57.0*	55.0*	19.7*	44.0*
70.0	57.0	32.0	70.0
53.0*	59.0*	21.6*	48.0*
69.0	58.0	34.0	74.0
56.0*	56.0*	21.1*	48.0*
78.0	52.0	31.0	74.0
57.0*	54.0*	19.3*	53.0*
74.0	53.0	27.3	81.0

For adhesive systems without primer:

I	II	III	IV
59.0*	51.0*	29.4*	49.0*
78.0	52.0	37.8	77.0
48.0*	44.0*	32.2*	59.0*
72.0	42.0	36.7	76.0
51.0*	42.0*	37.1*	55.0*
72.0	51.0	35.4	79.0
49.0*	54.0*	31.5*	54.0*
75.0	47.0	40.2	78.0
45.0*	47.0*	31.3*	49.0*
71.0	57.0	40.7	79.0
48.0*	56.0*	33.0*	58.0*
72.0	45.0	42.6	79.0

Perform the analysis of variance to determine whether the major source of variation is due to surface pretreatment or adhesive type. Use only the data marked with an asterisk.

8.27 Repeat Problem 8.26 using the unmarked data.

8.28 A psychology class performed an experiment to see if there is any correlation between amount of sleep and test-taking ability. Four students participated in the experiment. Each student varied his normal sleeping time from 4 h to 7 h and took a simple test the next day. The test occurred once a week for 4 weeks. Corresponding grades (out of 100%) are shown below. Does sleep time or student affect the grades?

Student	Grade following sleep time of (h):			
	4	5	6	7
1	78	79	84	89
2	84	82	84	91
3	80	84	86	83
4	71	73	73	77

8.29 A sophomore quantitative analysis student standardized three sulfide solutions by each of three methods: A (sodium nitroprusside), B (lead acetate), and c, the reference (modified calcium hydroxide). The data are given below.

Method	Result for solution:		
	1	2	3
A	5.21	5.18	5.22
B	5.30	5.23	5.12
C	5.18	5.38	5.09

Were the solutions different? Did the analysis procedure affect the results? Support your answer by a suitable AOV table and interpret the results.

8.30 Relative effectiveness has been measured for a catalyst at five different space velocities (V) for two different surface areas (A). The data follow.

A (m²/g)	V (s⁻¹) 1.5	2.4	3.2	4.6	7.3	$y_{i\cdot}$	$\sum_i y_{ij}^2$
1	3.84	3.06	2.42	2.06	1.63	13.01	36.8661
3	6.90	5.46	4.41	3.80	2.89	23.46	119.6618
$y_{\cdot j}$	10.74	8.52	6.83	5.86	4.52		
$\sum_i y_{ij}^2$	62.3556	39.1752	25.3045	18.6836	11.009		

Is catalyst effectiveness influenced by either space velocity or surface area? "Prove" your answer by stating and testing the appropriate null hypotheses at the $\alpha = 0.5$ level. Note that although the independent variables were arranged in log spacing to facilitate regression analysis (if warranted), that has nothing to do with the solution to this problem.

8.31 Four vertical elutriators were used to obtain samples of the concentration of cotton dust in the open-ended spinning room of a large textile mill. The results (in $\mu g/m^3$) are shown below. Each elutriator is located in a different part of the room; all have been calibrated to the OSHA standard. Is there any difference ($\alpha = 0.05$) between areas with regard to dust concentration?

VE₁	VE₂	VE₃	VE₄
182.6	174.3	182.0	181.7
173.4	178.5	182.1	183.4
190.1	180.0	184.6	180.6
178.6		180.9	
188.2			

8.32 The results of a laboratory determination for the percent purity of feed, distillate, and bottoms, with feed entering on tray 3 in a six-sieve-tray Scott fractional distillation column, are given below. Three samples were taken and analyzed in triplicate determinations for the feed, distillate, and bottoms.

Feed	Distillate	Bottoms
6.25	64.8	2.06
6.24	64.7	2.03
6.24	64.7	1.98
6.51	64.5	2.55
6.48	64.5	2.62
6.60	64.3	2.49
6.21	64.8	2.91
6.13	64.9	2.97
6.17	64.8	2.94

Does subsampling make a significant difference in the results?

8.33 The Visbreaker unit was fed three kinds of heavy oil: 30% hydrotreated, 50% hydrotreated, and 70% hydrotreated. Samples of the flash tower bottoms were taken each day for 4 days. Distillation at 10 mmHg was used to analyze the initial boiling point ($°F$) of the bottoms product. The distillation was carried out four times for each sample.

Sample	Result for oil treatment:		
	30%	50%	70%
1	425	457	510
	431	462	507
	436	460	500
	433	455	505
2	431	460	500
	423	456	510
	427	463	495
	429	465	498
3	428	482	505
	437	476	511
	436	480	506
	431	475	513
4	433	470	513
	435	476	505
	425	467	507
	430	465	510

Prepare the analysis of variance for these data and interpret the results.

8.34 In the gas absorption experiment in our laboratory, 1-L samples of the effluent (cleaned) gas are taken for ammonia analysis at equilibrium for each set of operating conditions. Four 40-mL aliquots are taken from each 1-L sample bag. Each aliquot is contacted with 10 mL of double-distilled, CO_2-free water. The water is analyzed for ammonia by specific ion electrode. For the past semester, the following results, in ppm ammonia, were obtained for an absorber water flow rate of 8 gal/hr and a total gas flow rate of 4 ft^3/min (70°F, 14.7 psia).

	% Ammonia entering gas phase						
	4	8	12	16	20	$y_{i\cdot\cdot}$	$y_{i\cdot\cdot}^2$
	12	21	37	48	60		
	13	24	37	42	62	725	525,625
	18	26	31	50	57		
	17	20	33	46	71		
	14	25	36	43	66		
	15	27	34	49	64	728	529,984
	13	22	31	51	59		
	10	24	39	41	65		
	11	19	36	49	63		
	14	23	37	52	61	734	538,756
	12	21	34	48	58		
	19	28	32	47	70		
$y_{\cdot j\cdot}$	168	280	417	566	756		
$y_{\cdot j\cdot}^2$	28,224	78,400	173,889	320,356	571,536		

$$\sum_i y_{\cdot j\cdot}^2 = 1{,}172{,}405 \qquad \sum_i y_{i\cdot\cdot}^2 = 1{,}594{,}365$$

$$y_{\cdots} = 2187 \qquad SS_M = 79{,}716.15$$

$$\sum_i \sum_j y_{ij\cdot}^2 = 390{,}959 \qquad SS_T = 98{,}307$$

Does inlet ammonia concentration affect absorber efficiency?

8.35 For one physical chemistry experiment, the teaching assistant made up a stock buffer solution. He took three samples once a week thereafter to check the pH. Two determinations were made on each weekly sample.

	Determinations for week:		
Sample	1	2	3
1	5.2	4.9	5.0
	5.3	4.8	5.0
2	4.9	5.0	5.3
	5.2	5.0	5.0
3	4.8	5.2	5.1
	5.1	5.2	5.2

Did the solution pH vary with time?

8.36 Invertase (β-D-fructofuranoside fructohydrolase) activity has been studied in the developing embryonic chicken. Duodenal loop tissue obtained from embryonic chicks ranging from 13 days to hatching was homogenated. Two subsamples from each homogenated tissue sample were then assayed for specific invertase activity. It has been proposed that invertase activity neither increases nor decreases from day 13 to day 15. Do the following invertase activity data (mg glucose released/mg protein/hr) support this opinion?

	Activity at day:		
Sample	13	14	15
1	56.8, 55.4	60.5, 60.0	62.1, 62.0
2	40.3, 41.7	59.8, 59.2	63.2, 63.0
3	54.3, 55.5	57.6, 56.9	59.8, 60.2
4	55.2, 55.7	59.3, 58.7	63.4, 63.1
5	59.2, 58.9	62.0, 61.8	67.3, 67.7

8.37 In a biochemistry experiment, the absorbance of each sample was measured at varying enzyme concentrations and times with the following results.

	Absorbance at enzyme concentration:		
Time (min)	100%	50%	25%
15	0.1549	0.0862	0.1024
30	0.3010	0.1427	0.1192
45	0.5230	0.2596	0.1612
60	0.6020	0.3870	0.2076

Is the reaction affected by either variable? Use the $\alpha = 0.02$ level.

8.38 Once a sizable pressure drop was obtained across the fixed-bed reactor, the antifoulant was injected at 800 ppm(v). Three material balances were conducted on the product wax at three different temperatures and flow rates. Since the finished product is to be in contact with food, the product must meet FDA specifications for food-grade wax. A determining test for the purity of the wax is the ASTM color test. FDA requires the hydrofinished wax to have a maximum color of 1.5. The plant data from the three material balances regarding the ASTM color test for the hydrofinished wax are given below for an antifoulant concentration of 800 ppm(v).

Flow rate (gal/h)	Result at temperature:		
	300°F	650°F	550°F
360	8.0	2.0	4.0
45	8.1	2.2	4.5
200	8.0	2.0	4.0

Does the flow rate have an effect on the ASTM color? Does temperature?

8.39 Two types of distillation are being studied in an organic chemistry lab: simple and fractional. The two tests are run separately following a completely randomized procedure. Before the experiment was begun, the volumes of distillate to be collected were randomized. Then, when these values are reached, the time is recorded. The distillate is then returned to the residue flask, allowed to cool, and the procedure is repeated for the next randomly selected volume. The following results have been placed in tabulated form.

	Distillation time	
Volume of distillate	Simple	Fractional
4	30	49
8	58	96
12	74	142
16	103	168
20	159	174
24	192	260
28	214	328

Is there a statistical difference in the time required for distillation between the two types of distillation?

8.40 The stress–strain relationships and breaking points of silicone sealants used for glazing in high-rise buildings have been evaluated. These sealants have the ability to flex slightly when put under strain. This ability allows glass used for exterior curtain walls to shift or flex without cracking under varying wind loads. Acid rain is a serious environmental problem affecting the failure of such sealants. The following data for four sealant formulations were obtained under different simulated acid rain concentrations for two types of sealants, low and high modulus. The same high stress level was used for all tests.

For low-modulus sealants:

	Formulation			
pH	1	2	3	4
6.9	1853	1767	2027	1800
6.8	1826	1760	1947	1777
6.7	1824	1710	1888	1693
6.6	1763	1713	1767	1700
6.5	1793	1663	1642	1654
6.4	1754	1623	1592	1600
5.0	1558	1520	1554	1554
4.9	1554	1513	1412	1525

For high-modulus sealants:

pH	Formulation			
	1	2	3	4
6.8	5729	5500	6002	4701
6.6	5527	5001	5523	4633
6.4	5423	4432	5331	4001
6.2	4001	3876	4873	3863
6.0	3327	3202	4432	3443
5.8	3130	3200	3823	2873
5.6	2902	2867	3002	2001

(a) Using two separate two-way analyses of variance, evaluate the effects of formulation and pH on the number of sine-wave stress–strain cycles to failure for low- and high-modulus sealants.

(b) The cycle rate for high-modulus sealants was much lower than for the low-modulus formulations to avoid overheating during testing. Can a valid three-way analysis of variance be performed on these data? If so, do it and interpret the results. If not, explain your reasons fully.

8.41 Local velocity data have been taken at different points along a straightening vane following a high-speed fan. Steady-state conditions have been reached, as there is no measurable change in the temperature of the air, and the local velocities are invariant with time for the constant fan speed. Four observations of the velocity (ft/min) have been made for each position.

1	2	3	4	5	6
600	450	350	550	550	400
650	450	300	550	600	400
600	350	400	550	600	450
600	400	350	550	575	450

Is the local velocity affected by the point at which measured?

REFERENCES

1. Astarita, G., G. Marrocci, and G. Palumbo, Non-Newtonian flow along inclined plane surfaces, *I&EC Fund.* **3**: 333 (1964).*

2. Auerbach, M. E., E. L. Bauer, and F. C. Naehod, Spectrophotometer wavelength reliability, *Ind. Qual. Control* **20**: 45–48 (1964).

3. Bancroft, T. A., *Topics in Intermediate Statistical Methods*, Vol. 1, Iowa State University Press, Ames, Iowa (1968), Ch. 8.

4. Conover, W. J., *Practical Nonparametric Statistics*, 2nd ed., Wiley, New York (1980), Ch. 5.

5. Dana, R., H. S. Bayer, and G. W. McElrath, Color and shade control of ceramic tile, *Ind. Qual. Control* **21**: 609–614 (1965).

6. Harper, J. C., and A. F. El Sahrigi, Thermal conductivities of gas-filled porous solids, *I&EC Fund.* **3**: 318 (1964).

7. Hensel, S. L., and R. E. Treybal, Air–water contact: adiabatic humidification of air with water in a packed tower, *Chem. Eng. Prog.* **48**: 362–370 (1952).

8. Ito, A., and K. Hayashi, Polyoxymethylene via radiation, *Hydrocarb. Proc.* **47**(11): 197–202 (1968).

9. Lemus, F., A mixed model factorial experiment in testing electrical connectors, *Ind. Qual. Control.* **17**(6): 12–16 (1960).

10. Mouradian, G., A statistical approach to design review, *Ind. Qual. Control* **22**: 516–520 (1966).

11. Scheffé, H., A method for judging all contrasts in the analysis of variance, *Biometrika* **40**: 87–104 (1953).

9

Regression Analysis

9.1 INTRODUCTION

Regression is a highly useful statistical technique for developing a quant-itative relationship between a dependent variable and one or more independent variables. It utilizes experimental data on the pertinent variables to develop a numerical relationship showing the influence of the independent variables on a dependent variable of the system.

Throughout engineering, regression may be applied to correlating data in a wide variety of problems ranging from the simple correlation of physical properties to the analysis of a complex industrial system. If nothing is known from theory about the relationship among the pertinent variables, a function may be assumed and fitted to experimental data on the system. Frequently, a linear function is assumed. In other cases where a linear function does not fit the experimental data properly, the engineer might try a polynomial or exponential function.

9.2 SIMPLE LINEAR REGRESSION

In the simplest case the proposed functional relationship between two variables is

$$Y = \beta_0 + \beta_1 X + \varepsilon. \tag{9.1}$$

In this model Y is the dependent variable, X the independent variable, and ε a random error (or residual) which is the amount of variation in Y not accounted for by the linear relationship. The parameters β_0 and β_1, called the *regression coefficients*, are unknown and are to be estimated. The variable X is *not* a random variable but takes on fixed values. It will be assumed that the errors ε are independent and have a normal distribution with mean 0 and variance σ^2, regardless of what fixed value of X is being considered. Taking the expectation of both sides of Eq. (9.1), we have

$$E(Y) = \beta_0 + \beta_1 X, \tag{9.2}$$

where we note that the expected value of the errors is zero.

In the simple linear regression model, the variable X can be taken to be a random variable, in which case Eq. (9.2) is written as

$$E(Y|X) = \beta_0 + \beta_1 X. \tag{9.3}$$

In this representation $E(Y|X)$ is the mean or expected value of Y given a fixed value of X. The mean $E(Y|X)$ is a conditional mean, that is, the mean of Y given X. This conditional mean can be written as $E(Y|X) = \mu_{Y|X}$. Equation (9.3) is called the *regression of Y on X*.

Under Eq. (9.1) the ε are normally distributed and the random variable Y has a normal distribution with mean $\beta_0 + \beta_1 X$ and variance σ^2. Under (9.3), the random variable Y, at a given value of X, has a normal distribution with mean $\beta_0 + \beta_1 X$ and variance σ^2. The main distinction between (9.1) and (9.3) is that under (9.1) the X values are fixed (nonrandom) and repeated values of Y can often be obtained for some X values, whereas under (9.3), X and Y have a joint distribution, and if X has a continuous distribution (such as the normal), repeated Y values for a given X may not be available in the sample. The model in (9.1) is termed a *simple linear model*, and the one in (9.3) is called a *simple linear regression model*. The estimation of β_0 and β_1 is the same under (9.1) and (9.3). Thus we will use the two terms for models (9.1) and (9.3) interchangeably. However, in discussing the correlation between X and Y in Section 9.8, we assume model (9.3); that is, X and Y have a joint distribution.

To estimate the relationship between Y and X suppose that we have n observations on Y and X, denoted by $(X_1, Y_1), (X_2, Y_2), \ldots, (X_n, Y_n)$. By Eqs. (9.1) and (9.3) we can write the assumed relationship between Y and X as

$$Y = E(Y|X) + \varepsilon, \tag{9.4}$$

where by Y on the left-hand side of Eq. (9.4) is meant Y, given X. The aim here is to estimate β_0 and β_1 and thus $E(Y|X)$ or Y in terms of the n observations. If $\hat{\beta}_0$ and $\hat{\beta}_1$ denote estimates of β_0 and β_1, an estimate of $E(Y)$ is denoted by $\hat{Y} = \hat{E}(Y) = \hat{\beta}_0 + \hat{\beta}_1 X$. Thus each observed Y_i can be written as

$$Y_i = \hat{Y}_i + e_i, \qquad i = 1,2,\ldots,n,$$

where $\hat{Y}_i$ is the estimate of $E(Y_i)$ and e_i is the estimate of ε_i. Therefore, if $E(Y)$ is a linear relationship,

$$Y_i = \beta_0 + \beta_1 X_i + \varepsilon_i = \hat{\beta}_0 + \hat{\beta}_1 X_i + e_i, \qquad i = 1,2,\ldots,n. \qquad (9.5)$$

The two equations (9.5) are illustrated in Fig. 9.1. The point (X_i, Y_i) denotes the ith observation. The "true" error or residual is $Y_i - (\beta_0 + \beta_1 X_i)$, the difference between the observed Y_i and the true unknown value $\beta_0 + \beta_1 X_i$. The observed residual e_i is $Y_i - (\hat{\beta}_0 + \hat{\beta}_1 X_i) = Y_i - \hat{Y}_i$, which is the difference between the observed Y_i and the estimated $\hat{Y}_i = \hat{\beta}_0 + \hat{\beta}_1 X_i$. The quantity $\hat{Y} = \hat{\beta}_0 + \hat{\beta}_1 X$ is commonly called the *predicted value* of Y resulting from the estimated regression line.

The problem is now to obtain estimates $\hat{\beta}_0$ and $\hat{\beta}_1$ from the sample for the unknown parameters β_0 and β_1. This can best be done by the method of *least squares*. This method minimizes the sum of squares, $\sum_{i=1}^{n} e_i^2 = SS_E$, of the differences between the predicted values and the experimental values for the dependent variable. The method is based on the principle that the best

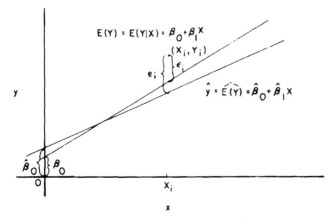

FIGURE 9.1 True and estimated regression lines.

estimators of β_0 and β_1 are those that minimize the sum of squares due to error, SS_E. The error sum of squares is

$$SS_E = \sum_{i=1}^{n} e_i^2 = \sum_{i=1}^{n} (Y_i - \hat{Y}_i)^2, \tag{9.6}$$

$$SS_E = \sum_{i=1}^{n} (Y_i - \hat{\beta}_0 - \hat{\beta}_1 X_i)^2. \tag{9.7}$$

To determine the minimum, the partial derivative of the error sum of squares with respect to each constant ($\hat{\beta}_0$ and $\hat{\beta}_1$ for this model) is set equal to zero to yield

$$\frac{\partial(SS_E)}{\partial \hat{\beta}_0} = \frac{\partial}{\partial \hat{\beta}_0}\left[\sum_{i=1}^{n} (Y_i - \hat{\beta}_0 - \hat{\beta}_1 X_i)^2\right] = 0, \tag{9.8}$$

$$\frac{\partial(SS_E)}{\partial \hat{\beta}_1} = \frac{\partial}{\partial \hat{\beta}_1}\left[\sum_{i=1}^{n} (Y_i - \hat{\beta}_0 - \hat{\beta}_1 X_i)^2\right] = 0. \tag{9.9}$$

These equations are called *normal equations*, indicating that they were obtained from the least-squares differentiation. Carrying out the differentiation, we obtain

$$n\hat{\beta}_0 + \hat{\beta}_1 \sum_i X_i = \sum_i Y_i, \tag{9.10}$$

$$\hat{\beta}_0 \sum_i X_i + \hat{\beta}_1 \sum_i X_i^2 = \sum_i X_i Y_i, \tag{9.11}$$

where all the summations go from $i = 1$ to $i = n$. The solutions to these normal equations are

$$\hat{\beta}_0 = \bar{Y} - \hat{\beta}_1 \bar{X}, \tag{9.12}$$

$$\hat{\beta}_1 = \frac{\sum_i (X_i - \bar{X})(Y_i - \bar{Y})}{\sum_i (X_i - \bar{X})^2}. \tag{9.13}$$

This solution for estimating β_0 and β_1 is called the least-squares solution. These estimates are used to give the regression equation

$$\hat{Y} = \hat{\beta}_0 + \hat{\beta}_1 X, \tag{9.14}$$

which is an estimate of the true linear relationship between Y and X.

The estimator $\hat{\beta}_1$ can also be written in the form

$$\hat{\beta}_1 = \frac{\sum_i X_i Y_i - n\bar{X}\bar{Y}}{\sum_i X_i^2 - n\bar{X}^2},$$

which is often more useful for computation. It can be shown that $\hat{\beta}_0$ and $\hat{\beta}_1$ are unbiased for β_0 and β_1; that is, $E(\hat{\beta}_0) = \beta_0$ and $E(\hat{\beta}_1) = \beta_1$.

The error sum of squares can be written as

$$SS_E = \sum_i (Y_i - \hat{\beta}_0 - \hat{\beta}_1 X_i)^2$$

$$= \sum_i [(Y_i - \bar{Y}) - \hat{\beta}_1 (X_i - \bar{X})]^2$$

$$= \sum_i (Y_i - \bar{Y})^2 - 2\hat{\beta}_1 \sum_i (Y_i - \bar{Y})(X_i - \bar{X}) + \hat{\beta}_1^2 \sum_i (X_i - \bar{X})^2.$$

The middle term above becomes $-2\hat{\beta}_1^2 \sum_i (X_i - \bar{X})^2$ on multiplying and dividing it by $\sum_i (X_i - \bar{X})^2$. Thus

$$SS_E = \sum_i (Y_i - \bar{Y})^2 - \hat{\beta}_1^2 \sum_i (X_i - \bar{X})^2. \tag{9.15}$$

The first term on the right-hand side of Eq. (9.15) is the total corrected sum of squares, SS_{TC}, of the Y's. The linear relationship between X and Y accounts for a reduction of $\hat{\beta}_1^2 \sum_i (X_i - \bar{X})^2$ in SS_{TC}. That is, if there is no linear relationship between X and Y (i.e., $\beta_1 = 0$), then $\sum_i (Y_i - \bar{Y})^2 = SS_E$. If there is a linear relationship between X and Y, then SS_E (or SS_{TC}) is reduced by an amount $\hat{\beta}_1^2 \sum_i (X_i - \bar{X})^2$, which is called the sum of squares due to regression and is denoted by SS_R. Equation (9.15) can be written as

$$SS_E = SS_{TC} - SS_R$$

or

$$SS_T \equiv \sum_i Y_i^2 = n\bar{Y}^2 + SS_E + SS_R$$

$$= SS_M + SS_E + SS_R.$$

Thus regression analysis may be looked on as the process of partititioning the total sum of squares, SS_T, into three parts: (1) the sum of squares due to the mean, SS_M, plus (2) the sum of squares due to error, SS_E (or deviations about the regression line), plus (3) the sum of squares due to regression, SS_R. Another way of stating this result is that each Y value is made up of three parts (or partitioned into three segments), each one leading to the corresponding sum of squares. That is,

$$Y_i = \bar{Y} + (\hat{Y}_i - \bar{Y}) + (Y_i - \hat{Y}_i), \qquad i = 1,2,\ldots,n.$$

Figure 9.2 shows the partition of Y in graphical form. It should be noted that the estimated regression line always passes through the point $(\bar{X}, \bar{Y})$. This is obvious from $\hat{Y} = \hat{\beta}_0 + \hat{\beta}_1 X = \bar{Y} - \hat{\beta}_1 \bar{X} + \hat{\beta}_1 X$.

It can be shown that $SS_E/(n-2)$ is an unbiased estimator of σ^2. Furthermore, SS_E/σ^2 has a χ^2-distribution with $n-2$ degrees of freedom.

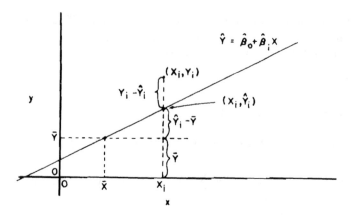

Figure 9.2 Partitioning of total sum of squares in simple linear regression.

There are various hypotheses that are of interest in relation to the simple linear regression model. One of these is $H_0: \beta_1 = \beta_1'$. The special case $H_0: \beta_1 = 0$ (i.e., $\beta_1' = 0$) is perhaps the most important since if $\beta_1 = 0$, there is no linear relationship between X and Y.

To make inferences regarding the parameter β_1 we need to know the distribution of its estimator $\hat{\beta}_1$. The estimator $\hat{\beta}_1$ can be written as a linear combination of $Y_1, Y_2, \ldots, Y_n$. That is,

$$\hat{\beta}_1 = \sum_i \left[\frac{X_i - \bar{X}}{\sum_j (X_j - \bar{X})^2} \right] Y_i.$$

From this it can be shown that $\hat{\beta}_1$ has a normal distribution with mean β_1 and variance $\sigma_{\hat{\beta}_1}^2 = \sigma^2 / \sum_i (X_i - \bar{X})^2$, keeping in mind that the X's are not random variables. From this it follows (see Section 6.6) that the random variable

$$T = \frac{\hat{\beta}_1 - \beta_1}{\hat{\sigma}_{\beta_1}} = \frac{\hat{\beta}_1 - \beta_1}{s_{\beta_1}} = \frac{\hat{\beta}_1 - \beta_1}{\{[SS_E/(n - 2)]/\sum_i (X_i - \bar{X})^2\}^{1/2}} \qquad (9.16)$$

has a t-distribution with $n - 2$ degrees of freedom, which can be used to make inferences relative to β_1. Note in Eq. (9.16) that σ^2 is estimated by $MS_E = SS_E/(n - 2)$. The term s_{β_1} is the estimate of the standard deviation of $\hat{\beta}_1$, the slope of the "fitted" regression line.

Example 9.1 An intermediate step in a reaction process is A → B. While this reaction is carried out at atmospheric pressure, the temperature varies from 1°C to 10°C. As a beginning step in the optimization of this process, the relation between conversion of A and temperature must be obtained. Pilot-plant studies have provided the following data (X = temperature, Y = yield):

x	y
1	3
2	5
3	7
4	10
5	11
6	14
7	15
8	17
9	20
10	21

The proposed model is $Y = \beta_0 + \beta_1 X + \varepsilon$. From these data we obtain the following:

$$\sum x_i = 55.0, \qquad \sum y_i = 123.0, \qquad \sum x_i y_i = 844.0,$$
$$\sum x_i^2 = 385.0, \qquad \sum y_i^2 = 1855.0, \qquad (\sum x_i)(\sum y_i) = 6765.0,$$
$$(\sum x_i)^2 = 3025.0, \qquad (\sum y_i)^2 = 15{,}129.0,$$
$$\bar{x} = \frac{\sum x_i}{n} = 5.5, \qquad \bar{y} = \frac{\sum y_i}{n} = 12.3,$$

$$\sum (x_i - \bar{x})(y_i - \bar{y}) = \sum x_i y_i - \frac{\sum x_i \sum y_i}{n} = 167.5,$$

$$\sum (x_i - \bar{x})^2 = \sum x_i^2 - \frac{(\sum x_i)^2}{n} = 82.5,$$

$$\sum (y_i - \bar{y})^2 = \sum y_i^2 - \frac{(\sum y_i)^2}{n} = 342.1,$$

$$\hat{\beta}_1 = \frac{\sum (x_i - \bar{x})(y_i - \bar{y})}{\sum (x_i - \bar{x})^2} = 2.0303,$$
$$\hat{\beta}_0 = \bar{y} - \hat{\beta}_1 \bar{x} = 1.1333.$$

(Remember that lowercase letters denote values of random variables, hence lowercase y's in the computations above. The variable X is not a random variable, but we use the same convention for X also.)

To test the usefulness of the regression line, we must find the standard deviation of the line. This is done in the following manner. Our regression model is

$$Y_i = \beta_0 + \beta_1 X_i + \varepsilon_i, \qquad i = 1,2,\ldots,n, \tag{9.17}$$

where the ε_i are normally and independently distributed with mean 0 and variance $\sigma_\varepsilon^2 = \sigma^2$, which is the same for each X. The subscript ε is used because σ_ε^2 is the variance of the errors. If we assume that the failure of the assumed model to fit is solely a function of errors, the mean square for deviation about regression (the residual mean square) can be used as an estimate of σ^2. The estimator for σ^2 is $\hat{\sigma}^2 = \text{MS}_E = \text{SS}_E/(n-2)$, or

$$s^2 = \hat{\sigma}^2 = \sum_i \frac{(Y_i - \hat{Y}_i)^2}{n-2} = \frac{\sum_i (Y_i - \bar{Y})^2 - \hat{\beta}_1 \sum_i (X_i - \bar{X})(Y_i - \bar{Y})}{n-2}. \tag{9.18}$$

Having determined the variance estimate $\hat{\sigma}^2$, it is a simple matter to obtain an estimate of the variance of the statistic $\hat{\beta}_1$, calculated in the regression analysis. The estimated variance of the regression coefficient $\hat{\beta}_1$ is

$$s_{\hat{\beta}_1}^2 = \frac{s^2}{\sum_{i=1}^{n} (X_i - \bar{X})^2}, \tag{9.19}$$

since

$$\sigma_{\hat{\beta}_1}^2 = \frac{\sigma^2}{\sum_{i=1}^{n} (X_i - \bar{X})^2}.$$

The sum of squares of deviations from the regression line is, according to Eq. (9.15),

$$\sum_i e_i^2 = \sum_i (y_i - \bar{y})^2 - \hat{\beta}_1^2 \sum_i (x_i - \bar{x})^2 = 2.025.$$

The mean-square deviation is s^2 given by

$$\hat{\sigma}^2 = s^2 = \frac{\sum_i e_i^2}{n-2} = 0.2531.$$

Note that we have reduced the degrees of freedom for s^2 by 2 to account for fitting the mean and the regression coefficient $\hat{\beta}_1$ for these data. The standard deviation of experimental data from the line is s, where

$$\hat{\sigma} = s = \sqrt{s^2} = 0.5031.$$

The standard deviation of the slope of the regression line is

$$s_{\hat{\beta}_1} = \frac{s}{[\sum_i (x_i - \bar{x})^2]^{1/2}} = 0.05539.$$

To test the regression of Y on X, that is, $H_0: \beta_1 = 0$, we use the t-statistic in Eq. (9.16) with $\beta_1 = 0$. The computed t value is

$$T = \frac{\hat{\beta}_1}{s_{\hat{\beta}_1}} = 36.655.$$

From Table IV, $t = 2.306$ with 8 degrees of freedom at a 0.05 significance level. Since the calculated t-value falls above the tabular value $t_{8,0.975} = 2.306$, the hypothesis that $\beta_1 = 0$ is rejected.

Suppose that theoretical considerations lead us to believe that $\beta_1 = 2.000$. To test whether or not the proposed β_1 is valid, we use the t-statistic in Eq. (9.16) with $\beta_1 = 2.0$, which is computed as

$$T = \frac{\hat{\beta}_1 - \beta_1}{s_{\hat{\beta}_1}} = \frac{2.0303 - 2.000}{0.05539} = 0.5470.$$

Since this value of t is less than the tabular value of $t_{8,0.975} = 2.306$, the hypothesis that $\beta_1 = 2.000$ is accepted.

9.2.1 Interval Estimation in Simple Linear Regression

Prior to determining confidence intervals or determining test procedures to be used, we recall three assumptions in the model $Y = \beta_0 + \beta_1 X$:

1. The independent variable X is a fixed variable whose values can be observed without error.
2. For any given value of X, Y is normally distributed with mean $\mu_{Y|X} = \beta_0 + \beta_1 X$ and variance $\sigma_{Y|X}^2 = \sigma^2$.
3. That the variance can be represented as $\sigma_\varepsilon^2 = \sigma^2$, which is the same for each X.

The estimator for σ^2, as mentioned previously, is $\hat{\sigma}^2 = MS_E = SS_E/(n-2)$.

Estimation and testing of the parameter β_1 was discussed in Section 9.2. Three other parameters of interest are the intercept β_0, the true mean, $\mu_{Y|X} = E(Y)$, of Y given a value of X, and the true predicted value, $\hat{Y}_i$, corresponding to a fixed value of X. The variances of the estimators of these three parameters can be shown to be

$$\sigma^2_{\hat{\beta}_0} = \left[\frac{1}{n} + \frac{\bar{X}^2}{\sum_i (X_i - \bar{X})^2} \right] \sigma^2. \tag{9.20}$$

$$\sigma^2_{\hat{\mu}_{Y|X}} = \left[\frac{1}{n} + \frac{(X - \bar{X})^2}{\sum_i (X_i - \bar{X})^2} \right] \sigma^2, \tag{9.21}$$

and

$$\sigma^2_{\hat{Y}} = \left[1 + \frac{1}{n} + \frac{(X - \bar{X})^2}{\sum_i (X_i - \bar{X})^2} \right] \sigma^2. \tag{9.22}$$

The estimators of the variances in (9.20), (9.21), and (9.22) are obtained by replacing σ^2 with $\hat{\sigma}^2 = SS_E/(n - 2)$ as defined by Eq. (9.18) to give

$$s^2_{\hat{\beta}_0} = \left[\frac{1}{n} + \frac{\bar{X}^2}{\sum_i (X_i - \bar{X})^2} \right] \hat{\sigma}^2, \tag{9.23}$$

$$s^2_{\hat{\mu}_{Y|x}} = \left[\frac{1}{n} + \frac{(X - \bar{X})^2}{\sum_i (X_i - \bar{X})^2} \right] \hat{\sigma}^2, \tag{9.24}$$

and

$$s^2_{\hat{Y}} = \left[1 + \frac{1}{n} + \frac{(X - \bar{X})^2}{\sum_i (X_i - \bar{X})^2} \right] \hat{\sigma}^2. \tag{9.25}$$

These variance estimators may be used in testing hypotheses about the unknown parameters, or they may be used to provide interval estimates of the same unknown parameters. As noted earlier, in linear regression problems the estimator of greatest interest is the slope $\hat{\beta}_1$. This, of course, is an estimator of β_1. To provide a $(1 - \alpha)100\%$ confidence interval of β_1, we compute

$$\left. \begin{array}{c} L \\ U \end{array} \right\} = \hat{\beta}_1 \mp s_{\hat{\beta}_1} t_{n-2, 1 - \alpha/2} \tag{9.26}$$

for a given significance level α, where $s_{\hat{\beta}_1}$ is given in Eq. (9.19), and L and U represent the lower and upper confidence limits, respectively. The confidence limits follow quite easily by considering the random variable given in Eq. (9.16).

If a confidence interval on β_0 is needed, we have only to compute

$$\left.\begin{array}{c} L \\ U \end{array}\right\} = \hat{\beta}_0 \mp s_{\hat{\beta}_0} t_{n-2,1-\alpha/2}. \tag{9.27}$$

where $s_{\hat{\beta}_0}^2$ is given in Eq. (9.23).

A $(1 - \alpha)100\%$ confidence interval on $\mu_{Y|X} = \beta_0 + \beta_1 X$ is given by

$$\left.\begin{array}{c} L \\ U \end{array}\right\} = \hat{Y} \mp s_{\hat{\mu}_{Y|X}} t_{n-2,1-\alpha/2}, \tag{9.28}$$

where $s_{\hat{\mu}_{Y|X}}^2$ is given in Eq. (9.24) and $\hat{Y} = \hat{\beta}_0 + \hat{\beta}_1 X$. The limits L, U yield a $(1 - \alpha)100\%$ confidence interval on the mean value of Y given a value of X.

Occasionally, it is desired to predict an individual Y value, Y', associated with a given X value. This can be done by using $\hat{Y}$ to predict an individual value other than a mean value. In this case the prediction interval is provided by

$$\left.\begin{array}{c} L \\ U \end{array}\right\} = \hat{Y} \mp s_{\hat{Y}} t_{n-2,1-\alpha/2}, \tag{9.29}$$

where $s_{\hat{Y}}^2$ is given in Eq. (9.25).

It should be noted that the confidence interval on $\mu_{Y|X}$ is shorter than the corresponding interval on a predicted Y since the latter takes into account the variability of individual Ys. This comes from the fact that $s_{\hat{Y}}^2 = s_{\hat{\mu}_{Y|X}}^2 + \hat{\sigma}^2$, from which we see that $s_{\hat{Y}}^2 > s_{\hat{\mu}_{Y|X}}^2$. Furthermore, the intervals on $\mu_{Y|X}$ and Y' are shortest when $X = \bar{X}$. This follows from considering Eqs. (9.24) and (9.25).

It is quite conceivable that we may at some time wish to determine a confidence region for the simultaneous estimation of β_0 and β_1. To do this we use the fact that the random variable

$$F = \frac{n(\hat{\beta}_0 - \beta_0)^2 + 2n\bar{X}(\hat{\beta}_0 - \beta_0)(\hat{\beta}_1 - \beta_1) + (\hat{\beta}_1 - \beta_1)^2 \sum_i X_i^2}{2\hat{\sigma}^2} \tag{9.30}$$

has an F-distribution with $v_1 = 2$ and $v_2 = n - 2$ degrees of freedom. Thus

$$\mathcal{P}(F \le F_{v_1,v_2,1-\alpha}) = 1 - \alpha,$$

and using the definition of F in Eq. (9.30) we obtain the two-dimensional confidence region given by the interior of the ellipse

$$\frac{n(\hat{\beta}_0 - \beta_0)^2 + 2n\bar{X}(\hat{\beta}_0 - \beta_0)(\hat{\beta}_1 - \beta_1) + (\hat{\beta}_1 - \beta_1)^2 \sum_i X_i^2}{2\hat{\sigma}^2} = F_{v_1,v_2,1-\alpha}. \tag{9.31}$$

Equation (9.31) is a second-degree equation in β_0 and β_1. The hypothesis $H_0: \beta_0 = \beta_0'$ and $\beta_1 = \beta_1'$ is rejected if the left-hand side of Eq. (9.31) evaluated at $\beta_0 = \beta_0'$ and $\beta_1 = \beta_1'$ exceeds $F_{v_1,v_2,1-\alpha}$.

Example 9.2 For the data in Example 9.1, construct 95% confidence intervals on (a) β_0, (b) β_1, (c) $\mu_{Y|X=4}$, and (d) predicted value, Y', when $X = 4$.

Solution   (a) The standard error of $\hat{\beta}_0$ is, by Eq. (9.23),

$$s_{\hat{\beta}_0} = \left[\frac{1}{10} + \frac{(5.5)^2}{82.5} \right]^{1/2} (0.5031)$$

$$= (0.10 + 0.3667)^{1/2}(0.5031)$$

$$= 0.3437.$$

Thus 95% confidence limits on β_0 are, by Eq. (9.27),

$$1.1333 \pm t_{8,0.975}s_{\hat{\beta}_0},$$

$$1.1333 \pm 2.306(0.3437),$$

$$1.1333 \pm 0.7926,$$

$$0.3407,\quad 1.9259.$$

(b) The standard error of $\hat{\beta}_1$ is $s_{\hat{\beta}_1} = 0.05539$, from Example 9.1. Thus by Eq. (9.26) the 95% confidence limits on β_1 are

$$2.0303 \pm 2.306(0.05539),$$

$$2.0303 \pm 0.1277,$$

$$1.9026,\quad 2.1580.$$

(c) The standard error of $\hat{\mu}_{Y|X}$ is, from Eq. (9.24),

$$s_{\hat{\mu}_{Y|X=4}} = \left[\frac{1}{10} + \frac{(4 - 5.5)^2}{82.5} \right]^{1/2} (0.5031)$$

$$= (0.1 + 0.0273)^{1/2}(0.5031)$$

$$= 0.3568(0.5031)$$

$$= 0.1795,$$

and by Eq. (9.28), the 95% confidence limits for $\mu_{Y|X}$ are

$$[1.1333 + 2.0303(4)] \mp 0.1795(2.306),$$

$$9.2545 \mp 0.4139,$$

$$8.8406,\quad 9.6684.$$

(d) From Eq. (9.25) we have

$$s_{\hat{Y}} = (1 + 0.1273)^{1/2}(0.5031)$$
$$= 1.0617(0.5031)$$
$$= 0.5341$$

and the 95% confidence limits on Y' are, from Eq. (9.29),

9.2545 $\mp$ 2.306(0.5341),

9.2545 $\mp$ 1.2316,

8.0229, 10.4861.

9.2.2 Hypothesis Testing in Simple Linear Regression

Sometimes it is desired to determine whether the estimated slope $\hat{\beta}_1$ is significantly different from some hypothetical value of β_1, say β_1'. In this case the hypothesis to be tested is $H_0: \beta_1 = \beta_1'$. The appropriate test statistic, as illustrated in Example 9.1, is

$$T = \frac{\hat{\beta}_1 - \beta_1'}{s_{\hat{\beta}_1}}. \tag{9.32}$$

A popular value for β_1' is 0, as this reflects the hypothesis that Y is independent of X. The critical region for testing $H_0: \beta_1 = \beta_1'$ against $H_A: \beta_1 \neq \beta_1'$ is

$$|T| = \left| \frac{\hat{\beta}_1 - \beta_1'}{s_{\hat{\beta}_1}} \right| > t_{n-2, 1 - \alpha/2}.$$

Critical regions for one-sided alternatives can easily be obtained.

Other test procedures in simple linear regression are concerned with such hypotheses as (1) $H_0: \beta_0 = \beta_0'$, (2) $H_0: \mu_{Y|X=X_0} = \mu_0$, and (3) $H_0: \beta_0 = 0$. Appropriate test procedures, statistics, and rejection regions are given in Table 9.1.

Example 9.3 For the data in Example 9.1, test the hypotheses (a) $H_0: \beta_0 = 1$, (b) $H_0: \beta_1 \leq 1.8$ against $H_A: \beta_1 > 1.8$, (c) $H_0: \mu_{Y|X=4} = 9$, and (d) $H_0: \beta_0 = 1$ and $\beta_1 = 1.8$. Use $\alpha = 0.05$.
Solution (a) From Table 9.1 we see that we need to compute $T = (\hat{\beta}_0 - 1)/s_{\hat{\beta}_0}$ to test $H_0: \beta_0 = 1$. We have

$$T = \frac{1.1333 - 1}{0.3437}$$
$$= 0.3878.$$

TABLE 9.1   Test Procedures in Simple Linear Regression

H_0	Statistic	Rejection region
$\beta_0 = \beta_0'$	$T = \dfrac{\hat{\beta}_0 - \beta_0'}{s_{\hat{\beta}_0}}$	
$\beta_1 = \beta_1'$	$T = \dfrac{\hat{\beta}_1 - \beta_1'}{s_{\hat{\beta}_1}}$	$T \geq t_{n-2,1-\alpha/2}$ or $T \leq t_{n-2,\alpha/2}$
$\mu_{Y\mid X = X_0} = \mu_0$	$T = \dfrac{\hat{\beta}_0 + \hat{\beta}_2 X_0 - \mu_0}{s_{\hat{y}}}$	
$\beta_0 = \beta_0'$ and $\beta_1 = \beta_1'$	$F = [n(\hat{\beta}_0 - \beta_0')^2 + 2n\bar{X}(\hat{\beta}_0 - \beta_0')(\hat{\beta}_1 - \beta_1') + (\hat{\beta}_1 - \beta_1')^2 \sum X^2]/2\hat{\sigma}^2$	$F \geq F_{2,n-2,1-\alpha}$
$\beta_1 = 0$	$F = MS_R/MS_E$	$F \geq F_{1,n-2,1-\alpha}$

Since $0.3878 < t_{8,0.975} = 2.306$, we accept $H_0: \beta_0 = 1$. This conclusion is also obtained from the confidence interval in Example 9.2(a) since $\beta_0' = 1$ is contained in that confidence interval.

(b) The hypothesis $H_0: \beta_1 \leq 1.8$ is one-sided and the pertinent statistic is $T = (\hat{\beta}_1 - 1.8)/s_{\hat{\beta}_1}$. We have

$$T = \frac{2.0303 - 1.8}{0.05539} = 4.1578.$$

Since $4.1578 > t_{8,0.95} = 1.860$, we reject $H_0: \beta_1 \leq 1.8$.

(c) The confidence interval in Example 9.2(c) contains $\mu_{Y\mid X = 4} = 9$; therefore, we accept $H_0: \mu_{Y\mid X = 4} = 9$.

(d) To test $H_0: \beta_0 = 1$ and $\beta_1 = 1.8$ we need to evaluate the F-statistic defined by Eq. (9.30), which is also in Table 9.1. Since $n = 10$, $\hat{\beta}_0 = 1.1333$, $\beta_0' = 1$, $\bar{x} = 5.5$, $\hat{\beta}_1 = 2.0303$, $\beta_1' = 1.8$, $\sum x^2 = 385.0$, and $\hat{\sigma}^2 = 0.2351$, we have

$$F = \frac{10(0.1333)^2 + 20(5.5)(0.1333)(0.2303) + (0.2303)^2(385.0)}{2(0.2531)}$$

$$= \frac{0.1777 + 3.3769 + 20.4197}{0.5062}$$

$$= 47.36.$$

Since $47.36 > F_{2,8,0.95} = 4.46$, we reject the joint hypothesis that $\beta_0 = 1$ and $\beta_1 = 1.8$.

9.2.3 Inverse Prediction in Simple Linear Regression

The equation $Y = \hat{\beta}_0 + \hat{\beta}_1 X$ may sometimes be used to estimate the unknown value of X associated with an observed value of Y. The procedure is as follows. We compute

$$\hat{X} = \frac{Y_0 - \hat{\beta}_0}{\hat{\beta}_1},\tag{9.33}$$

for which Y_0 is the observed value of Y from which we desire to estimate the associated value of X. The confidence interval for the true but still unknown X value is obtained from

$$\left.\begin{matrix}L\\U\end{matrix}\right\} = \bar{x} + \frac{\hat{\beta}_1(y_0 - \bar{y})}{\lambda} \mp \frac{t\hat{\sigma}}{\lambda}\left[c(y_0 - \bar{y})^2 + \lambda\left(\frac{n+1}{n}\right)\right]^{1/2},\tag{9.34}$$

where

$$c = \frac{1}{\sum_i (X_i - \bar{X})^2},\tag{9.35}$$

$$\lambda = \hat{\beta}_1^2 - ct^2\hat{\sigma}^2 = \hat{\beta}_1^2 - t^2 s_{\hat{\beta}_1}^2,\tag{9.36}$$

and

$$t = t_{n-2,\alpha/2}.\tag{9.37}$$

If as frequently occurs, several, say m, values of Y are associated with the same unknown value of X, Eqs. (9.33) and (9.34) are modified as follows:

$$\hat{x} = \frac{\bar{y}_0 - \hat{\beta}_0}{\hat{\beta}_1},\tag{9.38}$$

$$\left.\begin{matrix}L\\U\end{matrix}\right\} = \bar{x} + \frac{\hat{\beta}_1(\bar{y}_0 - \bar{y})}{\lambda} \mp \frac{t}{\lambda}\hat{\sigma}'\left[c(\bar{y}_0 - \bar{y})^2 + \lambda\left(\frac{n+m}{nm}\right)\right]^{1/2},\tag{9.39}$$

where

$$\bar{y}_0 = \frac{\sum_{i=1}^{m} y_{0i}}{m},\tag{9.40}$$

$$(\hat{\sigma}')^2 = \frac{(n-2)\hat{\sigma}^2 + \sum_{i=1}^{m}(y_{0i} - \bar{y}_0)^2}{n+m-3},\tag{9.41}$$

$$t = t_{n+m-3,\alpha/2},\tag{9.42}$$

and λ and c are as defined by Eqs. (9.35) and (9.36), respectively. If the number of Y values m associated with the unknown X is small by comparison to the total number of observations, an approximate solution may be obtained by using $\hat{\sigma}^2$ rather than $(\hat{\sigma}')^2$ in Eq. (9.39).

Example 9.4   Using the data in Example 9.1, construct a 99% confidence interval for the true value of X corresponding to the observed value $Y = Y_0 = 12$.

Solution   The confidence limits are given in Eq. (9.34). The required quantities are $\bar{x} = 5.5$, $\hat{\beta}_1 = 2.0303$, $y_0 = 12$, $\bar{y} = 12.3$, $\sum x_i^2 = 385.0$, $t = t_{8,0.995} = 3.355$, and $\hat{\sigma}^2 = 0.2531$. From these quantities we obtain $\lambda = 4.1147$. Thus the 99% confidence limits for X are

$$5.5 - 0.1480 \mp 0.4102(4.5264)^{1/2},$$

$$5.352 \mp 0.8727,$$

$$4.4793, \quad 6.2247.$$

9.2.4   Analysis of Variance in Simple Linear Regression

The mean square due to error, $MS_E = SS_E/(n-2)$, is an unbiased estimator for σ^2 because $E(MS_E) = \sigma^2$ regardless of whether or not the hypothesis $H_0: \beta_1 = 0$ is true. It can be shown that the expected value of the mean square due to regression, $MS_R = SS_R/(1)$, is a biased estimator for σ^2 unless $\beta_1 = 0$. This can be shown by

$$E(MS_R) = E\left(\frac{SS_R}{1}\right)$$

$$= \sigma^2 + \beta_1^2 \sum (X_i - \bar{X})^2 > \sigma^2.$$

These two expected mean squares suggest the use of the ratio

$$F = \frac{MS_R}{MS_E}$$

in testing $H_0: \beta_1 = 0$. This ratio has an F-distribution with $v_1 = 1$ and $v_2 = n - 2$ degrees of freedom if $H_0: \beta_1 = 0$ is true. Thus to test $H_0: \beta_1 = 0$ one rejects H_0 if $F > F_{1,n-2,1-\alpha}$. Since MS_R and MS_E both estimate σ^2 under $H_0: \beta_1 = 0$, one rejects H_0 if F is significantly larger than 1 since $E(MS_R) > \sigma^2$ when $H_0: \beta_1 = 0$ is not true. One should observe that $H_0: \beta_1 = 0$ can also be tested by using $T = (\hat{\beta}_1 - 0)/s_{\hat{\beta}_1}$ and the t-statistic with $n - 2$ degrees of freedom. In fact, it can be shown that T^2 is an F-statistic with 1 and $n - 2$ degrees of freedom, so that using F is equivalent to using T or T^2. The F-test for $H_0: \beta_1 = 0$ is the bottom entry in Table 9.1.

The analysis of variance in simple linear regression is summarized in Table 9.2.

TABLE 9.2 Analysis of Variance for Simple Linear Regression

Source	d.f.	SS	MS	EMS
Due to mean	1	$n\bar{Y}^2$	—	—
Due to regression (β_1)	1	$\hat{\beta}_1 \sum_i (X_i - \bar{X})(Y_i - \bar{Y}) = SS_R$	$MS_R = SS_R$	$\sigma^2 + \beta_1^2 \sum_i (X_i - \bar{X})^2$
Error or residual	$n - 2$	$\sum_i (Y_i - \hat{Y}_i)^2 = SS_E$	$MS_E = SS_E/(n - 2)$	σ^2
Total	n	$\sum_i Y_i^2 = SS_T$		

The AOV table for the data in Example 9.1 is given below.

Source	d.f.	SS	MS	EMS
Mean	1	1512.900	—	—
β_1	1	340.075	340.1	$\sigma^2 + 82.5\beta_1^2$
Error	8	2.025	0.2531	σ^2
Total	10	1855.000	—	—

The F value for testing $H_0: \beta_1 = 0$ is $F = 340.075/0.2531 = 1343.6$. Since $F_{1,8,0.95} = 5.32$, the hypothesis is rejected. The computed T value in Example 9.1 for testing the same hypothesis was $T = 36.654$. The two tests are equivalent with $T^2 = (36.654)^2 = 1343.5 \doteq F = 1343.6$, the difference being due to rounding error.

The use of the SAS system for simple linear regression is quite easy. (This program, as all the others for regression models has been taken with permission from *SAS for Linear Models*, SAS Institute, Cary, NC, 1981, 231 pp.) In this procedure, we introduce several new uses of the SAS general linear model routine.

```
(put initial JCL cards/statements here)
DATA;
INPUT depvar LOCATION indepvar LOCATION;
CARDS;
(put data here according to INPUT format)
PROC PRINT;
PROC PLOT;
PLOT depvar * indepvar;
PROC GLM;
MODEL depvar = indepvar;
OUTPUT OUT = NEW P = YHAT R = RESID;
PROC PLOT;
PLOT depvar * indepvar YHAT * indepvar = 'P'/OVERLAY;
PLOT RESID * indepvar/VREF = 0;
TITLE 'SIMPLE LINEAR REGRESSION OF indepvar ON depvar';
(put final JCL cards/statements here)
```

Some comments are in order. As usual, the dependent (depvar) variable must be listed first in the PROC PLOT command. This is also true for the MODEL statement. There, as elsewhere, "indepvar" refers to the independent variable. The MODEL statement instructs the SAS system to apply the least-squares routine to the INPUT data set. The OUTPUT statement tells the SAS system to create a new data set containing the independent variable and the predicted values of the dependent variable. The name appearing after OUT = is the name you have chosen for the new data set (NEW for this example). The name (YHAT in this example) appearing after P = in the OUTPUT statement names the predicted variable in the new set. The predicted values (YHAT) are found by inserting the values of the independent variable into the regression equation. The name appearing after R = in the OUTPUT statement assigns a name to the variable which contains the residuals in the new data set. A residual is the original value of the dependent variable minus its value as predicted corresponding to some particular value of the independent variable. If an observation is missing, no residual will be calculated, but the YHAT will contain the value predicted by the regression model for the corresponding value of the independent variable.

The second PLOT statement asks for the original and new data sets to be plotted on the same graph by use of the /OVERLAY instruction. The predicted (YHAT) values are labeled P by enclosing them in apostrophes so you can distinguish them from the original data (Fig. 9.3). The last PLOT statement tells the SAS system to plot the residuals (RESID) versus the independent variable. The statement also instructs the SAS system to draw a horizontal line at 0 on the vertical scale (/VREF = 0) (Fig. 9.4). This PLOT will show you whether you had the correct regression model. The values plotted should be randomly scattered about the reference line. If they show any recognizable trend, the mathematical expression for that shape should be added to a new trial model.

The use of the SAS program where "depvar" = CV and "indepvar" = NRE is shown in Example 9.5.

Example 9.5 The discharge coefficient C_v of venturi flowmeters in the turbulent regime should be constant at 0.98. Data as shown in the SAS output below were otained for $18{,}600 < N_{Re} < 39{,}130$. Does C_v vary with the Reynolds number, N_{Re}?

Solution Using the program above with the variables as defined, the following output was obtained.

SAS

OBS	CV	NRE
1	1.190	18600
2	0.849	24120
3	0.804	26130
4	0.945	29480
5	0.856	32170
6	1.030	33510
7	1.010	39130

SAS

GENERAL LINEAR MODELS PROCEDURE

DEPENDENT VARIABLE: CV

SOURCE	DF	SUM OF SQUARES	MEAN SQUARE	F VALUE
MODEL	1	0.00307222	0.00307222	0.15
ERROR	5	0.10474064	0.02094813	PR > F
CORRECTED TOTAL	6	0.10781286		0.7175

SOURCE	DF	TYPE I SS
NRE	1	0.00307222

SOURCE

NRE

PARAMETER	ESTIMATE	T FOR H0: PARAMETER=0	PR > |T|	STD ERROR OF ESTIMATE
INTERCEPT	1.05212932	4.05	0.0098	0.25982499
NRE	-3.3519014E-06	-0.38	0.7175	0.00000875

From the value of T_{β_0}, we conclude that the intercept is nonzero. For $H_0: \beta_0 = 0.98$, we have

$$T_{\beta_0} = \frac{\beta_0 - \hat{\beta}_0}{s_{\hat{\beta}_0}} = \frac{0.98 - 1.0521293}{0.25982499} = -0.278,$$

so we accept H_0 and tentatively conclude that the experiment verifies theory as C_v is statistically ($\alpha = 0.05$) = 0.98.

To complete this example, we must show that C_v does not vary with N_{Re}. To do this, note that $T_{\beta_1} = -0.38$ for $H_0: \beta_1 = 0$. We thus accept this hypothesis also and conclude that for the flow-rate range examined, C_v does not vary with N_{Re}.

9.2.5 Lack of Fit

The test for lack of fit of the regression model breaks up the residual sum of squares into a sum of squares for lack of fit and an experimental

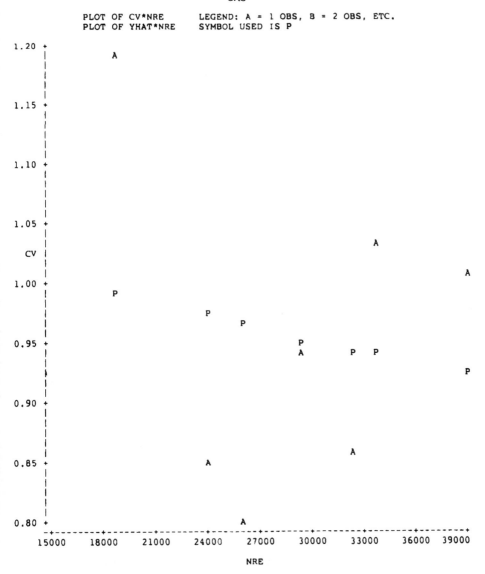

SAS

```
          PLOT OF CV*NRE        LEGEND: A = 1 OBS, B = 2 OBS, ETC.
          PLOT OF YHAT*NRE      SYMBOL USED IS P
```

FIGURE 9.3   Comparison of observed and predicted values of the venturi coefficient CV as a function of the Reynolds number, NRE.

SAS

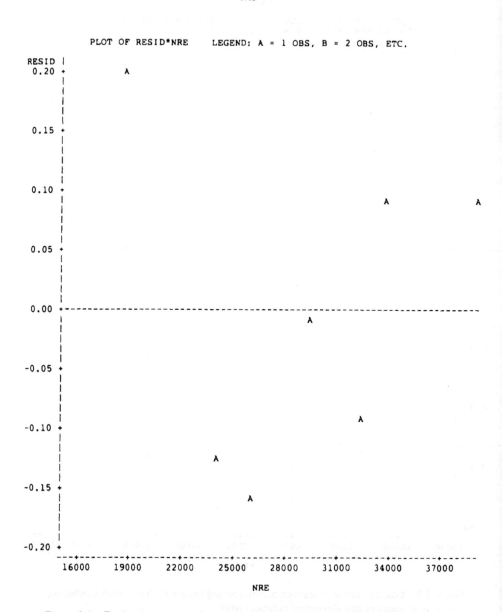

Figure 9.4 Evaluating residuals from linear regression of NRE on CV.

error sum of squares. This can be done only if we have some values of X for which we have more than one value for Y.

Suppose that of the n X's there are k distinct X's, where $k < n$, which occur with frequencies $n_1, n_2, \ldots, n_k$, where $n_1 + n_2 + \cdots + n_k = n$. The sum of squares of the n_i Y's corresponding to an X_i,

$$\sum_{j=1}^{n_i} (Y_j - \bar{Y})^2 = \sum_{j=1}^{n_i} Y_j^2 - n_i \bar{Y}^2 = (\text{SS}_E)_i,$$

is a sum of squares due to experimental error. The pooled sum of squares due to experimental error is denoted by

$$\text{SS}_E = \sum_{i=1}^{k} (\text{SS}_E)_i \qquad (9.43)$$

with $\sum_{i=1}^{k} (n_i - 1) = n - k$ degrees of freedom. The mean square due to experimental error is $\text{MS}_E = \text{SS}_E/(n - k)$. The usual residual sum of squares is $\text{SS}_{\text{Res}} = \sum_{i=1}^{n} (Y_i - \hat{Y}_i)^2$ with $n - 2$ degrees of freedom. The lack-of-fit sum of squares is $\text{SS}_{\text{L.F.}} = \text{SS}_{\text{Res}} - \text{SS}_E$ with $n - 2 - (n - k) = k - 2$ degrees of freedom. The critical region in testing H_0: lack of fit is $F > F_{k-2, n-k, 1-\alpha}$, where

$$F = \frac{\text{SS}_{\text{L.F.}}/(k - 2)}{\text{SS}_E/(n - k)} = \frac{\text{MS}_{\text{L.F.}}}{\text{MS}_E}$$

has an F-distribution with $k - 2$ and $n - k$ degrees of freedom.

9.2.6 Regression Through a Point

The least-squares regression technique will not necessarily produce an equation that will pass through some predetermined point. Such an occurrence is a necessity in many physical situations: at zero flow, the float in a rotameter should be at zero and the pressure drop across an orifice or other flowmeter should also be zero. If no liquid is flowing down a gas absorber, its efficiency cannot be anything other than zero. Given a particular point (X_p, Y_p) through which a line must pass, how can we find the coefficients which will cause that to occur? The model in question is now

$$Y_i - Y_p = \beta(X_i - X_p) + \varepsilon_i, \qquad i = 1, 2, \ldots, n$$

for which the normal equations must be solved as before. For the case of a linear model, the fitted equation must pass through (X_p, Y_p) instead of $(\bar{X}, \bar{Y})$. The normal equations for this case yield

$$\hat{\beta} = \frac{\sum_i (X_i - X_p)(Y_i - Y_p)}{\sum_i (X_i - X_p)^2}.$$

Example 9.6 The calibration data for the feed rotameter of the pilot plant distillation column are as follows:

Reading, Y	0	10	20	30	40	50	60	70	80	90	100
Flow rate, X (mL/min)	0	22	40	61	86	103	120	143	164	186	210

$$\sum X_i Y_i = 79{,}510, \qquad \sum Y_i = 550, \qquad \bar{Y} = 50,$$
$$\sum X_i = 1135, \qquad \sum X_i^2 = 164{,}251, \qquad \bar{X} = 103.181818,$$
$$X_p = 0, \qquad Y_p = 0, \qquad \hat{\beta} = 0.484076.$$

The corresponding standard least-squares fit of these data yields $\hat{\beta}_1 = 0.482821$, $\hat{\beta}_0 = 0.181665$. The difference is due, of course, to the fact that in the first case, the line was forced through the origin.

9.3 TESTING EQUALITY OF SLOPES

Suppose that a series of n observations of pairs (X, Y) can be partitioned into k groups with n_i pairs in the ith group such that $n_1 + n_2 + \cdots + n_k = n$. Assume a simple linear regression model for each group of observations with a common error variance for all groups. In this section we describe a procedure for testing the hypothesis $H_0 : \beta_1 = \beta_2 = \cdots = \beta_k$, the equality of the k slopes of k regression lines.

Consider first the case $k = 2$, that is, two groups $(X_{11}, Y_{11}), (X_{12}, Y_{12})$, $\ldots, (X_{1n_1}, Y_{1n_1})$ and $(X_{21}, Y_{21}), (X_{22}, Y_{22}), \ldots, (X_{2n_2}, Y_{2n_2})$. The estimates of β_1 and β_2 are, from Eq. (9.13),

$$\hat{\beta}_i = \frac{\sum_{j=1}^{n_i} (X_{ij} - \bar{X}_i)(Y_{ij} - \bar{Y}_i)}{\sum_{j=1}^{n_i} (X_{ij} - \bar{X}_i)^2}, \qquad i = 1, 2,$$

where $\bar{X}_i$ and $\bar{Y}_i$ are the means for the ith group. The variances of $\hat{\beta}_1$ and $\hat{\beta}_2$ are

$$s_{\hat{\beta}_1}^2 = \frac{\hat{\sigma}^2}{\sum_j (X_{1j} - \bar{X}_1)^2} \quad \text{and} \quad s_{\hat{\beta}_2}^2 = \frac{\hat{\sigma}^2}{\sum_j (X_{2j} - \bar{X}_2)^2}.$$

The variance of $\hat{\beta}_1 - \hat{\beta}_2$ is $s_{\hat{\beta}_1}^2 + s_{\hat{\beta}_2}^2$. Since $E(\hat{\beta}_1 - \hat{\beta}_2) = \beta_1 - \beta_2 = 0$ under $H_0 : \beta_1 = \beta_2$, and since $\hat{\beta}_1 - \hat{\beta}_2$ has a normal distribution with mean $\beta_1 - \beta_2$ and variance $\sigma_{\hat{\beta}_1}^2 + \sigma_{\hat{\beta}_2}^2$, the statistic

$$T = \frac{\hat{\beta}_1 - \hat{\beta}_2}{\hat{\sigma}\sqrt{1/\sum_j (X_{1j} - \bar{X}_1)^2 + 1/\sum_j (X_{2j} - \bar{X}_2)^2}} \tag{9.44}$$

has a t-distribution with $n_1 - 2 + n_2 - 2 = n_1 + n_2 - 4$ degrees of freedom, where

$$\hat{\sigma}^2 = \frac{(SS_E)_1 + (SS_E)_2}{n_1 + n_2 - 4}.$$

The terms $(SS_E)_1$ and $(SS_E)_2$ denote the error sums of squares corresponding to the two sets of data; that is,

$$(SS_E)_i = \sum_j (Y_{ij} - \bar{Y}_i)^2 - \hat{\beta}_i^2 \sum_j (X_{ij} - \bar{X}_i)^2, \qquad i = 1,2,$$

according to Eq. (9.15). To test the hypothesis $H_0: \beta_1 = \beta_2$, the procedure is to use the statistic in (9.44) with rejection region $T > t_{v,1-\alpha/2}$ or $T < t_{v,\alpha/2}$, where $v = n_1 + n_2 - 4$ degrees of freedom.

Example 9.7 Using the results of Problem 9.18, compare the slopes of the two calibration curves.

Solution The slopes $\hat{\beta}_{1,i}$ and their standard errors $s_{\hat{\beta}_{1,i}}$ are

$$\hat{\beta}_{1,1} = -0.13081677, \qquad s_{\hat{\beta}_{1,1}} = 0.00536172,$$
$$\hat{\beta}_{1,2} = -0.13072794, \qquad s_{\hat{\beta}_{1,2}} = 0.00483616.$$

From Eq. (9.44), which can be written as

$$T = \frac{\hat{\beta}_{1,1} - \hat{\beta}_{1,2} - (\beta_{1,1} - \beta_{1,2})}{\sqrt{s_{\hat{\beta}_{1,1}}^2 + s_{\hat{\beta}_{1,2}}^2}}.$$

we calculate $T = -0.0123$ with $n_1 + n_2 - 4 = 18$ degrees of freedom for this example. As $t_{18,0.975} = 2.101$, the null hypothesis of equal slopes is accepted.

In the case of k groups, let

$$(X_{i1}, Y_{i1}), (X_{i2}, Y_{i2}), \ldots, (X_{in_i}, Y_{in_i})$$

respresent the data for the ith group. The estimates of the slopes $\beta_1, \beta_2, \ldots, \beta_k$ are given by

$$\hat{\beta}_i = \frac{\sum_j (X_{ij} - \bar{X}_i)(Y_{ij} - \bar{Y}_i)}{\sum_j (X_{ij} - \bar{X}_i)^2}, \qquad i = 1, 2, \ldots, k.$$

The corresponding slope obtained by treating all the individual groups as one large group is $\hat{\beta}_p$, where

$$\hat{\beta}_p = \frac{\sum_{i=1}^k \sum_{j=1}^{n_i} (X_{ij} - \bar{X})(Y_{ij} - \bar{Y})}{\sum_{i=1}^k \sum_{j=1}^{n_i} (X_{ij} - \bar{X})^2},$$

where $\bar{X}$ and $\bar{Y}$ denote the overall means using all $n = \sum_{i=1}^{k} n_i$ values. The error sum of squares for the ith group is

$$(SS_E)_i = \sum_j (Y_{ij} - \bar{Y}_i)^2 - \hat{\beta}_i^2 \sum_j (X_{ij} - \bar{X}_i)^2, \tag{9.45}$$

and the pooled error sum of squares, denoted by S_1, is

$$\begin{aligned} S_1 &= \sum_{i=1}^{k} (SS_E)_i \\ &= \sum_{i=1}^{k} \left[\sum_j (Y_{ij} - \bar{Y}_i)^2 - \hat{\beta}_i^2 \sum_j (X_{ij} - \bar{X}_i)^2 \right], \end{aligned} \tag{9.46}$$

with $\sum_{i=1}^{k} (n_i - 2) = \sum_i n_i - 2k$ degrees of freedom. An unbiased estimate of σ^2 is furnished by

$$\hat{\sigma}^2 = \frac{S_1}{\sum_i n_i - 2k}.$$

The total sum of squares based on all k data sets collectively can be partitioned as

$$\begin{aligned} \sum_{i=1}^{k} \sum_{j=1}^{n_i} (Y_{ij} - \bar{Y})^2 = S_1 + \sum_{i=1}^{k} \sum_{j=1}^{n_i} {}^{\prime}\hat{\beta}_i - \hat{\beta}_p)^2 (X_{ij} - \bar{X}_i)^2 \\ + \hat{\beta}_p^2 \sum_{i=1}^{k} \sum_{j=1}^{n_i} (X_{ij} - \bar{X}_i)^2. \end{aligned} \tag{9.47}$$

The term S_1 is the pooled error sum of squares given by Eq. (9.46) and the second term, denoted by S_2, is the sum of squares due to differences between group slopes. The statistic

$$F = \frac{S_2/(k-1)}{S_1/(\sum_{i=1}^{k} n_i - 2k)}$$

has an F-distribution with $(k-1)$ and $(\sum_{i=1}^{k} n_i - 2k)$ degrees of freedom if $\beta_1 = \beta_2 = \cdots = \beta_k$. Thus the critical region for $H_0: \beta_1 = \beta_2 = \cdots = \beta_k$ is $F > F_{v_1, v_2, 1-\alpha}$, where $v_1 = k - 1$ and $v_2 = \sum_{i=1}^{k} n_i - 2k$. The sum S_2 can be written as

$$S_2 = \sum_i \sum_j (\hat{\beta}_i - \hat{\beta}_p)^2 (X_{ij} - \bar{X})^2 \tag{9.48a}$$

$$= \sum_{i=1}^{k} \hat{\beta}_i^2 A_i^2 - \hat{\beta}_p^2 \sum_{i=1}^{k} A_i, \tag{9.46}$$

where

$$A_i = \sum_{j=1}^{n_i} (X_{ij} - \bar{X}_i)^2.$$

9.4 MULTIPLE LINEAR REGRESSION

Matrix algebra is readily adaptable to multiple linear regression, where the model to be fitted is

$$Y = \beta_0 + \beta_1 X_1 + \beta_2 X_2 + \cdots + \beta_p X_p + \varepsilon. \tag{9.49}$$

The assumptions are the same as for the simple linear model except that now we have p independent variables. The same discussion of Eqs. (9.1) and (9.3) in simple linear regression is pertinent to the model (9.49). That is, we can consider (9.49) or the model

$$E(Y|X_1, \ldots, X_p) = \beta_0 + \beta_1 X_1 + \beta_2 X_2 + \cdots + \beta_p X_p,$$

which considers $\beta_0 + \beta_1 X_1 + \beta_2 X_2 + \cdots + \beta_p X_p$ to be the conditional mean of Y, given $X_1, X_2, \ldots, X_p$. The assumptions are the same as for the simple linear model (9.1) and the simple linear regression model (9.3) except that now we have p independent variables.

To obtain the least-squares estimates for the β_i, we must again minimize the error sum of squares. As with simple linear regression, we have n observations on Y, $X_1, X_2, \ldots, X_p$, and the error sum of squares is

$$\begin{aligned} \mathrm{SS_E} &= \sum_i e_i^2 = \sum_i (Y_i - \hat{Y}_i)^2 \\ &= \sum_i (Y_i - \hat{\beta}_0 - \hat{\beta}_1 X_{1i} - \hat{\beta}_2 X_{2i} - \cdots - \hat{\beta}_p X_{pi})^2, \end{aligned} \tag{9.50}$$

which is minimized by setting $\partial(\mathrm{SS_E})/\partial\hat{\beta}_i = 0$ to get the following system of normal equations:

$$\begin{aligned} n\hat{\beta}_0 + \hat{\beta}_1 \sum X_{1i} + \hat{\beta}_2 \sum X_{2i} + \cdots + \hat{\beta}_p \sum X_{pi} &= \sum Y_i, \\ \hat{\beta}_0 \sum X_{1i} + \hat{\beta}_1 \sum X_{1i}^2 + \hat{\beta}_2 \sum X_{1i} X_{2i} + \cdots + \hat{\beta}_p \sum X_{1i} X_{pi} &= \sum X_{1i} Y_i, \\ \hat{\beta}_0 \sum X_{2i} + \hat{\beta}_1 \sum X_{1i} X_{2i} + \hat{\beta}_2 \sum X_{2i}^2 + \cdots + \hat{\beta}_p \sum X_{2i} X_{pi} &= \sum X_{2i} Y_i, \\ \vdots \qquad\qquad \vdots \qquad\qquad \vdots \qquad\qquad\qquad \vdots \qquad & \\ \hat{\beta}_0 \sum X_{pi} + \hat{\beta}_1 \sum X_{1i} X_{pi} + \hat{\beta}_2 \sum X_{2i} X_{pi} + \cdots + \hat{\beta}_p \sum X_{pi}^2 &= \sum X_{pi} Y_i, \end{aligned} \tag{9.51}$$

where all the summations go from $i = 1$ to $i = n$. To obtain the estimates $\hat{\beta}_0, \hat{\beta}_1, \ldots, \hat{\beta}_p$ one needs to solve the system (9.51) of $p + 1$ linear equations, in the unknowns $\hat{\beta}_0, \hat{\beta}_1, \ldots, \hat{\beta}_p$. In the simple linear case we had two equations in two unknowns. A much easier approach to the normal equations is found from matrix algebra, where the $\hat{\beta}_i$ are determined by solving $X'X\hat{\beta} = X'Y$ which is actually the system (9.51) in matrix form.

In the case of p independent variables the sum of squares due to error is $SS_E = \sum_i (Y_i - \hat{Y}_i)^2$, where $\hat{Y}_i = \hat{\beta}_0 + \hat{\beta}_1 X_1 + \cdots + \hat{\beta}_p X_p$. It can be shown that SS_E can be written as

$$SS_E = \sum_i (Y_i - \hat{Y}_i)^2$$
$$= \sum_i (Y_i - \bar{Y})^2$$
$$- \left[\hat{\beta}_1 \sum_i (X_{1i} - \bar{X}_1)(Y_i - \bar{Y}) + \cdots + \hat{\beta}_p \sum_i (X_{pi} - \bar{X}_p)(Y_i - \bar{Y}) \right],$$

where $\bar{X}_j = \sum_{i=1}^n X_{ji}/n$. The sum enclosed in brackets is the sum of squares due to regression and thus

$$SS_E = SS_{TC} - SS_R$$

or

$$SS_{TC} = SS_E + SS_R,$$

where

$$SS_R = \hat{\beta}_1 \sum_i (X_{1i} - \bar{X}_i)(Y_i - \bar{Y}) + \cdots + \hat{\beta}_p \sum_i (X_{pi} - \bar{X}_p)(Y_i - \bar{Y}).$$

$$(9.52)$$

A hypothesis of great interest is $H_0: \beta_1 = \beta_2 = \cdots = \beta_p = 0$. To test this hypothesis, Table 9.3 is prepared. An unbiased estimate of σ^2 is furnished by $MS_E = \sum_i (Y_i - \hat{Y})^2/(n - p - 1)$. The hypothesis is tested by calculating $F = MS_R/MS_E$ and comparing it with $F_{p,(n-p-1),1-\alpha}$. The critical region is $F > F_{p,n-p-1,1-\alpha}$.

Another hypothesis of interest is $H_0: \beta_k = 0$, where β_k is a particular β. To test this hypothesis we need to know the distribution of the statistic $(\hat{\beta}_k - \beta_k)/s_{\hat{\beta}_k}$, where $s_{\hat{\beta}_k}^2$ denotes an estimate of the variance of $\hat{\beta}_k$. It can be

TABLE 9.3 Analysis of Variance for Multiple Linear Regression

Source	d.f.	SS	MS
Regression	p	SS_R	$MS_R = SS_R/p$
Error	$n - p - 1$	SS_E	$MS_E = SS_E/(n - p - 1)$
Total	$n - 1$	SS_{TC}	

shown from the theory regarding multiple regression [10] that $\hat{\beta}_{k-1}$ is unbiased for β_{k-1} and furthermore, that the variance of $\hat{\beta}_{k-1}$ is given by $c_{kk}\sigma^2$, where c_{kk} is the kkth element of the matrix $S^{-1} = C$. (It would prove instructive for the reader to check this out for the simple regression model.) In Ref. 10 it is shown that $c_{kk}\sigma^2$ is an unbiased estimator for $\sigma^2_{\hat{\beta}_{k-1}}$. Finally, the random variable $(\hat{\beta}_{k-1} - \beta_{k-1})/\hat{\sigma}\sqrt{c_{kk}}$ has a t-distribution with $(n - p - 1)$ degrees of freedom. Thus to test $H_0: \beta_{k-1} = 0$, we use the statistic $T = \hat{\beta}_{k-1}/\hat{\sigma}\sqrt{c_{kk}}$ and the rule is to reject H_0 if the computed T value is such that $T > t_{n-p-1,1-\alpha/2}$ or $T < t_{n-p-1,\alpha/2}$.

Example 9.8 It is necessary to relate the percent gas absorbed in a tower, Y, to the gas temperature, X_1, and the vapor pressure of the absorbing liquid, X_2. The postulated model is

$$Y = \beta_0 + \beta_1 X_1 + \beta_1 X_2 + \varepsilon.$$

The following data are available:

X_1 (°F)	X_2 (mmHg)	Y (% absorbed)
78.0	1.0	1.5
113.5	3.2	6.0
130.0	4.8	10.0
154.0	8.4	20.0
169.0	12.0	30.0
187.0	18.5	50.0
206.0	27.5	80.0
214.0	32.0	100.0

From these data the following values are calculated:

$\sum x_{1i} = 1251.5$, $\quad \sum x_{2i} = 107.4$, $\quad \sum x_{1i}x_{2i} = 20{,}359.29$,

$\sum x_{1i}^2 = 211{,}344.25$, $\quad \sum x_{2i}^2 = 2371.339$, $\quad \sum x_{1i}\sum x_{2i} = 134{,}411.0$,

$(\sum x_{1i})^2 = 1{,}566{,}252.25$, $\quad (\sum x_{2i})^2 = 11{,}534.75$, $\quad \sum y_i = 297.5$,

$\sum x_{1i}y_i = 57{,}478.0$, $\quad \sum x_{2i}y_i = 6921.70$, $\quad \sum y_i^2 = 20{,}338.25$,

$\sum x_{1i}\sum y_i = 372{,}321.25$, $\quad \sum x_{2i}\sum y_i = 31{,}951.49$,

$\bar{x}_1 = 156.4375$, $\quad \bar{x}_2 = 13.4249$, $\quad \bar{y} = 37.1875$

The corresponding corrected sums of squares are

$$\sum_i (x_{1i} - \bar{x}_1)^2 = 15{,}562.75,$$

$$\sum_i (x_{2i} - \bar{x}_2)^2 = 929.4951,$$

$$\sum_i (x_{1i} - \bar{x}_1)(x_{2i} - \bar{x}_2) = 3557.913,$$

$$\sum_i (x_{1i} - \bar{x}_1)(y_i - \bar{y}) = 10{,}937.844,$$

$$\sum_i (x_{2i} - \bar{x}_2)(y_i - \bar{y}) = 2927.763,$$

$$\sum_i (y_i - \bar{y})^2 = 9274.969.$$

The regression coefficients may now be found by solving the system

$$8\hat{\beta}_0 + 1251.5\hat{\beta}_1 + 107.4\hat{\beta}_2 = 297.5,$$
$$1251.5\hat{\beta}_0 + 211{,}344.25\hat{\beta}_1 + 20{,}359.29\hat{\beta}_2 = 57{,}478.0,$$
$$107.4\hat{\beta}_0 + 20{,}359.29\hat{\beta}_1 + 2371.339\hat{\beta}_2 = 6921.7.$$

The solution is $\hat{\beta}_1 = -0.13840$, $\hat{\beta}_2 = 3.6796$, and $\hat{\beta}_0 = 9.4398$. It can be shown that, in general,

$$\hat{\beta}_1 = \frac{s_{22}s_{1y} - s_{12}s_{2y}}{s_{11}s_{22} - s_{12}^2},$$

$$\hat{\beta}_2 = \frac{s_{11}s_{2y} - s_{12}s_{1y}}{s_{11}s_{22} - s_{12}^2},$$

and

$$\hat{\beta}_0 = \bar{Y} - \hat{\beta}_1\bar{X}_1 - \hat{\beta}_2\bar{X}_2,$$

where $s_{22} = \sum (X_{2i} - \bar{X}_2)^2$, $s_{1y} = \sum (X_{1i} - \bar{X}_1)(Y_i - \bar{Y})$, $s_{12} = \sum (X_{1i} - \bar{X}_1) \times \Phi X_{2i} - \bar{X}_2)$, and similarly for s_{11} and s_{2y}.

The resulting regression equation is

$$\hat{Y} = 9.4398 - 0.13840X_1 + 3.67961X_2.$$

In this example the sum of squares due to regression is

$$SS_R = \hat{\beta}_1 \sum (x_{1i} - \bar{x}_1)(y_i - \bar{y}) + \hat{\beta}_2 \sum_i (x_{2i} - \bar{x}_2)(y_i - \bar{y})$$

$$= -0.13840(10{,}937.84) + 3.67961(2927.762)$$

$$= 9259.22194.$$

The sum of squares due to error is thus

$$SS_E = \sum_i (y_i - \hat{y})^2$$

$$= \sum_i (y_i - \bar{y})^2 - SS_R$$

$$= 9274.96875 - 9259.22194$$

$$= 15.7468.$$

The hypothesis $H_0: \beta_1 = \beta_2 = 0$ is tested by using the F-ratio $F = MS_R/MS_E$. The computed F value is $f = 4629.6107/3.1494 = 1470.02$, which is highly significant, so $H_0: \beta_1 = \beta_2 = 0$ is rejected. The AOV table summarizing our results is given below.

Source	d.f.	SS	MS	F
Regression	2	9259.22194	4629.6107	1470.02
Error	5	15.74681	3.1494	
Total	7	9274.96875		

If one wanted to test hypotheses involving β_1 or β_2, one would use the t-statistics given by

$$T = \frac{\hat{\beta}_i - \beta_i}{s_{\hat{\beta}_i}}, \qquad i = 1,2.$$

As mentioned previously, estimates of $\sigma_{\hat{\beta}_1}^2$ and $\sigma_{\hat{\beta}_2}^2$ are obtained by finding the inverse of the coefficient matrix on the left-hand side of system (9.51), which in our example is

$$8\hat{\beta}_0 + \hat{\beta}_1 \sum_i X_{1i} + \hat{\beta}_2 \sum_i X_{2i} = \sum_i Y_i,$$

$$\hat{\beta}_0 \sum_i X_{1i} + \hat{\beta}_1 \sum_i X_{1i}^2 + \hat{\beta}_2 \sum_i X_{1i}X_{2i} = \sum_i X_{1i}Y_i,$$

$$\hat{\beta}_0 \sum_i X_{2i} + \hat{\beta}_1 \sum_i X_{1i}X_{2i} + \hat{\beta}_2 \sum_i X_{2i}^2 = \sum_i X_{2i}Y_i.$$

The estimates of $\sigma_{\hat{\beta}_1}^2$ and $\sigma_{\hat{\beta}_2}^2$ are then given by $s_{\hat{\beta}_1}^2 = c_{22}\hat{\sigma}^2$ and $s_{\hat{\beta}_2}^2 = c_{33}\hat{\sigma}^2$, where c_{22} and c_{33} are the second and third diagonal elements in the inverse of

$$\begin{pmatrix} 8 & 1{,}251.5 & 107.4 \\ 1251.5 & 211{,}344.25 & 20{,}359.29 \\ 107.4 & 20{,}359.29 & 2{,}371.339 \end{pmatrix} = \mathbf{S}.$$

In our example $c_{22} = 5.1446 \times 10^{-4}$ and $c_{33} = 8.6137 \times 10^{-3}$. Thus $s_{\hat{\beta}_1}^2 = 5.1446 \times 10^{-4} \cdot \sigma^2 = 5.1446 \times 10^{-4}(3.149) = 0.001620$. Similarly, $s_{\hat{\beta}_2}^2 = 0.02717$. To test the hypothesis $H_0: \beta_1 = 0$ the computed T value is $T = \hat{\beta}_1/s_{\hat{\beta}_1} = -0.1384/0.0402 = -3.443$. Since $-3.433 < t_{5,0.025} = -2.571$, $H_0: \beta_1 = 0$ is rejected at the 5% level of significance. To test $H_0: \beta_2 = 0$ the computed T value is $T = \hat{\beta}_2/s_{\hat{\beta}_2} = 3.67961/0.1648 = 22.328$. Since $22.328 > 2.571$, $H_0: \beta_2 = 0$ is likewise rejected at the 5% level of significance.

The use of the SAS system for multiple linear regression can be illustrated by the following example.

Example 9.9   In the production of ethylene glycol from ethylene oxide, the conversion of ethylene to ethylene oxide, Y, is a function of the activity X_1 of the silver catalyst and the residence time X_2. The following coded data are available:

Y	12.1	11.9	10.2	8.0	7.7	5.3	7.9	7.8	5.5	2.6
X_1	0	1	2	3	4	5	6	7	8	9
X_2	7	4	4	6	4	2	1	1	1	0

Calculate the regression coefficients.
Solution   The program is as follows.

```
(put initial JCL cards/statements here)
DATA;
INPUT X1   1   X2   3   Y   5-8;
CARDS;
(put data here according to INPUT format)
PROC PRINT;
PROC GLM;
MODEL Y = X1   X2;
OUTPUT OUT = NEW P = YHAT R = RESID;
PROC GLM;
MODEL YHAT = Y;
PROC PLOT;
PLOT YHAT * Y;
TITLE 'NOTE: IF MLR OK, GRAPH SLOPE = 1, INTERCEPT = 0';
(put final JCL cards/statements here)
```

Note that the INPUT statement can be adjusted to accommodate any number of variables of any size. The MODEL statement instructs the program to perform a multiple regression by the method of least squares, in this case using two independent variables. The PLOT statement in this program will result in a graph of YHAT, the predicted value, vs. Y, the original data (Fig. 9.5). If the regression has been significant, the resulting line should pass through the origin with slope of 1.0. To find out whether it does, we can look at F for the first MODEL and use PROC GLM for the YHAT vs. Y data to get the slope and intercept and their corresponding standard errors of estimate. The results follow.

SAS

GENERAL LINEAR MODELS PROCEDURE

DEPENDENT VARIABLE: Y

SOURCE	DF	SUM OF SQUARES	MEAN SQUARE	F VALUE
MODEL	2	65.68846912	32.84423456	16.53
ERROR	7	13.91153088	1.98736155	PR > F
CORRECTED TOTAL	9	79.60000000		0.0022

SOURCE	DF	TYPE I SS	F VALUE	PR > F
X1	1	63.71212121	32.06	0.0008
X2	1	1.97634791	0.99	0.3519

PARAMETER	ESTIMATE	T FOR H0: PARAMETER=0	PR > |T|	STD ERROR OF ESTIMATE
INTERCEPT	14.70755585	4.94	0.0017	2.97851835
X1	-1.20420499	-3.33	0.0125	0.36135254
X2	-0.46287779	-1.00	0.3519	0.46416575

From these results we see that the null hypothesis $H: \beta_1 = \beta_2 = 0$ is rejected at $\alpha = 0.05$. It appears, however, from the T_{β_2} value that β_2 is probably 0. From plotting the residuals $(Y_i - YHAT_i)$ vs. each of the independent variables, it appears that X_2^2 should have been included in the model rather than the X_2 term. Those plots are not shown.

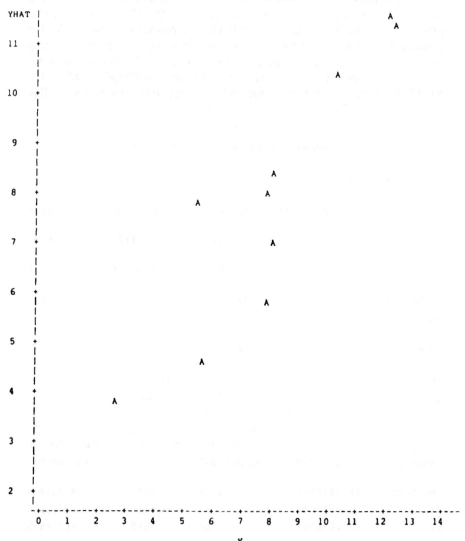

FIGURE **9.5** Comparison of observed (Y) and predicted (YHAT) values of ethylene oxide conversion as a result of multiple regression.

Continuing, let us examine the result of comparing the predicted (YHAT$_i$) and original (Y$_i$) conversion data.

SAS

GENERAL LINEAR MODELS PROCEDURE

DEPENDENT VARIABLE: YHAT

SOURCE	DF	SUM OF SQUARES	MEAN SQUARE	F VALUE
MODEL	1	54.20822833	54.20822833	37.77
ERROR	8	11.48024079	1.43503010	PR > F
CORRECTED TOTAL	9	65.68846912		0.0003

SOURCE	DF	TYPE I SS	F VALUE	PR > F
Y	1	54.20822833	37.77	0.0003

PARAMETER	ESTIMATE	T FOR H0: PARAMETER=0	PR > |T|	STD ERROR OF ESTIMATE
INTERCEPT	1.38066701	1.23	0.2551	1.12633543
Y	0.82523202	6.15	0.0003	0.13426845

From these results we see that we would accept $H_0: \beta_0 = 0$ and reject $H_0: \beta_1 = 0$ at $\alpha = 0.05$. From the standard error we calculate

$$T_{\beta_1} = \frac{-1 + 0.82523202}{0.13426845} = -1.302.$$

We thus accept the hypothesis that $\beta_1 = 1$ and conclude that the regression is probably satisfactory, even if the data do appear badly scattered.

9.5 POLYNOMIAL REGRESSION

In the case of polynomial or curvilinear regression, as given by the model,

$$Y = \beta_0 + \beta_1 X + \beta_2 X^2 + \cdots + \beta_p X^p + \varepsilon, \tag{9.53}$$

there is only one independent variable: X. However, $p - 1$ other independent variables are defined as powers of X. The powers of X can be considered as $W_1 = X$, $W_2 = X^2, \ldots, W_p = X^p$ and the model (9.53) is reduced to the multiple linear regression model given by Eq. (9.49) with independent variables $W_1, W_2, \ldots, W_p$. Thus the determination of the $\hat{\beta}_i$ in model (9.53) is done in the fashion of Section 9.4, the only difference being that to determine $W_1, W_2, \ldots, W_p$ one needs to know only X.

As an example, consider the quadratic model

$$Y = \beta_0 + \beta_1 X + \beta_2 X^2 + \varepsilon.$$

In this case $W_1 = X$ and $W_2 = X^2$ and the normal equations are, given n observations on X and Y,

$$n\hat{\beta}_0 + \hat{\beta}_1 \sum_i W_{1i} + \hat{\beta}_2 \sum_i W_{2i} = \sum_i Y_i,$$

$$\hat{\beta}_0 \sum_i W_{1i} + \hat{\beta}_1 \sum_i W_{1i}^2 + \hat{\beta}_2 \sum_i W_{1i} W_{2i} = \sum_i W_{1i} Y_i,$$

$$\hat{\beta}_0 \sum_i W_{2i} + \hat{\beta}_1 \sum_i W_{1i} W_{2i} + \hat{\beta}_2 \sum_i W_{2i}^2 = \sum_i W_{2i} Y_i.$$

However, $W_{1i} = X_i$ and $W_{2i} = X_i^2$. Thus the normal equations become

$$n\hat{\beta}_0 + \hat{\beta}_1 \sum_i X_i + \hat{\beta}_2 \sum_i X_i^2 = \sum_i Y_i,$$

$$\hat{\beta}_0 \sum_i X_i + \hat{\beta}_1 \sum_i X_i^2 + \hat{\beta}_2 \sum_i X_i^3 = \sum_i X_i Y_i. \qquad (9.54)$$

$$\hat{\beta}_0 \sum_i X_i^2 + \hat{\beta}_1 \sum_i X_i^3 + \hat{\beta}_2 \sum_i X_i^4 = \sum_i X_i^2 Y_i.$$

Equations (9.54) can be solved for $\hat{\beta}_0$, $\hat{\beta}_1$, and $\hat{\beta}_2$. Extensions to polynomials of higher degree are obvious and the solution follows in the same manner.

It should be pointed out that when one speaks of a *linear* model in regression the term *linear* means linear in the parameters $\beta_0, \beta_1, \ldots, \beta_p$ and not in the independent variable X. Thus the polynomial regression model in Eq. (9.53) is a linear model. However, the model $Y = \beta_0 e^{\beta_1 X}$ is not since Y is not linear in β_0 and β_1. The nonlinear model is discussed in Section 9.6.

Other examples of linear models (linear in the parameters) are

(1) $Y = \beta_0 + \beta_1 \log X + \beta_2 X^2 + \varepsilon.$
(2) $Y = \beta_0 + \beta_1 e^{-X} + \beta_2 X^{1/2} + \varepsilon.$
(3) $Y = \beta_0 + \beta_1 e^{-X_1} + \beta_2 X_2^2 + \beta_3 X_3 + \varepsilon.$

Models (1) and (2) involve one variable X but two independent variables defined in terms of X. Model (3) contains three independent variables and could be written as $Y = \beta_0 + \beta_1 W_1 + \beta_2 W_2 + \beta_3 W_3 + \varepsilon$, where $W_1 = e^{-X_1}$, $W_2 = X_2^2$, and $W_3 = X_3$.

Example 9.10   It is believed that the effect of temperature on catalyst activity is quadratic. Eight different temperatures (coded X data below) were used. The resulting activities are given as Y. The proposed model is

$$Y = \beta_0 + \beta_1 X + \beta_2 X^2 + \varepsilon.$$

The following data are available:

X	Y
2	0.846
4	0.573
6	0.401
8	0.288
10	0.209
12	0.153
14	0.111
16	0.078

If we let $W_1 = X$ and $W_2 = X^2$, the model reduces to the form

$$Y = \beta_0 + \beta_1 W_1 + \beta_2 W_2 + \varepsilon.$$

Following the same procedure as for multiple linear regression, the values of the β_i are estimated as

$$\hat{\beta}_0 = 1.05652,$$
$$\hat{\beta}_1 = -0.13114,$$
$$\hat{\beta}_2 = 0.00447.$$

The resulting regression equation is then

$$Y = 1.05652 - 0.13114X + 0.00447X^2.$$

In studies that involve only one independent variable it is generally advisable to consider a plot of the data. This is an easy task and it often suggests the type of model to be used. For example, the plot may suggest an absence of linearity in X in favor of a quadratic regression model; or it may suggest an exponential model $Y = \beta_0 e^{-\beta_1 X}$, which is discussed in Section 9.6.

The use of SAS procedures for quadratic regression follows the same pattern as for multiple linear regression with one change: a new variable, X^2, must be generated from the original data. Instructions for this must be placed between the INPUT and CARDS statements. The following example illustrates the procedure.

Example 9.11 Magnetic taconite is separated mechanically from a crushed ore slurry and rolled into marble-sized balls before charging to a pelletizing furnace. After burning in the furnace, the pellets must have a crushing strength of 8 to 10 lb. A binder, natural peat, is added during the balling process to increase the strength of the pellets. From the following data, propose an equation for the effect of peat concentration on crushing strength of taconite balls [20].*

Strength of taconite balls (lb)	Peat content (lb/ton of taconite)
3.6	0.0
9.8	4.0
14.7	8.0
16.2	12.0
16.0	16.0
15.5	20.0

Solution

(put initial JCL cards/statements here)

```
DATA;
INPUT STR 1-4  PEAT 6-9;
PEATSQ = PEAT * PEAT;
CARDS;
```

(put data here according to INPUT format)

```
PROC GLM;
MODEL STR = PEAT  PEATSQ;
OUTPUT OUT = NEW P = YHAT R = RESID;
PROC PRINT;
PROC PLOT;
PLOT STR * PEAT YHAT * PEAT = 'P'/OVERLAY;
```

(*Continued*)

```
PLOT RESID * PEAT/VREF = 0;
TITLE 'QUAD REGR OF PEAT CONC ON STRENGTH OF TACONITE';
PROC GLM;
MODEL YHAT = STR;
PROC PLOT;
PLOT YHAT * STR;
TITLE 'NOTE:  IF QUAD REGR OK, SLOPE = 1 AND INTERCEPT = 0';
```
(put final JCL cards/statements here)

Note that in all MODEL and PLOT statements, the dependent variable must be listed first. The form of the first PROC GLM and MODEL statements instruct the program to perform a quadratic least-squares regression. The second PROC GLM and MODEL statements produce a linear regression of the original data (STR_i) on the predicted ($YHAT_i$) values. The /OVERLAY command causes the original and predicted strength values to be plotted vs. peat concentration on the same graph. The last plot statement produces a graph of the residuals ($RESID_i = STR_i - YHAT_i$) vs. the peat concentration (PEAT). If the quadratic model was appropriate for the data set, the residuals should be randomly scattered about the reference line (/VREF = 0). The output is shown below.

SAS

DEP VARIABLE: STR

SOURCE	DF	SUM OF SQUARES	MEAN SQUARE	F VALUE	PROB>F
MODEL	2	125.102	62.551202	175.225	0.0008
ERROR	3	1.070929	0.356976		
C TOTAL	5	126.173			

DEP MEAN	12.633333	ADJ R-SQ	0.9859	
C.V.	4.729352			

VARIABLE	DF	PARAMETER ESTIMATE	STANDARD ERROR	T FOR H0: PARAMETER=0	PROB > |T|
INTERCEP	1	3.739286	0.541508	6.905	0.0062
PEAT	1	1.771696	0.127339	13.913	0.0008
PEATSQ	1	-0.060156	0.006111548	-9.843	0.0022

From the F-test and the t-tests of the individual coefficients we accept the regression as significant. The correlation index R^2 seems to confirm this.

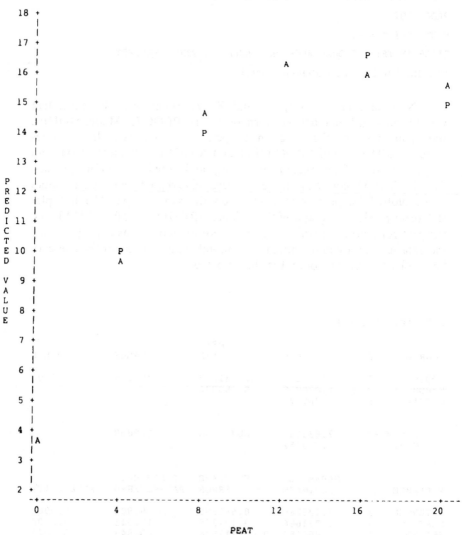

FIGURE **9.6** Comparison of original crushing strength values (A) with those predicted (P) by quadratic regression of peat concentration.

Figure 9.6 shows the original and predicted strength values as a function of peat concentration. From this graph the regression looks even more reasonable. Figure 9.7 shows the residuals $(STR_i - YHAT_i)$ vs. peat concentration. As the residuals are randomly scattered about the reference line, this is even more convincing evidence that the regression model used was the correct one.

Finally, we can use the $YHAT_i$ values generated by the OUTPUT statement in yet another way: to find the regression of strength (STR_i) on $YHAT_i$. As shown in the results from that PROC GLM routine, a linear model is appropriate for comparing the observed and predicted strength values. From the t-tests, $T_{\beta_0} = 0.174$, which leads us to accept the hypothesis that this model does pass through the origin. Using the standard error of estimate s_{β_1} to find $T_{\beta_1} = 0.185$, we similarly accept the null hypothesis that the slope is 1.0, as indicated in Fig. 9.8.

SAS

QUAD REGR OF PEAT CONC ON STRENGTH OF TACONITE

DEP VARIABLE: YHAT   PREDICTED VALUE

SOURCE	DF	SUM OF SQUARES	MEAN SQUARE	F VALUE	PROB>F
MODEL	1	124.041	124.041	467.267	0.0001
ERROR	4	1.061839	0.265460		
C TOTAL	5	125.102			

VARIABLE	DF	PARAMETER ESTIMATE	STANDARD ERROR	T FOR H0: PARAMETER=0	PROB > |T|
INTERCEP	1	0.107229	0.616468	0.174	0.8704
STR	1	0.991512	0.045869	21.616	0.0001

Example 9.12 The National Weather Service (NWS) calculates the wind-chill temperature from the following equation:

$$CHILTEMP = 0.0817(3.71\sqrt{V} + 5.81 - 0.25V)(T - 91.4) + 91.4$$

where V is the windspeed, mph, and T is the actual dry-bulb temperature, °F. The *Minneapolis Star-Tribune* has published the following "data" from an unknown source for its readers. To use these data, CHILTEMP = 29°F at $V = 25$ mph. 10°F, etc.

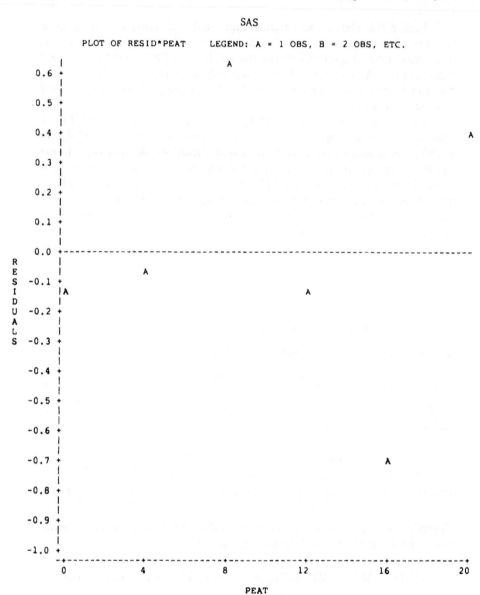

FIGURE 9.7   Evaluation of residuals from quadratic regression for random scatter.

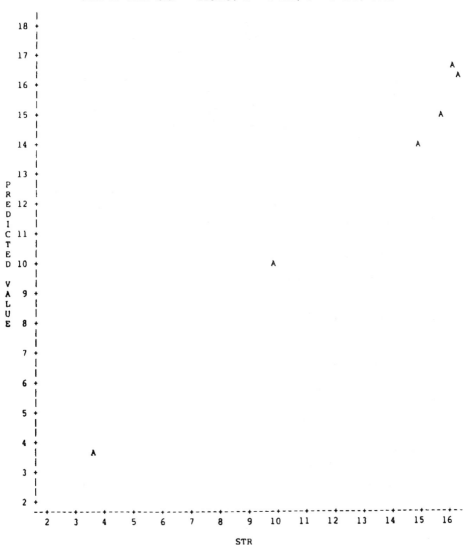

SAS

NOTE: IF QUAD REGR OK, SLOPE = 1 AND INTERCEPT = 0

PLOT OF YHAT*STR   LEGEND: A = 1 OBS, B = 2 OBS, ETC.

FIGURE 9.8 Comparison of predicted (YHAT) and observed (Y) values of crushing strength.

Dry-bulb temp. (°F)	CHILTEMP (°F) for wind speed (mph):							
	5	10	15	20	25	30	35	40
30	27	16	9	4	1	−2	−4	−5
20	16	3	−5	−10	−15	−18	−20	−21
10	6	−9	−18	−24	−29	−33	−35	−37
0	−5	−22	−31	−34	−44	−49	−52	−53
−10	−15	−34	−45	−53	−59	−64	−67	−69
−20	−26	−46	−58	−67	−74	−79	−82	−84
−30	−36	−58	−72	−81	−88	−93	−97	−100
−40	−47	−71	−85	−95	−103	−109	−113	−115

Develop a suitable equation from the *Star-Tribune* data and use it to evaluate the NWS equation for CHILTEMP.

Solution The lazy student will use the model

$$Y_{ij} = \beta_0 + \beta_1 X_1 + \beta_2 X_2 + \varepsilon_{ij},$$

as shown in the following SAS program.

(put initial JCL cards/statements here)

```
DATA TEMP;
INPUT Y 1-4 X1 6-8 X2 10-11;
LABEL Y = 'WINDCHILL TEMPERATURE'
      X1 = 'ACTUAL AIR TEMPEARATURE'
      X2 = 'WIND SPEED';
CARDS;
```

(insert data here according to input format)

```
PROC PRINT;
TITLE 'WINDCHILL TEMPERATURES';
PROC GLM;
TITLE 'MULTIPLE LINEAR REGRESSION ON WINDCHILL TEMPEATURE DATA';
MODEL Y = X1 X2;
OUTPUT OUT = NEW     P = YHAT    R = RESID;
PROC GLM;
TITLE ' EVALUATION OF REGRESSION MODEL';
MODEL YHAT = Y;
PROC PLOT;
PLOT YHAT*Y;
TITLE 'IF MLR OK, GRAPH SLOPE = 1 AND INTERCEPT = 0';
```

(put initial JCL cards/statements here)

The results are shown in the next AOV table.

General Linear Models Procedure

Dependent Variable: Y   WINDCHILL TEMPERATURE

Source	DF	Sum of Squares	Mean Square	F Value	Pr > F
Model	2	81007.8869	40503.9435	911.86	0.0001
Error	61	2709.5506	44.4189		
Corrected Total	63	83717.4375			

R-Square	C.V.	Root MSE	Y Mean
0.967635	-15.35435	6.66475	-43.40625

Source	DF	Type I SS	Mean Square	F Value	Pr > F
X1	1	65604.2411	65604.2411	1476.95	0.0001
X2	1	15403.6458	15403.6458	346.78	0.0001

Source	DF	Type III SS	Mean Square	F Value	Pr > F
X1	1	65604.2411	65604.2411	1476.95	0.0001
X2	1	15403.6458	15403.6458	346.78	0.0001

Parameter	Estimate	T for H0: Parameter=0	Pr > |T|	Std Error of Estimate
INTERCEPT	-5.950892857	-3.23	0.0020	1.84502686
X1	1.397321429	38.43	0.0001	0.03635918
X2	-1.354166667	-18.62	0.0001	0.07271836

This two-term multiple linear regression model produced results (model F-test and parameter t-tests) which indicate that this abbreviated model might be correct. In addition to the AOV shown above, R^2 for the model was 0.967635, indicating that the abbreviated model accounted for about 96.7% of the variability of the data. In this situation use of R^2 as a criterion for accepting the model leads to an erroneous decision.

When the predicted windchill values ($YHAT_i$) were plotted (Fig. 9.9) against the original values (Y_i), the result was a broad band, not the expected single line. At this point, new variables are defined between the INPUT and CARDS lines as follows:

ROOTV = SQRT(V);

TROOTV = T*ROOTV;

TV = T*V;

The model statement is written as

MODEL CHILT = TROOTV T TV ROOTV V;

to produce the following (second) AOV table for which $R^2 = 0.999596$.

General Linear Models Procedure

Dependent Variable: CHILT

Source	DF	Sum of Squares	Mean Square	F Value	Pr > F
Model	5	83683.6327	16736.7265	28715.76	0.0001
Error	58	33.8048	0.5828		
Corrected Total	63	83717.4375			

R-Square	C.V.	Root MSE	CHILT Mean
0.999596	-1.758826	0.76344	-43.40625

Source	DF	Type I SS	Mean Square	F Value	Pr > F
TROOTV	1	68533.2857	68533.2857	99999.99	0.0
T	1	318.7922	318.7922	546.96	0.0001
TV	1	119.3266	119.3266	204.73	0.0001
ROOTV	1	14217.3195	14217.3195	24393.13	0.0001
V	1	494.9088	494.9088	849.13	0.0001

Source	DF	Type III SS	Mean Square	F Value	Pr > F
TROOTV	1	100.71942	100.71942	172.81	0.0001
T	1	60.61442	60.61442	104.00	0.0001
TV	1	35.15944	35.15944	60.32	0.0001
ROOTV	1	1477.54546	1477.54546	2535.07	0.0001
V	1	494.90877	494.90877	849.13	0.0001

Parameter	Estimate	T for H0: Parameter=0	Pr > |T|	Std Error of Estimate
INTERCEPT	46.90080605	42.56	0.0001	1.10194748
TROOTV	0.30277375	13.15	0.0001	0.02303227
T	0.47917309	10.20	0.0001	0.04698720
TV	-0.02052414	-7.77	0.0001	0.00264253
ROOTV	-27.19652334	-50.35	0.0001	0.54015459
V	1.80587612	29.14	0.0001	0.06197271

As $F_{\text{model}} = 28715.76$, which is greater than $F_{5,58,0.95} \approx 2.376$, we reject the null hypothesis that the proposed model is invalid and conclude that it may be correct. By comparing T_{β_i} for each term in the revised five-term model to the NWS model vs. $t_{58,0.975} = 2.00173$, we have

$$T_{\beta_0} = -\frac{46.90080605 - 48.0145229}{1.10194748} = -1.01068,$$

$$T_{\beta_1} = \frac{0.303107 - 0.30277375}{0.02303227} = 0.01447,$$

$$T_{\beta_2} = \frac{0.47917309 - 0.474677}{0.04698720} = 0.09569,$$

$$T_{\beta_3} = \frac{-0.02052414 - (-0.020425)}{0.00264253} = -0.03752,$$

$$T_{\beta_4} = \frac{-27.19652334 - (-27.7039798)}{0.54015459} = 0.93946,$$

$$T_{\beta_5} = \frac{1.80587612 - 1.866845}{0.06197271} = -0.98380.$$

The parameter values in the NWS equation are accepted as probably correct.

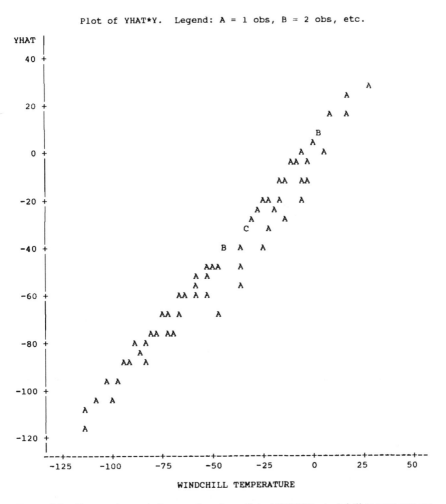

Plot of YHAT*Y.   Legend: A = 1 obs, B = 2 obs, etc.

Figure 9.9 Comparison of observed and predicted (YHAT) windchill temperatures as a result of two-term multiple linear regression.

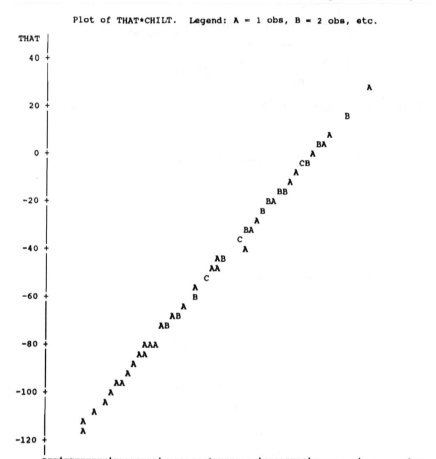

Plot of THAT*CHILT. Legend: A = 1 obs, B = 2 obs, etc.

FIGURE 9.10 Comparison of the observed (CHILT) and predicted (THAT) windchill temperatures as a result of five-term linear regression.

When the results of the comparison plot (Fig. 9.10) generated by use of the revised model are examined, it becomes obvious that the quality of the regression has been greatly improved: instead of a band, this plot shows the anticipated line. The results of the PROC GLM for the MODEL THAT = CHILT; show that

$$\text{THAT}_i = -0.01752728 + 0.999596207.$$

Evaluating the intercept and slope, we find that

$$T_{\beta_0=0} = \frac{-0.01752728 - 0}{0.14415901} = -0.1216$$

and

$$T_{\beta_1=1} = \frac{0.99959620 - 1}{0.00255151} = -0.1583.$$

When the T values are compared to $t_{62,0.975} \approx 1.99898$, the hypotheses of zero intercept and unit slope are accepted.

General Linear Models Procedure

Dependent Variable: THAT

Source	DF	Sum of Squares	Mean Square	F Value	Pr > F
Model	1	83649.8416	83649.8416	99999.99	0.0
Error	62	33.7911	0.5450		
Corrected Total	63	83683.6327			

	R-Square	C.V.	Root MSE	THAT Mean
	0.999596	-1.700800	0.73825	-43.40625

Source	DF	Type I SS	Mean Square	F Value	Pr > F
CHILT	1	83649.8416	83649.8416	99999.99	0.0

Source	DF	Type III SS	Mean Square	F Value	Pr > F
CHILT	1	83649.8416	83649.8416	99999.99	0.0

Parameter	Estimate	T for H0: Parameter=0	Pr > |T|	Std Error of Estimate
INTERCEPT	-.0175272811	-0.12	0.9036	0.14415901
CHILT	0.9995962037	391.77	0.0001	0.00255151

9.6 TRANSFORMATION OF DATA IN REGRESSION ANALYSIS

The basic assumptions in regression analysis are that the errors are independently and normally distributed with mean zero and variance σ^2 and, moreover, that the errors are homogeneous over the region of interest. If the error distribution is not normal or if the error variance is not homogeneous, it may be necessary to transform the dependent variable to attain desirable properties. When taking a transformation to linearize a given nonlinear model, it may also be useful to determine the variance of the transformed variables. These ideas are now discussed briefly.

9.6.1 Propogation of Error

We have seen that if $Y = a_1 X_1 + a_2 X_2 + \cdots + a_n X_n$, where the a's are constants and $X_1, X_2, \ldots, X_n$ are independent random variables, the variance of Y is $\sigma^2 = a_1^2 \sigma_{X_1}^2 + a_2^2 \sigma_{X_2}^2 + \cdots + a_n^2 \sigma_{X_n}^2$.

More generally, let $Y = f(X_1, X_2, \ldots, X_n)$ be some function of the independent random variables $X_1, X_2, \ldots, X_n$. The variance of Y is given, approximately, by

$$\sigma_Y^2 \cong \left(\frac{\partial f}{\partial X_1}\right)^2 \sigma_{X_1}^2 + \left(\frac{\partial f}{\partial X_2}\right)^2 \sigma_{X_2}^2 + \cdots + \left(\frac{\partial f}{\partial X_n}\right)^2 \sigma_{X_n}^2, \tag{9.55}$$

where the partial derivatives are evaluated at $X_1 = \mu_1, \ldots, X_n = \mu_n$. Equation (9.55) can be verified very easily by using the first $(n + 1)$ terms of the Taylor expansion of the function f of $X = (X_1, \ldots, X_n)$ about the point $\mu = (\mu_1, \mu_2, \ldots, \mu_n)$ where $\mu_i = E(X_i)$. We have

$$f(X_1, X_2, \ldots, X_n) \cong f(\mu_1, \mu_2, \ldots, \mu_n) + \sum_{i=1}^{n} \left(\frac{\partial f}{\partial X_i}\bigg|_{X_i = \mu_i}\right)(X_i - \mu_i).$$

Since the X_i are independent, the variance of the right-hand side yields Eq. (9.55), as required.

Equation (9.55) is an approximation. In many cases it adequately describes the total variance (precision) of the function f, and furthermore, it dictates which variable makes the largest contribution to the variance of the function. The partial derivatives in Eq. (9.55) are evaluated at $X_i = \mu_i$, $i = 1, 2, \ldots, n$, where $\mu_i = E(X_i)$.

As a special case, let $Y = f(X) = a_1 X_1 + a_2 X_2 + \cdots + a_n X_n$. We have $\partial f / \partial X_i = a_i$, $i = 1, 2, \ldots, n$. Thus

$$\sigma_Y^2 = a_1^2 \sigma_{X_1}^2 + a_2^2 \sigma_{X_2}^2 + \cdots + a_n^2 \sigma_{X_n}^2,$$

which we know holds exactly.

For another example suppose that $Z = \ln Y$. Since $\partial Z / \partial Y = 1/Y$ the variance of Z is $\sigma_Z^2 = (1/\mu_Y)^2 \sigma_Y^2$. If $\mu_Y = E(Y)$ is not known, $\sigma_Z^2 = (1/\bar{Y})\sigma_Y^2$, and if an estimate, $\hat{\sigma}_Y^2$, of σ_Y^2 is available, $\hat{\sigma}_Z^2 = (1/\bar{y})^2 \hat{\sigma}_Y^2$.

9.6.2 On Transforming the Data

Two reasons for transforming data are (1) lack of normality of error structure, and (2) lack of homogeneity in error structure. Lack of normality could occur when the tolerance on one side of the set point differed from the other side in reading an instrument. In other words, the error distribution could be skewed instead of bell-shaped. A heterogeneous error structure could occur in cases where the magnitude of the error depends on the region of values of the independent variable. For example, data consisting of values read from an instrument whose precision varies with the level of the reading are of this type.

Transformations to adjust for nonhomogeneity of the error structure are performed on the dependent variable. Sinibaldi et al. [19] discuss a method of Box and Cox [2] that is used in transforming the dependent variable to develop models for engineering applications. In many cases these transformations not only simplify the model but yield data that follow the underlying basic assumptions.

In some instances there may be a significant lack of fit in the model being used (see Section 9.3). Box and Tidwell [3] present methods of transforming the independent variables so that the resulting model is in as simple a form as possible. For further details, see Ref. 3.

9.6.3 Useful Transformations

A nonlinear model that occurs quite frequently is

$$Y = \beta_0 e^{\beta_1 X} \varepsilon'. \tag{9.56}$$

If β_1 is positive, Eq. (9.56) is termed an *exponential growth curve*; if negative, an *exponential decay curve*. Both have many applications in engineering and the basic sciences.

The problem in fitting a nonlinear model is that the resulting normal equations are not linear in the $\hat{\beta}$'s, and consequently, there is no clear-cut solution. There do exist iterative procedures for solving such systems of equations; however, we will not discuss that approach here. The interested reader is referred to Ref. 8.

The model $Y = \beta_0 e^{\beta_1 X} \varepsilon'$ is usually handled by using the log transformation and "linearizing" the model to yield the model

$$\ln Y = \ln \beta_0 + \beta_1 X + \varepsilon.$$

Letting $Z = \ln Y$, $\alpha_0 = \ln \beta_0$, and $\alpha_1 = \beta_1$, the model thus reduces to the linear model

$$Z = \alpha_0 + \alpha_1 X + \varepsilon. \tag{9.57}$$

Using the linear model (9.57), estimates $\hat{\alpha}_0$ and $\hat{\alpha}_1$ are obtained. From these one obtains the estimates $e^{\hat{\alpha}_0}$ and $\hat{\alpha}_1$. Now $\hat{\alpha}_0$ and $\hat{\alpha}_1$ are the least-squares estimates of α_0 and α_1; however, $e^{\hat{\alpha}_0}$ and $\hat{\alpha}_1$ are not necessarily the least-squares estimates of β_0 and β_1, the original parameters. Nevertheless, fitting the exponential model is commonly done by first linearizing and then doing the least-squares estimation.

Another nonlinear model that often occurs is the simple exponential model

$$Y = \beta_0 \beta_1^X \varepsilon'. \tag{9.58}$$

Using the natural logarithm transformation, we have the model

$$Z = \alpha_0 + \alpha_1 X + \varepsilon, \tag{9.59}$$

where $Z = \ln Y$, $\alpha_0 = \ln \beta_0$, and $\alpha_1 = \ln \beta_1$. Estimates of β_0 and β_1 can be obtained by fitting the linearized model (9.59).

One should be careful in using transformations such as the above, since if it is assumed that the original variable is normally distributed, the transformed variable may not be. The homogeneity of variance property may be violated similarly. Frequently, however, the original assumption of normality may not be justified and the transformed variables may have a distribution closer to normal.

This linearization process involves two assumptions: the experimental error is much smaller than the values of the dependent variable, and the resulting error structure is uniform. Even if the original error distribution is uniform, the errors of the transformed equation will not be. If the dependent variable covers a wide range, the assumption of uniform errors in the linearly transformed model will not be valid. The best way to check the results of a linearizing transformation is to perform the regression, test the model and all its parameters in the transform domain, reconvert to the original domain, use the parameters as they appear in the original domain to predict the values of the dependent variable ($\hat{Y}_i$), and compare the original (Y_i) and the predicted variables. The last two steps of this approach can be done by using the model

$$\hat{Y}_i = \hat{\beta}_0 + \hat{\beta}_1 Y,$$

where the expected values of $\hat{\beta}_0$ and $\hat{\beta}_1$ are 0 and 1, respectively.

Other nonlinear models that can be linearized are shown below with their corresponding linearized forms.

(1) Multiplicative: $Y_i = \alpha X_{1i}^\beta X_{2i}^\gamma X_{3i}^\delta \varepsilon_i'$ $\tag{9.60}$

$$Z_i = \ln Y_i = \ln \alpha + \beta \ln X_{1i} + \gamma \ln X_{2i}$$
$$+ \delta \ln X_{3i} + \varepsilon_i \tag{9.61}$$

(2) Exponential: $Y_i = \exp(\beta_0 + \beta_1 X_{1i} + \beta_2 X_{2i} + \varepsilon_i)$ $\tag{9.62}$

$$Z_i = \ln Y_i = \beta_0 + \beta_1 X_{1i} + \beta_2 X_{2i} + \varepsilon_i \tag{9.63}$$

(3)   Reciprocal: $Y_i = \dfrac{1}{\beta_0 + \beta_1 X_{1i} + \beta_2 X_{2i} + \varepsilon_i}$ (9.64)

$$Z_i = \frac{1}{Y_i} = \beta_0 + \beta_1 X_{1i} + \beta_2 X_{2i} + \varepsilon_i \tag{9.65}$$

(4) $Y_i = \dfrac{1}{1 + \exp(\beta_0 + \beta_1 X_{1i} + \beta_2 X_{2i} + \varepsilon_i)}$ (9.66)

$$Z_i = \ln\left(\frac{1}{Y_i} - 1\right) = \beta_0 + \beta_1 X_{1i} + \beta_2 X_{2i} + \varepsilon_i \tag{9.67}$$

Example 9.13   Calibration data for the specific ion electrode used for analysis of ammonia in liquid-phase samples from the pilot-scale gas absorber in the unit operations laboratory are given below. The concentration data are in gmol NH_4^+/L solution. The electrode response is in millivolts and becomes more positive as the concentration decreases.

NH_4^+ (mol/L)	MV
0.0001	− 125
0.00025	− 130
0.0005	− 131, − 142
0.001	− 138
0.0025	− 143
0.005	− 151, − 153
0.01	− 153
0.025	− 160
0.05	− 161
0.1	− 172
0.25	− 190
0.5	− 258

Unfortunately, this method of determining composition (and thus absorption efficiency, HTU, K_{OG}, etc.) is valid only in the straight-line portion of the curve. Find and evaluate a suitable equation for the calibration curve for this set of laboratory data in the form of log(liquid-phase NH_4^+ composition) = f(millivolt reading) and demonstrate in at least three ways that your equation is valid for use when working up data from your equipment.

Solution   The SAS program for the solution of this regression problem is as follows.

(put initial JCL cards/statements here)

```
DATA CALIB;
INPUT MV 1-4 CONC 6-12;
LNCONC=LOG10(CONC);
CARDS;
```

(put data here according to INPUT format)

```
PROC PRINT;
PROC PLOT;
PLOT LNCONC*MV;
TITLE 'EFFECT OF RESPONSE(MV) ON CONCENTRATION';
PROC GLM;
MODEL LNCONC=MV;
OUTPUT  OUT=NEW  P=CHAT  R=RESID;
PROC PRINT;
PROC PLOT;
PLOT LNCONC*MV CHAT*MV='P'/OVERLAY;
TITLE 'COMPARISON OF ORIGINAL AND PREDICTED VALUES';
PROC PLOT;
PLOT RESID*MV/VREF=0;
TITLE 'CHECKING RESIDUALS FROM SEMILOG REGRESSION';
PROC GLM;
MODEL CHAT=LNCONC;
PROC PLOT;
PLOT CHAT*LNCONC;
TITLE 'IF SEMILOG MODEL OK, (INTERCEPT, SLOPE)=(0,1)';
```

(put final JCL cards/statements here)

As seen in Fig. 9.11, the original data appear to have considerable scatter. From the first PROC GLM, the following analysis of variance was obtained. Comparison of F_{model} to $F_{1,10,0.95} = 4.96$ shows that the initial null hypothesis, that of an ineffective model, should be rejected. Further analysis of the individual parameters of this model indicates that neither β_0 nor β_1 is likely to be 0. Further confirmation of the usefulness of the semilogarithmic calibration equation is found on the overlay plot (Fig. 9.12) and the "shotgun" nature of the residual plot (Fig. 9.13).

When the predicted and original values are compared using the second PROC GLM, the results show that the intercept is likely to be zero and the slope is probably 1. These values are expected for an effective regression model.

$$T_{\beta_0=0} = -0.67,$$

$$T_{\beta_1=1} = \frac{0.9516525769 - 1}{0.06783063} = -0.7128.$$

The predicted and experimental values of the ammonia concentrations (as their logarithms) are shown in Fig. 9.14.

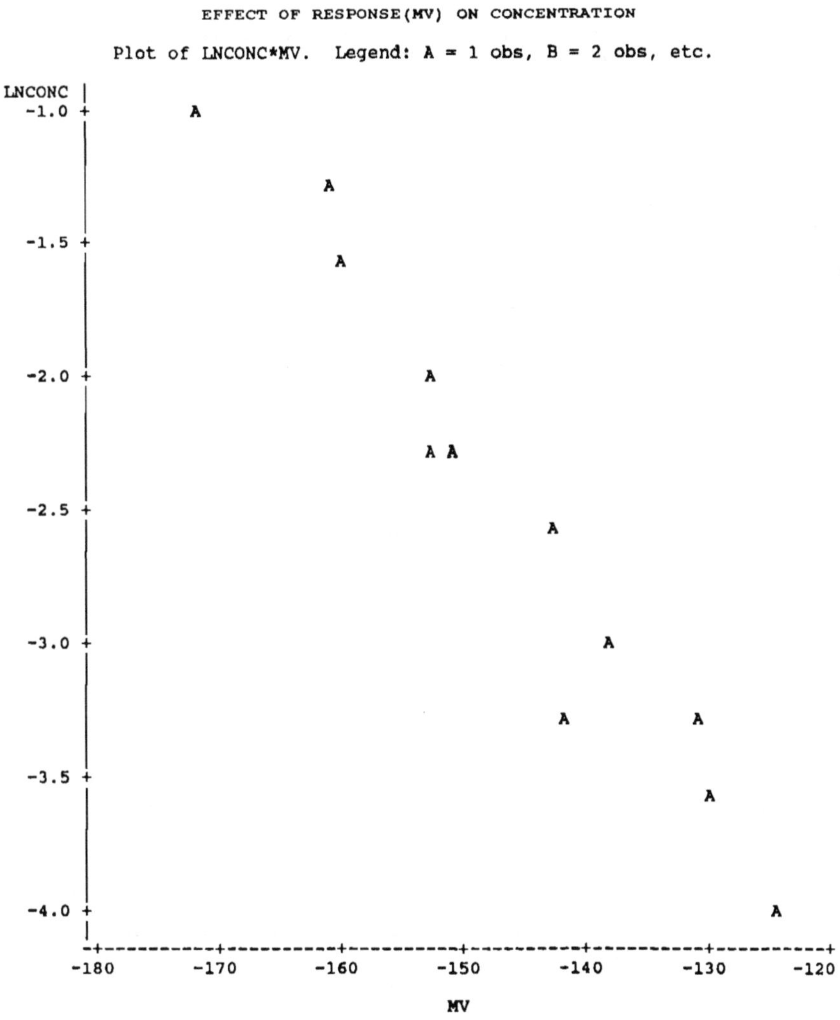

EFFECT OF RESPONSE(MV) ON CONCENTRATION

Plot of LNCONC*MV.   Legend: A = 1 obs, B = 2 obs, etc.

```
LNCONC |
  -1.0 +         A



  -1.5 +               A

                       A


  -2.0 +                    A

                              A A

  -2.5 +                         A


  -3.0 +                            A

                              A           A

  -3.5 +                                  A



  -4.0 +                                         A
       |
      -+----------+----------+----------+----------+----------+----------+
     -180       -170       -160       -150       -140       -130       -120
                                     MV
```

FIGURE **9.11**   Effect of concentration on electrode response (plotted in calibration curve style).

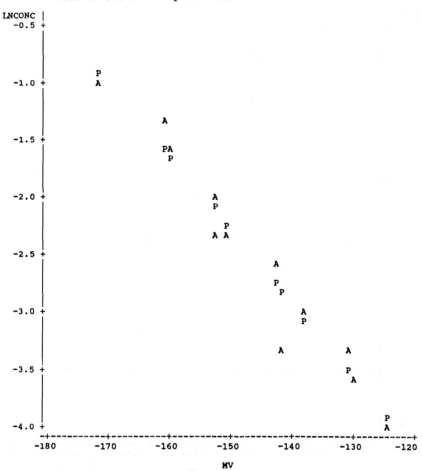

```
           Plot of LNCONC*MV.   Legend: A = 1 obs, B = 2 obs, etc.
           Plot of CHAT*MV.     Symbol used is 'P'.

LNCONC |
  -0.5 +
       |
       |
       |
       |
       |            P
  -1.0 +            A
       |
       |
       |
       |                     A
  -1.5 +
       |                   PA
       |                   P
       |
       |
  -2.0 +                        A
       |                        P
       |                          P
       |                        A A
  -2.5 +
       |                             A
       |                           P
       |                            P
  -3.0 +                                 A
       |                                 P
       |
       |                             A        A
  -3.5 +                                    P
       |                                    A
       |
       |
       |                                            P
  -4.0 +                                            A
       -+---------+---------+---------+---------+---------+---------+---------+
        -180     -170     -160     -150     -140     -130     -120
                                    MV
```

NOTE: 2 obs hidden.

FIGURE 9.12 Comparison of observed and predicted values.

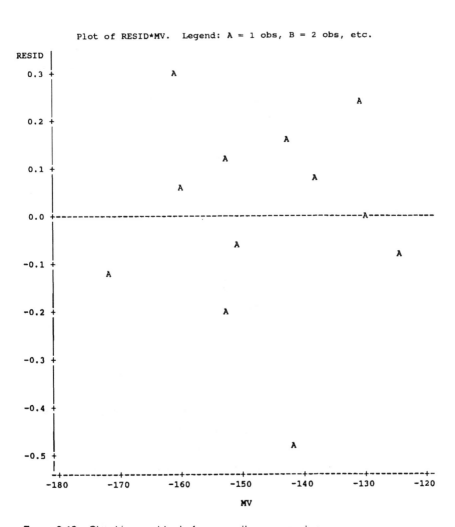

FIGURE 9.13 Checking residuals from semilog regression.

Dependent Variable: LNCONC

Source	DF	Sum of Squares	Mean Square	F Value	Pr > F
Model	1	9.34825155	9.34825155	196.84	0.0001
Error	10	0.47492529	0.04749253		
Corrected Total	11	9.82317684			

R-Square	C.V.	Root MSE	LNCONC Mean
0.951653	-8.627578	0.21793	-2.525944

Source	DF	Type I SS	Mean Square	F Value	Pr > F
MV	1	9.34825155	9.34825155	196.84	0.0001

Source	DF	Type III SS	Mean Square	F Value	Pr > F
MV	1	9.34825155	9.34825155	196.84	0.0001

Parameter	Estimate	T for H0: Parameter=0	Pr > |T|	Std Error of Estimate
INTERCEPT	-12.02319275	-17.69	0.0001	0.67984932
MV	-0.06479078	-14.03	0.0001	0.00461807

General Linear Models Procedure

Dependent Variable: CHAT

Source	DF	Sum of Squares	Mean Square	F Value	Pr > F
Model	1	8.89628768	8.89628768	196.84	0.0001
Error	10	0.45196387	0.04519639		
Corrected Total	11	9.34825155			

R-Square	C.V.	Root MSE	CHAT Mean
0.951653	-8.416434	0.21259	-2.525944

Source	DF	Type I SS	Mean Square	F Value	Pr > F
LNCONC	1	8.89628768	8.89628768	196.84	0.0001

Source	DF	Type III SS	Mean Square	F Value	Pr > F
LNCONC	1	8.89628768	8.89628768	196.84	0.0001

Parameter	Estimate	T for H0: Parameter=0	Pr > |T|	Std Error of Estimate
INTERCEPT	-.1221228911	-0.67	0.5174	0.18199595
LNCONC	0.9516525769	14.03	0.0001	0.06783063

When using a transformed model that is linear in the "new parameters," the assumptions pertinent to the new model need to be checked very carefully. This is especially true if any inferences are to be made on the parameters in the transformed model. If the nonlinear model, such as Eq. (9.56), is used directly, the problem is to determine the values of $\hat{\beta}_0$ and $\hat{\beta}_1$ that minimize

$$\text{SSE} = \sum_{i=1}^{n} (Y_i - \hat{\beta}_0 e^{\hat{\beta}_1 X_i})^2.$$

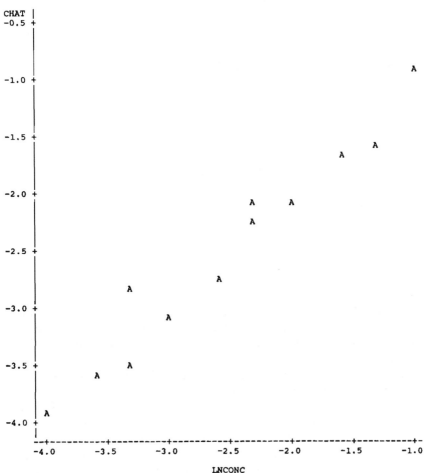

IF SEMILOG MODEL OK, (INTERCEPT, SLOPE)=(0,1)

Plot of CHAT*LNCONC. Legend: A = 1 obs, B = 2 obs, etc.

FIGURE 9.14 Comparison of predicted (CHAT) and measured (LNCONC) values of ammonia concentration.

This process is termed *nonlinear least squares*. If the derivatives $\partial(SS_E)/\hat{\beta}_0$ and $\partial(SS_E)/\hat{\beta}_1$ are set equal to zero, nonlinear equations in $\hat{\beta}_0$ and $\hat{\beta}_1$ result. Such equations are not solved as easily as the linear forms [Eqs. (9.10) and (9.11)]. However, computer software (e.g., PROC NLIN) is available that can be used to solve such equations (*SAS for Linear Models*, SAS Institute, Cary, NC, 1981, 231 pp.). It should also be noted that the estimates obtained from the transformed model will not be the same as those obtained from the original model by nonlinear least squares.

A frequently encountered linear (in the parameters β_i) model is the asymptotic growth curve, expressed by

$$Y_i = \beta_0 - \beta_1 e^{-X_i} + \varepsilon_i. \tag{9.68}$$

Another model is the sigmoid or logistic growth curve so common in biological and medical research:

$$Y_i = \frac{\beta_0}{1 + \beta_1 e^{X_i}} + \varepsilon_i. \tag{9.69}$$

Other nonlinear models are the Antoine equation,

$$Y_i = \beta_0 + \beta_1 e^{-\beta_2 X_i} + \varepsilon_i, \tag{9.70}$$

and mixed models such as

$$Y_i = \beta_0 + \beta_1 X_i + \beta_2 \beta_3^{X_i} + \varepsilon_i. \tag{9.71}$$

Unfortunately, these models cannot be linearized by a transformation of variables. Least-squares fits may be obtained by employing the appropriate software to yield estimates of the parameters by nonlinear least squares.

Example 9.14 The Gurney–Lurie charts give the relations between fractional temperature change accomplished (dimensionless) $(T_s - T)/(T_s - T_0)$ and the dimensionless group $kt/\rho C_p r^2$ for different values of the reciprocal Nusselt number k/hr and the fractional radius (dimensionless) x/r for cylinders in unsteady-state conduction. These relations consist of families of essentially straight lines for varying k/hr values for set levels of x/r when $(T_s - T)/(T_s - T_0)$ is plotted vs. $kt/\rho C_p r^2$ on semilog paper. For the case of $x/r = 0.6$ and $k/hr = 1$, find the correct mathematical expression for the following data:

$(T_s - T)/(T_s - T_0)$	$kt/\rho C_p r^2$
0.002	3.82
0.005	3.36
0.01	2.91
0.02	2.49
0.05	1.92
0.1	1.49
0.2	1.05
0.5	0.67

Solution For relations such as this, which are straight or essentially so on semilog paper, a model of the form $Y = \beta_0 \beta_1^X \varepsilon'$ is the proper starting point. The substitutions $Z = \log Y$, $\alpha_0 = \log \beta_0$, and $\alpha_1 = \log \beta_1$ are made to give a revised simple linear model $Z = \alpha_0 + \alpha_1 X + \varepsilon$. The following quantities are calculated:

$$\bar{z} = -1.500000, \qquad \bar{x} = 2.213749,$$
$$\sum z = -12.000000, \qquad \sum x = 17.70990,$$
$$\sum zx = -33.015480, \qquad \sum x^2 = 48.008040,$$
$$\sum z^2 = 22.737470, \qquad (\sum x)^2 = 313.64350,$$
$$\sum z \sum x = -212.519800, \qquad (\sum z)^2 = 143.99980.$$

Solution of the normal equations for the revised model gives the expression

$$\hat{Z} = 0.122229 - 0.73279X,$$

which is then converted to the original form to obtain the final result as

$$\hat{Y} = 1.3250(0.18501)^X.$$

Example 9.15 An orifice meter has been calibrated by varying the water flow rate over the range 1 to 10 ft/sec. The data are given below, where U = flow rate (ft/sec) and DELTAP = pressure drop, inches of mercury. The theoretical relationship is

$$U_i = C_0 \sqrt{\frac{2g_c}{(1 - \beta^4)\rho}} \sqrt{-\Delta P_i} \, \varepsilon_i'$$
$$= 0.61(2.49690198)\sqrt{-\Delta P_i} \, \varepsilon_i',$$

where $-\Delta P_i$ = in. water, the term in parentheses is the first square-root term, and the unit conversion terms are evaluated for water (68°F) flowing through a 0.625-in. orifice in a 1.025-in.-ID pipe.

The model may be approximated as

$$\log_{10} U_i = 0.18273133 + 0.5 \log_{10}(\text{DELTAP}_i) + \varepsilon_i.$$

Are the experimental data consistent with theory?

Solution   We will use the SAS program for simple linear regression after transformation of variables to evalute the data. The SAS program follows.

(put initial JCL cards/statements here)

```
DATA;
INPUT U 1-2  DELTAP 4-5;
Z = LOG10(U);
W = LOG10(DELTAP);
CARDS;
    1     5
    2     8
    3    10
    4    12
    5    14
    6    16
    7    18
    8    19
    9    23
   10    24
PROC PLOT;
PLOT U * DELTAP;
TITLE 'EFFECT OF PRESSURE DROP ON FLOW RATE';
PROC GLM;
MODEL Z = W;
OUTPUT OUT = NEW P = YHAT R = RESID;
PROC PRINT;
PROC PLOT;
PLOT Z * W YHAT * W = 'P'/OVERLAY;
TITLE 'COMPARISON OF ORIGINAL AND PREDICTED DATA';
PLOT RESID * W;
TITLE 'CHECKING RESIDUALS FOR RANDOM SCATTER';
```

(put final JCL cards/statements here)

The results follow.

SAS

EFFECT OF PRESSURE DROP ON FLOW RATE

GENERAL LINEAR MODELS PROCEDURE

DEPENDENT VARIABLE: Z

SOURCE	DF	SUM OF SQUARES	MEAN SQUARE	F VALUE
MODEL	1	0.90685273	0.90685273	1379.88
ERROR	8	0.00525757	0.00065720	PR > F
CORRECTED TOTAL	9	0.91211030		0.0001

SOURCE	DF	TYPE I SS	F VALUE	PR > F
W	1	0.90685273	1379.88	0.0001

PARAMETER	ESTIMATE	T FOR HO: PARAMETER=0	PR > |T|	STD ERROR OF ESTIMATE
INTERCEPT	-1.01226115	-22.18	0.0001	0.04563522
W	1.47534307	37.15	0.0001	0.03971662

From the F-test ($\alpha = 0.05$) we accept the regression as significant. Examining the T values for testing the β_i values, we conclude that both are nonzero. We also calculate

$$T_{\alpha_0} = \frac{-1.01226115 - 0.18273133}{0.04563522} = -18.177$$

and

$$T_{\alpha_1} = \frac{1.47534307 - 0.5}{0.03971662} = 24.557.$$

From the calculated T values we see that we must reject $H_0: \alpha_0 = 0.18273133$ and $H_0: \alpha_1 = 0.5$. The first result leads us to the conclusion that as g_c and β are constants, the orifice discharge coefficient C_0 is not 0.61 for this experiment. The second result leads us to doubt the well-accepted square-root dependency of flow rate on pressure drop. Note that all tests of hypotheses have been made in the "log domain," as that is how the data were handled to obtain the regression coefficients. In this case, the high (0.994^+) value of R^2 obtained by the methods described in Section 9.7 is totally misleading: the experimental data do not confirm the expected values. Figure 9.15 shows the observed (A) and predicted (P) values of $\log_{10} U_i$ or Z_i vs. $\log_{10}(\text{DELTAP}_i)$ or W_i. The agreement appears quite good. Do not be misled by the apparent shape of the residuals plot (Fig. 9.16), as the vertical scale there is expanded about eightfold compared to that in Fig. 9.15.

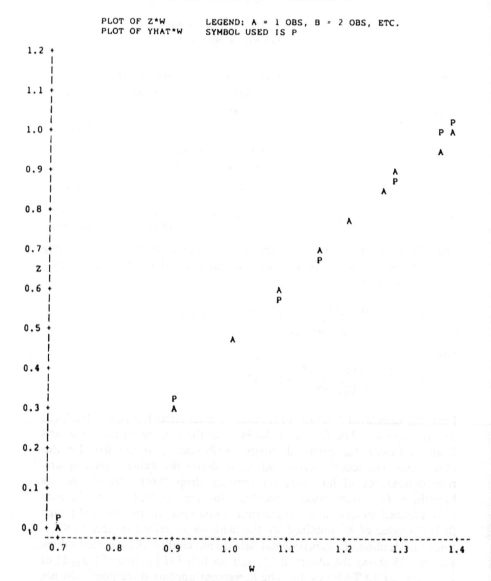

FIGURE 9.15   Comparison of observed (*A*) and predicted (*P*) values of Z ($= \log_{10} U$) as a function of W ($= \log_{10}$ DELTAP).

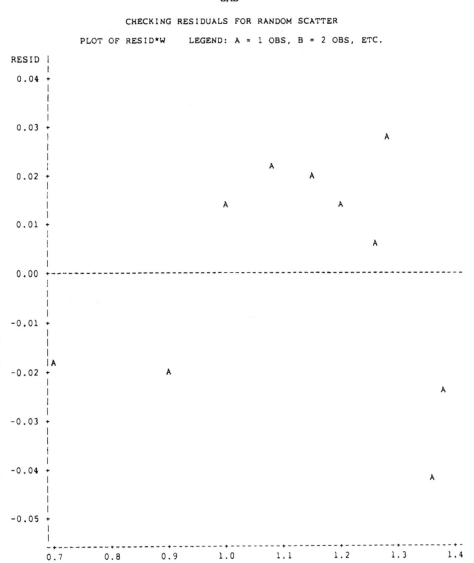

SAS

CHECKING RESIDUALS FOR RANDOM SCATTER

PLOT OF RESID*W   LEGEND: A = 1 OBS, B = 2 OBS, ETC.

FIGURE 9.16 Checking residuals from log–log model for random scatter.

Example 9.16 The vapor pressure of water absorbed on silica gel can be expressed as a function of the vapor pressure of pure water for various gel loadings in spacecraft humidity water recovering systems. For the water loading of 0.1 lb water/lb dry silica gel, the following data were obtained:

p^*, absorbed H_2O (in.Hg, abs)	p^*, pure H_2O (in.Hg, abs)
0.038	0.2
0.080	0.4
0.174	0.8
0.448	2.0
1.43	6.0
5.13	20
9.47	35

A plot of the p data on log–log paper yields a straight line, so an equation of the form $Y = \beta_0 X^{\beta_1} \varepsilon'$ is assumed as a model. If we let $Z = \log Y$, $\alpha_0 = \log \beta_0$, and $W = \log X$, we can reduce this to the simple linear case, $Z = \alpha_0 + \alpha_1 W + \varepsilon$. The following quantities are then calculated:

$$\bar{z} = -0.254785, \qquad \bar{w} = 0.390065,$$
$$\sum z = -1.783492, \qquad \sum w = 2.730458,$$
$$\sum z^2 = 5.400247, \qquad \sum w^2 = 5.429265,$$
$$\sum zw = 3.950120, \qquad (\sum w)^2 = 7.455401,$$
$$(\sum z)^2 = 3.3180843, \qquad \sum z \sum w = -4.869750.$$

Solution of the normal equations for $\hat{\alpha}_0$ and $\hat{\alpha}_1$ gives

$$Z = -0.670018 + 1.06452W,$$

which, on conversion to the original form, gives the desired result as

$$Y = 0.21379 X^{1.06452}.$$

9.7 NONLINEAR REGRESSION

Nonlinear regression can be used to determine the model parameters when no linearizing transformation can be found. This procedure finds the parameter values that minimize the error sum of squares in a nonlinear least-squares routine. The method is to take the partial derivatives of the model with respect to each of the parameters and set the resulting equations equal to zero. Because the model is nonlinear, the result of this least-squares procedure is a set of nonlinear equations that must be solved simultaneously by any of several methods, such as Gauss–Newton, Marquardt, steepest-descent, or multivariant secant.

The SAS software library contains a routine, PROC NLIN, which is often useful for nonlinear models. Two problems may be associated with this procedure. They are developing an empirical model when the correct form cannot be deduced or derived from first principles and estimating the range for each parameter in the iteration procedure.

There are four mandatory and two optional steps associated with PROC NLIN. The first required step is the assignment of a distinct name for each parameter to be estimated. Next, the model to be used (limited to one independent variable) must be specified with a starting value for each parameter to be estimated. Finally, the partial derivatives of the model parameters must be provided. The solution method must also be specified.

The optional criteria involve imposition of limits on the ranges of the parameters to be used in the iterative solutions. The convergence criterion may also be specified. Because the values of the parameters are unknown, a trial-and-error approach is often useful. A large range that is "certain" to contain the correct parameter value may be selected. Then a large step size for the iteration may be used to "home in" on the probable ranges of the parameters. Those areas may then be evaluated in more detail using smaller steps and greatly reduced ranges for the searches.

One caution is necessary. If the message "CONVERGENCE ASSUMED" appears on the output, it should be interpreted as "not converged at all," unless it can be demonstrated that the model, with the supposedly converged values of the parameters, can essentially reproduce the original data and that the residuals are small and randomly distributed. These "quality of fit" demonstrations have already been demonstrated in Examples 9.5 and 9.10.

Example 9.17 The vapor explosion apparatus (volume = 1.8 L) has been used to determine the explosion pressure as a function of concentration. The region under the "dome" is the flammable region. The lower and upper flammable limits (LFL and UFL, respectively) are the concentrations where the dome curve intersects the 0 bar explosion pressure line.

Explosion pressure (bar)	Flammable gas concentration (vol % in air)
0.1	2.3
0.2	7.8
1.4	2.7
2.0	7.7
2.6	2.7
3.2	7.6
4.05	3.0
5.0	7.4
6.1	7.2
7.3	7.0
7.5	3.5
8.0	3.9, 6.5
8.3	4.15
8.85	5.85

Find the equation describing the dome and the estimated values of the LFL and the UFL. What are the 95% confidence limits on each?
Solution

(put initial JCL here)

```
DATA SOTB;
INPUT PRES 1-10 CONC 11-20;
CARDS;
  0.1       2.3
  0.2       7.8
  1.4       2.7
  2.0       7.7
  2.6       2.7
  3.2       7.6
  4.05      3.0
  5.0       7.4
  6.1       7.2
  7.3       7.0
  7.5       3.5
  8.0       3.9
  8.0       6.5
  8.3       4.15
  8.85      5.85
PROC PLOT;
TITLE 'COMPARISON OF EXPLOSION PRESSURE RESULTS';
PLOT PRES*CONC;
PROC NLIN BEST = 10  METHOD = GAUSS;
PARMS BO = 1 TO 10 BY 1 B1 = 1 TO 10 BY 1
B2 = 1 TO 10 BY 1;
TITLE 'QUADRATIC MODEL FOR EXPLOSION PRESSURE';
MODEL PRES = BO-(B1*CONC**2)+(2*B1*B2*CONC)-(B1*B2**2);
DER.BO = 1;
DER.B1 = (-CONC**2)+(2*B2*CONC)-(B2**2);
```
(Continued)

```
DER.B2 = (2*B1÷CONC)-(2*B1*B2);
OUTPUT OUT = OUT1   P = PHAT   R = RESID1;
PROC PLOT;
TITLE 'COMPARISON OF ORIGINAL AND PREDICTED
VALUES FOR EXPLOSION PRESSURE';
PLOT PRES*CONC PHAT*CONC = 'P'/OVERLAY;
PLOT RESID1*CONC/VREF = 0;
PROC GLM;
MODEL PHAT = PRES;
PROC PLOT;
PLOT PHAT*PRES;
```

(put final JCL here)

The appearance of the original data (Fig. 9.17) is that of a truncated parabola that may have been rotated slightly clockwise. The parabola has

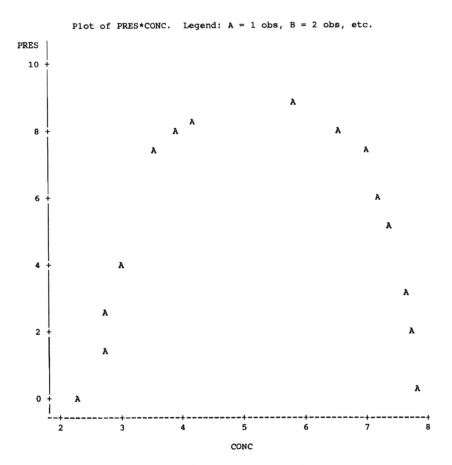

Plot of PRES*CONC.   Legend: A = 1 obs, B = 2 obs, etc.

Figure 9.17 Explosion pressure (Y) vs. volume percent flammable gas in air (X).

also been displaced from the (0,0) origin. For that reason, the general formula for a conic is used to develop the model

$$Y = \beta_0 - \beta_1 X^2 + 2\beta_1\beta_2 X - \beta_1\beta_2^2,$$

which is used in the program as the MODEL statement. The partial derivatives are inserted in the program by DER. statements and are

$$\frac{\partial Y}{\partial \hat{\beta}_0} = 1, \quad \frac{\partial Y}{\partial \hat{\beta}_1} = -X^2 + 2\hat{\beta}_2 X - \hat{\beta}_2^2 \quad \text{and} \quad \frac{\partial Y}{\partial \hat{\beta}_2} = 2\hat{\beta}_1 X - 2\hat{\beta}_1\hat{\beta}_2.$$

For the following example, the Gauss–Newton iteration procedure has been used. The "BEST = 10" option after PROC NLIN causes the 10 best values of the $\hat{\beta}_i$ and the corresponding residual (error) sums of squares to be printed. When convergence is reached, the converged values of the $\hat{\beta}_i$ are printed with the message "CONVERGENCE CRITERION MET."

QUADRATIC MODEL FOR EXPLOSION PRESSURE

Non-Linear Least Squares Grid Search   Dependent Variable PRES

BO	B1	B2	Sum of Squares
9.000000	1.000000	5.000000	26.638663
10.000000	1.000000	5.000000	35.408663
8.000000	1.000000	5.000000	47.868663
7.000000	1.000000	5.000000	99.098663
10.000000	1.000000	6.000000	173.568662
9.000000	1.000000	6.000000	177.598662
6.000000	1.000000	5.000000	180.328663
8.000000	1.000000	6.000000	211.628662
7.000000	1.000000	6.000000	275.658662
10.000000	2.000000	5.000000	277.754150

Non-Linear Least Squares Iterative Phase
Dependent Variable PRES   Method: Gauss-Newton

Iter	BO	B1	B2	Sum of Squares
0	9.000000	1.000000	5.000000	26.638663
1	10.212756	1.263864	5.257062	9.634820
2	10.261400	1.263864	5.203394	8.422128
3	10.265041	1.263864	5.203394	8.421929
4	10.265041	1.263864	5.203394	8.421929

NOTE: Convergence criterion met.

The analysis-of-variance table is then printed for the converged solution. The AOV table contains the asymptotic standard errors of the parameter estimates and the corresponding asymptotic 95% confidence limits for the parameters. The word *asymptotic* indicates that an asymptotic approach routine was used in meeting the convergence criterion.

Non-Linear Least Squares Summary Statistics   Dependent Variable PRES

Source	DF	Sum of Squares	Mean Square
Regression	3	477.95307079	159.31769026
Residual	12	8.42192921	0.70182743
Uncorrected Total	15	486.37500000	
(Corrected Total)	14	134.99100000	

Parameter	Estimate	Asymptotic Std. Error	Asymptotic 95 % Confidence Interval Lower	Upper
BO	10.26504066	0.46056645860	9.2615510707	11.268530245
B1	1.26386434	0.09451792472	1.0579271722	1.469801513
B2	5.20339400	0.04151463643	5.1129412481	5.293846760

Asymptotic Correlation Matrix

Corr	BO	B1	B2
BO	1	0.8826488623	-0.100377387
B1	0.8826488623	1	-0.092410654
B2	-0.100377387	-0.092410654	1

Figure 9.18 shows the comparison of the original and predicted values of the explosion pressure (*PRES*) as a function of the volume percent (*CONC*) of the flammable gas in air. The agreement of the experimental and the predicted values is good for the limbs of the parabola, but not of such high quality where the parabola was flattened to give the dome shape. Nevertheless, the residuals (Fig. 9.19) exhibit a random pattern.

General Linear Models Procedure

Dependent Variable: PHAT

Source	DF	Sum of Squares	Mean Square	F Value	Pr > F
Model	1	118.672576	118.672576	195.37	0.0001
Error	13	7.896495	0.607423		
Corrected Total	14	126.569071			

R-Square	C.V.	Root MSE	PHAT Mean
0.937611	16.10275	0.77937	4.8400000

Source	DF	Type I SS	Mean Square	F Value	Pr > F
PRES	1	118.672576	118.672576	195.37	0.0001

Source	DF	Type III SS	Mean Square	F Value	Pr > F
PRES	1	118.672576	118.672576	195.37	0.0001

Parameter	Estimate	T for H0: Parameter=0	Pr > |T|	Std Error of Estimate
INTERCEPT	0.3019618890	0.79	0.4434	0.38197337
PRES	0.9376111800	13.98	0.0001	0.06708001

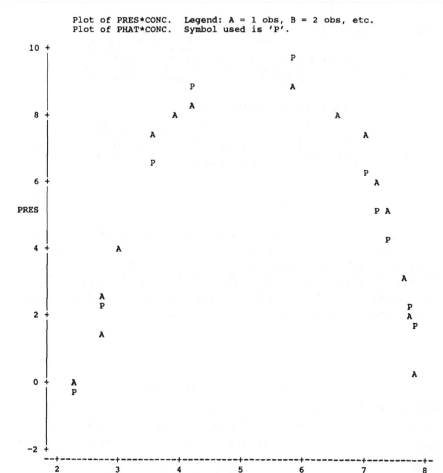

NOTE: 5 obs hidden.

Figure 9.18 Comparison of experimental and predicted values.

Evaluation of the predicted vs. original explosion pressures (Fig. 9.20) seems to indicate that the original model is not correct. If the point where $CONC = 0.1$ were in error, the rest of this plot would not be approximated as a straight line but as an exponential. When the original data were divided at $CONC = 5.0$ vol % flammable gas and the data fit by halves, much better results were obtained. For the left half $(2.3 \leq X \leq 4.15)$, the result was

$$PRES = -26.45540815 + 14.90300865CONC - 1.55608695(CONC)^2$$

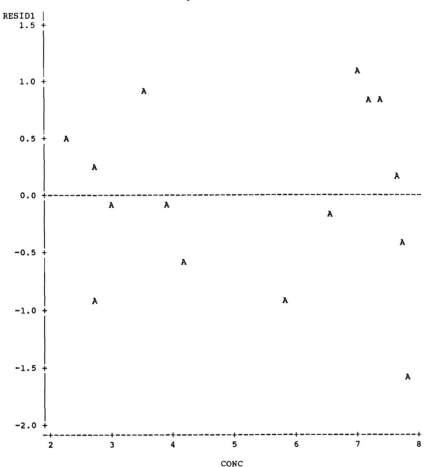

Plot of RESID1*CONC.   Legend: A = 1 obs, B = 2 obs, etc.

FIGURE **9.19**   Checking residuals for random scatter.

for which $R^2_{LH} = 0.979526$. For the right half ($5.85 \leq CONC \leq 7.8$), the result was

$$PRES = 152.0598270 - 0.0155122e^{CONC} - 50.4395405CONC$$
$$+ 4.5948842CONC^2,$$

for which $R^2_{RH} = 0.99808$. Splitting the datas provide a convenient way to avoid problems introduced by the rotation of the axes in this data set. The improvement in the quality of fit obtained by this approach is shown in Figs. 9.21 and 9.22. The values for LFL and UFL are 2.34 and 7.82 vol %, respectively, compared to the true values of 2.2 and 7.8 vol %.

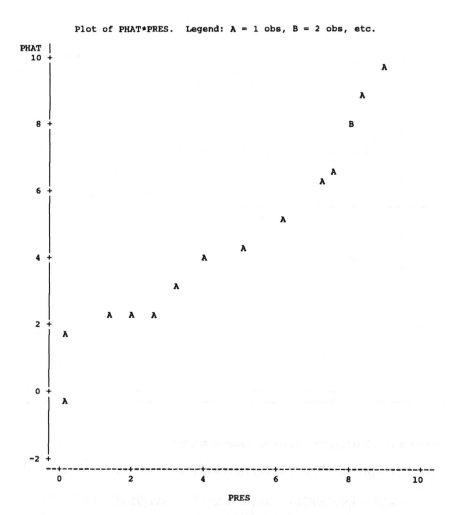

Figure 9.20 Comparison of predicted (PHAT) and experimental (PRES) explosion pressures.

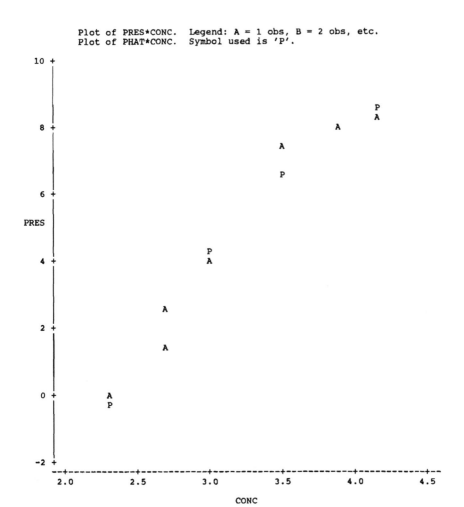

Plot of PRES*CONC. Legend: A = 1 obs, B = 2 obs, etc.
Plot of PHAT*CONC. Symbol used is 'P'.

NOTE: 3 obs hidden.

Figure 9.21 Comparison of predicted and experimental explosion pressures: left half.

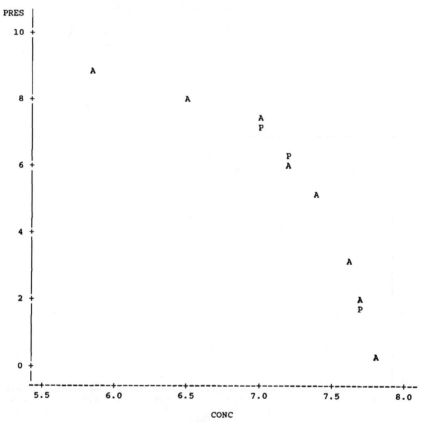

```
             Plot of PRES*CONC.   Legend: A = 1 obs, B = 2 obs, etc.
             Plot of PHAT*CONC.   Symbol used is 'P'.

PRES |
  10 +
     |
     |
     |            A
   8 +                            A
     |                              A
     |                              P
     |                                P
   6 +                                A
     |                                   A
     |
   4 +
     |
     |                                      A
     |
   2 +                                        A
     |                                        P
     |
     |
   0 +                                          A
     |
     --+------------+------------+------------+------------+------------+--
      5.5          6.0          6.5          7.0          7.5          8.0
                                     CONC
```

NOTE: 5 obs hidden.

FIGURE 9.22 Comparison of predicted and experimental explosion pressures:
right half.

9.8 CORRELATION ANALYSIS

Having determined that a relationship exists between variables, the next question which arises is that of how closely the variables are associated. The statistical techniques that have been developed to measure the amount of association between variables are called *correlation methods*. A statistical analysis performed to determine the degree of correlation is called a *correlation analysis*. The statistic used to measure correlation is the *correlation coefficient*. The correlation coefficient measures how well the regression equation fits the experimental data. As such, it is closely related to the standard error of estimate, $\hat{\sigma}$.

The correlation coefficient R should exhibit two characteristics:

1. It should be large when the variables are closely associated and small when there is little association.
2. It must be independent of the units used to measure the variables.

An effective correlation coefficient that exhibits these two characteristics is the square root of the fraction of the sum of squares of deviations of the original data from the regression curve that has been accounted for by the regression. This is a justifiable definition since the closeness of the regression curve to the data points is reflected in how much of the total corrected sum of squares, SS_{TC}, is accounted for by the sum of squares due to regression, SS_R. From Eq. (9.15) we have the identity $SS_{TC} = SS_E + SS_R$. If $SS_E = 0$ or $SS_R = SS_{TC}$, the data points fall on the regression curve, which fits the data perfectly. On the other hand, if $SS_R = 0$, there is no reduction in SS_{TC} due to the relation between X and Y, or we say that there is no relation (i.e., zero correlation) between X and Y. In view of this we define the correlation coefficient in terms of the proportional reduction in the error sum of squares accounted for by the regression of Y on X. The precise definition is

$$R^2 = \frac{SS_R}{SS_{TC}} = \frac{SS_{TC} - SS_E}{SS_{TC}} = 1 - \frac{SS_E}{SS_{TC}}.$$

As the $SS_E \le SS_{TC}$, R^2 will lie between 0 and 1. If the regression curve is a poor fit of the experimental data, R^2 will be close to zero.

9.8.1 Correlation in Simple Linear Regression

For the simple linear regression model, $Y = \beta_0 + \beta_1 X + \varepsilon$, the sum of squares due to regression is

$$SS_R = \hat{\beta}_1^2 \sum_i (X_i - \bar{X})^2.$$

Thus for the simple linear model we have

$$R^2 = \frac{SS_R}{SS_{TC}}$$
$$= \frac{\hat{\beta}_1^2 \sum_i (X_i - \bar{X})^2}{\sum_i (Y_i - \bar{Y})^2}$$
$$= \frac{[\sum_i (X_i - \bar{X})(Y_i - \bar{Y})]^2}{\sum_i (X_i - \bar{X})^2 \sum_i (Y_i - \bar{Y})^2}, \tag{9.72}$$

since $\hat{\beta}_1 = \sum_i (X_i - \bar{X})(Y_i - \bar{Y})/\sum_i (X_i - \bar{X})^2$. From Eq. (9.72) we have

$$R = \frac{\sum_i (X_i - \bar{X})(Y_i - \bar{Y})}{[\sum_i (X_i - \bar{X})^2 \sum_i (Y_i - \bar{Y})^2]^{1/2}}.$$

The correlation index for the simple linear model is usually denoted by r^2 rather than R^2. However, we shall use R^2, reserving r^2 to denote a particular value of the statistic R^2. From Eq. (9.72) it is obvious that the correlation coefficient associated with simple linear regression may readily be obtained from the statistics already calculated in the regression analysis. The values of R^2 lie between 0 and 1 and hence $-1 \leq R \leq 1$. A value of $R = 0$ means that there is no correlation between the variables and a value of either $+1$ or -1 implies perfect correlation. We say that there is perfect positive or negative correlation according as $r = +1$ or -1.

Example 9.18 Referring to the data of Example 9.1 and using Eq. (9.72), we calculate the simple linear correlation index as

$$r^2 = \frac{(167.5)^2}{82.5(342.1)} = 0.994,$$

indicating that the regression equation accounts for 99.4% of the variability of the data about $\bar{x}$. Since $\sum_i (x_i - \bar{x})(y_i - \bar{y}) = 167.5$, $r = +\sqrt{0.994} = 0.996$. This means that X and Y are positively correlated. That means that as X increases or decreases, the corresponding values of Y increase or decrease, accordingly. This also implies that the slope of the regression line is positive. In this example the value of the correlation coefficient is quite high, $r = 0.996$, indicating a "strong" linear relationship.

From the foregoing discussion it is seen that r^2 (or R^2) is just the proportion of the variation in $\sum_i (Y_i - \bar{Y})^2$ accounted for by the linear relationship between Y and X (or the X's). The simple correlation coefficient is defined as $r = \sqrt{r^2}$, $-1 \leq r \leq 1$. This term implies that X and Y have a joint distribution; that is, we are assuming model (9.3). In this case we can use the term *correlation coefficient* between X and Y and covariance of X and Y, since X is also a random variable.

If we denote the *sample covariance* of X and Y by s_{XY}, β_1 can be found from

$$\hat{\beta}_1 = \frac{s_{XY}}{s_X^2} \qquad\qquad (9.73)$$

and r can be found from

$$r = \frac{s_{XY}}{s_X s_Y}. \qquad\qquad (9.74)$$

The quantity $100r^2$ is the percentage of the corrected sum of squares that is accounted for by the simple linear regression $\hat{Y} = \hat{\beta}_0 + \hat{\beta}_1 X$. If this model does not account for enough of the variation in the data, a new regression equation should be used. Frequently, the form of the regression equation that will prove most effective in handling the data can be determined from the shape of the curve obtained from an appropriate plot of the data.

We have discussed linear regression assuming that the dependent variable Y, given a value of X, has a normal distribution with mean $\mu_{Y|X}$ and variance σ^2. Suppose that X is also a random variable. Then we have a two-dimensional random variable (X, Y) which has a bivariate probability density function $f(x,y)$. If it is assumed that $f(x,y)$ is the *bivariate normal* density function, the simple correlation coefficient r is an estimator for the population correlation coefficient ρ, which is defined by

$$\rho = E\left[\frac{(X - \mu_X)(Y - \mu_Y)}{\sigma_X \sigma_Y}\right]$$

$$= \frac{\sigma_{XY}}{\sigma_X \sigma_Y},$$

where $\sigma_{XY} = E[(X - \mu_X)(Y - \mu_Y)]$ is the population covariance of X and Y.

If the variable X is also a random variable, it has some variance which implies that the values of X are not necessarily observed or measured without error. The probability that X is not measured without error must be taken into consideration. The problem of estimating the parameters involved under these circumstances has not been resolved in general, so we will not pursue it here. Thus we will assume that X can be observed or measured without error.

In sampling from a bivariate normal population it is of interest to test the hypothesis $H_0: \rho = 0$, that is, that there is no linear relation between X and Y. It can be shown that if $\rho = 0$,

$$\frac{R - \rho}{S_R} = \frac{R}{S_R} = \frac{\hat{\beta}_1}{S_{\hat{\beta}_1}}. \qquad\qquad (9.75)$$

We already know [Eq. (9.16)] that $T = \hat{\beta}_1/s_{\hat{\beta}_1}$ has a t-distribution with $(n-2)$ degrees of freedom. Thus, in testing $H_0: \rho = 0$ against $H_A: \rho \neq 0$, the rejection region is $T \leq t_{n-2,\alpha/2}$ or $T \geq t_{n-2,1-\alpha/2}$, where $T = R/s_R$. Actually,

$$s_R = \sqrt{\frac{1-r^2}{n-2}}.$$

It should be noted that testing $H_0: \rho = 0$ is the same as testing $H_0: \beta_1 = 0$.

If $H_0: \rho = 0$ is rejected, the question of how good the linear relation is remains. Thus one may want to test a hypothesis of the type $H_0: \rho = \rho_0$, where $\rho_0 \neq 0$. To this end one can use the statistic

$$Z_r = \frac{1}{2} \ln \frac{1+R}{1-R},$$

which has been shown to have an approximate normal distribution with mean $\mu_{Z_r} = \frac{1}{2} \ln[(1+\rho)/(1-\rho)]$ and variance $\sigma_{Z_r}^2 = 1/(n-3)$. Thus the critical region in testing $H_0: \rho = \rho_0$ against $H_A: \rho \neq \rho_0$ is

$$Z > z_{1-\alpha/2} \quad \text{or} \quad Z < z_{\alpha/2},$$

where $Z = (Z_r - \mu_{Z_r})/\sigma_{Z_r}$ is the standard normal random variable (deviate) as discussed in Section 3.3.6.

Example 9.19   For the data in Example 9.1, test the hypothesis $H_0: \rho \geq 0.95$ against $H_A: \rho < 0.95$. Use $\alpha = 0.01$.

Solution   From Example 9.18 the sample correlation coefficient is $r = 0.997$. The statistic $Z_r = \frac{1}{2} \ln[(1+R)/(1-R)]$ when computed is

$$z_r = 0.5 \ln \frac{1.997}{0.003}$$
$$= 0.5(6.501)$$
$$= 3.250.$$

If $\rho = \rho_0 = 0.95$, then $\mu_{Z_r} = 0.5 \ln(1.95/0.05) = 1.832$ and $\sigma_{Z_r}^2 = 1/(n-3) = \frac{1}{7} = 0.1429$. Thus the computed Z value is $z = (3.250 - 1.832)/0.1429 = 9.923$, and since $z_{0.01} = -2.33$, we accept H_0.

9.8.2 Correlation in Multiple Linear Regression

In multiple linear regression, where the model is

$$Y = \beta_0 + \beta_1 X_1 + \beta_2 X_2 + \cdots + \beta_p X_p + \varepsilon, \tag{9.76}$$

the correlation index R^2 is, according to Eq. (9.52),

$$R^2 = \frac{SS_R}{SS_{TC}}$$

$$= \frac{\hat{\beta}_1 \sum_i (X_{1i} - \bar{X}_1)(Y_i - \bar{Y}) + \cdots + \hat{\beta}_p \sum_i (X_{pi} - \bar{X}_p)(Y_i - \bar{Y})}{\sum_i (Y_i - \bar{Y})^2}.$$

$$(9.77)$$

Equation (9.77) is analogous to Eq. (9.72). The coefficient R as defined by Eq. (9.77) is called the *multiple correlation coefficient* and can be viewed as being the simple correlation coefficient between Y and $\hat{Y}$, where $\hat{Y} = \hat{\beta}_0 + \hat{\beta}_1 X_1 + \cdots + \hat{\beta}_p X_p$.

One of the most common faults in regression analysis is failure to plot the data before postulating a model. Consider the following data, obtained from an experiment designed to measure the velocity profile in a tube. The location X is measured in the radial direction; the flow rate Y is measured in the longitudinal direction.

X	-2	-1	0	1	2
Y	4.2	0.9	0	1.1	4.0

Some students actually fitted these data to a simple linear model. Needless to say, they were quite surprised to find that $r^2 = 0.00027$. The data actually described a nearly perfect parabola. Had the data been plotted first, the proper model would have been obvious. The data would have then been fitted with a quadratic model and the resulting regression equation would have been

$$Y = -0.01714 - 0.02000X + 1.02857X^2.$$

R^2 for the quadratic model is 0.99754; that is, the square of the simple correlation coefficient between Y and $\hat{Y}$ can be viewed as being 0.99754.

Example 9.20 The correlation index R^2 may be obtained for the data of Example 9.8 as a means of determining the goodness of fit of the regression equation already estimated. Equation (9.77) is used to give

$$R^2 = \frac{\hat{\beta}_1 \sum_i (X_{1i} - \bar{X}_1)(Y_i - \bar{Y}) + \hat{\beta}_2 \sum_i (X_{2i} - \bar{X}_2)(Y_i - \bar{Y})}{\sum_i (Y_i - \bar{Y})^2},$$

where

$$\sum_i (y_i - \bar{y})^2 = \sum_i y_i^2 - \frac{(\sum_i y_i)^2}{n} = 9274.97.$$

Therefore,

$$r^2 = \frac{-0.13840(10{,}937.84) + 3.67961(2927.762)}{9274.97}$$

$$= 0.9983,$$

indicating that the proposed model fits the data exceptionally well and the regression equation in fact accounts for 99.9% of the variability of the experimental data about the mean $\bar{y}$. If the correlation coefficient between any two of the independent variables is high, say at least 0.8, then generally no great increase in information can be obtained by using multiple correlation for both variables as opposed to simple linear correlation for using only one of them. For example, using the data of Example 9.8, the simple correlation coefficients are $r^2_{X_1X_2} = 0.8271$, $r^2_{X_1Y} = 0.7686$, and $r^2_{X_2Y} = 0.9938$. Since R^2, as shown by Example 9.8, is 0.9983, very little was gained by adding X_1 to the correlation. However, using Y and both X_1 and X_2 increases the correlation from 0.7686 to 0.9983 over using Y and X_1. In Example 9.8 it was seen that $H_0: \beta_1 = 0$ was rejected at the 5% significance level. The hypothesis $H_0: \beta_2 = 0$ was also rejected at the 5% significance level.

One warning should be repeated here: High correlation coefficients can be obtained in models that leave very little or no physical meaning. It is good practice to plot the data, whenever possible, and select trial models according to the results of the plot. Frequently, the data will have to be plotted in several ways before a good (i.e., reasonable and usable) model is apparent. Consider the following example.

Example 9.21   All too often we encounter data that are nonlinear on semilog plots. One good example of this is the O'Connell [16] correlation between overall column efficiency E and the product of relative volatility α and liquid viscosity μ for multicomponent distillation systems. From the points below (estimated from O'Connell's original curve), find a mathematical relationship between E and $\alpha\mu$.

E	$\alpha\mu$
100	0.05
90	0.082
80	0.13
70	0.24
60	0.46
50	0.91
40	2.19
30	7.64

Solution When these data were plotted on rectangular coordinate paper, a quadratic model was suggested, which yielded

$$E = 85.6529 - 31.7893(\alpha\mu) + 3.21634(\alpha\mu)^2, \tag{A}$$

for which $R^2 = 0.8404$. As the model did not account for a large enough portion of the data variation, the data were plotted on semilog paper. From that, two models seemed obvious. The fitted results were

$$E = 52.6465 - 32.4268 \log(\alpha\mu), \tag{B}$$

for which $R^2 = 0.9743$ and

$$E = 47.1390 - 28.8853 \log(\alpha\mu) + 9.54988[\log(\alpha\mu)]^2, \tag{C}$$

for which $R^2 = 0.9916$. Although inclusion of the quadratic term in $\log(\alpha\mu)$ raised R^2, there are no known instances of $[\log(X)]^2$ having any real significance in a physical model. Therefore, equation (C) was discarded as useless.

One student suggested reversing the axes. This was done and gave

$$\log E = 1.86758 - 0.0582711(\alpha\mu), \tag{D}$$

for which $R^2 = 0.7023$. As we seemed to be headed in the wrong direction, the data were plotted on log–log paper and a nearly straight line was obtained. The data were then correlated using an appropriate model, which gave the result

$$\log E = 1.69014 - 0.242014 \log(\alpha\mu), \tag{E}$$

for which $R^2 = 0.9996$. At this point, the question was raised: "Which relation should we use, (B) or (E)?" This is almost like asking, "What is the best way to give change for a dollar?" While (E) may be slightly better than (B), it is not as easy to use because it involves $\log E$ rather than E itself. In this example, one would choose model (B) since it appears to be more useful than model (E). In similar circumstances, a choice is made based on usefulness. If the choice were between (A) and either (B) or (E), (A) would not be selected because of the lower R^2 value.

An alternative approach would be to break the curve into a series of segments that can be approximated as linear and using the techniques of Example 9.4. When this was done in two sections, $\alpha\mu \leq 2$ and $\alpha\mu > 2$, the results were, respectively,

$$E = 0.5117 - 0.14751 \log(\alpha\mu) \qquad \text{for } \alpha\mu \leq 2$$

and

$$E = 0.44916 - 0.07220 \log(\alpha\mu) \qquad \text{for } \alpha\mu > 2.$$

Potentially the errors encountered can be severe when taking this approach. At the breakpoint in this case, the error is a maximum. The error excursions in the center 60% of each segment are fairly small. The errors in the outer 20% of the range of each line are so large that the lines, in the vicinity of the breakpoint, are virtually useless for estimation purposes in the initial stages of bubble cap tower design. This illustrates an all too common failing on the part of the beginning statistics practitioner: laziness. Granted it is easier to segment the curve and solve the resulting pair of simple linear regressions. As we see in this example, the predictions of efficiency as it depends on the $\alpha\mu$ product are worthless. It is far better to expend the additional effort required to obtain a continuous curve that permits adequate predictions.

9.9 STEPWISE REGRESSION

The situation arises occasionally when several independent variables affect the outcome of an experiment. In that case you will want to find an empirical equation relating them that simultaneously satisfies two criteria: has a high value of R^2 and uses only the most important independent variables. This can be handled in two ways by use of PROC STEPWISE or PROC RSQUARE (*SAS User's Guide: Statistics*, Version 5 edition, SAS Institute, Cary, NC, 1985, 956 pp.). PROC STEPWISE has five methods of retaining individual terms in the model: forward selection, backward elimination, and so on. More discussion of these techniques can be found in Draper and Smith [8] and Daniel and Wood [7]. Stepwise regression requires the assumption that there is only one best equation and that the procedure will find it. This, of course, presupposes that the "right" independent variables were selected for the model. Another problem arises when there is a high correlation between independent variables: difficulty in interpreting the results. The general forward selection procedure begins by using the independent variables one at a time, comparing the corresponding reduction in the error sum of squares with some preset criterion ($\alpha = 0.15$ is the default for PROC STEPWISE) and then either retaining or rejecting the term. PROC STEPWISE is a modified forward selection method in which the F value for each term in the model is calculated, compared to the corresponding tabular value of F, and rejected if it is not significant at the preset significance level. Then the next term is added to the model and the process is repeated.

As the name implies, backward elimination starts with the full model and eliminates the terms one at a time based on their lack of contribution to the reduction in the error sum of squares. It should be noted that all independent variables not "read in" by INPUT and CARDS statements must be generated in the program. Such generations (X^2, $\sqrt{X}$, log X, etc.) must be placed between the INPUT and CARDS statements.

PROC RSQUARE constructs regression analyses for all possible combinations of the independent variables. The output includes the variables used and their associated values of R^2 for all possible combinations of the independent variables in the MODEL statement. The seemingly "best" models based on the R^2 values and compatibility with physical theory can hopefully be identified for further scrutiny or experimentation.

Example 9.22 This example, adapted from data given by Ostle,* demonstrates the use of PROC STEPWISE and PROC RSQUARE. The size (Y values measured in microns) of bainite grains was determined at different time intervals (X values, seconds) for a particular steel alloy undergoing isothermal transformation to bainite.

X (s)	Y (μm)	X (s)	Y (μm)
1	17	8	64
2	21	9	80
3	22	10	86
4	27	11	88
5	36	12	92
6	49	13	94
7	56		

Examine these data to find the most reasonable regression model.
Solution

(put initial JCL cards/statements here)

```
DATA TIME;
INPUT X1   1-2   Y   4-5;
X2 = X1*X1;
X3 = X1*X1*X1;
CARDS;
```

---

* Data reprinted from B. Ostle and R. W. Mensing, *Statistics in Research*, 3rd ed., copyright 1975 by the Iowa State University Press, Ames, Iowa, by permission.

(enter data here according to INPUT format)

```
PROC PRINT;
PROC PLOT;
PLOT Y*X1;
TITLE 'EFFECT OF ISOTHERMAL TRANSFORMATION TIME ON GRAIN SIZE';
PROC STEPWISE DATA = TIME;
MODEL Y = X1  X2  X3;
TITLE 'EFFECTIVENESS OF STEPWISE REGRESSION';
```

(put final JCL cards/statements here)

Note that data set should be named in both the DATA and PROC STEPWISE statements. The results of PROC STEPWISE are shown below.

```
                                SAS

        STEPWISE REGRESSION OF TRANSFORMATION TIME IN GRAIN SIZE

             STEPWISE REGRESSION PROCEDURE FOR DEPENDENT VARIABLE Y

 STEP 1      VARIABLE X1 ENTERED           R SQUARE = 0.96977720

                    DF          SUM OF SQUARES     MEAN SQUARE       F     PROB>F

 REGRESSION          1          10177.58791209  10177.58791209   352.96   0.0001
 ERROR              11            317.18131868     28.83466533
 TOTAL              12          10494.76923077

                  B VALUE          STD ERROR       TYPE II SS        F     PROB>F

 INTERCEPT       3.96153846
 X1              7.47802198        0.39803546   10177.58791209   352.96   0.0001
 -----------------------------------------------------------------------------
```

From the analysis of variance, we see that only the linear term was significant in the regression with the preset $\alpha = 0.15$ (default level if another option is not specified).

To continue this example, replace the PROC STEPWISE statement by PROC RSQUARE. Again, the data set is named in the DATA and PROC RSQUARE statements. The results are shown below.

```
        USE OF PROC RSQUARE ON TTT - GRAIN SIZE DATA

 N =     13      REGRESSION MODELS FOR DEPENDENT VARIABLE Y

 NUMBER IN      R-SQUARE        VARIABLES IN MODEL
   MODEL

     1          0.80322270      X3
     1          0.91122065      X2
     1          0.96977720      X1
 -------------------------------------------------
     2          0.97003083      X1 X2
     2          0.97135430      X1 X3
     2          0.99130555      X2 X3
 -------------------------------------------------
     3          0.99523836      X1 X2 X3
 -------------------------------------------------
```

This shows that a cubic model may be warranted for the data. These results do not conflict with those from PROC STEPWISE because of the default significance level for entering a new variable. We recommend plotting the data first, then follow with PROC RSQUARE, and finally, with the appropriate regression model using PROC GLM. The results from PROC GLM for this data set are shown below.

SAS

DEP VARIABLE: Y

SOURCE	DF	SUM OF SQUARES	MEAN SQUARE	F VALUE	PROB>F
MODEL	3	10444.797	3481.599	627.035	0.0001
ERROR	9	49.972278	5.552475		
C TOTAL	12	10494.769			

VARIABLE	DF	PARAMETER ESTIMATE	STANDARD ERROR	T FOR H0: PARAMETER=0	PROB > |T|
INTERCEP	1	21.727273	3.595816	6.042	0.0002
X1	1	-5.839577	2.141837	-2.726	0.0234
X2	1	2.343781	0.348835	6.719	0.0001
X3	1	-0.113345	0.016421	-6.903	0.0001

From the AOV for the entire model, the calculated F is much greater than $F_{3,9,0.95}$. We also note, however, that all the terms in the model are significant based on the t-tests.

Figures 9.23 and 9.24, respectively, illustrate another way of examining the regression. The predicted values ($YHAT_i$) should fall on or very near the observed Y_i values on an OVERLAY plot. Such is the case here. The residuals ($YHAT_i - Y_i$) should be randomly scattered about the reference line and they are. If the regression is valid, a plot of YHAT vs. Y should have a slope of 1.0 and pass through the origin. Figure 9.25 shows that this is so for this example.

Further examination of the results using

```
PROC GLM;
MODEL Y = X1  X2  X3;
OUTPUT OUT = NEW P = YHAT R = RESID;
PROC GLM;
MODEL YHAT = Y;
```

illustrates another way to test the effectiveness of the cubic regression: the intercept and slope should be 0 and 1, respectively. The results follow.

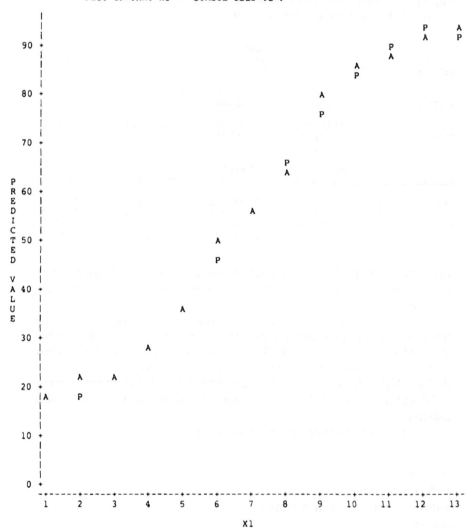

FIGURE **9.23** Comparison of grain growth predicted (P) by cubic regression with observed (A) values.

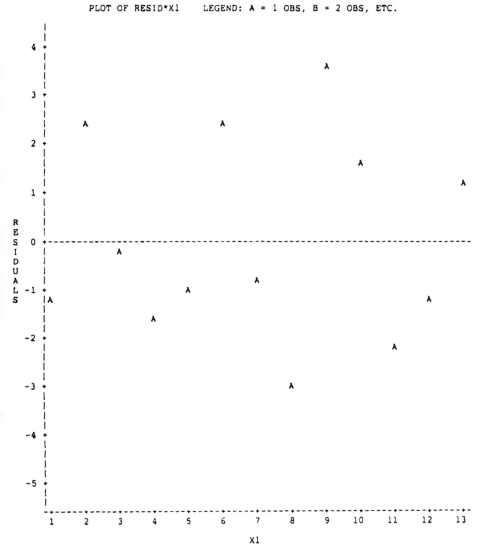

SAS

CHECKING RESIDUALS FOR RANDOM SCATTER

PLOT OF RESID*X1   LEGEND: A = 1 OBS, B = 2 OBS, ETC.

FIGURE 9.24 Checking residuals from cubic regression for random scatter.

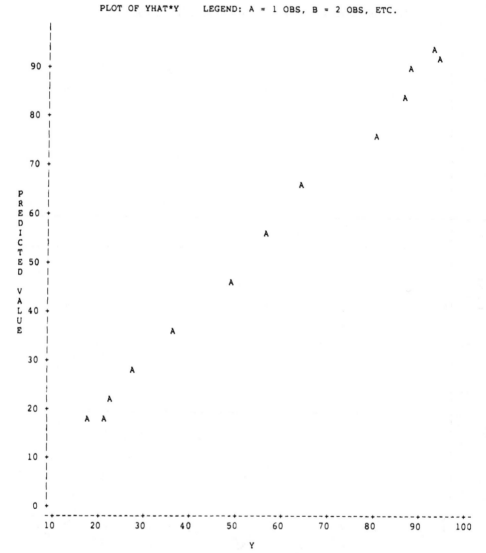

SAS

COMPARISON OF CALCULATED AND ORIGINAL DATA

PLOT OF YHAT*Y   LEGEND: A = 1 OBS, B = 2 OBS, ETC.

FIGURE 9.25   Evaluation of slope and intercept for grain size data.

SAS

EFFECTIVENESS OF CUBIC REGRESSION OF TIME ON GRAIN SIZE

DEP VARIABLE: YHAT   PREDICTED VALUE

SOURCE	DF	SUM OF SQUARES	MEAN SQUARE	F VALUE	PROB>F
MODEL	1	10395.063	10395.063	2299.130	0.0001
ERROR	11	49.734328	4.521303		
C TOTAL	12	10444.797			

VARIABLE	DF	PARAMETER ESTIMATE	STANDARD ERROR	T FOR HO: PARAMETER=0	PROB > |T|
INTERCEP	1	0.268117	1.309089	0.205	0.8415
Y	1	0.995238	0.020756	47.949	0.0001

From the T value for β_0 (intercept), we accept the null hypothesis that the intercept of the $\hat{Y}$ vs. Y model is zero. Calculating $T_{\beta_1} = (\beta_1 - \hat{\beta}_1)/S_{\beta_1}$, which is

$$T_{\beta_1} = \frac{1 - 0.995238}{0.020756} = 0.229,$$

and comparing it with $t_{11,0.975} = 2.201$, we see that there is no reason to believe that the slope is not 1, as seen in Fig. 9.25.

The use of R^2 as a criterion for evaluating empirical models is illustrated by the following example.

Example 9.23 The activity coefficients γ for toluene (T) and n-octane (O) vs. their corresponding mole fractions X in a binary mixture are shown below.

γ_T	γ_O	X
1.1757	1.2972	0.07
1.1492	1.2647	0.10
1.1168	1.2411	0.15
1.0965	1.2162	0.20
1.0944	1.1513	0.30
1.0522	1.0972	0.40
1.0387	1.0593	0.50
1.0221	1.0290	0.60
1.0116	1.0116	0.70
1.0039	1.0053	0.80
1.0028	1.0023	0.90
1.0000	1.0000	1.00

Find the most promising empirical models by use of PROC RSQUARE.
Solution   Using the usual linearizing transformations to logarithms, we
create

$$U = \text{LOG10(GAMMAT)}$$
$$W = \text{LOG10(XT)}.$$

The results of using PROC RSQUARE are shown below.

SAS

N=   12   REGRESSION MODELS FOR DEPENDENT VARIABLE GAMMAT

NUMBER IN MODEL	R-SQUARE	VARIABLES IN MODEL
1	0.70560351	XTSQ
1	0.88134528	XT
1	0.98552199	W
2	0.97805571	XT XTSQ
2	0.98562837	XT W
2	0.98584669	XTSQ W
3	0.98738233	XT XTSQ W

Figure 9.26 shows the original data. Initial trial models come immediately
to mind: quadratic in XT, $\log_{10} \gamma_T$ vs. XT, $\log_{10} \gamma_T$ vs. $\log_{10} XT$. As seen in
Fig. 9.27, the log–log model produced an approximately linear relation.
When we look at the OVERLAY plot (Fig. 9.28) we see that the linearizing
transformation used was appropriate: the predicted values γ_T differ only
slightly from the original values. This is even more obvious when we look
at the residuals plot (Fig. 9.29): the lack of a regular shape for the graph
indicates again that our selection of model was appropriate. The analysis of
variance below shows that the log–log model is significant ($F \cong 659.7$) at
$\alpha = 0.05$. Checking the residuals and comparing the predicted values to the
original data is the only reasonable way to evaluate transformed models
after obtaining favorable AOV results for them.

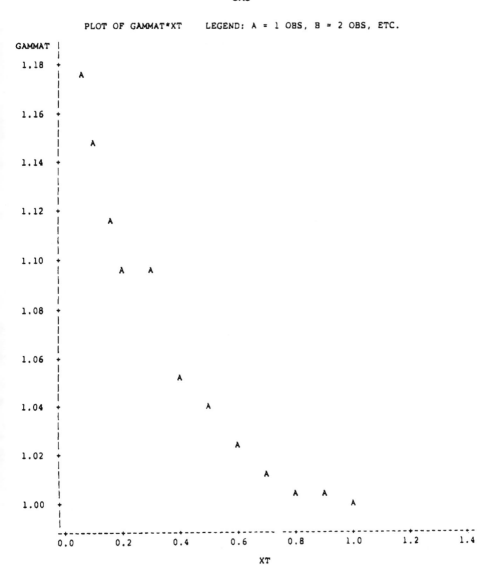

FIGURE 9.26   Activity coefficient vs. mole fraction for toluene.

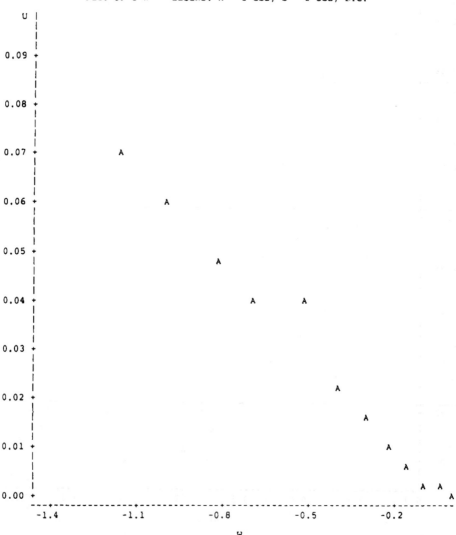

FIGURE 9.27   Effect of log–log transformation on activity coefficient data for toluene.

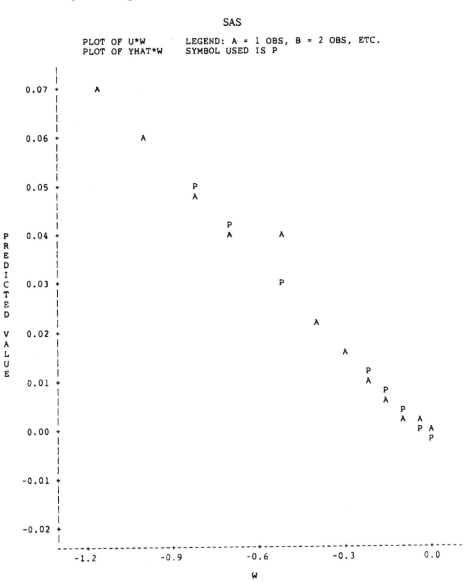

FIGURE **9.28** Comparison of observed (A) and predicted (P) toluene activity coefficients after log–log transformation.

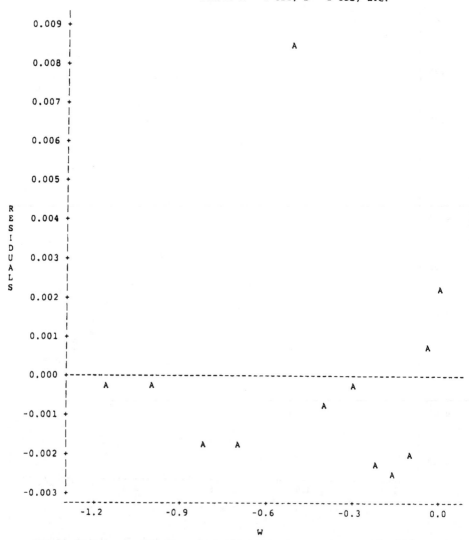

FIGURE 9.29   Residuals from log–log transformation of toluene activity coefficients.

SAS

DEP VARIABLE: U

SOURCE	DF	SUM OF SQUARES	MEAN SQUARE	F VALUE	PROB>F
MODEL	1	0.006597756	0.006597756	659.698	0.0001
ERROR	10	0.0001000118	.00001000118		
C TOTAL	11	0.006697767			

VARIABLE	DF	PARAMETER ESTIMATE	STANDARD ERROR	T FOR H0: PARAMETER=0	PROB > |T|
INTERCEP	1	-0.00230662	0.001435714	-1.607	0.1392
W	1	-0.063023	0.002453747	-25.685	0.0001

Problems

9.1 The relation between the heat capacity of liquid sulfuric acid and temperature is as follows:

Heat capacity (cal/g °C)	0.377	0.389	0.396	0.405	0.446	0.458
Temperature (°C)	50	100	150	200	250	300

The proposed model is

$$C_p = C_{po} + at + \varepsilon.$$

(a) By use of simple linear regression, evaluate C_{po} and a.
(b) What portion of the data does your equation explain?
(c) Are C_{po} and a significantly different from zero? Prove your answer.

9.2 In a study to determine the relationship between the incidence of sandspur and "goat-head" clumps on the intramural field behind our building, Y, and the number of times the grass is cut each year, X, the results of a 4-year study showed: $\bar{x} = 20$, $\bar{y} = 22$, $\sum_i (x_i - \bar{x})^2 = 225$, $\sum_i (y_i - \bar{y})^2 = 414$, $\sum_i (x_i - \bar{x})(y_i - \bar{y}) = 180$, and $n = 32$.

(a) What is the estimated population regression coefficient?
(b) Test the hypothesis that $\beta = 1$ for the model $Y_i = \alpha + \beta X_i + \varepsilon_i$.

9.3 For the data presented in Problem 8.25, determine the relationship between separation force and pulling angle. Test the resulting equation if it is believed that the true relation is

force (lb$_f$) = 40 + 0.75 (angle, in degrees).

9.4   Obtain the form of the linear relationships between blue colorant and lightness (L) for each level of yellow colorant from the data in Problem 8.22. Plot your results. Show the 98% confidence limit for each such relationship graphically.

9.5   The data for the calibration curve for the colorimetric analysis of fluoride ions in aqueous solution by the Lacroix and Labalade [13] method are as follows:

Concentration in cuvet (mg F$^-$/mL)	Percent transmittance		
	Trial 1	Trial 2	Average
0.000608	80.3	80.3	80.30
0.001216	80.5	80.4	80.45
0.001824	80.9	81.0	80.95
0.002432	81.2	81.8	81.50
0.003040	81.6	82.0	81.80
0.003648	82.9	82.5	82.70
0.004256	83.0	83.1	83.05
0.004864	83.9	84.0	83.95
0.005472	84.0	84.0	84.00
0.006080	85.0	84.8	84.90

The data can be expressed as

$$\%T = 79.46637 + 865.2737C,$$

where %T is the percent transmittance and C is the fluoride concentration in mg F$^-$/mL. The equation above is difficult to use for determining the concentration of an unknown from its %T. Find, by the least-squares technique, the desired relation between C and %T, test the validity of the regression constants, and give the 95% confidence interval on concentration for an unknown sample having 82.1%T.

9.6   Use the following data from the packed 6 ft × 3 in. gas absorber (water flow rate = 4 gal/min) to find the loading and flooding points mathematically as a result of suitable regression analyses.

$-\Delta P$	L/G	$-\Delta P$	L/G
0.10	21.7	0.75	5.8
0.13	19.0	0.90	5.0
0.12	17.2	1.05	4.6
0.20	13.0	1.10	4.0
0.30	11.4	1.22	3.6
0.35	9.6	1.35	3.4
0.50	8.5	1.55	3.2
0.60	7.8	1.70	2.6
0.65	7.0		

Compare the results with the corresponding L/G, moles water per mole gas, vs. pressure drop ($-\Delta P$, in H_2O) results you obtain by plotting the data. Comment on any differences.

9.7 A preliminary experiment has been run to check the working condition of the steam kettle in Ch.E. B-4. The experiment is carried out at atmospheric pressure with the steam pressure varying from 1 to 10 psig. Let X be the steam pressure in psig and Y be the temperature in °C.

X	Y
1	5
2	7
3	9
4	12
5	13
6	16
7	17
8	19
9	21
10	23

Find the relationship between steam pressure and the temperature in the center of the kettle at $t = 1$ min.

9.8 The following data were obtained two months apart for the relationship between refractive index and composition of mixtures of benzene and 1-propanol. Different lots of materials were used in each case.

| Benzene (mL) | 1-Propanol (mL) | Refractive index | |
		Run 1	Run 2
10	1	1.4837	1.4842
10	2	1.4724	1.4716
10	4	1.4650	1.4651
10	6	1.4571	1.4555
10	8	1.4430	1.4420
10	10	1.4385	1.4368
10	14	1.4263	1.4244
10	20	1.4108	1.4100
10	30	1.3941	1.3940
10	50	1.3892	1.3888
10	100	1.3816	1.3819

(a) Determine the relationship between these variables.

(b) Test each term at the $\alpha = 0.01$ level.

Support your answer statistically.

9.9 In an effort to work out a program for control of stack loss of a relatively valuable chemical, data were obtained on stack losses and suspected related independent variables for 139 days. Data on stack loss, water temperature, and air temperature are given below, where

$$y = 10 \text{ (stack loss in g/m}^3 - 3.0)$$
$$X_w = \text{water temperature in °C (°C} - 20\text{°C)}$$
$$X_a = \text{air temperature in °F (°F} - 50\text{°F)}.$$

Test	y	X_w	X_a
1	−20	−6	−11
2	−9	1	−3
3	−18	1	−2
4	−6	−1	−1
5	−3	2	−3
6	−5	−5	−10
7	−20	2	−20
8	+19	5	+14
9	−4	−1	−14
⋮	⋮	⋮	⋮
139			

$$\sum y = 1479, \qquad \sum x_w y = 12{,}797,$$
$$\sum x_w = 406, \qquad \sum y^2 = 99{,}077,$$
$$\sum x_a = 1, \qquad \sum x_w^2 = 3456,$$
$$\sum x_w x_a = 3533, \qquad \sum x_a^2 = 10{,}525,$$
$$\sum x_a y = 17{,}340, \qquad n = 139.$$

An independent investigation was made to determine the reproducibility of the stack loss measurement. Ten measurements at identical conditions were made and the data coded to correspond to y given above. From the replicate data $[y = 10 \text{ (stack loss in g/m}^3 - 3.0)]$, $\sum y = 42$, and $\sum y^2 = 1560$.

(a) Considering only the relationship between stack loss and water temperature:

 (1) Determine the best linear relationship between the variables and test its significance.

 (2) Establish 95% confidence limits of the regression equation for predicting stack loss at water temperatures of 15 to 30°C.

(b) Determine whether the addition of air temperature in the regression equation significantly improves the correlation.

(c) Do you feel that your best expression accounts adequately for the variation in stack losses? If not, what do you suggest as the next move?

9.10 The irritant factor Y of polluted air can be determined as a function of the concentrations of SO_2 and NO_2 in the atmosphere. The following data are available, where X_1 = parts NO_2 per 10 million parts of air and X_2 = parts SO_2 per 100 million parts of air.

Y	X_1	X_2
65	10	12.5
72	12	15
82	15	18
95	16	21
110	19	26
122	21	30
125	25	35
130	28	40

Determine the irritant factor as a function of X_1 and X_2.

9.11 Liquid Murphree plate efficiencies, E_L, in distillation are dependent on the closeness of approach to equilibrium on that plate. Equilibrium can be upset by changes in feed location or temperature. The following data were obtained for the methanol–water system using a 3-in.-ID column with six sieve plates, a total condenser, and a partial reboiler. For a constant external reflux ratio, the following data were obtained for E_L for plate 3. The feed and flow rates were constant in all cases.

Feed plate	Efficiency at feed temperature (°F):			
	90	105	125	150
2	7.5	41.2	51.6	65.2
3	54.0	62.5	84.2	100
4	29.0	48.6	41.0	36.0
5	12.3	14.6	18.3	8.3

Do either of these variables affect the Murphree efficiency?

9.12 In an investigation of the effects of carrier gas velocity and support particle diameter on retention time in gas chromatography, the following data have been obtained.

Y	X_1	X_2
6.5	76	81
8.4	92	96
9.7	106	108
11.4	120	150
13.4	136	159
15.6	159	173
18.7	200	214
20.0	218	230

Using multiple linear regression, develop a relationship among these three variables. Let Y be the retention time in minutes, X_1 be the carrier gas velocity in mL/min, and X_2 be the average particle diameter in μm.

9.13 Given the following data from the performance tests on a degassing tower:

X_1	X_2	Y
14	1.00	46
15	1.25	51
16	3.00	69
17	3.25	74
18	4.00	80
19	5.25	82
20	5.50	97

$$\sum_i (x_{1i} - \bar{x}_1)^2 = 28.0,$$

$$\sum_i (x_{1i} - \bar{x}_1)(x_{2i} - \bar{x}_2) = 22.50,$$

$$\sum_i (x_{2i} - \bar{x}_2)^2 = 18.7143,$$

$$\sum_i (x_{1i} - \bar{x}_1)(y_i - \bar{y}) = 226,$$

$$\sum_i (x_{2i} - \bar{x}_2)(y_i - \bar{y}) = 183.86,$$

$$\sum_i (y_i - \bar{y})^2 = 1915.43,$$

where X_1 is the liquor rate in hundreds of gallons per hour, X_2 the air velocity (ft/sec) and Y the percent dissolved CO removed:

(a) Calculate the regression equation between these variables.
(b) Test each term for significance at the 5% level.
(c) Give the 95% confidence interval for the true value of Y estimated from your regression equation for $x_1 = 18.5$ and $x_2 = 4.5$.

9.14 Consider the breakthrough curve below obtained in an investigation of the adsorbents and/or catalysts best suited for the reduction or removal of NO_2 from space cabin atmospheres.

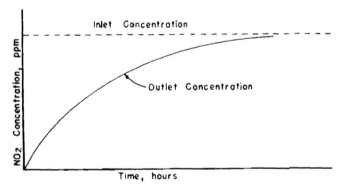

The adsorption is most likely of the Langmuir type. As such, the model describing the adsorption process is quadratic in form. Describe how you would find the regression coefficients involved for this model. Be specific and complete.

9.15   In the determination of the calibration curve for the analysis of ionic fluoride in mg F^-/mL of solution (the Y variable) vs. $\%T$ (the X variable), the data in Problem 9.5 were refit to the simple linear model $Y = \beta_0 + \beta_1 X + \varepsilon$, from which the following calculated results are available:

$$\sum_i x = 1647.198, \quad \left(\sum_i y\right)^2 = 0.004473, \quad \sum_i (x_i - \bar{x})^2 = 46.8125,$$

$$\sum_i x^2 = 135,710, \quad \sum_i xy = 5.561005, \quad \sum_i (y_i - \bar{y})^2 = 0.000061,$$

$$\left(\sum_i x\right)^2 = 2,713,264, \quad \sum_i x \sum_i y = 110.1646, \quad \sum_i (x_i - x)(y_i - y) = 0.052773.$$

$$\bar{x} = 82.53993, \qquad \bar{y} = 0.003344,$$

$$\sum_i y = 0.066880, \qquad \sum_i y^2 = 0.00285,$$

(a) Find the equation of the best line through these data.
(b) What portion of the variability in the data is accounted for by your equation?
(c) What range of fluoride concentration is expected if $\%T = 82.0$?

9.16   In a heat-transfer experiment, data were obtained on the Nusselt, Prandtl, and Reynolds numbers in the form

$$Y = \ln\left[\frac{N_{Nu}/10^3}{N_{Pr}^{0.4}}\right],$$

$$X = \ln \frac{N_{Re}}{10^5}$$

by Burrows [4]* in order to evaluate

$$N_{Nu} = C(N_{Re})^B(N_{Pr})^K.$$

Devise the best straight line through the following points, plot it, and show the 99% confidence limit on the line.

---

* Data reprinted by permission; copyright American Society for Quality Control.

Y	X
−0.3929	0.5417
−0.9978	0.0060
−0.5986	0.3542
−0.8564	−0.2303
−1.6724	−1.0524
−1.7430	−1.0496
−2.1144	−1.6100
−0.5211	0.5429
−1.0167	−0.0240

9.17 For the batch saponification of ethyl acetate a second-order rate equation was assumed to describe the rate of reaction. To establish that the second-order equation was a correct assumption, specific reaction-rate constants were calculated at various times during the reaction. They are as follows:

Time (s)	$k_2 \times 10^2$ (L/g mol·s)
69	0.7948
124	0.6572
179	0.7950
253	0.8157
305	0.9062
420	0.9249
600	0.8584
727	0.8653
902	0.8620
1201	0.7885

If the values for k_2 are nearly identical, it can be assumed that the second-order equation explained the kinetic data satisfactorily. Is it reasonable to believe that the second-order equation is the correct equation?

9.18 Two supposedly identical Brooks model R-2-65-5 rotameters with 316 stainless-steel spherical floats were calibrated for helium service at 20 psig input, 74°F. The data are as follows. (A dash means that no flow rate reading was taken. This does *not* mean that the flow rate at that scale reading was zero.)

Scale reading (mm)	Flow rate (mL He/min) for:	
	Tube 1	Tube 2
10	9.2	8.4
10	9.5	8.6
14	—	12.6
16	15.0	—
19	—	18.6
20	21.3	18.0
25	29.4	27.0
30	41.0	36.8
30	40.5	36.3
35	54.0	49.1
35	55.0	50.0
40	68.0	64.8
40	68.8	64.8
45	86.0	81.3
45	88.1	82.0
50	103.2	100.4
50	104.6	97.0
55	124.0	114.7
55	123.1	117.0
60	144.0	134.0
60	145.3	136.0
60	145.1	134.0

Let Y be the flow rate and X the scale reading.

(a) Obtain the regression coefficients for the quadratic model involved for each rotameter.

(b) Calculate the r^2 for each regression eqation from part (a).

(c) Compare the regression equations at the 0.05 significance level by comparing the individual regression coefficients: $\hat{\beta}_0$ with $\hat{\beta}'_0$, $\hat{\beta}_1$ with $\hat{\beta}'_1$, and $\hat{\beta}_2$ with $\hat{\beta}'_2$, where the $\hat{\beta}_i$ are for tube 1 and the $\hat{\beta}'_i$ are for tube 2.

9.19 Over the Reynolds number range $15{,}000 < N_{Re} < 40{,}000$, the following data were measured (f_{exp}) and predicted solely as a function of N_{Re} by the Konakov [17] equation (f_K) for the Fanning friction factor.

N_{Re}	f_{exp}	f_K
16,500	0.0285	0.0269
17,610	0.0298	0.0265
17,890	0.0224	0.0264
18,340	0.0289	0.0262
19,400	0.0275	0.0258
21,990	0.0239	0.0251
25,700	0.0265	0.0241
27,970	0.0231	0.0236
28,940	0.0270	0.0230
31,210	0.0182	0.0230
38,230	0.0107	0.0220

From the Konakov model, we have

$$f_K = 0.0858882 - 0.0060812 \ln N_{Re}.$$

Find the corresponding equation for the experimental data and compare those results with the predicted values.

9.20 In a recent experiment to determine the distillation curve for gasoline, acetone, and various mixtures of the two, the data below were obtained. The procedure of ASTM Standard D-86 was used.

	Temperature (°F)		
	Gasoline	Acetone	Acetone
Initial boiling-point:	95	128	133
Endpoint: % Recovered	368	140	142
5	110	132	133
10	117	132	133
20	134	134	134
30	158	134	134

(*continued*)

	Temperature (°F)		
	Gasoline	Acetone	Acetone
Initial boiling-point:	95	128	133
Endpoint: % Recovered	368	140	142
40	182	134	134
50	204	134	135
60	218	134	136
70	247	136	138
80	274	138	138
90	321	139	140
95	368	140	142
% Recovered	95	98	97
% Total recovery	97	99	98
% Loss	3	1	2

	Temperature (°F)		
	50% Gasoline 50% acetone	30% Gasoline 70% acetone	70% Gasoline 30% acetone
Initial boiling-point:	110	125	135
Endpoint: % Recovered	345	270	319
5	115	127	135
10	120	130	135
20	123	132	135
30	126	132	135
40	126	135	138
50	135	135	150
60	140	135	168
70	148	136	225
80	238	138	260
90	302	142	319
95	345	270	—
% Recovered	95	96	93
% Total recovery	96	97	95
% Loss	4	3	5

Using the two determinations for acetone as a measure of experimental error, statistically show how the addition of acetone to gasoline affects the ASTM distillation curve.

9.21 Calibration data for the ammonium electrode for effluent gas sample analysis are:

NH$_3$ (ppm)	MV
5	+ 13.8
13	0
20	− 10.6
34	− 20.0
50	− 26.7
90	− 35.0
150	− 45.5
250	− 56.0
500	− 66.8
1000	− 79.1

Find a suitable equation for the calibration curve in the form of $\log(NH_3$ concentration$) = f$(millivolt reading from the electrode).

9.22 Nitric acid is utilized to dissolve uranium metal. It has been determined that impurity concentration in the metal has a direct effect on its dissolution rate in HNO_3. Data for the dissolution of two samples of uranium, one containing 400 to 500 ppm carbon and one containing < 35 ppm carbon, are given below.

Dissolution time (h)	Amount dissolved (mg/cm^3) for impurity:	
	400–500 ppm	< 35 ppm
0.0	0.0	0.0
10.0	8.0	5.5
20.0	18.5	10.0
30.0	33.0	18.5
40.0	57.0	25.0
50.0	88.5	36.0

Plot the data and determine the best equations for the resultant curves [12].*

9.23 During air pollution studies concerning nitrogen oxides and hydrocarbon vapors, it has been determined that the use of chromite catalysts will catalytically promote the removal of NO in the presence of reducing agents such as CO. Removal is presumed to follow the reaction

$$2CO + 2NO \xrightarrow{\text{cat}} 2CO_2 + N_2.$$

Results of several runs given below for copper chromite indicate a reduction of catalyst effectiveness with time.

Time (min)	% CO removal
4.0	99.0
8.0	100.0
12.0	99.5
16.0	98.0
20.0	95.0
24.0	82.0
28.0	51.0
32.0	23.0

Plot the data and determine an equation for the effect of time on percent CO removal [18].†

9.24 For fluids being heated or cooled inside tubes, the following relation has been found empirically:

$$\frac{h}{C_p G} = 0.023 \left(\frac{\mu}{DG}\right)^{0.2} \left(\frac{k}{C_p \mu}\right)^{0.7}$$

or more simply, $N_{St} = 0.023 N_{Re}^{-0.2} N_{Pr}^{-0.7}$. This relation may be expressed graphically by plotting $N_{St} N_{Re}^{0.2}$ vs. N_{Pr} on log–log paper to yield a straight line for such fluids as water, air, acetone, ethanol, and *n*-butane. Data from

---

these and many other systems have been lumped together and the best straight line "eyeballed" for most practical purposes. You, however, need to use this relation in a computer program for neopentane. Before you can do so, you need proof that neopentane behaves in a similar manner. From the data below, determine whether or not the given functional relation will hold for your system.

$N_{St}N_{Re}^{0.2}$			N_{Pr}
0.007,	0.0073,	0.0077	6
0.0045,	0.0050,	0.0049	10
0.0034,	0.0036,	0.0032	20
0.0024,	0.0028,	0.0027	30
0.0020,	0.0021,	0.0023	40

9.25 The overall heat-transfer coefficient U (Btu/hr ft^2 °F) shows a log–log dependence on v_{max} (ft/sec) in heat exchangers with parallel banks of finned tubes. Typical data are listed below.

U			v_{max}
8.4,	7.9		300
9.7,	9.1		500
9.7,	9.9		600
10.4,	10.1,	10.1	800
11.3,	11.6,	11.4	1000
12.9,	14.0,	13.8	1200
14.8,	16.0,	15.1	1500

For these data, determine the relation between U and v_{max}. Also show that any constants you obtain are nonzero. What error can you expect if using your relation to predict the value of U corresponding to a v_{max} of 900 ft/sec?

9.26 In recent research by Graham [9] on the removal of SO_2 from air by reaction with MnO_2, the following data were obtained at 650°F. Run 12-18-1 used 30 MnO_2 pellets with an initial loaded container weight of

1.4188 g and an inlet concentration of 3620 ppm SO_2. Run 12-18-2 used 20 pellets with an initial loaded container weight of 1.3935 g. The tare weight in both cases was 1.3576 g. Times are in seconds, weights are for the container + pellets.

Run 12-18-1:

Time	Wt	Time	Wt
0	1.4188	2040	1.4305
178	1.4200	2145	1.4310
366	1.4210	2250	1.4315
382	1.4215	2340	1.4320
462	1.4220	2455	1.4325
520	1.4225	2556	1.4330
588	1.4230	2668	1.4335
640	1.4235	2779	1.4340
716	1.4240	2880	1.4345
863	1.4245	2980	1.4350
1003	1.4250	3085	1.4355
1314	1.4270	—	1.4360
1389	1.4275	3285	1.4365
1504	1.4280	3390	1.4370
1595	1.4285	3530	1.4375
1702	1.4290	3650	1.4380
1825	1.4295	3771	1.4385
1920	1.4300	3882	1.4390

Run 12-18-2:

Time	Wt	Time	Wt
0	1.3935	761	1.3975
150	1.3950	880	1.3980
244	1.3955	985	1.3985
335	1.3960	1111	1.3990
518	1.3965	1240	1.3995
635	1.3970	1347	1.4000

(continued)

Time	Wt	Time	Wt
1460	1.4005	3040	1.4049
1588	1.4010	3120	1.4052
1724	1.4015	3220	1.4055
1828	1.4018	3340	1.4059
1935	1.4021	3512	1.4063
2075	1.4025	3605	1.4066
2176	1.4028	3727	1.4069
2336	1.4032	3860	1.4073
2470	1.4035	3980	1.4077
2610	1.4039	4325	1.4082
2783	1.4043	4445	1.4085
2905	1.4046		

(a) Obtain the breakthrough curves for these runs by plotting percent weight gain vs. time on arithmetic and semilog paper.
(b) Use nonlinear regression techniques to obtain the equations of these curves.
(c) Compare the regression coefficients at the $\alpha = 0.05$ level and comment on the effect of the change in number of MnO_2 pellets.

9.27 The SO_2 removal studies in Problem 9.26 were extended to include runs at 800°F, 31 psig, 20 pellets per run. For runs 1-13-3 and 1-13-6, 20 pellets MnO_2 each were used. Tare weight was 1.3885 g for each. In run 1-13-3, the filled container weighed 1.3918 g; for run 1-13-6, it was 1.3938 g. The data are as follows:

Run 1-13-3:

Time	Wt	Time	Wt
0	1.3885	406	1.3955
75	1.3905	480	1.3965
128	1.3915	530	1.3970
170	1.3920	600	1.3975
232	1.3930	665	1.3980
281	1.3940	778	1.3995
360	1.3950	870	1.4005

(*continued*)

Time	Wt	Time	Wt
960	1.4015	1665	1.4065
1010	1.4020	1904	1.4080
1190	1.4035	2020	1.4085
1264	1.4040	2140	1.4090
1345	1.4045	2410	1.4100
1433	1.4050	2875	1.4110
1578	1.4060		

Run 1-13-6:

Time	Wt	Time	Wt
0	1.3937	1050	1.4100
150	1.3975	1124	1.4105
190	1.3985	1200	1.4115
245	1.3995	1300	1.4125
310	1.4005	1380	1.4130
366	1.4015	1442	1.4135
461	1.4030	1546	1.4140
526	1.4040	1620	1.4145
600	1.4050	1774	1.4150
672	1.4055	1879	1.4165
850	1.4080	1980	1.4170
917	1.4085	2050	1.4175
955	1.4090	2160	1.4180

SO_2 inlet concentration was 7540 ppm for both runs.

    (a)   Plot the breakthrough curves for these runs and determine statistically their mathematical form.

    (b)   Is there a significant difference in these results and those of run 12-18-2? How would you determine whether the difference is due to differences in MnO_2 pellets, SO_2 inlet concentration, or temperature?

9.28 Conley and Valint [6]* monitored the continuous oxidative degradation of poly(ethyl acrylate) at various temperatures. They determined the activation energy of the degradation reaction(s) as a function of temperature in a novel manner. The absorbance of infrared energy at 3.35 μm was measured at each temperature as a function of time. Those data were correlated with polymer weight loss so as to eliminate differences in film thickness between samples. The rate constants for the degradation (k) were obtained from the relationship between percent polymer remaining on the specimen holders and time.

Time (min)	Absorbance	% wt remaining
0	0.243	100
15	0.221	91
30	0.204	84
45	0.195	80
60	0.170	70
75	0.158	65
90	0.146	60
105	0.125	51
120	0.113	46
135	0.107	44

Temperature (°C)	Rate constant k (h^{-1})
180	0.64
190	1.28
200	9.3
210	11.9
220	27.4
230	221.6
240	472.2

---

* Data reprinted by permission of John Wiley & Sons, Inc.; copyright 1965 by John Wiley & Sons, Inc..

From the temperature dependence of $(-\log k)$, the activation energy of degradation was found to be 37 kcal/g mol. Assume that the variation in weight loss as measured by absorption at $3.35\,\mu m$ is the same for all degradation temperatures. What confidence limit can be placed on their values of the rate constants? What is the expected variance in the activation energy at any temperature within the range studied?

9.29–9.41   Calculate the correlation coefficients for Problems 9.1 to 9.13.

9.42   The grades in Phys. Chem. 347 and Ch.E. 330 (an applied solid-state chemistry course dealing with the physical properties of materials of all types) are as follows:

x_1 = Phys. Chem. 347	x_2 = Ch.E. 330
86	92
78	89
94	83
85	75
94	82
90	96
89	84
74	72

$$\sum x_1 = 670, \qquad \sum x_2^2 = 57{,}272,$$
$$\sum x_2 = 674, \qquad \sum x_1 x_2 = 56{,}663.$$
$$\sum x_1^2 = 56{,}514,$$

(a) What is the correlation coefficient for grades in these two courses?
(b) Develop an equation to predict Ch.E. 330 grades from the Phys. Chem. 347 grades above.
(c) Present and use two different techniques for testing the hypothesis that there is no relation between course grades.

9.43   Below are data obtained for the minimum bending radius for thin-walled (22- or 24-gauge) aluminum tubing to avoid crimping. The data are averages of 10 determinations per tube size.

(a) Find the appropriate relationship between bending radius and tube diameter.

Bending radius (in.)	Tube diameter (in.)
0.91	1/8
1.73	3/16
2.41	1/4
3.64	5/16
3.81	3/8
4.96	1/2
6.51	5/8
7.42	3/4
8.71	7/8
9.79	1

(b) What is the best estimate of the change in bending radius as the tube diameter is varied from $\frac{3}{16}$ in. to $\frac{7}{8}$ in.?

(c) What portion of the data is accounted for by your relationship?

9.44 Bennett and Franklin [1] present data* on the effect of annealing temperature on the density of a borosilicate glass with high silica content. For these data they obtained a cubic equation in terms of the orthogonal polynomial coefficients.

Annealing temperature (°C)	Density (g/cm^3)
450	2.23644
475	2.23534
500	2.23516
525	2.23574
550	2.23513
575	2.23674
600	2.23748
625	2.23902
650	2.23985
675	2.24005
700	2.24085
725	2.24120
750	2.24218

---

* Data reprinted by permission of John Wiley & Sons, Inc..

(a) Rework this problem to obtain the actual relationship between the variables involved. What part of the variation in the density does your regression equation explain?

(b) After plotting the data and your equation it appears that the relationship involved is, at least over the range of variables studied, part of a sine wave. Using appropriate transformations, fit the data to this new model. Transform the results of the fitted curve as necessary to the form of the original sine model. What portion of the variation in data does this approach provide? Which technique would you recommend for future use?

9.45 In an experiment on quantitative analysis by gas chromatography, the data below were obtained.

Composition (vol %)	EtFo peak height	Avg.	Bz peak height	Avg.
100% Bz	—	—	49.8	49.8
20% EtFo	28.0		45.6	
20% EtFo	28.3	28.166	46.6	46.166
20% EtFo	28.2		46.8	
40% EtFo	41.5		40.2	
40% EtFo	41.8	41.533	41.7	41.133
40% EtFo	41.3		41.5	
60% EtFo	51.5		28.2	
60% EtFo	51.5	51.900	27.0	27.733
60% EtFo	52.7		28.0	
80% EtFo	58.1		27.9	
80% EtFo	58.5	58.560	28.6	28.166
80% EtFo	59.1		28.0	
100% EtFo	63.3	63.3	—	—
Unknown	53		24.4	
Unknown	55	54.367	25.2	24.333
Unknown	55.1		23.4	

(a) Plot the two calibration curves on a single graph.

(b) Determine, using the least-squares technique, the best calibration curves to use in the analysis.

(c) One group of students obtained the following relation:

$$Y_1 = 0.050 + 186.706X_1 - 313.525X_1^2 + 312.482X_1^3 - 122.475X_1^4,$$
$$Y_2 = 1.761 + 136.333X_2 - 168.966X_2^2 + 81.516X_2^3,$$

where Y is the peak height and X is the weight fraction. Subscripts 1 and 2 denote ethyl formate (EtFo) and benzene (Bz), respectively. Comment on their solution with regard to form, model, correlation, and usefulness.

9.46 Chen et al. [5] studied heat transfer in the streamline flow of water through an annulus. For their data, fit an appropriate relation by the method of least squares. What portion of the variability in the data is accounted for by your equation?

9.47 Newton and Dodge [15]* studied the temperature dependence of thermodynamic equilibrium constant K on absolute temperature T (in Kelvin) for the synthesis of methanol from CO and H_2.

$1000/T$ (K^{-1})	log K
1.66	-4.15
1.73	-3.75
1.72	-3.65
1.75	-3.30
1.82	-3.10
1.81	-3.20
1.82	-3.00
1.82	-2.90
1.83	-2.95
1.88	-2.60
1.91	-2.70
1.91	-3.00
1.92	-2.30
2.05	-2.30
2.05	-2.15
2.05	-2.35

From their results, postulate the model involved and obtain the best possible estimates of the population parameters.

9.48 In the analysis of isopropanol and methyl ethyl ketone mixtures, a gas chromatographic procedure was used. Quantitative data were obtained and are shown below, where A represents isopropanol and B, methyl ethyl ketone.

W%	Peak height			Avg.
100.0 *A*	69.3	70.0	71.1	70.10
89.1 *A*	64.9	67.0	63.6	65.67
10.9 *B*	9.0	9.1	8.7	8.93
78.8 *A*	58.8	60.4		59.60
21.2 *B*	19.2	19.8		19.50
69.9 *A*	53.0	52.6		52.80
30.1 *B*	28.7	28.4		28.55
60.9 *A*	42.8	43.6		43.20
39.1 *B*	35.0	35.8		35.40
51.1 *A*	36.5	35.5		36.00
48.9 *B*	45.3	44.2		44.75
41.5 *A*	28.5	31.4	30.0	30.00
58.5 *B*	52.4	57.9	55.2	55.20
32.6 *A*	22.9	23.1		23.00
67.4 *B*	63.9	64.9		64.40
24.4 *A*	16.4	15.4	16.0	15.70
75.6 *B*	70.0	66.3	68.0	68.10
14.3 *A*	9.5	9.5		9.50
85.7 *B*	78.7	79.1		78.90
100.0 *B*	95.9	94.1	92.6	93.20
— *A*	44.1	43.5		43.80
— *B*	35.5	35.0		35.25

(a) What was the composition of the unknown sample?
(b) Give the 98% confidence limits for your answer.
(c) How good are these data? Support your answer by showing the analysis of variance on the calibration equations involved.

9.49 The overall heat-transfer coefficient U in evaporators is a function of the logarithm of the LMTD (labeled ΔT). For the following data, find the relation between U and Z where $Z = \log \Delta T$. Is the semilog model reasonable?

U	ΔT	Z
151.7	75.9	1.88024
186.2	84.25	1.92557
204.3	90.41	1.95622
221.7	97.04	1.98695
231.7	101.74	2.00749

$$\sum (U_i - \bar{U})(Z_i - \bar{Z}) = 6.3632296, \qquad \sum (U_i - \bar{U})^2 = 4013.728,$$
$$\sum (Z_i - \bar{Z})^2 = 0.01016400, \qquad \bar{U} = 199.12, \qquad \bar{Z} = 1.951294.$$

9.50 In a pilot-scale laboratory experiment, the composition (in mol %) of methanol from the internal and external liquid streams of a six-sieve-tray binary distillation column is determined by refractive index.

Mol % methanol	Refractive index
0	1.3330
5	1.3343
10	1.3370
18	1.3393
23	1.3415
31.5	1.3434
40.6	1.3428
52	1.3426
58	1.3415
65	1.3406
73	1.3382
80.5	1.3365
85	1.3347
89.6	1.3330
100	1.3302

Find a suitable regression equation for these data.

9.51 The number of diagnosed AIDS cases rose rapidly from 1980 through December 1987. Since then, as shown by the following data for sequential 6-month periods, the incidence appears to have been drastically reduced.

Time period	Incidence
Before 1981	240
Jan.–June 1981	240
July–Dec. 1981	320
Jan.–June 1982	540
July–Dec. 1982	920
Jan.–June 1983	1,420
July–Dec. 1983	2,040
Jan.–June 1984	2,700
July–Dec. 1984	3,800
Jan.–June 1985	5,000
July–Dec. 1985	7,500
Jan.–June 1986	8,600
July–Dec. 1986	11,000
Jan.–June 1987	13,500
July–Dec. 1987	14,520
Jan.–June 1988	17,500
July–Dec. 1988	17,800
Jan.–June 1989	18.520
July–Dec. 1989	18,000
Jan.–June 1990	18.480

Develop a suitable regression equation to express these data.

9.52 The increase in yield strength of age-hardened 96 wt % aluminum–4 wt % copper alloy rivets was measured as a function of elapsed time.

Increase (%)	Time (months)
0.29	6
0.52	7
0.79	8
1.25	9
1.81	10
2.61	11
4.25	12
7.38	13
11.30	14
18.12	15
28.12	16

Correlate the data by separate quadratic and exponential regressions and comment on the usefulness of those equations. What would you do now?

9.53 The reflectance loss of second-surface "common" glass mirrors increases with time due to the first-order surface degradation effect of exposure to windblown "dust" in dust storms. Extensive real-time data were used to develop simulation conditions in which an 8-h exposure is equivalent to a year of actual exposure on the southern High Plains of Texas. The following data for four samples show the reflectance loss as a function of cumulative hours of exposure.

% Loss	Time (months)
0	0
3, 3, 3, 2.9	8
4.8, 4.7, 4.8, 5	16
6, 6.1, 5.8, 5.9	24
7, 7, 7, 6.9	32
7.6, 7.5, 7.6, 7.4	40
8.2, 8.4, 8.6, 8.4	48
8.8, 8.7, 8.9, 8.8	56
9.6, 9.6, 9.5, 9.6	64
10, 10.2, 10.4, 9.8	72
10.5, 10.6, 10.5, 10.7	80

Develop and evaluate the appropriate exponential model for these data.

9.54 In preparation for making parabolic mirror cells for a solar-energy array, initial experiments used a sealed cloth supported over a short cylinder of 24-in. diameter. A vacuum of 0.6 psi was used to cause the "drumhead" to deflect. The deflection was measured along an arbitrarily selected diameter (i.e., from a traverse distance of 0 to 24 in.). These measurements were repeated 45 and 90 degrees from the original traverse by rotating the initial traverse points in a counterclockwise direction.

Material: plain cloth
Pressure: −0.60 psig

Traverse distance (in.)	$\dfrac{r}{R}$	Deflection (in.) for:		
		$\theta = 0°$	$\theta = 45°$	$\theta = 90°$
0	0	0.000	0.000	0.000
1	0.0833	0.131	0.127	0.128
2	0.1666	0.253	0.249	0.246
3	0.25	0.354	0.346	0.348
4	0.3333	0.443	0.434	0.445
5	0.4166	0.524	0.507	0.527
6	0.5	0.587	0.570	0.587
7	0.5833	0.637	0.624	0.640
8	0.6666	0.681	0.671	0.681
9	0.75	0.714	0.704	0.712
10	0.8333	0.739	0.725	0.733
11	0.9166	0.754	0.744	0.745
12	1.0	0.759	0.747	0.748
13	0.9166	0.754	0.740	0.739
14	0.8333	0.741	0.725	0.721
15	0.75	0.717	0.699	0.694
16	0.6666	0.684	0.665	0.658
17	0.5833	0.639	0.624	0.613
18	0.5	0.586	0.572	0.558
19	0.4166	0.515	0.508	0.494
20	0.3333	0.435	0.433	0.413
21	0.25	0.343	0.341	0.327
22	0.1666	0.245	0.236	0.228
23	0.0833	0.131	0.116	0.121
24	0	0.000	0.000	0.000

Use any of the traverse sets as assigned to find an equation describing the profile of the deflection of the simulated mirror surface in terms of the fractional radius, r/R, where r = distance from the outer edge to the traverse point and $R = 12$ in.

9.55 It has been claimed that OSHA reform legislation to increase effective workplace health and safety programs should significantly reduce workers' compensation (WC) costs which are said to have increased by 400% in a single decade. Consider the following data for 1974–1991:

Year	WC cost ($ billions)
1974	14.35
1975	15.3
1976	16.6
1977	17.25
1978	18.0
1979	19.35
1980	20.45
1981	21.35
1982	22.0
1983	23.6
1984	26.25
1985	29.1
1986	32.05
1987	35.1
1988	37.4
1989	40.5
1990	42.8
1991	46.6

(a) Comment on the validity of the claimed WC increase.
(b) Find a regression equation that adequately ($R^2 \geq 95\%$) represents the data.
(c) Estimate the 1993 WC costs. Compare your estimate to the 1993 actual WC cost as available from your local Social Security Administration, OSHA, or U.S. Department of Labor office.

9.56 The standard curve used in ASTM Standard E-119, *Fire Tests of Building Construction and Materials*, is defined by the following points:

Temperature (°F)	Elapsed Time
1000	5 min
1300	10 min
1550	30 min
1700	1 h
1850	2 h
2000	4 h
2300	8 h

What is the equation for this curve?

9.57 A series of experiments described in U.S. patent 4,450,145 obtained the loading values for SO_2 gas absorbed into a phosphate absorbent at $74°C \pm 1°C$. The inlet gas was saturated with water. The pH of the original absorbent was 4.4 ± 0.2. At equlibrium, the relationships between SO_2 concentration (vol %) are described by the following data.

SO_2 concentration	SO_2 loading	pH
2.0	5.50	3.22
1.66	5.53	3.07
1.38	4.86	3.18
1.16	4.41	3.18
0.82	3.77	3.28
0.61	3.20	3.40
0.32	2.56	3.50
0.25	2.11	3.80
0.15	1.31	3.84
0.10	1.28	3.86
0.05	0.96	4.00
0.04	0.57	4.12
0.03	0.57	4.17
0.02	0.38	4.18
0.01	0.24	4.12

Using nonlinear regression techniques, find equations relating each of these dependent variables to the inlet SO_2 concentration. Evaluate the validity of the equations by suitable methods.

9.58 Monthly average natural gas usage in millions of cubic feet per day have been reported for Lubbock, Texas, for 1993 as

Month	Usage
Jan.	1.336
Feb.	1.141
Mar.	1.035
Apr.	0.635
May	0.341

(continued)

Month	Usage
June	0.257
July	0.231
Aug.	0.224
Sept.	0.236
Oct.	0.315
Nov.	0.794
Dec.	1.104

(a) Find a suitable regression equation for these data and evaluate it in at least four ways at $\alpha = 0.05$.

(b) The January 1994 natural gas usage was 1.224 million cubic feet. Is that value reasonable, based on the 1993 data, or was Lubbock possibly enjoying a milder winter? Support your answer with all necessary statistical calculations.

9.59 In the determination of the number of transfer units (NTUs) by Mickley's [14] adaptation of the Merkel design equation, the data below were obtained for the integration.

i_G	$i_i - i_G$	i_G	$i_i - i_G$
30.8	4.4	36.0	7.5
31.2	4.6	36.4	7.6
31.6	4.8	36.8	8.2
32.0	5.0	37.2	8.8
32.4	5.1	37.6	9.4
32.8	5.2	38.0	10.0
33.2	5.8	38.4	10.6
33.6	6.1	38.8	11.2
34.0	6.6	39.2	11.8
34.4	6.8	39.6	12.4
34.8	7.0	40.0	13.0
35.2	7.2	40.4	13.6
35.6	7.2		

Prepare and execute a SAS program to plot $1/(i_i - i_G)$ vs. i_G so that an analytical integration can be obtained over the range $30.8 \le i_G \le 40.4$. The subscripts i and G refer to interface and bulk gas-phase conditions, respectively. The units of i are enthalpy, Bru/lb$_m$ bone dry air.

9.60 The overall heat-transfer coefficients (Btu/hr ft² °F) for use with a short-tube vertical evaporator have been evaluated as a function of glucose concentration (wt %) in aqueous solution from laboratory experiments. The data are shown below.

U (Btu/hr ft² °F)	X (wt % glucose)
1012	5
820	10
664	15
505	20
371	25
268	30
205	35
146	40
123	45

Select an appropriate model (or models) for these data by examining a plot of U vs. X, evaluate the model(s) by PROC STEPWISE or PROC RSQUARE, and then obtain the appropriate regression coefficients. Demonstrate in at least four ways that your model is suitable for these data.

9.61 The orifice discharge coefficient has been measured experimentally with the results shown below.

N_{Re}	C_0	N_{Re}	C_0
14,400	0.592	34,600	0.616
16,100	0.615	37,100	0.597
18,500	0.652	40,300	0.633
19,300	0.604	40,300	0.651
24,200	0.653	43,500	0.627
26,600	0.600	45,100	0.604
29,800	0.587	47,500	0.586
29,800	0.612	51,300	0.603
32,200	0.659	56,200	0.622
34,600	0.644	56,200	0.615

Are these data consistent with fluid-flow measurement theory? Support your answers with the appropriate tests of hypotheses.

9.62 In an attempt to verify experimentally the Colebrook and Altshul [11] equations for predicting the Fanning friction factor as a function of Reynolds number (N_{Re}), the following experimental and calculated results were obtained.

N_{Re}	f_C	f_A	f_{exp}
13,040	0.00772	0.00728	0.00698
14,380	0.00700	0.00709	0.00683
16,510	0.00678	0.00684	0.00703
18,010	0.00663	0.00688	0.00658
21,600	0.00633	0.00635	0.00591
22,790	0.00625	0.00625	0.00583
23,480	0.00620	0.00620	0.00539
25,250	0.00610	0.00608	0.00520

By the Colebrook equation, $f_C = 0.25[1.8 \log(N_{Re}/7)]^{-2}$. From the Altshul equation, $f_A = 0.0275(\varepsilon/D + 68/N_{Re})^{0.025}$ and ε was estimated as 1×10^{-5} in. The experimental values f_{exp}, were calculated from the Fanning equation:

$$f_E = 2g_c \frac{-\Delta P}{\rho v^2} \frac{L}{D}$$

using measured pressure drop and velocity data. For these results:

(a) Find an appropriate model for f_E as a function of N_{Re}.
(b) Compare your model to the Colebrook and Altshul models.
(c) Recommend, based on the comparison in part (b), a model for use in calculating friction losses in similar piping systems.

9.63 In a fluid-flow experiment, the pressure drop, ΔP_{cam} (lb_f/ft^2), was measured experimentally and logged by the Camille data acquisition system as a function of the Reynolds number, N_{Re}. The pressure drop for the valves and pipe fittings was calculated by the classic or 1-K method (ΔP_{1-K}) and by the newer 2-K method (ΔP_{2-K}). In the former method, the only criteria are fitting type and nominal pipe size. In the latter method, the effect of Reynolds number is also included. For the following data, which predictive method gives more realistic results?

N_{Re}	ΔP_{cam}	ΔP_{1-K}	ΔP_{2-K}
7,326	0.229	0.115	0.129
8,453	0.231	0.151	0.167
11,770	0.437	0.281	0.309
15,216	0.646	0.458	0.5
18,424	0.916	0.659	0.718
22,000	1.337	0.925	1.004
23,409	1.543	1.042	1.129
27,809	1.579	1.448	1.567
28,958	2.457	1.565	1.692
32,491	2.962	1.952	2.108
35,460	3.167	2.309	2.491
39,492	3.59	2.84	3.062
41,638	4.685	3.144	3.388
45,670	5.583	3.757	4.046
48,401	5.472	4.201	4.523
51,002	6.568	4.648	5.002
52,476	6.602	4.91	5.284
54,362	6.866	5.256	5.655
65,720	7.556	7.582	8.152
79,397	10.152	10.929	11.744

REFERENCES

1. Bennett, C. A., and N. L. Franklin, *Statistical Analysis in Chemistry and the Chemical Industry*, Wiley, New York (1954), p. 260.
2. Box, G. E. P., and D. R. Cox, An analysis of transformations, *J. Roy. Statist. Assoc.* **26**(2): 211–243 (1964).
3. Box, G. E. P., and P. W. Tidwell, Transformation of the independent variables, *Technometrics* **4**(4): 531–500 (1962).
4. Burrows, G. L., Interpreting straight line relationships, *Ind. Qual. Control* **15**(1): 15–16 (1958). Copyright American Society for Quality Control, Inc. Reprinted by permission.
5. Chen, C. Y., G. A. Hawkins, and H. L. Solberg, Heat transfer in annuli, *Trans. Am. Soc. Mech. Eng.* **68**: 99–106 (1946).
6. Conley, R. T., and P. L. Valint, Oxidative degradation of poly(ethyl acrylate), *J. Appl. Polym. Sci.* **9**: 785–797 (1965).
7. Daniel, C., and F. S. Wood, *Fitting Equations to Data*, Wiley-Interscience, New York (1971).
8. Draper, N. R., and H. Smith, *Applied Regression Analysis*, Wiley, New York (1966), Ch. 10.

9. Graham, R. R., Reactor performance prediction for the reaction of sulfur dioxide with manganese dioxide, unpublished Ph. D. dissertation, Library, Texas Tech University, Lubbock, Texas, 1969, pp. 84–87.
10. Graybill, F. A., *Introduction to Linear Statistical Models*, Vol. I, McGraw-Hill, New York (1961), Ch. 6.
11. Hooper, W. B., The two-K method predicts head losses in pipe fittings, *Chem. Eng.* **88**(17): 96–100 (1981).
12. Lacher, J. R., Dissolving uranium in nitric acid, *Ind. Eng. Chem.* **53**: 282–284 (1961).
13. Lacroix, S., and M. M. Labalade, *Anal. Chem. Acta.* **4**: 68 (1950).
14. Mickley, H. S., Design of forced draft air conditioning equipment, *Chem. Eng. Prog.* **45**: 739–745 (1949).
15. Newton, R. H., and B. F. Dodge, The equilibrium between hydrogen, formaldehyde, and methanol. II. The reaction $CO + 2H_2 \rightleftharpoons CH_3OH$, *J. Am. Chem. Soc.* **56**: 1287 (1934).
16. O'Connell, *Trans. Am. Inst. Chem. Eng.* **42**: 791 (1946).
17. Olujić, Ž., Compute friction factors fast for flow in pipes, *Chem. Eng.* **88**(25): 91–93 (1981).
18. Roth, J. F., Oxidation-reduction catalysis, *Ind. Eng. Chem.* **53**: 293 (1961).
19. Sinibaldi, F. J., Jr., T. L. Koehler, and A. H. Babis, Transformed data simplify and confirm math models, *Chem. Eng.* **71**(10): 139–146 (1971).
20. White, R. G., Pelletizing magnetic taconite concentrate, *Ind. Eng. Chem.* **53**: 215–216 (1961).

10

Statistical Process Control and Reliability

10.1 INTRODUCTION

In all production processes the product has *variability*. The variability may be in the quality of the product as measured by its density or color, or it may be in its yield as measured in units or by the percent of material produced in a given batch. A manufacturer wants to ensure that the production process is operating as intended. For this purpose control charts are used to monitor current production on a real-time basis.

Statistical process control (SPC) charts are a simple means of distinguishing between that part of the variability which is inherent in the process and that part which is due to abnormalities. One attempts to eliminate the abnormalities as much as possible. Much of the advantage to be obtained from the use of SPC charts is that the charts may be constructed such that a technician can readily use them. The experimenter or statistician need only construct the chart based on past history and check it periodically to detect when there is statistical evidence to rationally trigger management-induced change in a process.

In this chapter we apply the normal distribution and the central limit theorem to statistical process control. The interrelation between sampling and process control is considered briefly. There are many books that consider process control in detail. Consult the general references at the end of the chapter for further information on process control [2-7].

10.2 CONTROL CHARTS

Errors can be classified as random errors and systematic errors. *Random errors* or *random variation* is generally inherent in production process and due to chance causes, whereas *systematic errors* can be assigned to identifiable causes from any one of many sudden changes or abnormal situations. Statistical process control is a procedure for:

1. Measuring the amount of random variation inherent in a given process
2. Continually comparing the variation of a process continually with the past history of the random variation
3. Recognizing abnormal variation based on this past history
4. Locating the cause of this abnormal variation
5. Controlling or correcting the process to eliminate the abnormality

Figures 10.1, 10.2, and 10.3 are some typical examples of variability inherent in various processes. Figure 10.1 is a plot of the distribution of minor ingredients in a three-component mixture; Fig. 10.2 is the distribution of the acidity of a certain batch process; Fig. 10.3 shows the distribution of moisture in batches of dough. These figures are alike in that they have a single central hump dropping off to the horizontal axis on each side, and they are roughly symmetrical. Many examples of chance variation show similar behavior. This type of distribution can be approximated by the normal distribution.

No one imagines that the distribution of Figs. 10.1, 10.2, and 10.3, or similar examples of variation, are exactly normal, but the normal distribution can often be assumed as a close approximation to the actual model. Even when the actual distribution representing the chance or random error is not normal, the distribution of the means of samples from the nonnormal distribution will be approximately normal according to the central limit theorem given in Section 3.5.1.

Take a rectangular piece of cardboard about 11 in. long. Construct a line across its width dividing it into two equal halves. Construct inch-wide compartments with the middle compartment centered on the middle line. Number the compartments formed by these lines as follows:

-5	-4	-3	-2	-1	0	1	2	3	4	5

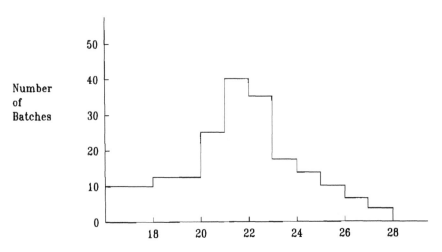

FIGURE **10.1** Distribution of minor ingredients in a three-component mixture.

Lay the cardboard on the floor and drop a dart from a height of 3 ft aiming to hit the middle line. Tabulate the compartments punctured for 10 drops of the dart. Repeat this 10 times.

Find the overall average compartment value for the 100 punctures by taking the average of the 10 sample averages. Plot this average value as a

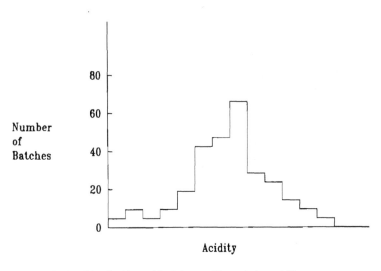

FIGURE **10.2** Distribution of batches with certain acidity.

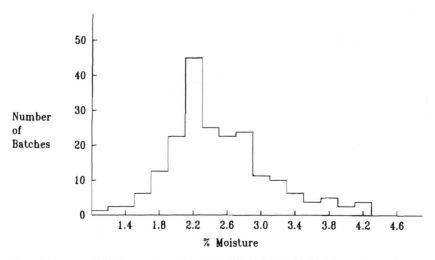

FIGURE 10.3 Distribution of batches of dough with a given moisture.

line horizontal to the abscissas, where the abscissas are the order of the
samples and ordinates are the average compartment values. Plot each of the
10 sample averages as ordinates versus the order of recording the data. The
data of Table 10.1 are plotted in this manner in Fig. 10.4. This plot gives a
picture of the data showing the central tendency, the spread of the data

TABLE 10.1 Results of Dart and Board Experiment

Sample	Mean compartment value ($n = 10$)
1	0.2
2	−0.3
3	−0.2
4	0.2
5	−0.2
6	−0.5
7	−0.1
8	0.4
9	0.3
10	−0.1
	Overmean −0.3
11	1.0

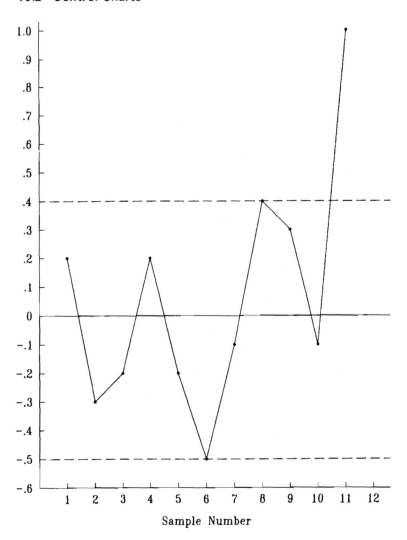

F<small>IGURE</small> **10.4** Control chart for dart and board experiment.

around the central value, and the information concerning the order in which the data were taken. As a crude approximation to the spread, the two dashed lines parallel to the center horizontal line are drawn through the sample averages farthest from the centerline.

Now, drop the dart an additional 10 times, but this time drop the dart from a height of 8 ft. Plot the mean of these 10 punctures at 11 on the control chart just constructed. The point at an abscissa value of 11 on Fig. 10.4 is the result of an experiment like the one just described. Notice that this value at 11 did not fall within the chance variation of the previous results but falls outside the control limits. Thus the change in the experiment average from 10 to 11 is revealed on the control chart of Fig. 10.4, and in addition to knowing that something changed, we also know when the change took place. On the usual chart the upper and lower control limits (UCL and LCL) are determined from the standard error of the mean if a normal distribution is assumed for the data rather than extreme values as in Fig. 10.4.

The major assumptions in statistical process control theory are (1) that the chance error as determined from the sample data follow a normal distribution (or for sample means, the sample size is large enough that the central limit theorem applies); and (2) that abnormal variation will be recognized as it will fall outside the statistically predicted spread of the data. In the following sections we look at statistical process control theory, specific applications, and types of control charts.

10.3   STATISTICAL PROCESS CONTROL

The hypothesis or assumption in statistical process control theory is that the data are statistically uniform and that any variation is due only to a constant system of chance causes; that is, the system is in statistical control. If the data are in statistical control, the stated hypothesis is accepted. If the data do not comply or are out of control, the hypothesis is rejected and the change in variation is assigned to a definite cause, even though it may not be mechanically identified. Therefore, it is termed an *assignable cause*.

The statistics used in statistical process control theory are the sample mean for the jth sample:

$$\bar{Y}_j = \frac{\sum_{i=1}^{n} Y_i}{n}, \qquad j = 1, 2, \ldots, N, \tag{10.1}$$

the sample variance:

$$S_j^2 = \frac{\sum_{i=1}^{n} (Y_i - \bar{Y}_j)^2}{n - 1}, \qquad j = 1, 2, \ldots, N, \tag{10.2}$$

the range:

$$R_j = Y_{\max} - Y_{\min}, \qquad j = 1, 2, \ldots, N, \tag{10.3}$$

the overall mean (which is the best estimate of the population mean)

$$\bar{\bar{Y}} = \frac{\sum_{j=1}^{N} \bar{Y}_j}{N}, \tag{10.4}$$

the variance of all the data:

$$S^2 = \frac{\sum_{i=1}^{Nn}(Y_i - \bar{\bar{Y}})^2}{Nn}, \tag{10.5}$$

the mean range:

$$\bar{R} = \frac{\sum_{j=1}^{N} R_j}{Nn}, \tag{10.6}$$

and the standard error of the sample means:

$$S_{\bar{Y}} = \frac{S}{\sqrt{n}}. \tag{10.7}$$

In Eqs. (10.1) to (10.7), the Y_i's are the sample values, n the sample size, Y_{max} the maximum value in a given sample, Y_{min} the minimum value in a given sample, and N the number of samples. Straightforward computation of S lumps all data into one population. This approach is incorrect if the process has changed during the control chart window. This bias on S can be substantially eliminated by using $\bar{R}$. To accomplish this, use $\bar{R}/d_2$ as an unbiased estimator of σ, where d_2 is given in Table 10.2. Thus one generally uses A_2, where $A_2 = 3/d_2\sqrt{n}$ to compute process control limits.

The control limits on a control chart are based on the normal distribution, so that 95% of the sample averages from a process that is in control should be within the limits $\bar{\bar{Y}} \pm 2s_{\bar{Y}}$ and 99.7% should be within the limits $\bar{\bar{Y}} \pm 3s_{\bar{Y}}$.

To determine the control limits for a control chart, one proceeds as follows:

1. Obtain samples at definite time intervals and measure the variable to be controlled.
2. Calculate each sample mean and sample variance.
3. After a number of sample means and variances have been determined, say 10, calculate the overall mean for these 10.
4. Calculate the standard deviation of the samples and the standard error of the mean of the sample averages.
5. Calculate control limits from the following equations:

   Warning limits: UWL, LWL $= \bar{\bar{Y}} \pm 2s_{\bar{Y}}$ (10.8)

   Control limits: UCL, LCL $= \bar{\bar{Y}} \pm 3s_{\bar{Y}}$. (10.9)

6. Plot the sample means as a function of time (the order of sampling) and draw a horizontal line at an ordinate value of the overall sample mean $\bar{\bar{Y}}$ and draw horizontal lines for the control limits.

TABLE 10.2   Parameters for Use with Control Charts

Sample size	A_2	d_2	D_3	D_4
2	1.880	1.128	0	3.276
3	1.023	1.693	0	2.575
4	0.729	2.059	0	2.282
5	0.577	2.326	0	2.115
6	0.483	2.534	0	2.004
7	0.419	2.704	0.076	1.924
8	0.373	2.847	0.136	1.864
9	0.337	2.970	0.184	1.816
10	0.308	3.078	0.223	1.777
11	0.285	3.173	0.256	1.744
12	0.266	3.258	0.284	1.719
13	0.249	3.336	0.308	1.692
14	0.235	3.407	0.329	1.671
15	0.223	3.472	0.348	1.652
16	0.212	3.532	0.364	1.636
17	0.203	3.588	0.379	1.621
18	0.194	3.640	0.392	1.608
19	0.187	3.389	0.404	1.596
20	0.180	3.735	0.414	1.586

Source: Ref. 1.

Example 10.1   Following are the viscosity data from a process. Each day four viscosities are measured and these four measurements constitute the sample ($n = 4$). Also listed are the range and mean of each day's viscosity measurements.

Day	Viscosities				Total	Average	Range
1	66	68	58	70	262	65.5	12
2	60	58	68	65	251	62.7	10
3	69	71	68	76	284	71.0	8
4	71	78	69	61	279	69.7	17
5	66	65	63	71	265	66.2	8
6	60	58	59	68	245	61.2	10
7	55	58	51	51	215	53.7	7
8	57	56	60	57	230	57.5	4

(*continued*)

Day	Viscosities				Total	Average	Range
9	65	65	66	63	259	64.7	3
10	64	73	63	60	260	65.0	13
11	63	63	60	62	248	62.0	3
12	61	57	66	59	243	60.7	9
13	55	56	49	56	216	54.0	7
14	58	62	54	61	235	58.7	8
15	54	49	64	72	239	59.7	23
16	63	71	59	57	250	62.5	14
17	55	54	63	46	218	54.5	17
18	54	64	51	46	215	53.7	18
19	44	41	47	49	181	45.2	8

$$s^* = \frac{\bar{R}}{d_2} = \frac{9.46}{2.059} = 4.594,$$

$$s_{\bar{Y}} = \frac{s}{\sqrt{n}} = \frac{6.38}{\sqrt{n}} = 3.19,$$

$$\bar{\bar{Y}} = \frac{932.3}{15} = 62.15.$$

Warning: UWL $= 62.15 + 2(3.19) = 68.5$

LWL $= 62.15 - 2(3.19) = 55.8$

Outer: UCL $= 62.15 + 3(3.19) = 71.7$

LCL $= 62.15 - 3(3.19) = 52.6.$

Construct a control chart based on the first 15 days. For the first 15 days $\bar{R} = 142/15 = 9.46$.

Solution The control chart for the average sample viscosities with the centerline at 62.15, the warning and control limits, and the sample averages for the 19 days is shown in Fig. 10.5. Consider the results of the example as plotted in Fig. 10.5. In the absence of abnormal sources of variability, 95%, or 19 of 20 sample averages should fall between the warning limits and 99.7%, or 997 of 1000 sample averages should fall between the control limits. Any point that does fall outside the control limits is a good indication of abnormal variation (i.e., definite cause). On the other hand, if only one point falls outside the warning limits, one would not be so likely to accept this as an

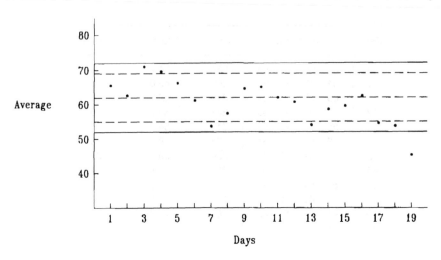

Figure 10.5   Control chart for viscosity averages of Example 10.1.

indication of abnormal variation. The probability of two successive points falling outside one of the warning limits is $(1/20)(1/20)$, or $1/400$; the probability of three in a row is $1/8000$. Therefore, when two or more consecutive points fall outside the warning limits, one suspects abnormal variation or some change in the average viscosity of the process. At days 3 and 4 there are two points above the warning limits. These points lead one to suspect abnormal variation. The problem was corrected and on day 5 the points are back inside the warning limits. The sample means on days 13, 17, 18, and 19 are below the warning limits. This large number of points below the limits (the last three consecutive) is a good indication that the process has gone out of control and that there has been a shift in the process average. Actually, one would interpret runs of seven or more points on one side of the mean as a possible shift in mean due probably to some systematic change in the system. The assumption of a given distribution inherently includes the assumption that the data will scatter randomly about the mean. If the data do not scatter about this mean, common sense tells you that something has changed. If one wishes to put a number on this, the probability of a run of points on one side may be calculated as was done previously for any control limit.

Example 10.2   Mrs. Smith decided to go on a diet: Since she would become discouraged if she weighed herself daily, she decided to weigh herself only twice a week. Her progress is as follows:

Week	First weighing	Second weighing	R	$\bar{x}$
1	140	141	1	140.5
2	138	137	1	137.5
3	138	136	2	137.0
4	135	136	1	135.5
5	133	133	0	133.0
6	138	139	1	138.5
7	137	139	2	138.0
8	134	133	1	133.5
9	131	130	1	130.5
10	128	126	2	127.0

$$\bar{R} = \frac{12}{10} = 1.2, \quad UCL_{\bar{x}} = \bar{x} + A_2\bar{R} = 135.1 + 1.880(1.2) = 137.36,$$

$$\bar{x} = 135.1, \quad LCL_{\bar{x}} = \bar{x} - A_2\bar{R} = 135.1 - 1.880(1.2) = 132.84.$$

Figure 10.6 shows the control charts for $\bar{x}$ and R. Note that weeks 1, 2, 6, 7, 9, and 10 are all outside the limits for the average. Can you discuss reasons for these out-of-control points?

Example 10.3 A basketball coach wants to study the average number of points scored by his post men in a ball game. It is suggested that a control chart be used.

Game	T	L	E	W	R	$\bar{x}$
1	22	7	23	3	20	13.75
2	13	14	22	6	16	13.75
3	15	13	16	4	12	12.00
4	13	9	6	6	7	8.50
5	3	6	12	2	10	5.75
6	9	1	10	1	9	5.25
7	8	4	14	4	10	7.50
8	15	5	13	3	12	9.00
9	14	6	14	6	8	10.00
10	12	8	10	9	4	9.75
11	16	7	15	2	14	10.00
12	8	16	14	4	12	10.50

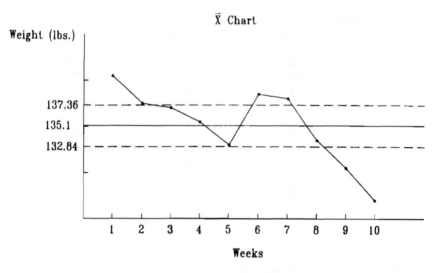

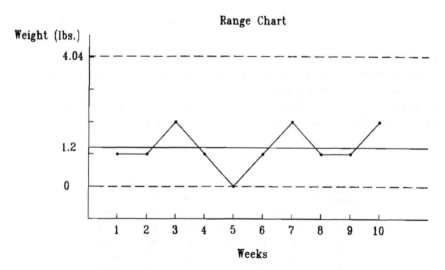

FIGURE 10.6 $\bar{x}$ and $\bar{R}$ charts.

$$\bar{x} = 9.65, \qquad \bar{R} = 11.17, \qquad n = 4,$$

$$d_2 = 2.059, \qquad \hat{\sigma} = \frac{\bar{R}}{d_2} = 5.42.$$

This is shown in Fig. 10.7.

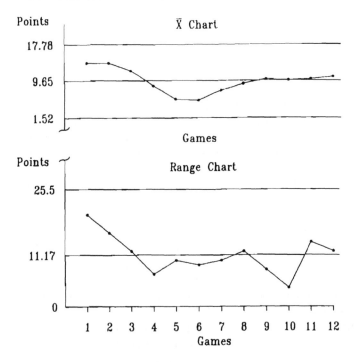

Figure **10.7** Range control chart.

The distribution of ranges may also be used to detect abnormal variation. The range chart is not so sensitive to variation as the chart for mean values. The range is not assumed to follow the normal distribution and obviously will vary more (on a relative basis) than the sample mean. The range chart, however, is useful in detecting unusual variation about the mean.

A standard set of symbols is used in statistical process control. The average chart is most often referred to as the $\bar{x}$ chart (X-bar chart) and the range chart as the R *chart*. The control limits are given by:

$$\text{LCL}_{\bar{x}}; \text{UCL}_{\bar{x}} = \bar{\bar{X}} \pm 3s_{\bar{x}} = \bar{\bar{X}} \pm \frac{3\bar{R}}{d_2 \cdot \sqrt{n}} = \bar{\bar{X}} \pm A_2\bar{R}, \qquad (10.10)$$

where the warning limits are the same except for the 2 in place of the 3 in Eq. (10.10) and A_2 is replaced by $2A_3/3$. For the range chart the upper and lower control limits are different and are given as

$$UCL_R = D_4 \bar{R}, \qquad\qquad\qquad (10.11)$$
$$LCL_R = D_3 \bar{R}, \qquad\qquad\qquad (10.12)$$

where D_3 and D_4 are factors computed from a normal distribution as a function of sample size and are listed in Table 10.2.

A control chart may be considered as a normal distribution in three dimensions where the third dimension is time or order. A control chart might be thought of as a normal distribution lying on its side, as shown in Fig. 10.8. The overall mean $\bar{x}$ is considered to be the population mean; the limits are lines through $2\sigma_{\bar{x}}$ and $3\sigma_{\bar{x}}$; the abscissa is the time or order scale.

Figure 10.9 shows the average chart for flare times with the corresponding normal distribution below them. The average (X-bar) chart (Fig. 10.9) shows that the mean of sample 42 is out of control; the lower part of the figure indicates this as a shift of the average of the normal distribution of flare times. The range chart (Fig. 10.10) shows that the range of sample 80 is out of control; the lower part of the figure indicates that this is a change in the spread of the normal distribution of flare times.

Statistical process control and the $\bar{X}$ and R charts are often used in production control. These procedures have been developed to help make a

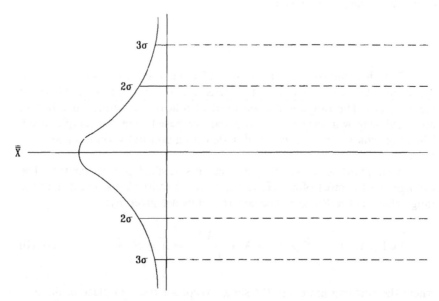

FIGURE **10.8** Control chart and normal distribution.

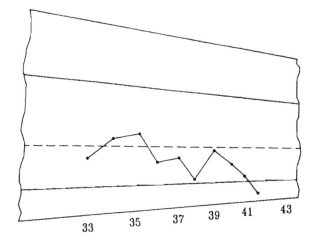

Sample Number

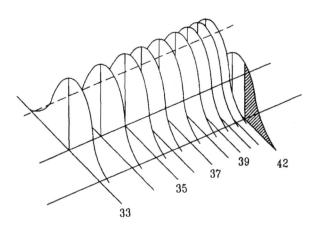

Figure 10.9   Mean out of control.

multitude of interchangeable parts so that a minimum of rejects are produced. We consider next a series of examples of the use and interpretation of control theory. The examples are from real situations, but they were selected in each case to illustrate a particular type of situation.

Example 10.4   Shift in Overall Average   During the operation of a method for recovering oil known as water flood in central Oklahoma it was necessary to meter the fluids produced from each of nine production wells in an injection-production well pattern. The pumps, which are positive-displacement rod pumps, deliver a constant volume of fluid on each stroke, so the

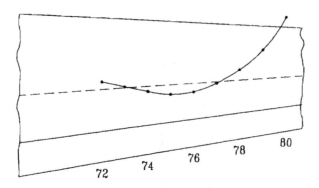

Sample Number

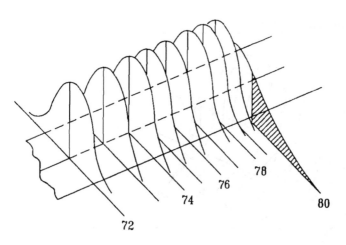

FIGURE **10.10** Range out of control.

simplest way to meter the fluid is to calibrate the pumps and simply count the strokes. (This is easily done with a mechanical counter attached to the polish rod.) The pumps were adjusted to pump a certain average number of strokes per day by using an on–off timer. The number of strokes per day was checked three times per day and the pumps were calibrated once a week. A control chart was kept of the pump strokes per 10-min period using the three 10-min checks each day as the sample. The following table gives the number of strokes per 10-min period for the three times each day, the range, the sum of the three samples, and the sample average.

Day	Pump strokes/10 min			Total	Range	Sample avg.
1	40	41	42	123	2	41
2	43	40	39	122	4	40.7
3	38	44	39	121	6	40.3
4	40	42	43	125	3	41.7
5	41	41	41	123	0	41
6	42	40	39	121	3	40.3
7	37	39	42	118	5	39.3
8	44	40	41	125	4	41.7
9	40	41	42	123	2	41
10	42	42	42	126	0	42
11	41	40	44	125	4	41.7
12	39	39	41	119	2	39.7
13	41	40	42	123	2	41
14	40	40	39	119	1	39.7
15	39	39	37	115	2	38.3
16	40	36	37	113	4	37.7
17	39	36	35	110	4	36.7
18	37	36	35	108	2	36
19	40	42	43	125	3	41.7
20	41	42	42	125	1	41.7

For the first 10 days: $\bar{\bar{X}} = \dfrac{\sum X}{Nn} = \dfrac{1227}{30} = 40.9,$

$$\bar{R} = \frac{\sum R}{10} = \frac{29}{10} = 2.9,$$

$\text{UCL}_{\bar{x}} = \bar{\bar{X}} + A_2\bar{R} = 40.9 + 1.023(2.9) = 43.9,$

$\text{LCL}_{\bar{x}} = \bar{\bar{X}} - A_2\bar{R} = 40.9 - 1.023(2.9) = 37.9,$

$\text{UCL}_R = D_4\bar{R} = 2.282(2.9) = 6.618,$

$\text{LCL}_R = D_3\bar{R} = 0(2.9) = 0.$

The $\bar{X}$ and R charts for the data of the table for 20 days are plotted in Fig. 10.11.

It is obvious that a shift in process average (i.e., a change in the stroke rate) started about day 15. The point on day 16 is below the lower control limits on the $\bar{X}$ chart. Days 17 and 18 confirmed that the stroke rate was out of control and the difficulty was finally located on day 18; the V-belt driving the pump had started slipping. The belt was tightened and day 19 is back in control. The range chart was in control during the entire time.

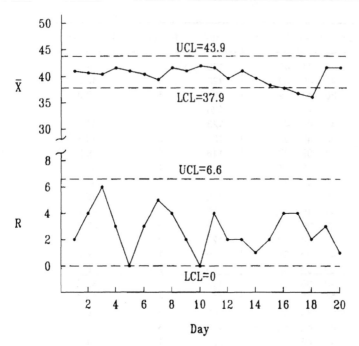

F<small>IGURE</small> **10.11**   Control charts for pump strokes.

Example 10.5   Irregular Shift in Overall Average; Change in Range   The porosity of several hundred sandstone plugs was to be measured by a service laboratory. The plugs were delivered in sets of 50 and the results reported for these 50 plugs before the next set was delivered. Two plugs of known porosity were always placed in each group of 50 as a means of checking on the reliability of the continuing measurements. The results of 16 sets of measurements of these two plugs are as follows:

Sample	Porosity (%)		Range	Sample avg.
1	24.0	24.1	0.1	24.05
2	23.8	23.9	0.1	23.85
3	24.2	23.9	0.3	24.05
4	24.5	24.0	0.5	24.25
5	23.8	23.7	0.1	23.75
6	24.2	24.6	0.4	24.40
7	22.1	22.0	0.1	22.05

(continued)

Sample	Porosity (%)		Range	Sample avg.
8	21.3	21.0	0.3	21.15
9	25.6	25.5	0.1	25.55
10	23.9	23.9	0.0	23.90
11	24.0	24.6	0.6	24.30
12	22.5	24.7	2.2	23.60
13	22.4	24.8	2.4	23.60
14	24.1	23.9	0.2	24.00
15	23.7	23.9	0.2	23.80
16	24.1	24.0	0.1	24.05

Previous measurements of these plugs had shown porosity of 24% and an average range of 0.5%.

From previous measurements: $\bar{\bar{X}} = 24\%$

$$\bar{R} = 0.5\%$$

$$(\text{UCL, LCL})_{\bar{X}} = 24 \pm (1.880)(0.5) = 24.9, 23.1,$$

$$\text{UCL}_R = (3.267)(0.5) = 1.6,$$

$$\text{LCL}_R = (0)(0.5) = 0.$$

Figure 10.12 gives the $\bar{X}$ and R charts for the measurements of the porosity of the two known plugs with the average and limits based on the measurement stated in the preceding paragraph. Measurements of samples 7, 8, and 9 on the $\bar{X}$ chart are obviously out of control and in an irregular fashion. Upon checking with the service laboratory it was found that a new technician was used to make the measurements starting on sample 7. Subsequent measurements after sample 9 were made with experienced technicians. The R chart is out of control on samples 12 and 13. Upon checking with the service company it was found that the drying oven used to dry the plugs had developed a faulty thermostat and the drying temperature was much lower than it should have been; the dry weight of the plugs varied more than usual as a result of this lower drying temperature. The faulty thermostat was fixed and the remaining data shown remained in control.

Example 10.6 Steady Trend in Average Figure 10.13 shows the $\bar{X}$ and R charts for the diameter of a part manufactured for the console typewriter on a computer. Limits on the chart were set from construction data (data not given) concerning the tolerance of the part and previous data on the manufacture of this part. Six days from the start of the chart in Fig. 10.13, the sample average diameter began to rise steadily until on day 11 the diameter was almost on the upper control limit. At this time the grinding tool

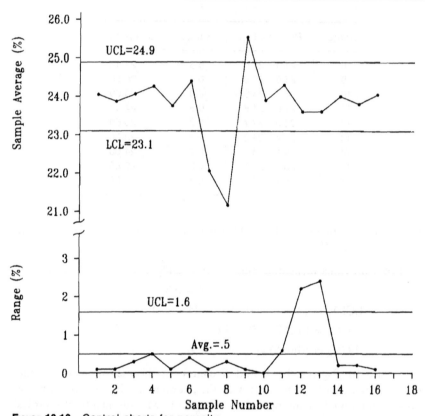

FIGURE **10.12**   Control charts for porosity.

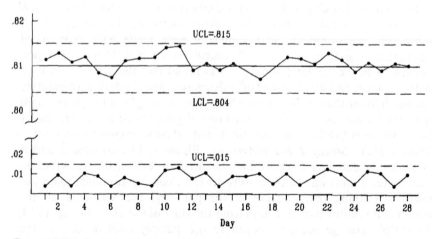

FIGURE **10.13**   Control charts for diameter.

used to manufacture the part was replaced; this tool had steadily worn down, causing the diameter of the part to increase. The steady trend in the average was caused by tool wear.

Example 10.7 Moving Average Limits Example 10.6 showed that tool wear will cause a manufactured part to go out of control. One may wish to take this known or expected change in average into account when setting up a control chart. The following data are measurements on parts from a machine where it is known that there is rapid tool wear. The specified dimensions for the part are 64.4 ± 0.30 mm. The subgroup sample size is 5.

Subgroup	X (mm)	R (mm)
1	64.15	0.10
2	64.17	0.15
3	64.22	0.10
4	64.30	0.14
5	64.32	0.09
6	64.35	0.10
7	64.34	0.12
8	64.38	0.04
9	64.41	0.08
10	64.42	0.13
11	64.51	0.10
12	64.56	0.09
13	64.55	0.11

Since we know that the subgroup (sample) average is not constant (rather, it is increasing) the "average" line and the control limits should slope as determined by the data. The trend line for the averages may be determined in any one of several ways, but the most common (see Chapter 9) is to fit a straight line to the data for the averages from the table by the method of least squares. The equation for the data of the table gives the following straight-line equation: $\bar{X} = 64.126 + 0.033x$, where x is the subgroup number. The upper and lower control limits are $A_2\bar{R}$ above and below the sloping trend line and parallel to it. In this case $A_2\bar{R} = 0.577(0.10) = 0.0577$. For this type of chart one must decide on the initial machine setting and when it is necessary to replace the tool. The tolerance spread for this particular part is $64.70 - 64.10 = 0.60$. The estimate of the standard deviation of the measurement is $\bar{R}/d_2 = 0.10/2.3236 = 0.043$. To be on the safe side the initial limits will be made twice the estimate of the standard deviation above the

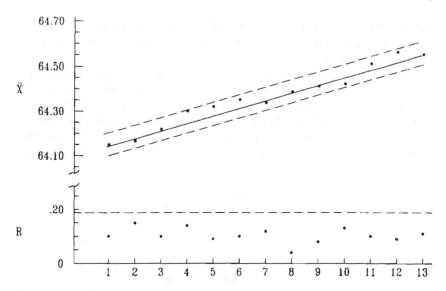

FIGURE 10.14   Control charts for data with a steady trend.

lower control limit. Thus the initial setup is made at $64.10 + 2(0.043) =$ 64.186; the tool is replaced (a new setup) when the trend line (the data on the trend line) reaches $64.70 - 3(0.043) = 64.571$. The control chart with a moving average and control limits for the data of the table is plotted in Fig. 10.14. This chart is interpreted the same as regular charts; any value falling outside the control limits is out of control. Since there is autocorrelation in this situation, the run rules are invalid. The range chart is the same as in the previous examples.

Example 10.8   Irregular Changes in Both Average and Spread   Figure 10.15 shows the $\bar{X}$ and R charts for the diameter of a part. The data plotted on the charts are out of control irregularly both in average and range. This situation often occurs at the startup of a process when all the problems have not been solved. Figure 10.16 shows $\bar{X}$ and R charts where the changes are also irregular. In this case these charts show data resulting from an inexperienced operator just starting on an existing process line.

Example 10.9   Changing Limits   Often, the sample size that is used for control chart measurements varies. During a test operation in a water-flood project a particular material was added to the water to help increase the flooding efficiency. This material mainly dissolved in the water, but some residue of particles less than 5 μm in size were suspended in the water. The resulting solutions were filtered to remove a majority of this material, because

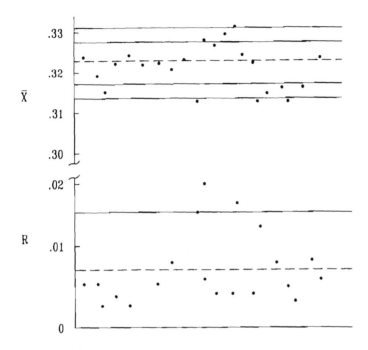

Figure 10.15 Control chart for diameter where both average and spread are irregular.

the suspended particles would build up on the injection well faces and cause the wells to plug. Previous tests had shown that the amount of suspended material could be measured by taking 10-mL samples of the filter solution and spinning them for 30 min in a centrifuge; then the length and breadth of the smear at the bottom of the tube were measured and the sum of the length and breadth used as a measure of the amount of suspended solids. The first 10 days were used to calculate the control limits. Although the procedure described was carried out at the field site efficiently, the method of measurement (i.e., the centrifuging) often broke some of the sample vials. Thus the number of measurements per sample could vary all the way from six (the maximum number of tubes that fit the centrifuge) to zero. In actual operation no more than three tubes were ever lost. Thus the method of measurement caused a varying sample size, and to take this into account a control chart with changing limits was used to control this particular operation. The following table gives the data from the first 15 days of operation of the process just described.

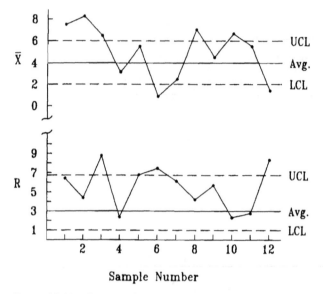

FIGURE 10.16 Control charts for new operator on an existing process line.

The sample values are the sum of the length and breadth (in millimeters) of the smear of particles found at the bottom of a 10-mL centrifuge tube after 30 min of centrifuging. There are as many as six values per sample.

Day	Length plus breadth (mm)						R	$\bar{X}$
1	5	5	6	5	5	6	1	5.3
2	5	6	7	—	7	8	3	6.4
3	4	5	5	6	—	—	2	5.0
4	5	5	6	6	7	5	2	5.7
5	5	6	7	8	4	5	4	5.8
6	—	—	—	7	5	5	2	5.7
7	5	5	5	5	5	5	0	5.0
8	6	—	7	5	6	8	3	6.4
9	6	8	7	—	—	5	3	6.5
10	6	7	7	5	5	6	2	6.0
11	4	5	5	6	6	5	2	5.2
12	—	—	4	4	4	3	1	3.8
13	9	9	8	10	9	—	2	9.0
14	8	9	10	11	11	9	3	9.7
15	5	6	5	5	—	6	1	5.4

The dashes indicate that for that value the centrifuging tube broke. The range of the samples and the sample average are also given in the table. The overall average and the average range for the first 10 days is calculated at the bottom of the table. Control limits for the $\bar{X}$ chart are computed for sample sizes of 3, 4, 5, and 6.

$$\bar{X} = \frac{291}{51} = 5.8,$$

$$\bar{R} = \frac{22}{10} = 2.2.$$

For sample sizes of 3, 4, 5, and 6, UCL, LCL $= \bar{\bar{X}} \pm A_2 \bar{R}$.

3: UCL, LCL $= 5.8 \pm (1.023)(2.2) = 8.1, 3.5$
4: UCL, LCl $= 5.8 \pm (0.729)(2.2) = 7.4, 4.2$
5: UCL, LCL $= 5.8 \pm (0.577)(2.2) = 7.1, 4.5$
6: UCL, LCL $= 5.8 \pm (0.483)(2.2) = 6.9, 4.7$

The average value and the varying control limits are plotted in Fig. 10.17 along with sample averages for the first 15 days. On day 12 the data fell below the lower control limit. In this case the system is not "out of control" because the lower the value, the better it is. Thus the upper limit is the one that we were concerned with. On the other hand, it may be an indication of the process going out of control as it did on days 13 and 14.

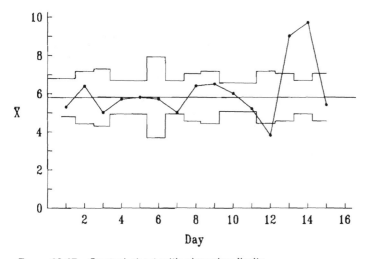

FIGURE **10.17** Control chart with changing limits.

The good filtering action of day 12 is probably due to an excess of material on the plate and frame filter. This excess gave very good filtering, but it also caused the pressure drop across the filter to rise. The filter sheets began to rupture on day 13 and the filter ceased to operate effectively. On day 14 the filter sheets were changed and the process was back in control on day 15.

In Example 10.9 we considered the problem of changing sample size and the resulting changing limits on the control chart. When the sample size is constant, it is often adequate to plot the sum of the sample values rather than the sample average, thus saving a computational step. For the data of the table in Example 10.4 we could have plotted the sum of the three pump stroke rates rather than the average. In this case the control limits are

$$CL = n(\bar{X} \pm A_2 \bar{R}).$$

The data are plotted in Fig. 10.18 as a sum chart. The interpretation is the same as that in Example 10.1.

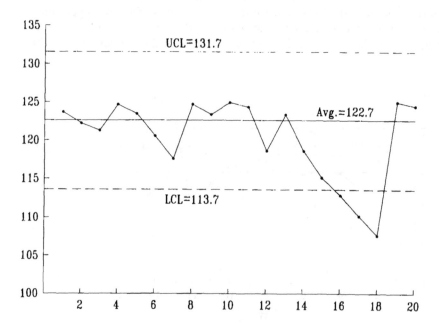

FIGURE 10.18   Sum chart for data of Example 10.4.

10.4 OTHER QUALITY-CONTROL PROCEDURES

The control charts up to this point use information about the process contained in the last plotted point and ignore any other past information. This observation has led to methods of incorporating past information about a process to maintain its output in control about a target value μ_0. One such procedure is to plot the cumulative sum (CUSUM) of deviations about the target value. The CUSUM control chart is thus formed by plotting the quantities $S_i = \sum_{k=1}^{i} (\bar{X}_k - \mu_0)$ against the sample number i, where $\bar{X}_k$ is the mean of the kth sample. As long as the process is in control, small positive and negative deviations tend to offset each other and the CUSUM fluctuates around zero. If the mean shifts upward to some value $\mu_1 > \mu_0$, say, then an upward or positive drift will develop in the CUSUM. Conversely, if the mean shifts downward to some $\mu_1 < \mu_0$, a downward or negative drift will develop in the CUSUM.

A formal procedure for determining whether a process is out of control using the CUSUM procedure involves placing a V mask on the CUSUM control chart so that upward or downward shifts can be determined. The V mask forms a visual frame of reference similar to the control limits in the ordinary control charts. The determination of the V mask is somewhat involved, and one should consult more advanced process control texts for further information on this approach. It suffices to point out that a CUSUM control chart is generally slow to detect large process shifts but can be designed to detect small shifts more effectively than the standard $\bar{X}$ charts when run rules are not used.

Other alternatives to the usual $\bar{X}$ and R charts have been developed when we are interested in detecting small shifts and only one observation is taken at each point. An exponentially weighted moving average (or EWMA) chart is easier to set up and operate than a CUSUM control chart. The exponentially weighted moving average is given as

$$z_t = \lambda \bar{X}_t + (1 - \lambda)z_{t-1}, \qquad 0 < \lambda < 1$$

where λ is a constant and the starting value required with the first sample at $t = 1$ is $z_0 = \bar{X}$.

Clearly, for $\lambda = 1$ the EWMA reduces to the usual $\bar{X}$ chart. For $\lambda < 1$, previous values of $\bar{X}$ are used but not given as much weight as the later values of $\bar{X}$. In fact, the weights $\lambda(1 - \lambda)^j$ are easily shown to decrease geometrically with the age of the sample mean. If $\lambda = 0.5$, the weight given to the current sample mean is 0.25 and the weights given to the previous means are 0.125, 0.0625, and so on.

If the $\bar{X}_i$ are independent random variables with variance σ^2/n, the variance of z_t is

$$\frac{\sigma^2}{n}\left(\frac{\lambda}{2-\lambda}\right)[1-(1-\lambda)^{2t}],$$

and as the sample number t increases, the variance of z_t increases to a limiting value

$$\sigma_z^2 = \frac{\sigma^2}{n}\frac{\lambda}{2-\lambda}.$$

Hence the upper and lower control limits are

$$\text{UCL} = \bar{\bar{X}} + 3\sigma\sqrt{\frac{\lambda}{(2-\lambda)n}}$$

and

$$\text{LCL} = \bar{\bar{X}} - 3\sigma\sqrt{\frac{\lambda}{(2-\lambda)n}},$$

respectively, if the sample number t is moderately large. Usually, σ must be estimated from an R chart, so the control limits become

$$\text{UCL} = \bar{\bar{X}} + A_2\bar{R}\sqrt{\frac{\lambda}{2-\lambda}},$$

$$\text{LCL} = \bar{\bar{X}} - A_2\bar{R}\sqrt{\frac{\lambda}{2-\lambda}},$$

where if only individual observations are taken, one uses the moving range of two successive observations to estimate the process variability. That is, $R_i = |X_i - X_{i-1}|$ and $\bar{R}$ is the mean of these. Run rules are not used with EWMA due to autocorrelation.

Space does not permit further detailed information concerning other statistical process control procedures. Further information on quality schemes can be found in the references.

10.5   ACCEPTANCE SAMPLING

Another important area of statistical quality control is acceptance sampling. When a batch of items is received by a purchaser, a decision must be made whether to accept or reject the batch. This batch is treated as a statistical population and a random sample inspected since inspection of every item is usually uneconomical, due to the time or cost involved. Thus the decision to reject a lot is made based on a random sample of items from the batch.

The inspected sample items from a batch are usually classified as defective or satisfactory. In case quantitative measurements are made to determine this, those items falling outside tolerance specifications are classified as *defectives*. A good batch is one that has a small proportion of defectives π. The decision to accept or reject the batch is based on the number of defectives found in the sample. Thus acceptance sampling is just an adaptation of classical hypothesis testing. The null hypothesis is $H_0: \pi \leq p_0$ (the lot is good), while the alternative is that $\pi > p_0$ (the lot is bad).

The value of p_0 is referred to as the *acceptable quality level* (AQL). As we know, two types of errors can occur when testing a hypothesis. We might reject a lot that is in fact acceptable, thus committing a type I error or fail to reject an unacceptable lot, thus committing a type II error. The probability of committing a type I error, α, is called the *producer's risk*, and the probability of committing a type II error, β, is called the *consumer's risk*. A sampling plan thus needs to be developed to have both α and β at small levels so that the consumer's risk and the producer's risk are in satisfactory balance.

To determine the decision rule, assume that the batch size is N, of which r are defective and the sample size to be taken without replacement is n. In an earlier chapter we saw that the number of sample defectives x is a random variable having a hypergeometric distribution.

$$f(x) = \frac{\binom{r}{x}\binom{N-r}{n-x}}{\binom{N}{n}},$$

where x is an integer between $\max[0, n - (N - r)]$ and $\min(n, r)$. Since a batch is acceptable if $x \leq c$, the constant is called the *acceptance number*. In this context the significance level is given by

$$\alpha = \mathscr{P}(\text{reject } H_0 | \pi = p_0)$$
$$= \mathscr{P}(x > c | \pi = p_0)$$

and

$$1 - \alpha = \mathscr{P}(x \leq c | \pi = p_0)$$
$$= \mathscr{P}(\text{accepting } H_0 | \pi = p_0)$$
$$= \mathscr{P}(A).$$

We thus study the probability of accepting the batch $\mathscr{P}(A)$ as a function of the true p_0 for a given n and c. A plot of these probabilities is called an *operating characteristic curve* (OC curve).

We now illustrate its calculation with a numerical example.

Example 10.10 A supplier ships batches of size 25 to another company. The receiving company has a sampling plan which states that 5 items are to be taken without replacement from each batch. If there are no defectives among these 5, the entire lot is accepted; otherwise, it is rejected. This is modeled as

$$OC(p_0) = \mathscr{P}(X = 0) = \frac{\binom{r}{0}\binom{25-r}{5}}{\binom{25}{5}},$$

where $p_0 = r/25$. Letting r be 0, 1, 2, ... we find that $OC(0) = 1$, $OC(0.04) = 0.8$, $OC(0.08) = 0.633$, $OC(0.12) = 0.496$, $OC(0.16) = 0.383$, and so on. From seeing these probabilities, either the supplier or the receiving company may find that this sampling plan is unsatisfactory. For instance, the receiving company may feel that the probability of accepting a batch with four defective $OC(0.16) = 0.383$ is too low from the supplier's point of view. Thus the sampling plan would need to be changed.

Generally, in practice, lot sizes are much larger than 25, and thus approximations to the hypergeometric probabilities are utilized for the situations where the batch size is large compared to the sample size. Normally, inspection is designed to protect the producer against the rejection of batches with percent defectives less than the AQL and protect the consumer from accepting batches with percent defectives greater than the AQL. Once a balance between these two interests is established, generally based on the quality history of batches being good, a more economical plan may be agreed upon. However, various sampling plans are generally carefully studied before deciding on an optimal plan for a particular problem.

Many handbooks are available with operating characteristic curves and related properties of various sampling plans. One of the most widely used sets of plans is Military Standard 105D (MIL-STD-105D). For more detailed treatment of sampling inspection plans and comparisons of possible competing plans, the interested reader can find further information in the general references.

10.6 RELIABILITY

In industry, the distribution of the time to failure, T, of a product (device, component, system, etc.) is of great importance. The failure process of a product is often very complex and it is difficult to understand its underlying physics. Describing the failure process mathematically may be even more difficult. Consequently, the distribution of the time to failure becomes very important because of its usefulness in providing a statistical summary of the time to failure of the product under study. Determining the distribution of T can be a very involved process, and selection of a life distribution is mostly an art. Even if life test data (observed failure times) are available, the proper choice of the distribution is still a problem. Distributions of T are often skewed (asymmetric), causing a sparcity of observations in the tail regions of the distributions that are used to model the data.

It has been recognized that the difficulties in selecting the appropriate distribution can be overcome if one can distinguish among the many available distributions on the basis of physical considerations. The concept of hazard rate provides a means of making such a distinction. Before discussing the concept of hazard rate, we discuss the related idea of failure rate.

In developing the concept of hazard rate, we assume that the time to failure, T, is a continuous random variable. The rate at which items fail in the interval $[t_1, t_2]$ is called the *failure rate* in the interval. This rate can be defined as the probability that a failure per unit time occurs in the interval given that a failure has not occurred by time t_1. Let the random variable, T, denote the time to failure of a given item under consideration. The failure rate in the interval $[t_1, t_2]$, $FR(t_2, t_1)$, is then

$$FR(t_2,t_1) = \frac{\mathscr{P}(t_1 \le T \le t_2 | T \ge t_1)}{t_2 - t_1} = \frac{\mathscr{P}(t_1 \le T \le t_2)}{\mathscr{P}(T \ge t_1)} \frac{1}{t_2 - t_1}$$
$$= \frac{F(t_2) - F(t_1)}{R(t_1)} \frac{1}{t_2 - t_1},$$

where the function $R(t_1) = 1 - F(t_1)$ is known as the *reliability function* (or *survival function*). The first factor is the conditional probability of failure in the interval $[t_1, t_2]$, given survival to time t_1, and the second factor is a dimensional characteristic used to express the conditional probability on a per-unit time basis. Although the failure rate is expressed as "failures per unit time," the unit "time" may be replaced by "cycles," "distance," and so on. The interval failure rate can be written as

$$FR(t_2,t_1) = \frac{\mathscr{P}(t_1 \le T \le t_2 | T \ge t_1)}{t_2 - t_1} = \frac{F(t_2) - F(t_1)}{t_2 - t_1} \frac{1}{R(t_1)}.$$

Letting $t_1 = t$ and $t_2 = t + \Delta t$, we have

$$FR(t, t + \Delta t) = \frac{F(t + \Delta t) - F(t)}{\Delta t} \frac{1}{R(t)}. \tag{10.13}$$

The *hazard rate* is defined as the limit of the interval failure rate as the length of the interval, Δt, goes to zero. Thus the hazard rate, $h(t)$, is the instantaneous failure rate at time t and is given, from (10.13), by

$$h(t) = \lim_{\Delta t \to 0} FR(t,t + \Delta t) = \frac{f(t)}{R(t)}. \qquad (10.14)$$

The hazard rate is also called the *hazard rate function* or simply the *hazard function*. We note here that given that an item has survived to time t, $h(t) \, dt$ is the probability that the item will fail during the interval $(t, t + dt)$. In functional terms, $h(t)$ is the rate of change of the conditional probability of failure given in time t. In comparison, $f(t) \, dt$ is the unconditional probability of failure in $(t, t + dt)$ and $f(t)$ is the rate of change of the regular (unconditional) probability of failure. The significance of the hazard function is that it describes the change in failure rate over the lifetime of a population of items.

A lifetime distribution can be chosen based on the type of hazard function one might "expect" given the population of items under study. A very popular hazard function, $h(t)$, is one that has a bathtub shape, shown in Fig. 10.19. This bathtub hazard function represents the operation of items in terms of three phases: (1) the initial (or break-in) phase, (2) the chance (random) phase, and (3) the wearout phase. During the break-in phase the hazard rate may be high due to defects in the material and/or manufacturing process. As the initial failures have taken place, the hazard rate becomes

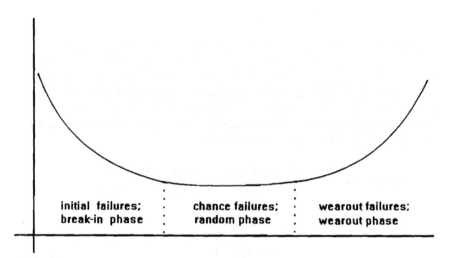

initial failures;   :   chance failures;   :   wearout failures;
break-in phase   :   random phase   :   wearout phase

FIGURE 10.19   Typical (bathtub) hazard rate curve.

nearly constant, due to failures occurring due to sudden stress or unpredict-
able operating conditions. In the wearout phase items fail due to aging,
deterioration, and an accumulation of shock and fatigue. As stated earlier,
it may be more convenient to choose a life distribution on the basis of features
of the hazard rate function rather than the shape of the life probability
density function (pdf) itself.

There is an important and useful relationship between the hazard
function and the reliability function $R(t) = \mathscr{P}(T \geq t) = 1 - F(t)$. Assume
that $R(0) = 1$. From (10.14) we have

$$h(x) = \frac{f(x)}{R(x)} = \frac{dF(x)/dx}{R(x)} = -\frac{dR(x)/dx}{R(x)}. \tag{10.15}$$

Therefore, integrating both sides of (10.15) yields

$$\int_0^t h(x)\,dx = -\int_0^t \frac{R'(x)}{R(x)}\,dx$$

$$= -\ln R(x)\Big|_0^t$$

$$= -[\ln R(t) - \ln R(0)]$$

$$= -\ln R(t).$$

Thus

$$e^{-\int_0^t h(x)\,dx} = R(t). \tag{10.16}$$

Since $dR(t)|\,dt = -f(t)$, we have, from (10.16),

$$f(t) = h(t)e^{-\int_0^t f(x)\,dx}.$$

The three functions $f(t)$, $R(t)$, and $h(t)$ are all related and one uniquely
determines the others.

Example 10.11   Consider the exponential pdf

$$f(t) = \tfrac{1}{4}e^{-t/4}, \qquad t \geq 0.$$

Find the reliability function and the hazard function (as functions of time t)
corresponding to the pdf $f(t)$.

Solution   The reliability function is

$$R(t) = \mathscr{P}(T \geq t)$$

$$= \int_t^\infty \tfrac{1}{4}e^{-x/4}\,dx$$

$$= -e^{-x/4}\Big|_t^x$$

$$= e^{-t/4}, \qquad t \geq 0.$$

The hazard function, from Eq. (10.15), is

$$h(t) = \frac{f(t)}{R(t)} = \frac{\frac{1}{4}e^{-t/4}}{e^{-t/4}} = \frac{1}{4}, \qquad t \geq 0.$$

Thus, for the exponential pdf of this example the hazard function is constant for all values of t. This means that for this exponential pdf all failures occurring are chance or random failures. See Problem 10.10 for another example involving the exponential pdf.

Example 10.12   A very popular hazard function is

$$h(t) = \alpha\beta t^{\beta - 1}, \qquad t > 0, \quad \alpha > 0, \quad \beta > 0.$$

This function has the property that if $\beta = 1$, the hazard function is constant, indicating random failures; if $\beta > 1$, the hazard function is increasing, indicating that failures are due to wearout; and if $\beta < 1$, the function is decreasing, indicating a malfunction or break-in failures. From (10.16) we have

$$R(t) = e^{-\int_0^t \alpha\beta x^{\beta - 1}\, dx} = e^{-\alpha t^\beta}.$$

Thus, by (10.14),

$$\begin{aligned} f(t) &= h(t)R(t) \\ &= \alpha\beta t^{\beta - 1}e^{-\alpha t^\beta}, \qquad \alpha > 0, \quad \beta > 0, \quad t > 0. \end{aligned}$$

This is a Weibull pdf, which differs from the form given in Eq. (3.39). The pdf in Eq. (3.39) involves a scale parameter λ. Both forms are used in practice.

10.6.1   Reliability of Series and Parallel Systems

Multiple-component systems are arrangements of components arranged in various configurations. Many systems have components (or subsystems) arranged in a "series" configuration or in a "parallel" configuration, or in a combination of the two.

A *series system* is the most common system and is the easiest to analyze. It consists of, say, n components operating in such a manner that the system operates successfully if and only if all the components operate successfully. That is, the system fails if any of its components fail. A series system of n components is depicted in Fig. 10.20. A series system cannot be more reliable

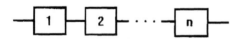

Figure 10.20   Series system.

than its least reliable component (weakest link concept). Let T be the time to failure of the system and $T_1, T_2, \ldots, T_n$ be the times to failure of the n components. Assuming that $T_1, T_2, \ldots, T_n$ are independent random variables, the reliability of the system is

$$
\begin{aligned}
R_s(t) &= \mathscr{P}(T > t) \\
&= \mathscr{P}(T_1 > t, T_2 > t, \ldots, T_n > t) \\
&= \mathscr{P}(T_1 > t)\mathscr{P}(T_2 > t) \cdots \mathscr{P}(T_n > t) \\
&= \prod_{i=1}^{n} \mathscr{P}(T_i > t) \\
&= \prod_{i=1}^{n} R_i(t),
\end{aligned}
$$

where $R_i(t) = \mathscr{P}(T_i > t)$ is the reliability of the ith component.

Example 10.13 Consider a series system of $n = 6$ components. If each component has a reliability of 0.95 at time t, the system reliability at time t is

$$
R_s(t) = (0.95)^6 = 0.735.
$$

This means that at time t the system has a 73.5% chance of operating.

Example 10.14 Suppose that a company is designing a series of $n = 6$ components and it wants the system reliability to be at least 0.95. If the reliability of each component is to be the same, what is the minimum reliability for each component?

Solution We want $R_s(t) = \prod_{i=1}^{6} R_i(t) = [R(t)]^6 \geq 0.95$, where $R(t)$, the reliability of each component, is

$$
R(t) = (0.95)^{1/6} = 0.9915.
$$

This means each component must have a probability of at least 0.9915 of operating successfully in order for the system to have a probability of 0.95 of operating successfully.

The reliability of a system of components is increased by arranging the components as a parallel system. Such a system, of n components (or subsystems) in parallel, operates successfully if k or more of the components ($k \leq n$) of the system operate successfully. If $k = 1$, the system is often called a *purely parallel system*, and if $k > 1$, it is called a *k-out-of-n system*. We will consider $k = 1$, or the purely parallel system. A parallel system is shown in Fig. 10.21. The parallel system operates successfully if at least one of the components operates successfully, or it fails if all the components fail. Thus

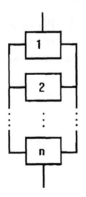

Figure 10.21   Parallel system.

the system reliability is

$$R_s(t) = \mathscr{P}(T \geq t)$$
$$= 1 - \mathscr{P}(T \geq t)$$
$$= 1 - \mathscr{P}(T_1 \geq t).\mathscr{P}(T_2 \geq t)\cdots\mathscr{P}(T_n \geq t)$$
$$= 1 - \prod_{i=1}^{n} \mathscr{P}(T_i < t)$$
$$= 1 - \prod_{i=1}^{n} [1 - R_i(t)], \tag{10.17}$$

where $R_i(t)$ is the reliability of the ith component.

Example 10.15   Consider a system of $n = 6$ components connected in parallel. If each component has a reliability of 0.80 at time t, the reliability of the system at time t is $R_s(t) = 1 - (1 - 0.80)^6 = 0.999936$. This shows how much more reliable parallel systems are over series systems.

A parallel system can be considered as a system of n components where one component operating means that the system is operating. When that component fails, the second is used; when the second fails, the third is used; until the last fails, at which time the system fails. A parallel system is thus a system with several components on *standby*.

Example 10.16   Consider the following system of $n = 7$ independent components. The system can be viewed as consisting of four subsystems connected in series. The reliabilities of the four subsystems are $R_I(t) = 0.99$, $R_{II}(t) = 1 - (1 - 0.9)^2 = 0.99$, $R_{III}(t) = 1 - (1 - 0.92)(1 - 0.80)(1 - 0.85) = 1 - (0.08)(0.20)(0.15) = 0.9976$, and $R_{IV}(t) = 0.95$. The system reliability is thus

$$R_s(t) = R_I(t)R_{II}(t)R_{III}(t)R_{IV}(t) = 0.9289.$$

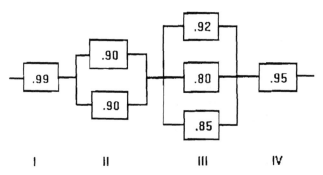

From the foregoing it is evident that a system with many independent components in series may have a very low system reliability even if each component alone is highly reliable. One way to increase system reliability is to replace single components with several similar components in a purely parallel configuration, although this type of redundancy could prove costly.

PROBLEMS

10.1 Breathe normally. Record the number of times you exhale in 1 min. Repeat the counting process five times.

   (a)   Plot the numbers on a graph, with the vertical axis being the numbers observed and the horizontal axis being the sample numbers.

   (b)   Draw two lines parallel to the horizontal axis, one passing through the highest count and one through the lowest count.

   (c)   Do one of the following: (1) some exercise for a short period of time, or (2) breathe deeply such as a doctor would expect you to do.

   (d)   Record the number of exhalations in 1 min. Plot the count on the chart in part (a) as the sixth sample, and note the difference between this count and the others.

   (e)   Would you say that the process is out of control as a result of the action taken in part (c)?

10.2 The following table gives coded values for the amount of fuel per second flowing through a velocity regulator valve in a missile.

Sample	X_1	X_2	X_3	X_4	X_5	Mean ($\bar{X}$)	Range (R)
1	10.9	9.2	11.0	10.2	9.9		
2	9.4	10.6	9.9	10.6	10.8		
3	9.5	9.8	9.8	9.6	10.2		
4	9.9	8.2	10.0	9.2	10.8		
5	12.2	9.8	10.5	9.9	11.1		
6	9.9	10.0	10.0	11.0	9.9		
7	10.8	11.3	11.6	10.8	11.1		
8	11.0	9.8	10.7	11.0	10.8		
9	10.4	10.2	10.3	10.3	10.7		
10	8.1	10.0	9.6	9.3	9.6		
11	10.4	9.78	10.5	10.0	11.2		
12	10.6	10.0	10.3	8.2	9.7		
13	10.5	10.5	10.6	8.4	11.2		
14	11.1	11.2	10.2	10.4	10.9		
15	11.2	11.0	11.2	9.9	11.4		
16	9.9	9.9	9.5	9.6	10.3		
17	10.5	12.6	11.0	11.0	11.1		
18	11.7	11.7	11.4	12.2	11.2		
19	10.6	11.9	11.6	9.2	11.4		
20	12.2	10.9	11.6	10.8	11.7		
Average						$\bar{X} =$	$\bar{R} =$

(a) Calculate the values of $\bar{X}$ and R and fill in the table.

(b) Construct the control charts for the data above (i.e., the $\bar{X}$ and R charts), showing the outer control limits.

(c) Plot the data on the control charts and determine if this process is always in control.

(d) Should inner limits have been used as a warning to the astronaut that the missile was starting to burn the fuel at too great or too slow a rate so that the appropriate measures could be taken?

10.3 Mr. Mean Moment joined the Raggy Rollers' bowling team. His bowling scores during the first 5 days were as follows:

1	2	3	4	5
100	110	111	115	125
114	160	135	140	130
170	123	150	132	140

(a)   Construct an $\bar{X}$ and an R chart showing control limits based on the results above.

(b)   Do the data indicate any improvement during the 5 days?

(c)   Is Mean Moment becoming a more consistent bowler? Is he an accurate bowler?

(d)   Would you want him on your team if your team average were 125? 135?

M and M, as he is called by his partners, returns for the second week of bowling. He has received some advice from an expert on how he might improve his bowling game. His scores for the second week are as follows:

6	7	8	9	10
110	125	130	140	157
166	170	125	144	150
180	185	185	160	142

(e)   Construct an $\bar{X}$ and an R chart showing control limits based on these new data (days 6 to 10).

(f)   Plot the means and ranges of these new data on the charts constructed in part (a). Would you say that he is now a more accurate bowler? Is he more consistent?

(g)   Construct an $\bar{X}$ and an R chart showing control limits based on all the data (days 1 to 10). Should the control limits constructed in parts (a), (e), or (g) be used to best describe the population of bowling scores?

After these two weeks, M and M decides to purchase his own bowling shoes and ball. (He has decided to stop bowling barefooted). His third week of bowling produced the following scores:

11	12	13	14	15
153	171	158	169	165
159	183	164	172	165
168	186	173	184	170

(h)   In which week was he most accurate on average?
(i)   In which game was he most accurate?
(j)   In which week was he most consistent?

10.4 A nail manufacturing firm recently started producing six-penny nails. For the first 6 days of production it was decided to check on the process by picking 20 nails from the assembly belt each day and recording the lengths (in inches) of the six-penny nails. (A six-penny nail is supposed to have an overall length of 2 in.) The following data were recorded for those 6 days.

	Day					
Sample	1	2	3	4	5	6
1	1.9993	2.0010	2.0003	2.0001	1.9996	2.0014
2	2.0000	2.0009	2.0015	1.9996	1.9997	2.0009
3	2.0015	1.9999	1.9996	2.0007	2.0003	2.0001
4	2.0009	2.0013	2.0013	2.0000	2.0010	1.9999
5	2.0010	2.0013	2.0012	2.0000	2.0013	1.9995
6	1.9999	1.9995	2.0015	1.9999	1.9998	2.0000
7	1.9996	2.0015	1.9996	2.0009	2.0018	1.9997
8	2.0013	1.9997	2.0000	2.0000	1.9999	2.0000
9	2.0014	2.0000	2.0010	2.0014	2.0010	2.0000
10	1.9995	2.0007	2.0016	2.0009	2.0015	1.9996
11	2.0000	2.9998	1.9997	2.0012	2.0000	2.0013
12	2.0001	2.0000	2.0001	1.9994	2.0010	2.0003
13	1.9997	2.0000	2.0005	2.0015	1.9995	2.0012
14	2.0003	2.0000	1.9995	1.9996	2.0008	2.0001
15	2.9998	2.0010	1.9996	2.0000	1.9996	2.0012
16	2.0000	2.0001	2.0001	1.9999	2.0000	2.0007
17	2.0001	2.0007	2.0003	2.0014	1.9998	2.0000
18	2.0007	2.0003	2.0008	2.0000	2.0011	1.9998
19	1.9996	1.9999	1.9999	1.9997	2.0000	1.9999
20	2.0010	2.0000	1.9995	1.9997	1.9997	2.0002

The manager then let the operation continue without worrying about control (he was not statistically oriented!) until receiving complaints from customers some 30 days later. He then started taking samples from the assembly belt to see if the complaints were justified. The following data resulted.

Sample	Day				
	37	38	39	40	41
1	1.9985	2.0010	2.0000	2.0013	2.0014
2	2.0007	1.9996	2.0000	2.0012	2.0009
3	1.9997	1.9999	2.0020	2.0015	2.0001
4	1.9989	2.0015	2.0007	1.9996	2.0017
5	2.0000	1.9999	2.0003	2.0023	1.9999
6	2.0010	2.0019	1.9999	2.0000	1.9995
7	2.0011	2.0003	2.0000	2.0010	2.0000
8	2.0015	2.0007	2.0001	2.0016	1.9997
9	1.9990	1.9998	1.9983	1.9997	2.0000
10	2.0003	2.0000	2.0007	2.0001	2.0000
11	2.0000	1.9985	2.0000	1.9982	1.9996
12	1.9995	1.9996	1.9999	2.0005	2.0013
13	1.9994	2.0000	2.0009	1.9995	2.0003
14	2.0007	1.9998	1.9994	1.9996	2.0012
15	2.0012	1.9995	1.9996	2.0001	2.0001
16	2.0000	2.0001	2.0014	2.0003	1.9980
17	2.0003	2.0003	2.0001	2.0008	2.0007
18	2.0008	2.0010	2.6015	1.9999	2.0012
19	2.0001	1.9996	1.9998	1.9995	2.0000
20	2.0001	2.0013	1.9998	2.0000	1.9998

(a) Compute $\bar{X}$ and R charts based on the first 6 days of operation.

(b) Plot the data above on the charts in part (a).

(c) Are the complaints about the average length of the six-penny nails justified?

(d) Are the complaints about the variation in the lengths of the nails justified? If so, on which day should this have become evident to the manager?

10.5 The House of Commons entertained a law regulating skirt lengths to 4 in. above the knees. They decided to have an official "investigator" observe and measure the lengths of skirts on a sample of women in Piccadilly Square. The following results were reported from selecting the first three women encountered immediately after 10 A.M. and also after 10 P.M.:

	M	T	W	Th	F	S	Su	M	T
A.M.	2.5	4.5	1.9	4.2	4.4	3.2	1.5	2.6	4.3
data	4.4	2.4	2.7	2.6	2.6	4.3	2.0	4.6	2.7
	2.5	2.4	4.1	2.2	2.4	2.3	3.3	3.1	2.3
P.M.	6.7	5.1	5.7	5.4	6.0	5.8	6.3	6.6	6.7
data	4.4	6.1	5.7	5.6	5.2	5.7	4.9	4.4	5.8
	4.6	4.1	3.6	3.6	4.1	3.8	4.1	4.6	3.8

(The House of Commons decided not to enact the law.)

(a) Construct $\bar{X}$ and R charts using all the data obtained each day for the first 5 days and plot $\bar{X}$ and R for the other 4 days, indicating on which days there is a significant difference.

(b) Follow the instructions in part (a) for the A.M. data only; for the P.M. data only. Do you believe that the A.M. and P.M. data differ significantly?

10.6 A toothbrush sales manager for a district in the Ozarks has kept up with the number of sales per month for the past 5 years for each of his salespeople. His top salespeople (of which there are 10) have had an average of 2550 sales per month with a standard deviation of 25 sales. A new sales technique involving C_s brand toothpaste seems promising. One of his top salespersons is instructed to use this technique. The resulting sales for the next 6-month period were as follows:

Month	1	2	3	4	5	6
Sales	2348	2714	2613	2492	2391	2701

(a) Plot these data on the control chart.

(b) Does the sales technique employing brand C_s toothpaste seem to be better than the usual technique employed by the sales personnel?

10.7 In a machining operation, measurements are taken on the diameters of cylinders. With the process in control the control limits for the mean were determined to be 31.36 and 31.26. The results of four new measurements on five different lots yielded the following data.

Lot	Measurements			
1	31.30	31.26	31.31	31.29
2	31.29	31.31	31.35	31.26
3	31.25	31.26	31.30	31.29
4	31.27	31.33	31.37	31.34
5	31.40	31.38	31.36	31.35

Specify which of these lots are determined to be in control.

10.8 Construct operating characteristic curves for the following sampling plans.

(a) $N = 50$, $n = 10$, $c = 2$
(b) $N = 50$, $n = 10$, $c = 4$
(c) Discuss the comparison of these two plans from a consumer's and a producer's point of view.

10.9 Do Problem 10.8 with batch sizes of 1000 using the appropriate approximation to the hypergeometric distribution.

10.10 The length of time T, in hours that a battery will hold its charge is exponential with parameter β; that is, $f(t) = (1/\beta)e^{-t\beta}$, $\beta > 0$, $t > 0$.

(a) Find the mean and variance of T.
(b) Find the cumulative distribution function of T.
(c) Find the reliability function of T.
(d) Find the hazard function of T.
(e) What is the failure rate at $t = 3$ h? At $t = 12$ h?
(f) Is the hazard function an increasing or a decreasing function? Is this reasonable from a practical point of view?

10.11 Let T have the Weibull pdf

$$f(t) = \alpha\beta t^{\beta-1}e^{-\alpha t^{\beta}}, \qquad t > 0, \quad \alpha > 0, \quad \beta > 0.$$

(a) Verify that $f(t)$ is a pdf.
(b) Find the mean of T.
(c) Find the variance of T.
(d) Find $G(x) = \mathscr{P}(X \leq x)$, where $X = \alpha T^{\beta}$.
(e) Use the result in part (d) to find the pdf of X.
(f) Find the hazard function for the Weibull pdf.

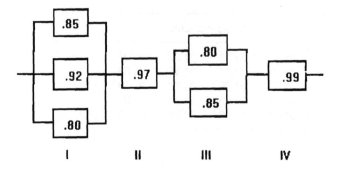

I   II   III   IV

10.12   A system consists of $n = 7$ components, as shown above.

(a)   Find the reliability of subsystems I and III.

(b)   Find the system reliability.

(c)   If II is replaced with two components in parallel, each with reliability 0.97, what is the reliability of the resulting subsystem II?

(d)   What is the reliability of the new system after making the changes in part (c).

(e)   Make a similar change in subsystem IV and compute the new system reliability.

   10.13   A system can consist of n identical and independent components in parallel, where each component has reliability 0.9. If it is desired to have a system reliability of at least 0.9999, how many components should be connected in parallel?

   10.14   If eight identical and independent components are connected in series, what must the reliability of each component be in order for the system to have reliability 0.98?

REFERENCES

1.   American Society of Testing Materials *ASTM Manual on Quality Control of Materials*, ASTM, Philadelphia (1951).

2.   Cowden, Dudley, J., *Statistical Methods in Quality Control*, Prentice Hall, Englewood Cliffs, NJ (1957).

3.   Deming, W. E., *Some Theory of Sampling*, Wiley, New York (1966).

4.   Dodge, H. F., and H. G. Romig, *Sampling Inspection Tables: Single and Double Sampling*, 2nd ed., Wiley, New York (1959).

5.   Enrick, N. L., *Quality Control and Reliability*, 5th ed. The Industrial Press, New York (1966).

6.   Grant, Eugene L., and Richard S. Levenworth, *Statistical Quality Control*, 4th ed., McGraw-Hill, New York (1972).

7.   Juran, J. M. (ed.), *Quality Control Handbook*, 2nd ed., McGraw-Hill, New York (1962).

11

Experimental Design

11.1 INTRODUCTION

It is important at this point to consider the manner in which the experimental data were collected, as this greatly influences the choice of the proper technique for data analysis. Before going any further, it is well to point out that the person performing the data analysis should be fully aware of several things:

1. What was to have been found out?
2. What is considered a significant answer?
3. How are the data to be collected, and what are the factors that influence the responses?

If an experiment has been designed or planned properly, the data will have been collected in the most efficient form for the problem being considered. *Experimental design* is the sequence of steps initially taken to ensure that the data will be obtained in such a way that their analysis will lead immediately to valid statistical inferences. Before a design can be chosen, the following questions must be answered:

1. How is the effect to be measured?
2. What factors influence the effect?
3. How many of the factors will be considered at one time?
4. How many replications (repetitions) of the experiment will be required?

5. What type of data analysis is required (regression, AOV, etc.)?
6. What level of difference in effects is considered significant?

The purpose of statistically designing an experiment is to collect the maximum amount of relevant information with a minimum expenditure of time and resources. It is important also to remember that the design of the experiment should be as simple as possible and consistent with the requirements of the problem. Three principles to be considered are replication, randomization, and control.

Replication is merely a complete repetition of the basic experiment. It provides an estimate of the magnitude of the experimental error. It also makes tests of significance of effects possible.

Randomization is the means used to eliminate bias in the experimental units and/or treatment combinations. If the data are random, it is safe to assume that they are independently distributed. Errors associated with experimental units that are adjacent in time or space will tend to be correlated, thus violating the assumption of independence. Randomization helps to make this correlation as small as possible so that the analyses can be carried out as though the assumption of independence were true.

An *experimental unit* is a unit to which a single treatment combination is applied in a single replication of the experiment. The term *treatment* (or treatment combinations) means the experimental conditions that are imposed on an experimental unit in a particular experiment. For example, in an experiment that studies a chemical process, the purpose may be to determine how temperature affects the response obtained from the process. Temperature would be a treatment. Another term, used interchangeably with treatment, is the term *factor*. This is especially true in the case of a factor that is quantitative, such as temperature. In such a case, the different values of the quantitative factor are referred to as *levels* of the factor. The factors are often called *independent variables*. In our example the different temperatures used in the experiment would be levels of the factor (or treatment) temperature. The one-way classification model studied in Chapter 8 can be referred to as a *one-factor* model. Similarly, the two-way classification model can be referred to as a *two-factor* experiment. Another example of a treatment is the amount of catalyst used in a chemical conversion process. A treatment combination could be in combination of a certain catalyst at a particular temperature. A treatment combination may consist of a single value of a single variable, or it may represent a combination of factors.

Experimental error, which has been mentioned previously, measures the failure of two identical experiments to yield identical results. The experimental error is composed of many things: measurement errors, observation errors, variation in experimental units, errors associated with the particular experiment, and all factors not being studied which could possibly have an effect on the outcomes of the experiment.

Occasionally, two or more effects in an experiment are *confounded*. That is, it is impossible to separate their individual effects when the statistical analysis of the data is performed. It might appear that confounding would be rigorously avoided, but this is not always true. In complex experiments where several levels of many effects are to be used, confounding often produces a significant decrease in the size of the experiment. Of course, the analysis of the data is more difficult and the effectiveness of the experiment may be seriously impaired if any data are missing.

Control refers to the way in which the experimental units in a particular design are balanced, blocked, and grouped. *Balancing* means the assignment of the treatment combinations to the experimental units in such a way that a balanced or symmetric configuration is obtained. For example, in a two-factor experiment involving a certain mechanical process in textile production, the two factors may be line tension and winding speed. If the same number of responses are obtained for each tension–speed combination, we say that the design is *balanced*. Otherwise, it is unbalanced or we say simply that data are missing. *Blocking* is the assignment of the experimental units to blocks in such a manner that the units within any particular block are as homogeneous as possible. *Grouping* refers to the placement of homogeneous experimental units into different groups to which separate treatments may be assigned.

Example 11.1 Consider the problem of testing the effect of electrolyte (HBF_4) concentration on the available power of $Mn_3(PO_4)_2$–Pb dry-charge batteries. Suppose that 20 batteries are available. In this case the experimental units are the batteries. Suppose that there are four concentrations of HBF_4 available: 100, 90, 80, and 70. The treatment (or factor) is $\%$ HBF_4 and the levels are 100, 90, 80, and 70. The 20 batteries can be grouped so that there are 5 batteries corresponding to each concentration of HBF_4. (The batteries for each concentration are chosen at random.)

Example 11.2 Suppose that an experiment is conducted to study the effect of fuel rate, burner angle, and steam/fuel ratio on the roof temperature of a steel furnace. In this case we have three factors (or treatments): (1) fuel rate, (2) burner angle, and (3) steam/fuel ratio. If two levels of each factor are used, we have what is commonly called a *three-factor factorial* with two levels of each factor, or a 2^3 factorial. In this case we must consider the added effect of the various combinations of two or three factors. These combination effects are called *interactions*.

In any experiment the number of replicates depends on the variance measured on a per experimental unit basis, the probability of committing both type I and type II errors, and the desired magnitude of the difference between two treatment means.

Consider the hypothesis $H_0: \mu_1 - \mu_2 = 0$. In this case, the statistic T_0 defined by

$$T_0 = \frac{\bar{X}_1 - \bar{X}_2 - 0}{\sqrt{2\sigma^2/r}} = \frac{\bar{X}_1 - \bar{X}_2 - \delta}{(2\sigma^2/r)^{1/2}} + \frac{\delta}{(2\sigma^2/r)^{1/2}}.$$

has a normal distribution with mean 0 and variance 1. We will write this more compactly by saying that "T_0 has an $N(0,1)$ distribution" or "T_0 is $N(0,1)$." The alternative is $H_A: \mu_1 - \mu_2 = \delta$. Under this alternative, we have

$$T_A = \frac{\bar{X}_1 - \bar{X}_2 - \delta}{(2\sigma^2/r)^{1/2}}, \tag{11.1}$$

which is $N(0,1)$. The distribution of $T_0 = (\bar{X}_1 - \bar{X}_2)/\sqrt{2\sigma^2/r}$ is shown in Fig. 11.1 for both hypotheses in the case when $\delta > 0$.

Suppose that the level of significance is fixed at α. Note that as $t_{1-\alpha}$ increases, type I error decreases, but type II error increases. For type I error $= \alpha$, we must have $(T_0 \geq t_{1-\alpha}|\delta = 0) = \alpha$. If we also specify $\beta =$ type II error, we must have

$$T_0 < t_{1-\alpha}|\delta = \left[T_A < t_{1-\alpha} - \frac{\delta}{(2\sigma^2/r)^{1/2}} \middle| \delta \right] = T_A < t_\beta|\delta = \beta,$$

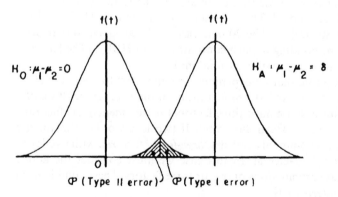

$H_0 : \mu_1 - \mu_2 = 0$ f(t)   f(t) $H_A : \mu_1 - \mu_2 = \delta$

$\mathcal{P}$ (Type II error) $\mathcal{P}$ (Type I error)

FIGURE 11.1 Error types for testing differences between treatment means.

according to Eq. (11.1). This implies that

$$\frac{\delta}{(2\sigma^2/r)^{1/2}} = t_{1-\alpha} - t_\beta. \tag{11.2}$$

As we usually do not know σ^2, we use S^2 as an estimator for it and determine the number of replicates required, r, by

$$r = \frac{2S^2}{\delta^2} \cdot (t_{1-\alpha} - t_\beta)^2. \tag{11.3}$$

The use of Eq. (11.3) to obtain the number of replications r required for a given probability $\mathscr{P}$ of obtaining a significant result requires a trial-and-error procedure, as the degrees of freedom for the t-values $t_{1-\alpha}$ and t_β both depend on r. For those cases where the true difference δ and the true standard error per unit $\sigma/\sqrt{r}$ are known, the number of replicates required for one- and two-tailed tests may readily be found in Table 11.1.

For the case where the number of replicates to be used is known, Eq. (11.3) may be rearranged and used to find the size of the true difference δ, which can be detected provided that the true standard error per unit, σ, or its corresponding estimate, s, is known. The true difference δ which can be detected with probability β for a given size α is given by

$$\delta = \sigma \sqrt{\frac{2}{r}} (t_{1-\alpha} - t_\beta). \tag{11.4}$$

The probability $\mathscr{P}$ of obtaining a significant result can be obtained from the formula

$$\mathscr{P} = 1 - \tfrac{1}{2}p_\beta, \tag{11.5}$$

where p_β is the probability corresponding to t_β in the t-table. The quantities t_β, $t_{1-\alpha}$, δ, σ, and r are the same as those used in Eqs. (11.3) and (11.4).

Example 11.3   In an experiment utilizing paired observations on colorimetric determination of Cr^{7+} at two different wavelengths, the estimate of the true standard error per unit was 8% of the overall mean. Ten replicate sets of data were obtained. The degrees of freedom available for estimating σ^2 was 24. What is the probability of obtaining a significant result when the true difference in means is 14% of the overall mean? Let the level of significance be $\alpha = 0.05$.

Solution   The estimate of the true standard error of the difference between the means, $\bar{d} = \bar{x}_1 - \bar{x}_2$, of the results obtained at different wavelengths is

$$\left(\frac{2s^2}{r}\right)^{1/2} = \sqrt{\frac{2(8)^2}{10}} = 3.578.$$

TABLE 11.1 Replications Required for a Given Probability of Obtaining a Significant Result for One-Class Data[a]

Upper figure: Test of significance at 5% level, probability 80%
Middle figure: Test of significance at 5% level, probability 90%
Lower figure: Test of significance at 1% level, probability 95%

True difference (δ) as percent of the mean	True standard error per unit (σ) as percent of the mean														
	2	3	4	5	6	7	8	9	10	11	12	14	16	18	20
						One-tailed tests									
5	3	6	9	13	19	25	33	41	50						
	4	7	12	18	26	35	45								
	7	13	22	33	47										
10	2	2	3	4	6	7	9	11	13	16	19	25	33	41	50
	2	3	4	5	8	9	12	15	18	22	26	35	45		
	3	4	7	9	13	17	22	27	33	40	47				
15	2	2	2	3	3	4	5	6	7	8	9	12	15	19	23
	2	2	3	3	4	5	6	7	9	10	12	16	21	26	31
	2	3	4	5	7	8	10	13	15	18	22	29	37	47	
20	2	2	2	2	2	3	3	4	4	5	6	7	9	11	13
	2	2	2	2	3	3	4	5	5	6	7	9	12	15	18
	2	3	3	4	4	5	7	8	9	11	13	17	22	27	33
25	2	2	2	2	2	2	3	3	3	4	4	5	6	7	9
	2	2	2	2	2	3	3	3	4	5	5	7	8	10	12
	2	2	3	3	3	4	5	6	7	8	9	11	14	18	22
30	2	2	2	2	2	2	2	2	3	3	3	4	5	6	7
	2	2	2	2	2	2	3	3	3	4	4	5	6	7	9
	2	2	2	3	3	3	4	4	5	6	7	8	10	13	15
						Two-tailed tests									
5	4	7	11	17	24	32	41								
	5	9	15	22	31	42									
	7	14	24	38											
10	2	3	4	5	7	9	11	14	17	20	24	32	41		
	2	3	5	7	9	12	15	18	22	27	31	42			
	3	5	7	11	14	19	24	30	37	45					

(*continued*)

TABLE **11.1** (*continued*)

True difference (δ) as percent of the mean	True standard error per unit (σ) as percent of the mean														
	2	3	4	5	6	7	8	9	10	11	12	14	16	18	20
15	2	2	3	3	4	5	6	7	8	10	11	15	19	24	29
	2	2	3	4	5	6	7	9	11	13	15	19	25	31	39
	3	3	4	6	7	9	12	14	17	21	24	33	42		
20	2	2	2	3	3	3	4	5	5	6	7	9	11	14	17
	2	2	2	3	3	4	5	6	7	8	9	12	15	18	22
	2	3	3	4	5	6	7	9	11	12	14	19	24	30	37
25	2	2	2	2	2	3	3	3	4	4	5	6	7	9	11
	2	2	2	2	3	3	4	4	5	5	6	8	10	12	15
	2	2	3	3	4	5	5	6	7	9	10	13	16	20	24
30	2	2	2	2	2	2	3	3	3	4	4	5	6	7	8
	2	2	2	2	2	3	3	3	4	4	5	6	7	8	11
	2	2	3	3	3	4	4	5	6	6	7	9	12	14	17

Source: Ref. 8. By permission of Butterworth & Co. (Publishers) Ltd.©
[a] In constructing the table, it was assumed that the number of degrees of freedom for error is $3(r-1)$; this would apply in a randomized blocks experiment with four treatments.

To test the significance of $\bar{d} = \bar{x}_1 - \bar{x}_2$, we use a two-tailed t-test with 24 degrees of freedom, for which $t = 2.064$. Therefore, if $\bar{d}$ is to be significant,

$$\bar{d} = \bar{x}_1 - \bar{x}_2 \geq \sqrt{\frac{2}{r}}\, st_{1-\alpha} = 3.578(2.064) = 7.385.$$

From Eq. (11.2), using the fact that we have a two-tailed test, we have

$$t_\beta = \frac{-\delta}{(2s^2/r)^{1/2}} + t_{1-\alpha/2}$$

$$= \frac{-14}{[2(8)^2/10]^{1/2}} + 2.064$$

$$= \frac{-14}{3.578} + 2.064$$

$$= -1.851.$$

From the t-table, Table IV, we find that the probability of T being less than -1.851 or greater than $+1.851$ is 0.08; that is, the probability of not finding a significant difference is 0.08. Thus the probability of finding a significant difference is $1 - 0.08 = 0.92$, which is also the power of the test. Consequently, the chances are quite high that a significant difference between the means of the two types of determinations would be observed in this experiment.

Example 11.4 In Example 11.3 suppose that $\alpha = 0.05$. How many replicates are required in order to have four out of five chances of obtaining a significant result if we wish to detect a true difference that is 15% of the mean?

Solution We have $\alpha = 0.05$, $\sigma = 8\%$ of the mean, $\delta = 15$, and the probability of obtaining a significant result is 80%. Thus since we have a two-tailed test, from Table 11.1 we see that $r = 6$ replicates are needed. The answer could also have been obtained by using Eq. (11.3).

In some experiments the experimental units fall into two categories, such as defective and nondefective. One might then be interested in comparing the percentages of units that fall into one of the classes under two different treatments. Table 11.2 gives the number of replications required in such experiments, where the units fall into one of two classes.

Example 11.5 Two different methods of manufacturing compression rings for $\frac{1}{4}$-in. copper tubing are being used. Method I produces about 20% defectives. If method II produces as low as 5% defectives, the manufacturer wishes to be fairly certain of obtaining a significant result in comparing the two methods. Suppose that the test is to be two-tailed at the 5% level and the manufacturer is satisfied with a probability of 0.80 of obtaining a significant result. How many rings are needed to carry out the test?

Solution We have

$$\mathcal{P}_1 = \text{smaller percentage of defectives} = 5\%$$
$$\delta = \text{size of difference to be detected}$$
$$= 20\% - 5\%$$
$$= 15\%.$$

From Table 11.2, for two-tailed tests we read the upper figure corresponding to $\mathcal{P}_1 = 5$ and $\delta = 15$ to obtain $r = 69$. The upper figure corresponds to $\alpha = 0.05$ and the given probability of 0.80. The size of each sample is 69, so that a total of 138 rings need to be obtained.

Various factors affect the size of an experimental unit. These are practical considerations that are used to determine the optimum amount of material to be studied or used in each of the optimum number of replicates, the nature of the experimental material, the nature of the treatments, and the cost of each experimental unit.

To sum up, the requirements for a good experiment are that the treatment comparisons should:

TABLE 11.2 Replications Required for a Given Probability of Obtaining a Significant Result for Two-Class Data

Upper figure: Test of significance at 5% level, probability 80%
Middle figure: Test of significance at 5% level, probability 90%
Lower figure: Test of significance at 1% level, probability 95%

$\mathscr{P}_1$ = smaller % success	$\delta = \mathscr{P}_2 - \mathscr{P}_1$ = larger minus smaller percentage of success													
	5	10	15	20	25	30	35	40	45	50	55	60	65	70
One-tailed tests														
5	330	105	55	35	25	20	16	13	11	9	8	7	6	6
	460	145	76	48	34	26	21	17	15	13	11	9	8	7
	850	270	140	89	63	47	37	30	25	21	19	17	14	13
10	540	155	76	47	32	23	19	15	13	11	9	8	7	6
	740	210	105	64	44	33	25	21	17	14	12	11	9	8
	1370	390	195	120	81	60	46	37	30	25	21	19	16	14
15	710	200	94	56	38	27	21	17	14	12	10	8	7	6
	990	270	130	77	52	38	29	22	19	16	13	10	10	8
	1820	500	240	145	96	69	52	41	33	27	22	20	17	14
20	860	230	110	63	42	30	22	18	15	12	10	8	7	6
	1190	320	150	88	58	41	31	24	20	16	14	11	10	8
	2190	590	280	160	105	76	57	44	35	28	23	20	17	14
25	980	260	120	69	45	32	24	19	15	12	10	8	7	—
	1360	360	165	96	63	44	33	25	21	16	14	11	9	—
	2510	660	300	175	115	81	60	46	36	29	23	20	16	—
30	1080	280	130	73	47	33	24	19	15	12	10	8	—	—
	1500	390	175	100	65	46	33	25	21	16	13	11	—	—
	2760	720	330	185	120	84	61	47	36	28	22	19	—	—
35	1160	300	135	75	48	33	24	19	15	12	9	—	—	—
	1600	410	185	105	67	46	33	25	20	16	12	—	—	—
	2960	750	340	190	125	85	61	46	35	27	21	—	—	—
40	1210	310	135	76	48	33	24	18	14	11	—	—	—	—
	1670	420	190	105	67	46	33	24	19	14	—	—	—	—
	3080	780	350	195	125	84	60	44	33	25	—	—	—	—
45	1230	310	135	75	47	32	22	17	13	—	—	—	—	—
	1710	430	190	105	65	44	31	22	17	—	—	—	—	—
	3140	790	350	190	120	81	57	41	30	—	—	—	—	—
50	1230	310	135	75	45	30	21	15	—	—	—	—	—	—
	1710	420	185	100	63	41	29	21	—	—	—	—	—	—
	3140	780	340	185	115	76	52	37	—	—	—	—	—	—

(*continued*)

TABLE 11.2 (*continued*)

$\mathscr{P}_1$ = smaller % success	$\delta = \mathscr{P}_2 - \mathscr{P}_1$ = larger minus smaller percentage of success													
	5	10	15	20	25	30	35	40	45	50	55	60	65	70
	Two-tailed tests													
5	420	130	69	44	31	24	20	16	14	12	10	9	9	7
	570	175	93	59	42	32	25	21	18	15	13	11	10	9
	960	300	155	100	71	54	42	34	28	24	21	19	16	14
10	680	195	96	59	41	30	23	19	16	13	11	10	9	7
	910	260	130	79	54	40	31	24	21	18	15	13	11	10
	1550	440	220	135	92	68	52	41	34	28	23	21	18	15
15	910	250	120	71	48	34	26	21	17	14	12	10	9	8
	1220	330	160	95	64	46	35	27	22	19	16	13	11	10
	2060	560	270	160	110	78	59	47	37	31	25	21	19	16
20	1090	290	135	80	53	38	28	22	18	15	13	10	9	7
	1460	390	185	105	71	51	38	29	23	20	16	14	11	10
	2470	660	310	180	120	86	64	50	40	32	26	21	19	15
25	1250	330	150	88	57	40	30	23	19	15	13	10	9	—
	1680	440	200	115	77	54	40	31	24	20	16	13	11	—
	2840	740	340	200	130	92	68	52	41	32	26	21	18	—
30	1380	360	160	93	60	42	31	23	19	15	12	10	—	—
	1840	480	220	125	80	56	41	31	24	20	16	13	—	—
	3120	810	370	210	135	95	69	53	41	32	25	21	—	—
35	1470	380	170	96	61	42	31	23	18	14	11	—	—	—
	1970	500	225	130	82	57	41	31	23	19	15	—	—	—
	3340	850	380	215	140	96	69	52	40	31	23	—	—	—
40	1530	390	175	97	61	42	30	22	17	13	—	—	—	—
	2050	520	230	130	82	56	40	29	22	18	—	—	—	—
	3480	880	390	220	140	95	68	50	37	28	—	—	—	—
45	1560	390	175	96	60	40	28	21	16	—	—	—	—	—
	2100	520	230	130	80	54	38	27	21	—	—	—	—	—
	3550	890	390	215	135	92	64	47	34	—	—	—	—	—
50	1560	390	170	93	57	38	26	19	—	—	—	—	—	—
	2100	520	225	125	77	51	35	24	—	—	—	—	—	—
	3550	880	380	210	130	86	59	41	—	—	—	—	—	—

Source: Ref. 3. Reprinted by permission of John Wiley and Sons, Inc.

1. Be free of systematic error
2. Be made sufficiently precise, which depends on:
    a. The intrinsic variability of the experimental material
    b. The intrinsic variability of the measuring devices and techniques
    c. The number of experimental units available
    d. The experimental design
3. Have a wide range of validity
4. Not be unnecessarily complicated, consistent with the desired results
5. Be easy to manipulate so as to obtain the standard error between two treatment combinations

11.2   RANDOM SEQUENCING IN EXPERIMENTAL DESIGNS

To prevent bias, all experiments should be carried out using proper randomization. This approach should minimize the effect of uncontrolled variables on the results (i.e., the dependent responses). In this book and others, the tabulated results used in example and homework problems give the impression that the data have been collected in some rigid pattern. As you will see later in the sections on Latin and Greco-Latin square designs, it is sometimes necessary to follow a pattern in assigning the treatments to experimental units and in conducting the experiment. Even so, randomization of the experimental sequence is still practiced to the maximum extent possible.

Consider the data of the one-way AOV experiment described in Example 8.1. The five individual determinations by each method were not collected one after another, and so on through all 20 analyses. Instead, the 20 samples of water were poured into separate flasks and stoppered. A table of random numbers was used to determine the order and method of analysis for each sample. The results were arranged in the order shown in Example 8.1 for ease of presentation.

All data for Example 8.4 were obtained during a single distillation experiment. Samples were drawn simultaneously from the three trays, divided, and analyzed. This procedure was repeated twice at the same operating conditions to yield the 18 refractive indices used for composition analysis. Randomization was introduced into this experiment in two ways. First, the order of analysis was randomized over the entire set of samples once all had been collected. Second, there is always some variability in the action on each individual tray even though the average temperature profile appears to remain constant within our ability to measure it.

In Example 8.5 the independent variables are lab group (fixed) and reflux ratio (random). The groups were randomly assigned the days on which they would perform the distillation experiment before the semester started. Although each group used the same four reflux ratios, the instructor randomized the order in which the reflux ratios were to be used when giving each group its laboratory assignment.

There are some situations in which complete randomization is not possible due to the physical nature of the process, time constraints, and so on. A pilot-scale cooling tower with horizontally oriented redwood slats as the tower fill presents a physical constraint to randomization. Although the water to be cooled is supposedly distributed uniformly over the top layer of slats, there is some channeling at low flow rates. If the combined effects of water and air flow rates are to be examined to produce a two-way AOV, laboratory time constraints require some adjustment of the experimental layout. The most informative approach would be a completely randomized design. Unfortunately, the degree of wetting of the redwood slats increases up to some threshhold water flow rate. The only way to dry the slats for the next run is to pull air through the tower at a high volume rate or to allow the slats to dry out slowly, a process that takes about a week. Thus, to evaluate the effects of water and air rates on tower performance, the order of water rates is first randomized. The air flow rates are randomized within each water rate for this model II situation. The air rates are re-randomized for each subsequent water rate.

How does this randomization process occur? First, list all the experimental conditions. This list is usually prepared in ascending or descending order. The choice does not matter. Next, use a table of random numbers or the random number function on your pocket calculator to determine the order in which the experiments will be run. When using a table of random numbers such as that shown in Table 11.3, start anywhere and move in any direction you like. Take the digits in groups of two, three, or four to create random numbers. Assign the first such number to the first treatment listed. Pick a direction in the table (vertical, horizontal, or diagonal), move a fixed number of digits, and start over. If by chance you get the same number as any number created previously, delete the repeated number and move on as described.

Let's use Example 8.5 to see how the randomization process occurs. Start in the fourth column, sixth row: 0 and move down two digits: 080. Next, move right three digits: 536. Next, move down three digits: 875. Now, move right three digits: 380. The lab groups are then assigned to the distillation experiment in the following order: group 1, group 4, group 2, and group 3, corresponding to ranking of the random numbers and assigning them to the group order:

Group	1	2	3	4
Random number	080	536	875	380

TABLE 11.3   Random Numbers

```
8  1  7  3  3  3  4  0  6  4  7  6  5  8  4  6  4  9  3  4  1  8
8  9  5  5  0  0  5  0  7  6  2  5  9  0  4  0  7  0  5  2  6  4
3  8  0  4  2  2  1  4  6  2  5  4  6  4  4  0  5  2  3  8  3  6
9  2  0  0  5  8  0  7  3  3  8  2  2  7  9  8  3  6  7  7  9  5
8  7  9  4  4  9  6  8  8  6  0  5  2  1  8  2  0  6  2  6  8  4
0  3  8  0  4  2  7  5  4  1  5  2  7  0  8  5  8  5  1  5  9  8
2  3  5  8  1  9  8  9  1  7  5  6  3  8  7  0  2  4  5  8  8  8
5  9  9  0  5  3  6  5  0  8  5  0  6  4  9  1  8  4  1  1  5  8
4  4  7  2  7  7  8  8  0  3  9  5  1  6  2  4  8  3  6  3  8  4
4  6  6  3  4  9  7  4  1  0  5  3  1  4  9  9  2  2  6  3  5  7
6  1  6  6  6  4  5  3  8  0  7  0  5  7  0  5  2  2  3  4  2  7
9  6  6  8  4  5  4  8  4  1  1  0  3  3  9  4  3  7  9  6  1  1
3  8  5  8  1  7  5  5  9  9  7  6  4  4  4  1  0  2  9  9  4  2
1  2  3  3  2  8  3  1  2  5  8  8  3  2  5  8  3  9  0  5  3  9
3  6  3  4  2  6  0  1  8  7  7  7  0  9  6  6  0  6  6  9  4  8
2  2  6  8  7  5  4  3  6  3  5  7  6  9  2  7  3  6  9  1  2  3
0  7  9  1  1  8  2  9  9  5  2  6  4  2  0  3  6  6  8  5  4  4
9  5  5  5  1  2  4  7  4  7  5  0  2  0  7  3  3  8  8  2  5  8
0  5  8  4  7  6  4  0  7  8  5  0  0  6  8  5  3  7  9  5  4  7
5  6  0  9  6  1  6  3  8  3  2  0  7  0  7  3  4  5  8  5  6  5
1  0  9  0  9  3  3  3  2  3  4  4  7  1  2  1  0  8  3  6  4  8
7  1  0  7  6  4  1  8  2  2  1  3  3  9  1  0  9  5  0  3  5  9
8  4  1  8  0  9  5  2  1  5  6  9  1  4  3  1  5  0  0  5  1  6
3  6  6  4  8  4  4  1  8  8  6  1  6  4  7  1  9  3  0  5  0  4
1  3  6  1  5  6  6  2  0  0  3  5  7  1  0  5  6  1  9  2  9  2
2  0  7  3  9  0  7  7  4  2  5  6  5  9  0  9  3  4  9  6  6  4
0  3  2  0  2  2  7  4  4  6  3  9  4  7  7  9  1  2  6  8  2  8
3  7  9  5  8  8  2  8  5  1  8  8  5  6  6  8  9  0  7  8  3  5
1  0  2  1  4  4  8  4  1  9  4  4  3  8  4  3  9  5  7  8  4  3
7  0  7  5  5  6  4  9  5  8  0  7  1  7  1  4  4  3  4  6  6  1
6  7  0  6  1  4  2  1  4  8  4  3  2  1  0  5  7  7  0  4  1  4
9  4  9  1  3  9  6  6  2  6  9  3  2  5  8  0  6  9  1  1  1  2
0  3  3  2  6  5  2  6  6  8  9  3  7  1  2  6  4  7  2  7  5  7
6  5  4  1  5  8  2  0  1  3  8  1  8  1  4  0  4  5  8  1  0  7
9  2  8  8  5  2  1  6  1  4  3  4  3  5  7  6  9  2  3  7  7  8
1  5  4  8  3  2  9  8  1  0  3  2  0  8  9  6  7  6  7  9  7  9
4  7  3  5  0  5  5  1  5  5  6  1  6  8  6  2  8  2  2  9  8  8
3  6  6  6  0  9  5  5  9  8  0  3  4  2  7  1  2  9  4  3  9  2
6  0  8  8  1  5  3  1  4  2  3  2  0  3  0  5  5  7  0  3  6  9
3  7  8  6  7  5  5  1  2  4  8  4  1  0  1  3  4  3  2  8  7  4
0  1  7  4  6  2  1  9  0  0  4  6  9  9  8  4  4  2  2  3  2  5
5  0  9  5  6  8  4  1  7  2  5  2  4  0  4  3  7  3  7  8  6  2
7  8  7  7  6  8  3  7  1  4  6  9  8  5  0  6  3  3  0  2  0  6
6  1  1  8  5  2  3  2  6  0  2  4  3  6  9  9  1  3  9  6  7  2
5  3  4  9  0  9  8  2  5  4  4  0  7  7  3  2  1  2  6  2  3  0
5  1  8  0  6  9  8  9  4  1  9  7  5  5  5  7  5  9  3  0  3  6
7  2  1  1  0  3  0  0  2  9  9  3  6  1  8  2  6  9  6  9  6  1
4  2  0  1  2  0  0  5  0  9  0  3  5  4  3  8  8  4  1  9  0  1
7  4  7  2  2  4  3  8  1  6  8  0  1  7  7  7  0  0  4  7  6  0
0  6  4  2  2  1  9  0  3  8  9  2  2  2  6  7  6  7  1  9  2  5
1  5  1  4  2  3  5  3  4  0  8  3  8  1  7  2  0  7  4  9  8  0
4  5  6  3  7  6  7  1  5  2  8  3  3  1  4  5  4  6  5  4  6  0
3  6  2  5  0  7  4  2  5  5  2  4  5  5  9  1  1  2  1  9  0  5
9  0  2  9  4  3  8  7  7  8  2  4  8  5  7  9  9  4  7  9  6  7
```

Source: R. M. Bethea and R. R. Rhinehart, *Applied Engineering Statistics*,
Marcel Dekker, New York (1991), p. 209.

To assign the reflux ratios (moles returned to moles of overhead
product), let's start at the second column of the seventeenth row: 7 and
move diagonally down and to the right in groups of three digits. The
resulting random numbers are 754, 631, 280, and 595. The reflux ratios are
assigned as

Reflux ratio		1	2	3	4
Random number		754	631	280	595

For this group, the order of reflux ratios to be used are 3, 4, 2, and 1. The randomization process is repeated for each group.

11.3 SOURCES OF ERROR

There are two important sources of experimental error: the variability of the material itself and poor experimental technique. The accuracy of the experiment can be increased by increasing the size of the experiment, by refining the experimental technique, and by handling the material so as to reduce the variability of the various effects. At this point, two terms often confused should be defined: *accuracy* is the approach to the true value of a datum point: *precision* is the repeatability of the results of a particular experiment. It should be noted here that when the degrees of freedom decrease, the limits of error for a true difference between treatment means are increased and the probability of getting a significant result is decreased accordingly.

Before starting into a discussion of experimental design, it is well to reiterate the two basic assumptions used in making an analysis of variance. These are that the treatment and environmental effects are additive and that the experimental errors are normally and independently distributed with mean zero and variance σ^2. This is usually written as $NID(0,\sigma^2)$. This is saying that the observations comprise a random sample.

11.4 COMPLETELY RANDOMIZED DESIGNS

In completely randomized designs, the treatments are allocated entirely by chance. The design is completely flexible. Any number of treatments or replications may be used. You may vary the replications from treatment to treatment and all available experimental material can be utilized. Several advantages of this method are that the statistical analysis of the results is simple even for the case of missing data. The relative loss of information due to missing data is less for the completely random design than for any other design. All variation among the experimental units goes into the experimental error.

The completely randomized design should be used when the experimental material is homogeneous or in anticipation of missing data points or in small experiments when an increase in accuracy from other designs does not outweigh the loss in experimental error degrees of freedom.

11.4.1 Analysis of Variance

The analysis of variance in the completely randomized design is based on the model

$$Y_{ij} = \mu + \tau_j + \varepsilon_{ij}, \qquad i = 1,2,\dots,n, \quad j = 1,2,\dots,t, \qquad (11.6)$$

where the assumptions inherent in the model are linearity, normality, additivity, independence, and homogeneity of variances among the treatments. The model equation (11.6) is actually the one-way AOV model discussed in Chapter 8. There are a total of nt experimental units, and n units are assigned to each of t treatments. There are two models with which we are concerned. Model I requires that $\sum_{j=1}^{t} \tau_j = 0$. This model is concerned only with the t treatments present in the experiment. Model II requires the τ_j to be random variables which are $NID(0,\sigma_\tau^2)$. In this case we are concerned with the population of treatments, of which only t (which are sampled at random) are present in the experiment.

The hypothesis of interest in model I, fixed effects, is $H_0: \tau_1 = \tau_2 = \cdots = \tau_t = 0$, that is, that of no treatment effect. In model II we are still interested in the hypothesis of no treatment effect; however, the τ_j are random variables with mean zero and variance σ_τ^2. In this case the hypothesis of no treatment effect is $H_0: \sigma_\tau^2 = 0$.

The analysis of variance for completely randomized designs with equal numbers of observations per treatment, n, is shown in Table 11.4. For the estimation of σ_τ^2 that is of interest in the case of model II, we use the estimator S_τ^2 given by

$$S_\tau^2 = \frac{MS_{Tr} - MS_E}{n}.$$

This follows from the expected mean squares in Table 11.4.

$$E(MS_{Tr}) = \sigma^2 + n\sigma_\tau^2 \qquad \text{and} \qquad E(MS_E) = \sigma^2,$$

which imply that $E[MS_{Tr} - MS_E)/n] = \sigma_\tau^2$.

To estimate the variance of the mean corresponding to the jth treatment, in the case of model I we use

$$\text{var}(j\text{th mean}) = \hat{V}(\bar{Y}_{\cdot j}) = S_{\bar{Y}}^2 = \frac{MS_E}{n}. \qquad (11.7)$$

This and other ideas were presented in Chapter 8 in discussing the one-way AOV model. A point estimate of the true mean effect of the jth treatment, $\hat{\mu}_j = \widehat{\mu + \tau_j}$, is

$$\hat{\mu}_j = \widehat{\mu + \tau_j} = \bar{y}_{\cdot j}. \qquad (11.8)$$

TABLE 11.4   Analysis of Variance for Completely Randomized Design with Equal
Numbers of Observations per Treatment

Source	d.f.	SS	EMS
Mean	1	$\left(\sum_i \sum_j Y_{ij}\right)^2 \Big/ tn = SS_M$	—
Among treatments	$t-1$	$\dfrac{\sum_{j=1}^{t}(\sum_{i=1}^{n} Y_{ij})^2}{n} - SS_M$ $= S_{Tr}$	Model I: $\sigma^2 + n\dfrac{\sum_j^t \tau_j^2}{t-1}$ Model II: $\sigma^2 + n\sigma_t^2$
Within treatments	$t(n-1)$	$\sum_i \sum_j Y_{ij}^2 - SS_M - SS_{Tr} = SS_E$	σ^2
Total	tn	$\sum_i \sum_j Y_{ij}^2 = SS_T$	

Example 11.6   A young engineer was assigned the task of testing the effect of
electrolyte ($\overset{.}{H}BF_4$) concentration on the available power of $Mn_3(PO_4)_2$–Pb
dry-charge batteries. He elected to perform the required experimental work
in a completely random design. The following data (A min) were obtained
from small-scale tests:

	% HBF$_4$		
40	30	20	10
62	61	58	56
64	60	60	54
64	61	59	54
62	61	58	55
63	62	58	55

(a) Prepare a complete AOV table for model I and model II
    interpretations and describe their differences.
(b) Are there any significant differences between treatments?
(c) What is the variance of the difference between the second and
    fourth treatment means?

Solution   The treatment sums and means are as follows:

$y_{.j}$:   315   305   293   274   for decreasing % HBF_4

$\bar{y}_{.j}$:   63.0   61.0   58.6   54.8   for decreasing % HBF_4.

Also,

$$y_{..} = \sum_i \sum_j y_{ij} = 1187, \qquad \bar{y} = \text{overall mean} = 59.35.$$

The sums of squares obtained from Table 11.4 are

$$SS_T = \sum_i \sum_j y_{ij}^2 = 70{,}647,$$

$$SS_M = \frac{(\sum_i \sum_j y_{ij})^2}{nt} = \frac{(1187)^2}{20} = 70{,}448.45,$$

$$SS_{Tr} = \frac{\sum_j (y_{.j}^2)}{n} - SS_M,$$

$$= \frac{(315)^2 + (305)^2 + (293)^2 + (274)^2}{5} - 70{,}448.45$$

$$= 186.55,$$

$$SS_E = SS_T - SS_M - SS_{Tr}$$

$$= 70{,}647 - 70{,}448.45 - 186.55 = 12.00$$

(a)   AOV table:

Source	d.f.	SS	MS	EMS	Model
Mean	1	70,448.45	70,448.45	—	—
Treatments	3	186.55	62.18	$\sigma^2 + \dfrac{5}{3} \displaystyle\sum_{j=1}^{4} \tau_j^2$	I
Error	16	12.0	0.75	$\sigma^2 + 5\sigma_\tau^2$	II
				σ^2	—
Total	20	70,647			

Model I assumes four fixed treatments applied to fixed experimental units in a random manner. Model II assumes assignment of each of four randomly selected treatments to five randomly selected experimental units.

(b) $H_0: \tau_j = 0$ (or $H_0: \sigma_\tau^2 = 0$).

$F_{3,16,0.99} = 5.29.$

$$F = \frac{MS_{Tt}}{MS_E} = \frac{62.18}{0.75} = 82.41.$$

As $F > 5.29$, we reject the null hypothesis and conclude that there is very likely a difference in the available power related to HBF_4 concentration in the electrolyte.

(c)   To calculate the variance of the difference between two treatment means, $2\sigma^2/r$, where $r =$ number of replicates (in this case, $r = 5$), we first estimate σ^2 by $S^2 = MS_E$. Therefore, an estimate of the variance of the difference between the second and fourth treatment means is $2(0.75)/5 = 0.20$.

11.4.2   Two-Way Analysis of Variance

Completely randomized designs may also be developed for two, three, four, or more treatments. Simple designs involving only two independent variables with the treatment combinations arranged to provide for the evaluation of interactions between those variables have been discussed in Section 8.4.2. The corresponding assumptions behind the four different two-way analysis-of-variance models and the resulting F-tests have been discussed in Section 8.4.3.

11.4.3   Two-Way Analysis of Variance with Subsampling

Although we did not show an example of subsampling in two-way analysis of variance in Chapter 8, the model is

$$Y_{ijk} = \mu + \alpha_i + \beta_j + \varepsilon_{ij} + \eta_{ijk}, \quad \begin{matrix} i = 1,2,\ldots,a, \\ j = 1,2,\ldots,b, \quad k = 1,2,\ldots,n. \end{matrix} \tag{11.9}$$

In subsampling, there are no independently repeated observations to yield an interaction term in the model. Instead, the samples from each of the treatment combinations are divided prior to analysis or testing. The subsamples provide an estimate of the sampling error independently of the experimental error. The corresponding calculations are similar to those presented in Tables 8.6 and 8.7 with a few significant changes. The experimental error sum of squares, SS_{EE}, is calculated as

$$SS_{EE} = SS_{AC} - SS_A - SS_B,$$

where

$$SS_{AC} = \frac{\sum \sum Y_{ij}^2}{n} - \frac{Y_{...}^2}{abn}.$$

The sum of squares for experimental error is thus $SS_{EE} = SS_{AC} - SS_A - SS_B$, where the sums of squares of the independent variables are calculated in the usual manner. The sampling error sum of squares, SS_{SE}, is the total corrected sum of squares, less the among-cells sum of squares, or $SS_{SE} = SS_T - SS_M - SS_{AC}$.

The true experimental error σ^2 is not directly separable from the sampling error σ_η^2. Hence the first F-test in the analysis of variance is the evaluation of $H_0: \sigma^2 = 0$ to determine which error is dominant. If $F = MS_{EE}/MS_{SE}$ is significantly greater than 1, the hypothesis that the sampling error is the dominant error is rejected as false. Regardless of the outcome of this first hypothesis test, the α_i and β_j variable effects are always evaluated against the dominant error as done previously in one-way analysis of variance with subsampling (Section 8.3.5). The F calculations used to estimate the significance of the independent variables proceed in a way analogous to those of Example 8.4. After stating the appropriate null hypotheses, the ratio of the mean squares for each term to the dominant error defines the value of the corresponding F. The expected mean squares for two-way analysis of variance with subsampling are shown in Table 11.5.

11.4.4 Three-Way Analysis of Variance

The model for three-way analysis of variance with interaction is

$$\left. \begin{array}{l} Y_{ijkl} = \mu + \alpha_i + \beta_j + \gamma_k + (\alpha\beta)_{ij} + (\alpha\gamma)_{ik} + (\beta\gamma)_{jk} + (\alpha\beta\gamma)_{ijk} + \varepsilon_{ijkl} \\ i = 1,2,\ldots,a, \quad j = 1,2,\ldots,b, \quad k = 1,2,\ldots,c, \quad l = 1,2,\ldots,n. \end{array} \right\}$$

(11.10)

The expected mean squares for the three-way analysis of variance model are presented in Table 11.6. The F-test ratios for the various treatments and interactions are given in Table 11.7. Each F-ratio is developed from the ratio of expected mean squares, which should be 1 if the null hypothesis that the treatment or interaction effect is insignificant is accepted as true. As an example, to test $H_0: \beta_1 = \beta_2 = \cdots = \beta_b = 0$ for a model III situation, if H_0 is true, then

$$F = \frac{MS_B}{MS_{BC}} \approx \frac{EMS_B}{EMS_{BC}} = \frac{\sigma^2 + na\sigma_{\beta\gamma}^2 + nac \sum_{j=1}^{b} \beta_j^2 / (b-1)}{\sigma^2 + na\sigma_{\beta\gamma}^2}.$$

TABLE 11.5   Expected Mean Squares in Two-Way Analysis of Variance with Subsampling in Completely Randomized Design

Source	Model I α,β: fixed	Model II α,β: random	Model III α: fixed, β: random	Model IV α: random, β: fixed
Mean Treatments	—	—	—	—
A	$\sigma_\eta^2 + n\sigma^2 + nb\sum_{i=1}^{a}\alpha_i^2/(a-1)$	$\sigma_\eta^2 + n\sigma^2 + nb\sigma_\alpha^2$	$\sigma_\eta^2 + n\sigma^2 + nb\sum_{i=1}^{a}\alpha_i^2/(a-1)$	$\sigma_\eta^2 + n\sigma^2 + nb\sigma_\alpha^2$
B	$\sigma_\eta^2 + n\sigma^2 + na\sum_{j=1}^{b}\beta_j^2/(b-1)$	$\sigma_\eta^2 + n\sigma^2 + na\sigma_\beta^2$	$\sigma_\eta^2 + n\sigma^2 + na\sigma_\beta^2$	$\sigma_\eta^2 + n\sigma^2 + na\sum_{j=1}^{b}\beta_j^2/(b-1)$
Experimental error	$\sigma_\eta^2 + n\sigma^2$	$\sigma_\eta^2 + n\sigma^2$	$\sigma_\eta^2 + n\sigma^2$	$\sigma_\eta^2 + n\sigma^2$
Sampling error	σ_η^2	σ_η^2	σ_η^2	σ_η^2

TABLE 11.6 Expected Mean Squares in Three-Way Analysis of Variance

Source	Model I α,β,γ: fixed	Model II α,β,γ: random	Model III α,β: fixed γ: random	Model IV α: fixed β,γ: random
Mean	—	—	—	—
Treatments				
A	$\sigma^2 + nbc \sum_{i=1}^{a} \alpha_i^2/(a-1)$	$\sigma^2 + n\sigma_{\alpha\beta\gamma}^2 + nc\sigma_{\alpha\beta}^2 + nb\sigma_{\alpha\gamma}^2 + nbc\sigma_{\alpha}^2$	$\sigma^2 + nb\sigma_{\alpha\gamma}^2 + nbc \sum_{i=1}^{a} \alpha_i^2/(a-1)$	$\sigma^2 + n\sigma_{\alpha\beta\gamma}^2 + nb\sigma_{\alpha\gamma}^2 + nc\sigma_{\alpha\beta}^2 + nbc \sum_{i=1}^{a} \alpha_i^2/(a-1)$
B	$\sigma^2 + nac \sum_{j=1}^{b} \beta_j^2/(b-1)$	$\sigma^2 + n\sigma_{\alpha\beta\gamma}^2 + nc\sigma_{\alpha\beta}^2 + na\sigma_{\beta\gamma}^2 + nac\sigma_{\beta}^2$	$\sigma^2 + na\sigma_{\beta\gamma}^2 + nac \sum_{j=1}^{b} \beta_j^2/(b-1)$	$\sigma^2 + na\sigma_{\beta\gamma}^2 + nac\sigma_{\beta}^2$
C	$\sigma^2 + nab \sum_{k=1}^{c} \gamma_k^2/(c-1)$	$\sigma^2 + n\sigma_{\alpha\beta\gamma}^2 + nb\sigma_{\alpha\gamma}^2 + na\sigma_{\beta\gamma}^2 + nab\sigma_{\gamma}^2$	$\sigma^2 + nab\sigma_{\gamma}^2$	$\sigma^2 + na\sigma_{\beta\gamma}^2 + nab\sigma_{\gamma}^2$
AB	$\sigma^2 + nc \sum_{i=1}^{a}\sum_{j=1}^{b} (\alpha\beta)_{ij}^2/(a-1)(b-1)$	$\sigma^2 + n\sigma_{\alpha\beta\gamma}^2 + nc\sigma_{\alpha\beta}^2$	$\sigma^2 + n\sigma_{\alpha\beta\gamma}^2 + nc \sum_{i=1}^{a}\sum_{j=1}^{b} (\alpha\beta)_{ij}^2/(a-1)(b-1)$	$\sigma^2 + n\sigma_{\alpha\beta\gamma}^2 + nc\sigma_{\alpha\beta}^2$
AC	$\sigma^2 + nb \sum_{i=1}^{a}\sum_{k=1}^{c} (\alpha\gamma)_{ik}^2/(a-1)(c-1)$	$\sigma^2 + n\sigma_{\alpha\beta\gamma}^2 + nb\sigma_{\alpha\gamma}^2$	$\sigma^2 + nb\sigma_{\alpha\gamma}^2$	$\sigma^2 + n\sigma_{\alpha\beta\gamma}^2 + nb\sigma_{\alpha\gamma}^2$
BC	$\sigma^2 + na \sum_{j=1}^{b}\sum_{k=1}^{c} (\beta\gamma)_{jk}^2/(b-1)(c-1)$	$\sigma^2 + n\sigma_{\alpha\beta\gamma}^2 + na\sigma_{\beta\gamma}^2$	$\sigma^2 + na\sigma_{\beta\gamma}^2$	$\sigma^2 + na\sigma_{\beta\gamma}^2$
ABC	$\sigma^2 + n \sum_{i=1}^{a}\sum_{j=1}^{b}\sum_{k=1}^{c} (\alpha\beta\gamma)_{ijk}^2/(a-1)(b-1)(c-1)$	$\sigma^2 + n\sigma_{\alpha\beta\gamma}^2$	$\sigma^2 + n\sigma_{\alpha\beta\gamma}^2$	$\sigma^2 + n\sigma_{\alpha\beta\gamma}^2$
Error	σ^2	σ^2	σ^2	σ^2

TABLE 11.7   *F*-Ratios for Testing Hypotheses in Three-Way Analysis of Variance

Source	Model I α,β,γ: fixed	Model II α,β,γ: random	Model III α,β: fixed γ: random	Model IV α: fixed β,γ: random
Mean	—	—	—	--
A	A/E	No exact test	A/AC	No exact test
B	B/E	No exact test	B/BC	B/BC
C	C/E	No exact test	C/E	C/BC
AB	AB/E	AB/ABC	AB/ABC	AB/ABC
AC	AC/E	AC/ABC	AC/E	AC/ABC
BC	BC/E	BC/ABC	BC/E	BC/E
ABC	ABC/E	ABC/E	ABC/E	ABC/E
E	—	—	—	—

If this null hypothesis is to be true, then $\sum_{j=1}^{b} \beta_j^2$ must be zero and the ratio of expected mean squares reduces to 1. To test this H_0, calculate $F = MS_B/MS_{BC}$ from the experimental data and compare the resulting value to the corresponding critical value of F with $(b-1)$ and $(b-1)(c-1)$ degrees of freedom.

In some cases, there is no exact test. In that situation, we resort to Satterthwaite's procedure. As an example, to test $H_0: \sigma_\alpha^2 = 0$ in a model II situation, we calculate

$$F \approx \frac{EMS_A}{EMS_{AB} + EMS_{AC} - EMS_{ABC}}$$

$$= \frac{\sigma^2 + n\sigma_{\alpha\beta\gamma}^2 + nc\sigma_{\alpha\beta}^2 + nb\sigma_{\alpha\gamma}^2 + nbc\sigma_\alpha^2}{(\sigma^2 + n\sigma_{\alpha\beta\gamma}^2 + nc\sigma_{\alpha\beta}^2) + (\sigma^2 + n\sigma_{\alpha\beta\gamma}^2 + nb\sigma_{\alpha\gamma}^2) - (\sigma^2 + n\sigma_{\alpha\beta\gamma}^2)}$$

$$= \frac{\sigma^2 + n\sigma_{\alpha\beta\gamma}^2 + nc\sigma_{\alpha\beta}^2 + nb\sigma_{\alpha\gamma}^2 + nbc\sigma_\alpha^2}{\sigma^2 + n\sigma_{\alpha\beta\gamma}^2 + nc\sigma_{\alpha\beta}^2 + nb\sigma_{\alpha\gamma}^2},$$

which reduces to 1 if H_0 is true. The degrees of freedom for the denominator are estimated as

$$\hat{v}_{denom} = \frac{(a_{AB}MS_{AB} + a_{AC}MS_{AC} + a_{ABC}MS_{ABC})^2}{(a_{AB}MS_{AB})^2/v_{AB} + (a_{AC}MS_{AC})^2/v_{AC} + (a_{ABC}MS_{ABC})^2/v_{ABC}}$$

where the a_i are integer constants. For this example, $a_{AB} = 1$, $a_{AC} = 1$, and $a_{ABC} = -1$.

An example of the calculations involved in a three-way AOV has been shown in Example 8.9 and Table 8.9 for a model I situation.

11.4.5 Four-Way Analysis of Variance

We now consider an example of a four-factor model, each factor at two levels.
Example 11.7 A completely randomized experiment was conducted to study the effect on the roof temperature Y of a steel furnace, of the following four factors, each at two levels:

F: fuel rate (α)
O: oxygen in waste gas (β)
B: burner angle (γ)
S: steam–fuel ratio (δ)

The model for this experiment is supposed to be adequately represented by

$$Y_{ijkl} = \mu + \alpha_i + \beta_j + \gamma_k + \delta_l + (\alpha\beta)_{ij} + (\alpha\gamma)_{ik} + (\alpha\delta)_{il}$$
$$+ (\beta\gamma)_{jk} + (\beta\delta)_{jl} + (\gamma\delta)_{kl} + \varepsilon_{ijkl}. \tag{11.11}$$

The key feature of this assumed model is, of course, that all population interactions of order higher than 1 are assumed to equal zero, which means that the corresponding single-degree-of-freedom mean squares are assumed to estimate only error.

The observed values of roof temperature were as follows, where subscript 1 refers to the low level and 2 to the high level of the variable involved:

$$y_{1111} = 1138, \quad y_{1211} = 1065, \quad y_{2111} = 1152, \quad y_{2211} = 1125,$$
$$y_{1112} = 1206, \quad y_{1212} = 1103, \quad y_{2112} = 1309, \quad y_{2212} = 1373,$$
$$y_{1121} = 1082, \quad y_{1221} = 1046, \quad y_{2121} = 1091, \quad y_{2221} = 1129,$$
$$y_{1122} = 1198, \quad y_{1222} = 1168, \quad y_{2122} = 1359, \quad y_{2222} = 1301.$$

(a) On the basis of these observations, prepare a complete AOV table. The sums of squares and expected mean squares for selected main effects and interactions are shown below as a guide.

$$SS_F = \frac{\sum_i y_{i\cdots}^2}{8} - \frac{y_{\cdots}^2}{16},$$

$$EMS_F = \sigma^2 + 8 \sum_i \alpha_i^2.$$

$$SS_{FO} = \frac{\sum_i \sum_j y_{ij\cdots}^2}{4} - \frac{\sum_i y_{i\cdots}^2}{8} - \frac{\sum_j y_{\cdot j\cdots}^2}{8} + \frac{y_{\cdots}^2}{16},$$

$$EMS_{FO} = \sigma^2 + 4 \sum_i \sum_j (\alpha\beta)_{ij}^2,$$

$$SS_{FOB} = \frac{\sum_i \sum_j \sum_k y_{ijk\cdot}^2}{2} - \frac{\sum_i \sum_j y_{ij\cdot\cdot}^2}{4} - \frac{\sum_i \sum_k y_{i\cdot k\cdot}^2}{4}$$

$$- \frac{\sum_j \sum_k y_{\cdot jk\cdot}^2}{4} + \frac{\sum_i y_{i\cdots}^2}{8} + \frac{\sum_j y_{\cdot j\cdot\cdot}^2}{8} + \frac{\sum_k y_{\cdot\cdot k\cdot}^2}{8}$$

$$- \frac{y_{\cdots}^2}{16},$$

$$EMS_{FOB} = \sigma^2,$$

$$SS_{FOBS} = \sum_i \sum_j \sum_k \sum_l y_{ijkl}^2 - \frac{\sum_i \sum_j \sum_k y_{ijk\cdot}^2}{2} - \frac{\sum_i \sum_k \sum_l y_{i\cdot kl}^2}{2}$$

$$- \frac{\sum_j \sum_k \sum_l y_{\cdot jkl}^2}{2} - \frac{\sum_i \sum_j \sum_l y_{ij\cdot l}^2}{2} + \frac{\sum_i \sum_j y_{ij\cdot\cdot}^2}{4}$$

$$+ \frac{\sum_i \sum_k y_{i\cdot k\cdot}^2}{4} + \frac{\sum_i \sum_l y_{i\cdot\cdot l}^2}{4} + \frac{\sum_k \sum_l y_{\cdot\cdot kl}^2}{4}$$

$$+ \frac{\sum_j \sum_l y_{\cdot j\cdot l}^2}{4} - \frac{\sum_i y_{i\cdots}^2}{8} - \frac{\sum_j y_{\cdot j\cdot\cdot}^2}{8} - \frac{\sum_k y_{\cdot\cdot k\cdot}^2}{8}$$

$$- \frac{\sum_l y_{\cdots l}^2}{8} + \frac{y_{\cdots}^2}{16},$$

$$EMS_{FOBS} = \sigma^2.$$

(b) Test the plausibility of the assumption regarding error by an F-test pitting the four degrees of freedom for second-order interactions against the single degree of freedom for the third-order interaction. A first-order interaction contains two factors, a second-order interaction contains three factors, and so on.

(c) If things turn out satisfactorily in (b), pool the five degrees of freedom for error as planned and test all 10 main effects and first-order interactions against this pooled error.

Solution   AOV table (asterisk indicates significant at 0.05 level):

Source	d.f.	SS (=MS as d.f. = 1)	EMS
F	1	43,368.06	$\sigma^2 + 8\sum_i \alpha_i^2$*
O	1	3,164.06	$\sigma^2 + 8\sum_j \beta_j^2$
B	1	588.06	$\sigma^2 + 8\sum_k \gamma_k^2$

(continued)

Source	d.f.	SS ($= $MS as d.f. $= 1$)	EMS
S	1	88,357.06	$\sigma^2 + 8 \sum_i \delta_i^{2*}$
FO	1	4,192.56	$\sigma^2 + 4 \sum_i \sum_j (\alpha\beta)_{ij}^2$
FB	1	232.56	$\sigma^2 + 4 \sum_i \sum_k (\alpha\gamma)_{ik}^2$
FS	1	15,687.56	$\sigma^2 + 4 \sum_i \sum_l (\alpha\delta)_{il}^{2*}$
OB	1	175.56	$\sigma^2 + 4 \sum_j \sum_k (\beta\gamma)_{jk}^2$
OS	1	52.56	$\sigma^2 + 4 \sum_j \sum_l (\beta\delta)_{jl}^2$
BS	1	1,743.06	$\sigma^2 + 4 \sum_k \sum_l (\gamma\delta)_{kl}^2$
FOB	1	1,743.06	σ^2
FOS	1	22.56	σ^2
FBS	1	588.06	σ^2
OBS	1	1,425.06	σ^2
FOBS	1	3,108.06	σ^2
	15	164,448.44	

(b) $\dfrac{MS_{FOB} + MS_{FOS} + MS_{FBS} + MS_{OBS}}{4} = \dfrac{3778.74}{4} = 944.685 = \mathscr{S}_1$

$\dfrac{\mathscr{S}_1}{MS_{FOBS}} = \dfrac{944.685}{3108.06} = 0.30394$, which is less than $F_{4,1,0.95} = 225$, so we accept H_0: interactions of order greater than 1 estimate error only.

(c) $\dfrac{MS_{FOB} + MS_{FOS} + MS_{FBS} + MS_{OBS} + MS_{FOBS}}{5} = \dfrac{6886.8}{5}$

$\qquad\qquad\qquad\qquad\qquad\qquad\qquad\qquad = 1377.3 = \mathscr{S}_2$

$F_{1,5,0.95} = 6.61$

$\dfrac{MS_F}{\mathscr{S}_2} = \dfrac{43368.06}{1377.3} = 31.48$, so we reject H_0: fuel rate is not significant.

$\dfrac{MS_O}{\mathscr{S}_2} = \dfrac{3164.06}{1377.3} = 2.297$, so we accept H_0: oxygen content in waste gas is not significant.

$$\frac{MS_B}{\mathscr{S}_2} = \frac{558.06}{1377.3} = 0.427,$$ so we accept H_0: burner angle is not significant.

$$\frac{MS_S}{\mathscr{S}_2} = \frac{88356.56}{1377.3} = 64.151,$$ so we reject H_0: steam/fuel ratio is not significant.

$$\frac{MS_{FO}}{\mathscr{S}_2} = \frac{4192.56}{1377.3} = 3.044,$$ so we accept H_0: interaction between fuel rate and oxygen content is not significant.

$$\frac{MS_{FB}}{\mathscr{S}_2} = \frac{232.56}{1377.3} = 0.169,$$ so we accept H_0: interaction between fuel rate and burner angle is not significant.

$$\frac{MS_{FS}}{\mathscr{S}_2} = \frac{15,687.56}{1,377.3} = 11.390,$$ so we reject H_0: interaction between fuel rate and steam ratio is not significant.

$$\frac{MS_{OB}}{\mathscr{S}_2} = \frac{175.56}{1377.3} = 0.127,$$ so we accept H_0: interaction between oxygen content and burner angle is not significant.

$$\frac{MS_{OS}}{\mathscr{S}_2} = \frac{52.56}{1377.3} = 0.038,$$ so we accept H_0: interaction between oxygen content and steam/fuel ratio is not significant.

$$\frac{MS_{BS}}{\mathscr{S}_2} = \frac{1743.06}{1377.3} = 1.266,$$ so we accept H_0: interaction between burner angle and steam/fuel ratio is not significant.

11.4.6 Nested Designs

In experimental designs such as the two- and three-way AOV designs, the factors are "crossed" with each other (i.e., there are observations present for each treatment combination). For example, in the two-way AOV design in Section 8.4.3, every level of factor A occurs with every level of factor B. There are situations where the factors are not crossed but factors may be "nested" inside one another.

In comparing four methods for doing a chemical analysis, suppose that a batch of a certain substance (chemical) is split into 60 parts, with 15 parts allocated to each method. For each method, three technicians will perform five replicate analyses on the chemical. The total number of observations obtained will be $4(3)(5) = 60$. If there are three different technicians for each method (i.e., a total of 15 technicians), the technician effect is nested within the method. If the same three technicians were used across methods, the method and technician factors would be crossed.

As another example, consider an experiment intended to measure a certain flow characteristic Y at a certain location in a whirlpool tank, as a function of I nozzle apertures. In this case there might be J runs for each nozzle aperture. That is, the run effect is nested within the aperture.

The model for two factors A and B, where *factor B is nested within factor A*, is

$$Y_{ijk} = \mu + \alpha_i + \beta_{ij} + \varepsilon_{ijk},$$
$$i = 1,2,\ldots,a, \quad j = 1,2,\ldots,b, \quad k = 1,2,\ldots,n, \quad (11.12)$$

where μ is the main effect, α_i the true effect of the ith level of A, β_{ij} the true effect of the jth level of factor B nested within the ith level of factor A, and ε_{ijk} is the random error corresponding to the kth unit for the jth level of B within the ith level of A. We assume that ε_{ijk} are $NID(0,\sigma^2)$. We further assume that the design is balanced.

The two-factor nested design is sometimes referred to as a *two-fold nested design*. The AOV sums of squares and associated degrees of freedom are the same for the twofold nested design as those in the AOV table for subsampling in a completely randomized design in Table 8.4. One main difference is that factors A and B may both be fixed in the twofold nested design, whereas only factor A may have that option in the completely randomized subsampling design. The expected mean squares for the twofold nested design when both A and B are fixed are

$$E(MS_A) = \sigma^2 + \frac{nb \sum_{i=1}^{a} \alpha_i^2}{I-1},$$

$$E(MS_{B(A)}) = \sigma^2 + \frac{n \sum_{i=1}^{a} \sum_{j=1}^{b} \beta_{ij}^2}{I(J-1)},$$

$$E(MS_E) = \sigma^2.$$

The F-ratio for testing the appropriate hypotheses in the twofold nested design are:

$$A \text{ and } B \text{ fixed}: F = \frac{MS_A}{MS_E}$$

$$A \text{ and } B \text{ random}: F = \frac{MS_A}{MS_{B(A)}}$$

$$A \text{ fixed, } B \text{ random}: F = \frac{MS_A}{MS_{B(A)}}$$

The case of A random and B fixed is rarely used and is thus not considered here.

If there are three factors (A, B, and C), C could be nested within B, and the combination of C and B could be nested within A. This would be a threefold nested design. Designs with more factors could be defined accordingly. Nested designs are also known as *hierarchical designs*. Furthermore, there are designs that contain factors, some of which are crossed and some of which are nested. For example, in a three-factor experiment, A and B could be crossed and that combination nested within C. Such designs are not covered in this book.

We now consider an example for a twofold nested design.

Example 11.8 Consider the following AOV table, computed on the basis of an experiment intended to measure a certain flow characteristic Y at a certain location in a whirlpool tank, as a function of I nozzle apertures:

Source	d.f.	MS	EMS
Among nozzle apertures	5	12,489	$\sigma_\varepsilon^2 + 3\sigma_\beta^2 + 30\sigma_\alpha^2$
Among runs, within nozzle apertures	54	3,339	$\sigma_\varepsilon^2 + 3\sigma_\beta^2$
Among measurements, within runs	120	627	σ_ε^2

(a) Write out the model equations that probably were assumed for this experiment.

(b) How many nozzle settings were used? Runs per setting? Measurements per run?

(c) State, as precisely as you can, what μ, α_i, β_{ij}, and ε_{ijk} signify for this experiment.

(d) Perform the F-test suggested by the AOV table of the null hypothesis that nozzle aperture has no effect on the flow characteristic Y.

(e) Perform the F-test of the null hypothesis that σ_β^2 is three times as large as σ_ε^2 (use a two-sided F-test).

(f) Give a CI for $\sigma_\beta^2/\sigma_\varepsilon^2$.

(g) Estimate $V[\bar{Y}_3..]$.

(h) Assuming that $Y_3.. = 193.7$, find a CI for $\mu_3 = \mu + \alpha_3$.

(i) Assume that the estimates of σ_ε^2 and σ_β^2 are exact. Assume that the overhead cost of a run is \$2.00 and that the cost of a measurement is 50 cents. Find the most efficient of all experimental plans costing as much as the given plan.

Solution

(a) $Y_{ijk} = \mu + \alpha_i + \beta_{ij} + \varepsilon_{ijk}$.

(b) $I = 6 =$ nozzle settings,

$J = 10 =$ number of runs per setting,

$K = 3 =$ measurements per run.

(c) $\mu =$ overall population mean of entire experiment,

$\alpha_i =$ contribution to Y_{ijk} due to treatment (nozzle settings),

$\beta_{ij} =$ contribution to Y_{ijk} due to run to run variations within any one treatment,

$\varepsilon_{ijk} =$ contribution to Y_{ijk} due to measurement variation in a run at any treatment level.

(d) $\mathrm{MS}_{n.a.}/\mathrm{MS}_{run} = 12{,}489/3339 = F = 3.7403$. $H_0: \sigma_\alpha^2 = 0$. $F > F_{5,54,0.95}$ is the test criterion, and as 3.7403 is $> F_{5,54,0.95} = 2.384$, we reject H_0.

(e) $H_0: \sigma_\beta^2/\sigma_\varepsilon^2 = 3$ or $H_0: \sigma_\beta^2 = 3\sigma_\varepsilon^2$.

$$\frac{\mathrm{MS}_{runs}/(\sigma_\varepsilon^2 + 3\sigma_\beta^2)}{\mathrm{MS}_{meas.}/\sigma_\varepsilon^2} = \frac{\mathrm{MS}_{runs}/10}{\mathrm{MS}_{meas.}} = F.$$

$$F = \frac{3339/10}{627} = \frac{333.9}{627} = 0.5325.$$

$F_{54,120,0.05} < F < F_{54,120,0.95}$ is the acceptance criterion. $1/F_{120,54,0.95} < F < F_{54,120,0.95}$ is found from tabular F-values as $1/1.488 < 0.5325 < 1.464$. which reduces to the partial inequality $0.672 \not< 0.5325 < 1.464$, so we reject H_0 and conclude that there is a significant variation due to replications (runs).

(f) $H_0: \sigma_\beta^2/\sigma_\varepsilon^2 = l$. Accept if

$$F_{54,120,0.05} < \frac{\mathrm{MS}_{runs}/(1 + 3l)}{\mathrm{MS}_{meas.}} < F_{54,120,0.95},$$

$$0.672 < \frac{5.32536}{1 + 3l} < 1.464.$$

For 90% CI limits, $5.32536/(1 + 3l) = 1.464$, so $l_L = 0.879$. and $5.32536 (1 + 3l) = 0.672$, so $l_U = 2.308$ to yield $0.879 < l < 2.308$ as the 90%₀ CI for l.

(g) Estimate $V[\bar{Y}_3..] = \mathrm{MS}_{runs}/JK = 3339/30 = 111.3$.

(h) $E[\bar{Y}_3..] = \mu_3 = \mu + \alpha_3$.

$$\mu_3: \bar{Y}_3 \pm t_{54,0.05}\sqrt{\frac{\mathrm{MS}_{runs}}{JK}},$$

$$\mu_3: 193.7 \pm t_{54,0.05}\sqrt{\frac{3339}{30}},$$

$$\mu_3: 193.7 \pm t_{54,0.05}\sqrt{111.3},$$

μ_3: 193.7 $\pm$ 2.0043(10.55),
μ_3: 193.7 $\pm$ 21.14,

172.6 < μ_3 < 214.8 is a 90% CI for μ_3.

(i) Cost/nozzle opening = \$2/run (10 runs) + (\$0.5/meas.)(3 meas./run) (10 runs) = \$35. Must minimize $\sigma_\beta^2/J + \sigma_\varepsilon^2/JK = \text{var}[\bar{Y}_3..] = 111.3$ subject to $2J + JK/2 = 35$.

$\sigma_\varepsilon^2 = 627$; $\sigma_\beta^2 = (3339 - 627)/3 = 2712/3 = 904$; $904/J + 627/JK = 111.3$ or $904K + 627 = 111.3JK$. Also, $2J + JK/2 = 35$ or $K = (70 - 4J)/J$, so

$$\frac{904(70 - 4J)}{J} + 627 = 111.3J\left(\frac{70 - 4J}{J}\right)$$

and $904(70 - 4J) + 627J = 111.3(70J - 4J^2)$ or $445.2J^2 - 10,780J + 63,280 = 0$. $dJ/dK = 0 = 890.4J - 10,780$, so $J = 10,780/890.4 = 12.107$ and $K = 1.782$. As fractional runs and measurements are impossible, use $J = 12$, $K = 2$ for best results costing no more than \$35.

11.5 RANDOMIZED COMPLETE BLOCK DESIGN

In the case of the randomized complete block design, the experimental units are divided into groups, each of which is a single trial or replication. Here the major variations are kept between blocks so as to keep the experimental error in each group small. Because of this grouping or blocking, as it is more often called, we get more accuracy than with the completely randomized design. In the randomized complete block design, any number of treatments or replications may be used. The statistical analysis is easy even for missing data. In the case where the error sum of squares for some treatments is larger than for other treatments, we can still get an unbiased estimate for testing any specific combination of treatment means. This is a decided advantage. It should be noted here that the replications provide unbiased comparisons of differences between replications. The method of using the randomized complete block design is to group the experimental units into blocks and then randomly assign the treatments within each block.

The basic assumption for a randomized complete block design with one observation per experimental unit is that the observations may be represented by the model given as

$$Y_{ij} = \mu + \tau_i + \beta_j + \varepsilon_{ij}, \qquad i = 1,2,\ldots,t, \quad j = 1,2,\ldots,b, \qquad (11.13)$$

where μ is the true mean effect, β_j the true effect of the jth block, and τ_i is the true effect of the ith treatment. The additional assumption made is that the ε_{ij} are $\text{NID}(0,\sigma^2)$.

11.5.1 Analysis of Variance, RCB

The analysis of variance for a randomized complete block design with one observation per experimental unit is given in Table 11.8. As in the case of the completely randomized design, either model I or model II may be assumed with respect to the β_j and τ_i. The hypothesis H_0: $\tau_i = 0$ is of primary importance and may be tested by computing

$$F = \frac{\text{MS for treatments}}{\text{MS}_E}. \qquad (11.14)$$

In Table 11.8 the B_j are the individual block sums $(Y_{.j})$ and the T_i are the individual treatment sums $(Y_{i.})$. If the value of F so computed by Eq. (11.14) exceeds $F_{v_1, v_2, 1 - \alpha}$, the hypothesis will be rejected and the conclusion is that there are significant differences among the treatments. The hypothesis H_0: $\beta_j = 0$, although of minor importance, may be tested in a similar manner using $F = (\text{MS for blocks})/\text{MS}_E$. However, due to the manner in which the experiment is set up, the hypothesis H_0: $\beta_j = 0$ should not be tested except as a check on the blocking of the experiment. The entire purpose of a RCB design is to reduce experimental error and get a more efficient test of H_0: $\tau_i = 0$, $i = 1,2,\ldots,t$. It is wise to check the efficacy of blocking by an

TABLE 11.8 Analysis of Variance of Randomized Complete Block Design with One Observation per Experimental Unit

Source	d.f.	SS	EMS
Mean	1	$\left(\sum_i \sum_j Y_{ij}\right)^2 / bt = \text{SS}_M$	—
Blocks	$b-1$	$\sum_j B_j^2/t - \text{SS}_M = \text{SS}_B$	$\sigma^2 + t \sum_j \beta_j^2 \ (b-1)$
Treatments	$t-1$	$\sum_i T_i^2/b - \text{SS}_M = \text{SS}_{Tr}$	$\sigma^2 + b \sum_i \tau_i^2 \ (t-1)$
Error	$(b-1)(t-1)$	$\sum_i \sum_j Y_{ij}^2 - \text{SS}_M - \text{SS}_B - \text{SS}_{Tr} = \text{SS}_E$	σ^2
Total	bt	$\sum_i \sum_j Y_{ij}^2$	

F-test on the blocks. If that test gives an insignificant *F*-value, the blocking was almost certainly faulty. In that case, the entire experiment should be repeated with more careful attention to assignment of the treatments to the experimental units.

Example 11.9 In an effort to determine the effect of temperature on degradation of a particular polymer proposed for use in manned spacecraft, five temperature levels were used in a randomized complete block design. The temperature effect was measured as ppm (parts per million) total hydrocarbons produced for samples of consistent mass. The tests were conducted using 5-g/L samples in each test chamber to simulate the weight/volume ratio expected in the actual mission. Five replicates were made at each temperature. Compute the AOV table.

Solution AOV table:

Source	d.f.	SS	MS	EMS
Blocks	4	10,000	2500	$\sigma^2 + \frac{5}{4}\sum_i \beta_i^2$
Temperature	4	11,200	2800	$\sigma^2 + \frac{5}{4}\sum_i \tau_i^2$
Error	16	24,000	1500	σ^2
	24			

The hypothesis $H_{0_1}: \tau_i = 0$ can be tested by comparing $F_{4,16,0.95}$ with $F = MS_{Tr}/MS_E$. We have

$$F_{4,16,0.95} = 3.01,$$

$$F = \frac{MS_{Tr}}{MS_E} = \frac{2800}{1500} = 1.866.$$

Since $F < F_{4,16,0.95}$, we accept H_{0_1}, that the treatments (temperatures) are not significant.

11.5.2 Missing Data, RCB

There are two common cases of missing data in the randomized complete block design: (1) a complete block is missing, and (2) the treatment is completely missing. In the first case, if at least two blocks remain, the analysis of variance proceeds as usual with one less block than originally thought to be available. If at least two treatments remain, we proceed in the regular manner.

A more common situation is the one in which one observation is missing. A correction for a single missing observation is made by assigning a value for the missing observation that minimizes the experimental error sum of squares when the standard analysis of variance is performed. For this technique an expression for the experimental sum of squares is obtained and differentiated with respect to the missing observation M, equated to zero and solved for M. The resulting estimate of the missing observation is

$$M = \frac{tT + bB - S}{(t-1)(b-1)},$$ (11.15)

where t is the number of treatments, b the number of blocks, T the $\sum Y$ with same treatment as M, B the $\sum Y$ in the same block as M, and S the sum of all actual observations. This value M is then entered into the table of data and the analysis proceeds as before, with one important exception.

The degrees of freedom associated with the experimental error and the total degrees of freedom must be decreased by one to account for the simulated datum point. After this is done, the treatment mean square calculated from the data with the estimated observation inserted is greater than the expected value of the experimental error mean square. Therefore, any test of hypothesis that does not correct for this fact will be biased. The correction for bias, Z, is made by decreasing the treatment sum of squares by an amount

$$Z = \frac{[B - (t-1)M]^2}{t(t-1)}$$ (11.16)

to give a new corrected treatment sum of squares which is then used in the analysis of variance.

Example 11.10   For the RCB data below, in which blocks are different Δt's and the treatments are different feeding arrangements to an evaporator system, test the hypothesis that the true treatment means are equal at the 5% level.

		Block		
Treatment	B_1	B_2	B_3	B_4
T_1	3.6	3.8	3.0	4.0
T_2	4.1	3.7	3.9	4.2
T_3	3.0	3.3	3.4	3.8
T_4	—	4.5	4.0	4.8

Solution A value for the missing datum point of steam economy may be estimated from Eq. (11.15). We have

$$M = \frac{tT + bB - S}{(t-1)(n-1)}$$

$$= \frac{4 \times 13.3 + 4 \times 10.7 - 57.1}{3 \times 3} = 4.32,$$

$$SS_T = \sum_i \sum_j y_{ij}^2 = 239.5924,$$

$$\sum_i \sum_j y_{ij} = 61.42,$$

$$SS_M = \frac{(\sum_i \sum_j y_{ij})^2}{bt} = \frac{(61.42)^2}{16} = 235.7760,$$

$$SS_B = \frac{\sum_j B_j^2}{t} - SS_M = \frac{946.4204}{4} - 235.7760 = 0.82908,$$

$$SS_{Tr} = \frac{\sum_i T_i^2}{b} - SS_M = \frac{952.8844}{4} - 235.7760 = 2.44508.$$

Due to the calculation of the missing datum point, the value of SS_{Tr} above must be corrected by subtracting the following correction term Z from it [see Eq. (11.16):

$$Z = \frac{[B - (t-1)M]^2}{t(t-1)}$$

$$= \frac{[10.7 - 3(4.32)]^2}{3(4)} = \frac{(-2.26)^2}{12} = \frac{5.0176}{12}$$

$$= 0.4256.$$

The new value of SS_{Tr} is then SS'_{Tr}, which is

$$SS'_{Tr} = SS_{Tr} - Z = 2.44508 - 0.4256 = 2.01948.$$

Also

$$SS_E = SS_T - SS_M - SS_B - SS_{Tr}$$
$$= 239.5924 - 235.7760 - 0.82908 - 2.44508$$
$$= 0.54224.$$

Analysis of variance:

Source	d.f.	SS	MS	f
Mean	1	235.7760	—	—
Blocks	3	0.82908	0.27636	0.4077
Treatments	3	2.01948	0.67316	9.932
Error	8	0.54224	0.06778	—
Total	15	239.1668		

To test the H_0 that there is no difference in treatment means, $F = MS'_{Tr}/MS_E$ will be compared with

$$F_{3,8,0.95} = 4.07.$$

Since $F_{treatments} > F_{3,8,0.95}$, the null hypothesis that the treatment means are equal is rejected at the 95% level and it may be said that there is a significant difference in steam economy due to different feeding arrangements.

Since $F_{blocks} < F_{3,8,0.95}$, the blocks do not have a significant effect on steam economy. This is not as it should be since the blocks (Δt's) should constitute a major source of variation in a RCB design. It is known that Δt changes affect steam economy significantly for a given feeding arrangement. The experiment might have been more informative with a feeding arrangement as the major source of variation (blocks) since it is known that the feeding arrangement affects the steam economy. Although no interaction tests were possible with this design, a certain feeding arrangement $-\Delta t$ combination might have been very significant in this experiment and could have affected the AOV accordingly.

11.5.3 Paired Observations, RCB

We can consider the method of paired observations, discussed in Section 7.7, as a randomized block with two blocks. In this case the test for equality of means $H_0: \mu_1 = \mu_2$ is obtained by computing the statistic

$$T = \frac{\bar{X}_1 - \bar{X}_2}{S_{\bar{x}_1 - \bar{x}_2}}, \tag{11.17}$$

where the acceptance region is given by

$$t_{n-1,\alpha/2} < T < t_{n-1,1-\alpha/2}. \tag{11.18}$$

It can be shown that the F-statistic, using $t = 2$, in the RCB design is the square of the T-statistic given in Eq. (11.17).

11.5.4 Subsampling in a Randomized Complete Block Design

When subsampling is used in a randomized complete block design, the corresponding statistical model is

$$Y_{ijk} = \mu + \tau_i + \beta_j + \varepsilon_{ij} + \eta_{ijk},$$
$$i = 1,2,\ldots,t, \quad j = 1,2,\ldots,b, \quad k = 1,2,\ldots,n, \quad (11.19)$$

and the corresponding analysis of variance is presented in Table 11.9. In the calculation for Table 11.9, B_j refers to the total of all observations in the jth block, and T_i is the total of all observations subjected to the ith treatment. The among-cells sum of squares used in calculating the sampling error sum of squares is given as

$$SS_{AC} = \frac{\sum_{i=1}^{t} \sum_{j=1}^{b} Y_{ij\cdot}^2}{n} - \frac{Y_{\cdot\cdot\cdot}^2}{btn}. \qquad (11.20)$$

Treatment effects can be tested as before by using an F-ratio, which is the mean square for treatments divided by the experimental error mean square,

TABLE 11.9 Analysis of Variance for Randomized Complete Block Design with Subsampling in the Experimental Units (Model I)

Source	d.f.	SS	EMS
Mean	1	$\left(\sum_i \sum_j \sum_k Y_{ijk}\right)^2 / btn = SS_M$	—
Blocks	$b-1$	$\sum_i B_j^2/tn - SS_M = SS_B$	$\sigma_\eta^2 + n\sigma^2 + tn \sum_j^b \beta_j^2/(b-1)$
Treatments	$t-1$	$\sum_i T_i^2/bn - SS_M = SS_{Tr}$	$\sigma_\eta^2 + n\sigma^2 + bn \sum_i^t \tau_i^2/(t-1)$
Experimental error	$(b-1)(t-1)$	$SS_{AC} - SS_B - SS_{Tr} = SS_{EE}$	$\sigma_\eta^2 + n\sigma^2$
Sampling error	$bt(n-1)$	$\sum_i \sum_j \sum_k Y_{ijk}^2 - SS_M - SS_{AC}$ $= SS_{EE}$	σ_η^2
Total	btn	$\sum_i \sum_j \sum_k Y_{ijk}^2 = SS_T$	

and comparing it with the tabular value of F, for the corresponding degrees of freedom and for a given significance level.

Example 11.11   In an experiment designed as a randomized complete block, the effect of cobalt concentration (% Co) on the tensile strength of steel was evaluated. Four levels of % Co were used. Three different crucibles were used in the alloying process. Each crucible was sampled once and the samples divided before analysis to give the following data, which have been presented in thousands of psi for convenience in making the calculations.

	Crucible		
% Co	1	2	3
1	49,50	44,45	53,56
2	60,62	53,56	64,65
3	64,67	63,65	74,78
4	71,75	65,67	76,80

It is believed that a reasonable model for this experiment is

$$Y_{ijk} = \mu + \tau_i + \beta_j + \varepsilon_{ij} + \eta_{ijk}, \qquad i = 1,2,3,4, \quad j = 1,2,3, \quad k = 1,2,$$

where i, j, and k correspond to % Co, crucible, and replication, respectively. The model then gives the observed tensile strengths Y_{ijk} in terms of the overall mean μ, the block or crucible effect β_j, and the treatment or % Co effect τ_i; and includes terms for experimental error, ε_{ij}, and subsampling error, η_{ijk}. For these data, compute the values for the AOV table.

Solution   Using the model for subsampling in randomized complete blocks (*SAS for Linear Models*), the following program is prepared.

(put initial JCL cards/statements here)

```
DATA ALLOY;
INPUT CRUC  1  CO  3  TENSTR  5-6;
CARDS;
```

(put data here according to INPUT format)

```
PROC PRINT;
PROC GLM;
CLASSES  CO  CRUC  CO(CRUC);
MODEL TENSTR = CO  CRUC  CO(CRUC);
TITLE 'EFFECT OF CRUCIBLE AND COBALT CONCENTRATION ON TENSILE
        STRENGTH';
```

(put final JCL cards/statements here)

SAS

EFFECT OF CRUCIBLE AND COBALT CONCENTRATION ON TENSILE STRENGTH

GENERAL LINEAR MODELS PROCEDURE

DEPENDENT VARIABLE: TENSTR

SOURCE	DF	SUM OF SQUARES	MEAN SQUARE	F VALUE
MODEL	11	2366.83333333	215.16666667	57.38
ERROR	12	45.00000000	3.75000000	PR > F
CORRECTED TOTAL	23	2411.83333333		0.0001

SOURCE	DF	TYPE I SS	F VALUE	PR > F
CRUC	2	485.33333333	64.71	0.0001
CO	3	1847.50000000	164.22	0.0001
CO(CRUC)	6	34.00000000	1.51	0.2552

The output consists of the AOV table shown here. The CO(CRUC) component in the MODEL statement results in the experimental error term. The ERROR term in the output is the sampling error term. We first compare experimental error to sampling error by

$$F = \frac{MS_{EE}}{MS_{SE}} = \frac{34.0/6}{45.0/12} = 1.511,$$

which is not significant, so $H_0: \sigma^2 = 0$ is accepted. When the treatments (% Co) are tested vs. sampling error as the major error source,

$$F_{Tr} = \frac{MS_{Tr}}{MS_{SE}} = \frac{1847.5/3}{45/12} = 164.22,$$

which is significant, so the hypothesis of no treatment effects ($H_0: \sum \tau_i^2 = 0$) is rejected. The same result is obtained if the two error terms are pooled to give

$$MS_{PE} = \frac{SS_{EE} + SS_{SE}}{df_{EE} + df_{SE}} = 4.3888$$

with $F_{Tr} = 140.32$ when using the pooled error term to increase precision.
Example 11.12 In a randomized complete block experiment intended to determine the effect of percent cobalt on steel tensile strength, three different percentages of cobalt were annealed in each of four ovens. The 12 resulting specimens were then divided into three samples each, and the resulting 36 samples tested for strength. The 36 strength values thus obtained were as follows (values are in thousands of psi):

	Oven			
% CO	1	2	3	4
1	50,47,55	43,42,51	49,53,57	45,49,55
2	64,59,55	48,55,62	66,63,65	67,63,60
3	60,66,69	59,60,70	67,83,75	63,72,70

For this example, the following calculations are made.

$$SS_M = \frac{Y_{...}^2}{btn} = 126,854.6944,$$

$$SS_T = \sum_{i=1}^{t} \sum_{j=1}^{b} \sum_{k=1}^{n} Y_{ijk}^2 = 129,995,$$

$$SS_B = \sum_{j=1}^{b} \frac{B_j^2}{tn} - SS_M = 450.30\overline{5},$$

$$SS_{Tr} = \sum_{i=1}^{t} \frac{T_i^2}{bn} - SS_M = 2007.0\overline{5},$$

$$SS_{AC} = \sum_{i=1}^{t} \sum_{j=1}^{b} \frac{Y_{ij.}^2}{n} - SS_M = 2519.63\overline{8},$$

$$SS_{EE} = SS_{AC} - SS_B - SS_{Tr} = 62.5777768,$$
$$SS_{SE} = SS_T - SS_M - SS_{AC} = 620.6666676.$$

Analysis of variance:

Source	d.f.	SS	MS
Mean	1	129,374.$\bar{3}$	—
Treatments (CO)	2	2,007.0$\bar{5}$	1,003.52$\bar{7}$
Blocks (ovens)	3	450.30$\bar{5}$	150.1018518
EE	6	62.5777768	10.42962947
SE	24	620	25.83$\bar{3}$
Total	36	129,995	

We first evaluate H_{0_1}: $\sigma^2 = 0$ by calculating $F_{EE} = MS_{EE}/MS_{SE} = 0.4037$, which is less than the critical value of $F_{6,24,0.95} = 2.51$. We accept H_{0_1} and conclude that the sampling is the dominant error source. As we have 24 degrees of freedom for the sampling error, there is no need to pool error sources. We proceed by calculating $F_{Tr} = MS_{Tr}/MS_{SE} = 38.8045$, which is greater than the critical value of $F_{2,24,0.95} = 3.40$. We thus reject H_{0_2}: $\sigma_t^2 = 0$ and conclude that the cobalt concentration probably affects tensile strength. Finally, we check the block effect just to be sure that we designed the experiment correctly. Here $F_B = MS_B/MS_{SE} = 5.8041$, which is greater than the critical value of $F_{3,24,0.95} = 3.01$. We conclude that blocks (ovens) constitute a major source of variation and we reject H_{0_3}: $\sum_j \beta_j^2 = 0$.

It is often advantageous when using randomized complete block designs to consider factorial treatment combinations. Two important cases are the two-factor case, for which the statistical model is given by

$$Y_{ijk} = \mu + \rho_i + \alpha_j + \beta_k + (\alpha\beta)_{jk} + \varepsilon_{ijk}, \qquad i = 1,2,\ldots,r,$$
$$j = 1,2,\ldots,a, \quad k = 1,2,\ldots,b, \qquad (11.21)$$

and the three-factor case, for which the statistical model is

$$Y_{ijkl} = \mu + \rho_i + \alpha_j + \beta_k + (\alpha\beta)_{jk} + \gamma_l + (\alpha\gamma)_{jl} + (\beta\gamma)_{kl} + (\alpha\beta\gamma)_{jkl} + \varepsilon_{ijkl},$$
$$i = 1,2,\ldots,r, \quad j = 1,2,\ldots,a, \quad k = 1,2,\ldots,b, \quad l = 1,2,\ldots,c. \qquad (11.22)$$

In these equations, μ is the true mean effect, ρ_i is the true effect of the ith replicate, and the various terms involving α_j, β_k, γ_l, $(\alpha\beta)_{jk}$, and so on, are the true effects of the several factors and their interactions. The ε's are then the true effects of the experimental units. The analysis of variance proceeds in a manner analogous to that used for factorial treatment combinations in the completely randomized design. The interested student is referred to Ostle and Mensing [5] for an analysis of multifactor factorials in a randomized complete block design with subsampling and several determinations per subsampling unit.

11.5.5 Nonparametric Methods and Randomized Complete Blocks

There is a popular and efficient nonparametric method that is useful in analyzing data that conform to a randomized complete block design. This procedure is referred to as the *Friedman test* (see Conover [4]). The procedure used in a randomized complete block is an extension of the paired *t*-test procedure discussed in Section 7.7 for comparing the means of two normal populations. The Friedman test is a nonparametric analog to the randomized complete block procedure.

Friedman Test

We are interested in comparing the effects of t treatments when the extraneous variation is controlled by blocking. The experimental units are divided into b blocks, each of size t. The elements within a block are chosen so that they are as alike as possible with respect to the extraneous variable. The t treatments are then assigned at random within each block.

To test the hypothesis H_0: equal treatment means, the observations within each of the b blocks ranked 1 to t (in increasing order). The rank assigned to any tied observations is the mean of the corresponding ranks. The rank total, T_i, for each of the t treatments is then computed. The Friedman test statistic is defined by

$$S = \frac{12}{bt(t+1)} \sum_{i=1}^{b} \left[T_i - \frac{b(t+1)}{2} \right]^2$$

$$= \frac{12}{bt(t+1)} \sum_{i=1}^{b} T_i^2 - 3b(t+1). \qquad (11.23)$$

It can be shown that when H_0 is true, $E(T_i) = b(t+1)/2$. Consequently, under H_0, the Friedman statistic is a measure of deviations of the computed rank totals from their expected value. The hypothesis H_0 is rejected for large values of the Friedman statistic. The exact distribution of S has been tabulated for small values of b and t. It is also known that S has an approximate chi-squared distribution with $t - 1$ degrees of freedom. At the level of significance α, H_0 is rejected if the computed value of S exceeds $\chi^2_{t-1, 1-\alpha}$.

Example 11.13 Consider a randomized complete block experiment (similar to that of Problem 11.21), where there are $t = 7$ treatment levels (feeding arrangements) and $b = 4$ blocks (different ΔTs). The data are

	Treatment						
	T_1	T_2	T_3	T_4	T_5	T_6	T_7
B_1	3.6	4.1	3.0	4.1	4.3	3.1	4.2
B_2	3.8	3.7	3.3	4.5	4.2	3.4	4.4
B_3	3.0	3.9	3.4	4.0	4.1	3.3	4.5
B_4	4.0	4.2	3.8	4.8	4.7	3.2	4.7

Ranking the observations within each block, we get

	Treatment						
	T_1	T_2	T_3	T_4	T_5	T_6	T_7
B_1	3	4.5	1	4.5	7	2	6
B_2	4	3	1	7	5	2	6
B_3	1	4	3	5	6	2	7
B_4	3	4	2	7	5.5	1	5.5
Rank totals	11	15.5	7	23.5	23.5	7	24.5

The computed value of the Friedman test statistic is

$$S = \frac{12}{4(7)(8)} \cdot [(11)^2 + \cdots + (24.5)^2] - 3(4)(8)$$

$$= 115.929 - 96$$

$$= 19.929.$$

Since $\chi^2_{6, 0.95} = 12.5916$, we reject the hypothesis of no treatment differences since $S = 19.929 > 12.5916$.

There are nonparametric methods available for determining which treatments differ significantly from one another. For further reading and information the reader is referred to Conover [4].

11.6 LATIN SQUARE DESIGNS

A design that is particularly useful in industrial experimentation is the Latin square design. We have seen how randomized blocks can be used to reduce experimental error by eliminating a source of variation in which there is no interest; that is, if a block effect exists, the corresponding test procedure used to test treatment effects is more sensitive to actual treatment differences. Also, the probability of a type II error is decreased. If there are two sources of variation to control, a Latin square provides a very good method of analyzing treatment differences. The Latin square can thus be viewed as a method of assessing treatment effects when a double type of blocking is imposed on the experimental units. In this sense we can say that the Latin square is an extension of the randomized complete block design.

Example 11.14 Consider an experiment involving study of the tire life of four brands of tires. There are 16 tires, four of each brand, available. Furthermore, four cars are available. The treatment in this case is tire brand. There are two other factors that effect tire life: (1) car type and (2) position of the tire on the car. Label the cars I, II, III, and IV and tire position as RF, LF, RR, and LR. The four cars can be set up as rows and the four positions as columns and the tires can then be assigned at random subject to the condition that each brand occur once in each row and each column. This example will be analyzed further in Example 11.15.

In a Latin square the two types of blocks are referred to as rows and columns. The differences among rows and among columns represent two sources of variation that the design is meant to remove. The experimental error per unit usually increases as the size of the square increases. This leads us to the conclusion that in a small Latin square such as a 3 × 3 or a 4 × 4, a substantial reduction in experimental error must be obtained over that obtainable with the completely randomized or randomized complete block design to counteract for the concomitant loss of degrees of freedom for error in the Latin square design. Latin squares are often used when different treatments are applied in sequence to the same experimental units. In this case the rows become successive periods or applications and the columns are the experimental units.

To form a Latin square design, assign one variable as rows, the second as columns, and the third as treatments. Then randomly assign values of the row and column variables as shown below.

	C_3	C_1	C_4	C_2
R_4				
R_1				
R_3				
R_2				

Each small "box" represents one experimental unit to which will be assigned (in this case of a 4 × 4 Latin square) one of the four treatments. For the sake of simplicity, start at the upper left corner and assign the treatments in any order across the first row or down the first column. The permute the treatments across (or down) the rest of the square. Shown below are four simple 4 × 4 Latin squares. Designation of the levels of the row and column variables have been omitted for clarity.

A	B	C	D
B	C	D	A
C	D	A	B
D	A	B	C

Column permutation

A	B	C	D
D	A	B	C
C	D	A	B
B	C	D	A

Row permutation

A	B	C	D
D	C	B	A
C	D	A	B
B	A	D	C

Helical permutation

A	B	C	D
B	A	D	C
C	D	A	B
D	C	B	A

Alternating forward
reverse permutation

Diagonal and other permutations are possible. It is not necessary to start with any particular order of treatments for the first row or column. The important feature to observe is that each row and each column contains each treatment once and only once.

11.6.1 Analysis of Variance for the Latin Square

The model for a general $m \times m$ Latin square with one observation per experimental unit is represented by

$$Y_{ij(k)} = \mu + \rho_i + \gamma_j + \tau_k + \varepsilon_{ij(k)},$$

$$i = 1,2,\ldots,m, \quad j = 1,2,\ldots,m, \quad k = 1,2,\ldots,m, \quad (11.24)$$

where $\sum_i \rho_i = \sum_j \gamma_j = \sum_k \tau_k = 0$ and the $\varepsilon_{ij(k)}$ are NID($0,\sigma^2$). The variables ρ_i, γ_j, and τ_k are the true effects associated with the ith row, jth column, and kth treatment. The reason for the parentheses around the k in Eq. (11.24) is that there are actually m^2 observations, not m^3 observations such as are present in the usual three-factor design with one replication. The analysis of variance for an $m \times m$ Latin square with one observation per experimental unit corresponding to the model given in Eq. (11.24) is shown in Table 11.10, where R_i, C_j, and T_k represent the indicated row, column, and treatment totals.

TABLE 11.10 Analysis of Variance for an $m \times m$ Latin Square with One Observation per Experimental Unit

Source	d.f.	SS	EMS
Mean	1	$\left(\sum_i \sum_j Y_{ij(k)} \right)^2 \Big/ m^2 = SS_M$ or $\left(\sum_i \sum_k Y_{i(j)k} \right)^2 \Big/ m^2 = SS_M$	
Rows	$m-1$	$\sum_i R_i^2/m - SS_M = SS_R$	$\sigma^2 + \dfrac{m}{m-1} \sum_i \rho_i^2$
Columns	$m-1$	$\sum_j C_j^2/m - SS_M = SS_C$	$\sigma^2 + \dfrac{m}{m-1} \sum_j \gamma_j^2$
Treatments	$m-1$	$\sum_k T_k^2/m - SS_M = SS_{Tr}$	$\sigma^2 + \dfrac{m}{m-1} \sum_k \tau_k^2$
Experimental error	$(m-1)(m-2)$	$\sum_i \sum_j \sum_k Y_{ij(k)}^2 - SS_M - SS_R$ $- SS_C - SS_{Tr}$	σ^2
Total	m^2	$\sum_i \sum_j \sum_k Y_{ij(k)}^2 = SS_T$	

It would appear that the Latin square gives us a very powerful technique for assessing treatment differences in the presence of two other sources of variation. This is true; however, a price must be paid and in this case the price is the assumption of no interactions in the model in Eq. (11.24). This assumption must be considered by the experimenter, and if in doubt it may be necessary to check it out by use of one of the models considered previously. Note that the Latin square calls for the same number of rows, columns, and treatments.

Example 11.15 Four brands (A, B, C, D) of tires were evaluated by a Latin square design in the following manner. Sixteen tires, four of each brand, were put on four cars (I, II, III, IV), in such a way that:

1. Each car had a tire from every brand.
2. Each brand was tried in a RF, LF, RR, and LR position.

3. The exact disposition of the 16 tires was determined by the random selection of a 4 × 4 Latin square, which turned out to be

	Position			
Car	RF	LF	RR	LR
I	A	B	C	D
II	B	D	A	C
III	C	A	D	B
IV	D	C	B	A

The cars were driven 5000 miles, and tire wear, as measured by a standard procedure, turned out to be

31	33	47	54
36	53	42	54
51	43	62	49
81	78	72	84

(a) What car had the hot-rod driver? Does the presence of the hot-rod driver invalidate the experiment?
(b) Test for the equality of tire effects, the equality of position effects, and the equality of car effects.

Solution

	Position								
Car	RF		LF		RR		LR		Row Σ
I	A	31	B	33	C	47	D	54	165
II	B	36	D	53	A	42	C	54	185
III	C	51	A	43	D	62	B	49	205
IV	D	81	C	78	B	72	A	84	315
Col Σ		199		207		223		241	870

We have

$$\frac{y^2_{\cdot\cdot(\cdot)}}{16} = \frac{(870)^2}{16} = 47{,}306.25,$$

$$\frac{\sum_j y^2_{\cdot j(\cdot)}}{4} = \frac{190{,}260.}{4} = 47{,}565,$$

$$\frac{\sum_i y^2_{i\cdot(\cdot)}}{4} = \frac{202{,}700.}{4} = 50{,}675,$$

$$\frac{\sum_k y^2_{\cdot\cdot(k)}}{4} = \frac{(200)^2 + (190)^2 + (230)^2 + (250)^2}{4},$$

$$= 47{,}875.$$

Also,

SS for cars: $\dfrac{\sum_i y^2_{i\cdot(\cdot)}}{4} - \dfrac{y^2_{\cdot\cdot(\cdot)}}{16} = 3368.75,$

SS for positions: $\dfrac{\sum_j y^2_{\cdot j(\cdot)}}{4} - \dfrac{y^2_{\cdot\cdot(\cdot)}}{16} = 258.75,$

SS for tires: $\dfrac{\sum_k y^2_{\cdot\cdot(k)}}{4} - \dfrac{y^2_{\cdot\cdot(\cdot)}}{16} = 568.75,$

$$SS_{TC} = 51{,}540. - 47{,}306.25 = \sum_i \sum_j y^2_{ij(k)} - \frac{y^2_{\cdot\cdot(\cdot)}}{16} = 4233.75,$$

$$SS_E = SS_{TC} - (SS_{cars} + SS_{pos} + SS_{tires}) = 37.50.$$

(a) Car IV had the hot-rod driver, as there was much more tire wear on that car than on any other. His presence does not invalidate the experiment, as we may consider him to be the "upper limit of wear" while the driver of car I may be the "lower limit of wear," cars II and III being somewhere in between. Anyway, the design tests each tire in each position by each driver and the AOV eliminates bias of tire position and car effects. The AOV table is given below.

Source of variation	d.f.	SS	MS
Blocks (cars)	3	3368.75	1122.917
Treatments (tires)	3	568.75	189.583
Positions	3	258.75	86.25
Exptl. error	6	37.50	6.25
Total corrected	15	4233.75	

(b) The hypotheses involving various effects and the corresponding decisions follow.

H_{0_1}: tire effects are equal

$$F = \frac{\text{tire MS}}{\text{MS}_E}; F_{3,6,0.95} = 4.76.$$

$$F = \frac{189.583}{6.25} = 30.33, \text{ so we reject } H_{0_1}.$$

H_{0_2}: position effects are equal

$$F = \frac{\text{pos. MS}}{\text{MS}_E}; F_{3,6,0.95} = 4.76.$$

$$F = \frac{86.25}{6.25} = 13.80, \text{ so we reject } H_{0_2}.$$

H_{0_3}: car effects are equal

$$F = \frac{\text{car MS}}{\text{MS}_E}; F_{3,6,0.95} = 4.76.$$

$$F = \frac{1122.917}{6.25} = 179.67, \text{ so we reject } H_{0_3}.$$

11.6.2 Missing Data, LS

If a single observation is missing in an experiment conducted according to an $m \times m$ Latin square, its value may be estimated by using

$$M = \frac{m(R + C + T) - 2S}{(m - 1)(m - 2)}, \tag{11.25}$$

where $R = \sum Y$ in same row as M, $C = \sum Y$ in same column as M, $T = \sum Y$ with same treatment as M, S is the sum of all actual observations, and m is the number of rows = number of columns = number of treatments. As before, the degrees of freedom associated with the experimental error must be decreased by one. The treatment sum of squares must be corrected for bias by subtracting a correction,

$$Z = \frac{[S - R - C - (m - 1)T]^2}{(m - 1)^2(m - 2)}, \tag{11.26}$$

from the calculated or original treatment sum of squares before hypothesis testing.

Example 11.16 A 4 × 4 Latin square design was used in a series of tests to determine the effect of water pressure, airflow rate, and the number of nozzles in operation on the overall efficiency of the scrubber-impinger for the control of particulate effluents from cotton lint cleaners. The results of these tests are tabulated in Table 11.11. The statistical evaluation of the data is shown in Table 11.12.

TABLE 11.11 Results for Testing the Effect of Water Pressure, Airflow, and Nozzle Operation on Efficiency

Airflow (cfm)	Nozzle operation	Efficiency
6400	N_A	95
	N_C	85
	N_D	67
	N_B	94
5120	N_A	90
	N_C	—[a]
	N_D	83
	N_B	96
2560	N_A	81
	N_C	80
	N_D	83
	N_B	95
4160	N_A	95
	N_C	88
	N_D	88
	N_B	88

[a] Power failure during first part of test.

The analysis of variance is summarized in Table 11.13. Each variable was then tested to determine whether it had a significant effect on efficiency at a significance level of $\alpha = 0.05$. The tabular F value that corresponds to this significance level is 4.76. As the calculated F values were all less than 4.76, the null hypothesis of no significant effect on efficiency was accepted for each variable over the range studied.

In this example, none of the major variables has any significant effect on performance of the air pollution control device. The most plausible explanation is that either the effects of the proper variables were not examined or that the ranges used in the experiment for these variables were too small.

TABLE 11.12   Latin Square for the Effect of Water Pressure, Airflow, and Nozzle Operation on Efficiency

| Nozzle operation | Collection efficiency, η, for airflow rate (cfm): | | | | |
	5120	2560	4160	6400	$\sum_i \eta_j$
N_A	90	81	95	95	361
N_C	90[a]	80	88	85	343
N_D	83	83	88	67	321
N_B	96	95	88	94	373
$\sum_i \eta_i$	359	339	359	341	1398

[a] Calculated efficiency.

TABLE 11.13   Analysis of Variance for Scrubber Tests

Source	d.f.	SS	MS	F
Mean	1	122,150.25	—	—
Nozzle operation	3	204.5	68.167	0.729
Airflow rate	3	90.75	30.25	0.323
Water pressure on	3	106.75	35.583	
nozzle taps		(98.75)[a]	(32.917)[a]	(0.352)[a]
Error	5[b]	467.75	93.55	—
Total	15[b]	122,823		

[a] Adjusted to account for bias from missing datum point.
[b] One d.f. subtracted for missing datum point.

11.7   GRECO-LATIN SQUARE

The concept of a Latin square design is easily extended to that of a Greco-Latin square. In this type of design equal numbers of treatments are tested on equal numbers of experimental units at different times or in different locations. The use of the Greco-Latin square results in a tremendous saving in time and money in that the total number of observations is greatly reduced. However, this design, as in the case of the Latin square, should be used with caution as no interactions are tolerated.

A Greco-Latin square can be obtained by superimposing one Latin square using Latin letters on another Latin square involving Greek letters in such a way that each Greek–Latin letter pair occurs once and only once. The construction of Greco-Latin squares has been studied and it has been shown that $n \times n$ Greco-Latin squares can be constructed provided that n is not 2 or 6.

Greco-Latin squares for the cases $n = 3$, 4, and 5 are given below:

Aα	Bγ	Cβ		Aα	Bγ	Cδ	Dβ
Bβ	Cα	Aγ		Bβ	Aδ	Dγ	Cα
Cγ	Aβ	Bα		Cγ	Dα	Aβ	Bδ
				Dδ	Cβ	Bα	Aγ

Aα	Bγ	Cε	Dβ	Eδ
Bβ	Cδ	Dα	Eγ	Aε
Cγ	Dε	Eβ	Aδ	Bα
Dδ	Eα	Aγ	Bε	Cβ
Eε	Aβ	Bδ	Cα	Dγ

In each case rows represent one variable, columns a second variable, Greek letters a third variable, and Latin letters represent treatments (a fourth variable). Thus a Greco-Latin square can be viewed as a design subject to three types of blocking. An $n \times n$ Greco-Latin square requires n^2 observations as compared with the n^4 observations that are required by a four-factor design where all factor-level combinations are present. The AOV table for a Greco-Latin square is similar to that of a Latin square with one added variable and will not be given here. It should be observed, however, that each of the four variables (factors) has $n - 1$ degrees of freedom associated with it, and consequently, the experimental error sum of squares has degrees of freedom $= n^2 - 1 - 4(n - 1) = (n - 1)(n - 3)$.

The model for a Greco-Latin square is

$$Y_{ij(kl)} = \mu + \rho_i + \gamma_j + \beta_k + \tau_l + \varepsilon_{ij(kl)}, \qquad i,j,k,l = 1,2,\ldots,n. \qquad (11.27)$$

such that

$$\sum \rho_i = \sum \gamma_j = \sum \beta_k = \sum \tau_l = 0.$$

Although i, j, k, and l range from 1 to n, only n^2 observations are present.

11.8 FACTORIAL EXPERIMENTS

We have considered a wide variety of experimental designs in Chapter 8 and in the present chapter. The type of design that has come to be called a factorial design has found great popularity in industrial investigations. We have already stated that the terms *independent variable, factor,* and *treatment* are often used interchangeably, although in the language of experimental design the term *factor* is preferred. Very often, it is desired to study various factors in a single experiment. In studying a chemical process it may be desired to study the factors pressure, temperature, and type of catalyst. Suppose further that one uses three different pressures, four temperatures, and two catalysts. This yields a three-factor experiment with three levels of the first factor, four levels of the second, and two levels of the third.

There are various reasons for using a factorial experiment. First, it allows us to study the interactions (or independence) of the various factors. For example, is there an added effect in using pressure in combination with several temperatures? Second is the gain in time and effort since all observations may be used to study the effects of all factors. For example, in the problem above, if one conducted three one-factor experiments, the observations in each experiment would yield information about the corresponding factor only and more experimental units would be needed to obtain the same accuracy as using the three-factor experiment. Finally, the results have wider application since each factor has been studied with varying levels of the other factors.

Since factorial experiments are used quite often, we now present some notation and topics pertaining to such experiments.

11.8.1 Main Effects

Consider a two-factor experiment with two levels of each factor. The two factors are denoted by A and B and the two levels of A and B are denoted by a_0, a_1 and b_0, b_1, respectively. The quantities $a_0 b_0, a_0 b_1, a_1 b_0,$ and $a_1 b_1$ denote the average response from the corresponding level combination as well as the particular combination. The experiment can be displayed as follows:

	a_0	a_1
b_0	$a_0 b_0$	$a_1 b_0$
b_1	$a_0 b_1$	$a_1 b_1$

The effect of factor A at level b_0 of factor B is given by $a_1 b_0 - a_0 b_0$. Similarly, the effect of factor A at level b_1 of factor B is given by $a_1 b_1 - a_0 b_1$. The *main effect* of factor A is then defined by

$$\frac{(a_1 b_0 - a_0 b_0) + (a_1 b_1 - a_0 b_1)}{2} = \frac{(a_1 - a_0)(b_1 + b_0)}{2} = [A]. \qquad (11.28)$$

In the same fashion, the main effect of B is given by

$$\frac{(a_1 + a_0)(b_1 - b_0)}{2} = [B]. \tag{11.29}$$

The effects of factor A at b_0 and b_1 will not be the same unless A and B are independent of each other, that is, unless A and B do not interact. A measure of interaction between A and B is given by

$$\frac{(a_1b_1 - a_1b_0) - (a_0b_1 - a_0b_0)}{2} = \frac{(a_1b_1 - a_0b_1) - (a_1b_0 - a_0b_0)}{2} = [AB]. \tag{11.30}$$

If there is no interaction, the terms in the numerators should be the same and the interaction would then be zero.

The effect of one factor at a given level of a second factor is sometimes called a *simple effect*. Thus $a_1b_0 - a_0b_0$ is a simple effect. A main effect of factor A is then the average simple effect averaged over all levels of the other factor. There is no interaction between A and B if all the simple effects corresponding to A are equal to the main effect. These ideas can be extended to higher-order factorial experiments [2,3,5,7].

The two- and three-way AOV models discussed in Chapter 8 correspond to factorial experiments. The model $Y_{ijk} = \mu + \alpha_i + \beta_j + (\alpha\beta)_{ij} + \varepsilon_{ijk}$ considered in Section 8.4.2 is a two-factor factorial with n replications. There are a levels of factor A and b levels of factor B. Tables 8.6 and 8.7 give the appropriate AOVs.

Example 11.17 The efficiency of multiple-effect evaporator systems can be expressed in terms of the steam economy, which is the ratio of the mass of total vapor (new steam) produced to the mass of the process steam used. As the number of effects increases, the steam economy increases. The following data were collected in a pilot-sized double-effect evaporator system as a 2^2 factorial experiment with $r = 3$ replications.

Replication	Feed = 6		Feed = 12		$Y_{i.}$
	Steam = 5	Steam = 10	Steam = 5	Steam = 10	
1	0.71	0.96	0.52	0.65	2.84
2	0.67	1.06	0.48	0.65	2.86
3	0.68	1.03	0.47	0.63	2.81
$Y_{.j}$	2.06	3.05	1.47	1.93	Variable totals
Feed rate, A	-1	-1	$+1$	$+1$	-1.71
Steam rate, B	-1	$+1$	-1	$+1$	$+1.45$
Interaction, AB	$+1$	-1	-1	$+1$	-0.53

For this system, the feed rates, A, were 6 and 12 lb_m/min, and the steam rates, B, were 5 and 10 lb_m/min. What is the significance of the variance components at the 95% confidence level? Note that the replicates have been used as blocks and that the data table has been rotated 90° counterclockwise in the display.

Solution   In this example we have arbitrarily assigned a value of -1 to the lower level of both variables and a value of $+1$ to the higher level of each variable. The analysis-of-variance table is developed using the following summations to determine the contribution of the variables and their interaction to the sum of squares for treatments.

Factorial effect	Multiplier for treatment total				Contribution to SS_{Tr}
	y_{00}	y_{10}	y_{01}	y_{11}	
A	-1	-1	$+1$	$+1$	$(\sum_{k=1}^{r}[A]_k)^2/4r$
B	-1	$+1$	-1	$+1$	$(\sum_{k=1}^{r}[B]_k)^2/4r$
AB	$+1$	-1	-1	$+1$	$(\sum_{k=1}^{r}[AB]_k)^2/4r$

We begin by determining $SS_M = Y_{..}^2/4r = 6.035008\bar{3}$ and

$$SS_T = \sum_i \sum_j \sum_k Y_{ijk}^2 = 6.4581.$$

The replication effect, SS_R, is calculated by $\sum_i Y_{i.}^2/4 - SS_M = 0.00031\bar{6}$. To calculate the effect totals, we use the lines of the multiplier section of the data table. The feed rate effect $= A = -2.06 - 3.05 + 1.41 + 1.93 = -1.71$. To obtain the steam rate effect $= B = -2.06 + 3.05 - 1.47 + 1.93 = 1.45$. Similarly, the interaction term $= AB = 2.06 - 3.05 - 1.47 + 1.93 = -0.53$. The resulting AOV table follows.

Source	d.f.	SS	MS
Mean	1	$6.035008\bar{3}$	
Replication	2	$0.00031\bar{6}$	
Feed rate, A	1	0.243675	0.243675
Steam rate, B	1	$0.175208\bar{3}$	$0.175208\bar{3}$
Interaction, AB	1	$0.023408\bar{3}$	$0.023408\bar{3}$
Error	6	0.007483335	0.001247223
Total	12	6.4851	

For this model II situation, we first test the interaction term (see Table 8.7) by calculating

$$F_{AB} = \frac{MS_{AB}}{MS_E} = 18.769,$$

which is larger than $F_{1,6,0.95} = 5.99$. Therefore, we reject the $H_0: \sigma^2_{\alpha\beta} = 0$ and proceed to test the A and B terms against the AB interaction. To evaluate the effect of the feed rate, we calculate $F_A = MS_A/MS_{AB} = 10.4098$, which when compared to $F_{1,1,0.95} = 161.$, proves to be insignificant. Similarly, $F_B = MS_B/MS_{AB} = 7.4848$ shows that for these pilot-scale data, neither main effect is significant. An appropriate response would be to repeat the entire experiment in the next semester, as the laboratory results do not agree with industrial observations.

11.8.2 Confounding

We discuss briefly an idea that is very useful, especially in factorial experiments. This is the idea of confounding. In working with randomized block experiments where each block is treated as a factor, it may not be possible to have a complete replicate of the experiment within one block. *Confounding* is used to reduce the size of blocks by sacrificing accuracy on the higher-order interactions. It reduces the experimental error by the use of more homogeneous blocks than those of complete replicates. The disadvantages of confounding are that it reduces the replications on the confounded treatment comparisons and increases the complexity of the calculations. If a term is confounded, it will be impossible to separate its effects from those of the effects of the blocks. The error term in a confounded factorial is made up of interactions between the treatments and the incomplete blocks. If confounding has been effective, the error for a third-order interaction (being composed of comparisons among blocks) will be greater than the error for the remainder of the experiments. The main effect of a factor will not be confounded; that is, main effects will be kept free of block effects if every block contains each level of the factor the same number of times.

With only one replication we cannot obtain an estimate of error from the interaction of treatments and blocks. If, however, certain higher-order interactions are negligible, their mean squares in the analysis of variance will behave exactly like components of the error mean square. Therefore, they can be used to provide an estimate of error. If some of these interactions happen to be quite large, the error mean square that is used as the experimental error mean square will overestimate the true error variance, and the fact that the interactions are large might not be discovered. Consider the case of the factorial design involving two levels of three different factors, as shown below.

Factorial effect	(1)	a	b	c	ab	ac	bc	abc
M (mean)	+	+	+	+	+	+	+	+
A	−	+	−	−	+	+	−	+
B	−	−	+	−	+	−	+	+
C	−	−	−	+	−	+	+	+
AB	+	−	−	+	+	−	−	+
AC	+	−	+	−	−	+	−	+
BC	+	+	−	−	−	−	+	+
ABC	−	+	+	+	−	−	−	+

a, b, c, and so on, are the second-level effects of corresponding factors A, B, C, and so on. (1) is used as the first level of all factors. The first level of any factor is indicated by a lack of the corresponding lowercase letter. In other words, bc in a 2^3 experiment means the first level of a and the second levels of b and c.

The simple effect of A at the second level of B is $ab - b$ and the simple effect of A at the first level of B is $a - (1)$. In like manner, the main effect of C is found from

$$4 \times \text{main effect of } C = c + ac + bc + abc - (1) - a - b - ab. \quad (11.31)$$

In a 2^n factorial experiment the interaction effect, say $(BCDE)$, is expressed by

$$(BCDE) = \frac{1}{2^{n-1}} [(a + 1)(b - 1)(c - 1)(d - 1)(e - 1)(f + 1)]. \quad (11.32)$$

In other words, write out the product $\prod_i (i - 1)$ for each letter i of the interaction and multiply by $\prod_i (i + 1)$ for each letter i not in the interaction. In this case, as before, the factorial main effects are A, B, C, and so on. The contribution of any effect to the sum of squares is the sum of squares due to that effect divided by $r2^n$, where r is the number of replications and n is the order of the factorial.

Example 11.18 One of your colleagues is interested in the effect of flow rate and temperature on the retention time for the gas chromatographic analysis of 1-nitropropane. Four flow rates (60, 90, 120, 150 mL H_2/min) and five temperatures (40, 50, 60, 70, 80°C) are to be investigated. Five columns are available, all having the same packing. To change from one flow rate to another requires about 2 h, while a temperature change can be accomplished in about 15 min. Consequently, a reason exists to do the five temperature treatments for each flow rate together.

(a) Give a description of how you would carry out the experiment physically. Five replications are needed.

(b) Give the model for the experiment, the consequent AOV, and a discussion.

Solution (a) This is a factorial (4 × 5) experiment with a total of 20 treatments. There are several ways in which the physical layout of those treatments can be approached, all of which are concerned with assigning the flow rates, temperatures, or replications to the five columns. One of the first decisions to be made is: "Do we want the same degree of sensitivity in estimating flow rate effects, and vice versa?" Because of the difficulty in changing flow rates, it is suggested that we may have to reduce the sensitivity of the test for flow rate effects and to compensate for this loss in precision by increasing the sensitivity of the test for temperature effects. Having decided to assign all five temperatures as subplots to a given main plot flow rate, we now have to decide whether to assign flow rate or replication to the five columns on the basis of sensitivity desired and information available from past experience. Let us assume that the five columns are five separate individual columns, in which case there is reason to suspect they may not be exactly alike and may not give the same results.

If we decide to assign flow rate to the columns, we can re-randomize these rates at the beginning of each replication, or we can maintain the same flow rate in a given column for all five replications. In the latter case, flow rate is completely confounded with column and it is not possible to calculate a valid error for testing flow rate effects. The only advantage of this is the experiment could be run in a shorter time, since no changes in flow rate would have to be made during the course of the experiment. In the first case, where flow rates are re-randomized at the beginning of each replication, the identity of the replications is not obvious, except with respect to time (which we assume has no effect on the results obtained). Because we have no reason for pairing replications, the AOV for this design would be as follows, where flow rates are considered main plots and temperatures are the subplots assigned to main plots in a split fashion:

Source	d.f.
Flow rates	3
Replications flow rates	16
Temperatures	4
$T \times FR$	12
Pooled error	64

(b) One would expect the design above to be less sensitive in testing flow rate effects because the error term has been inflated by an unidentified portion due to replication. An alternative design that is suggested for use pulls out this unidentified replication effect in such a way as to make replication correspond to a physical feature in the experiment. This is done by considering the columns as the replication and assigning flow rates and temperatures within a single column. The model for this design is

$$Y_{ijk} = \mu + \mathrm{FR}_i + R_j + \varepsilon_{ij} + T_k + (\mathrm{FR} \times T)_{ik} + \delta_{ijk},$$

$$i = 1,2,3,\ldots,n, \quad j = 1,2,\ldots,m, \quad k = 1,2,\ldots,h, \qquad (11.33)$$

for which the abbreviated analysis of variance is

Source	d.f.
Reps	4
Flow rates	3
Error a (ε_{ij})	12
Temperature	4
$T \times \mathrm{FR}$	12
Error b (δ_{ijk})	64

11.9 OTHER DESIGNS

11.9.1 Split-Plot Designs

In a two-factor experiment involving factors A and B, it may be desirable to get more precise information on factor B and its interaction AB with A than on factor A. For example, in a randomized complete block design, suppose that one has four levels of factor (treatment) A and four blocks. This is a one-factor experiment replicated over four blocks. Suppose there is a second factor B that is present in the experiment and that it is of primary importance. Moreover, its interaction with factor A is also important. Suppose further that factor B has two levels and that each level of B occurs with each level of A within each block. The configuration might look as follows:

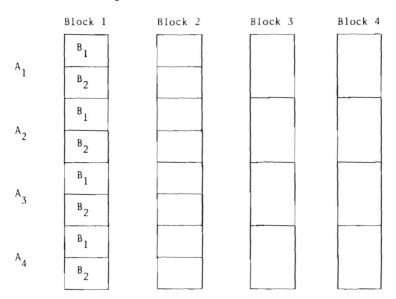

Such a design is called a *split-plot experiment* in randomized blocks. One can think of the main plots as being those that accommodate the levels of *A* and the subplots as those that accommodate the levels of *B*. Thus the main plots are "split" to accommodate the levels of the actually important factor *B*.

Example 11.19 In a plastics curing operation, temperature can readily be controlled at two levels and pressure at four levels. To speed up the experimental work needed to evaluate the effects of these two variables on the surface finish of the resulting products from the molding process. four molds can be used. In this experiment, the molds are the blocks. the whole plots are temperature level, and the subplots are pressure levels. This is done because the primary variable affecting surface finish and strength of the finished parts is pressure. Use of the split-plot design will give us some information about the temperature–pressure interaction.

The model for a split plot in randomized blocks is

$$Y_{ijk} = \mu + \alpha_i + \beta_j + \varepsilon_{ij} + \tau_k + (\alpha\tau)_{ik} + \delta_{ijk},$$
$$i = 1,2,\ldots,a, \quad j = 1,2,\ldots,r, \quad k = 1,2,\ldots,b. \qquad (11.34)$$

ε_{ij} are $\mathrm{NID}(0,\sigma_z^2)$, and δ_{ijk} are $\mathrm{NID}(0,\sigma_\tau^2)$, where β stands for blocks or replicates, α stands for main plot treatment, and τ stands for subplot treatment.

In split-plot designs certain main effects are confounded. This is in contrast to factorial experiments, where interaction effects are confounded. If the subplots are regarded as the experimental units, the subplot treatments (levels of factor B) are applied to blocks of two units (i.e., the main or whole plot). Therefore, the differences among these blocks (main plots) are confounded with the differences among the levels of the individual treatments. The number of degrees of freedom for experimental error mean square is smaller for whole-plot comparisons than for subplot comparisons. The average experimental error over all treatments is the same for randomized complete blocks as for split plots. Therefore, no net gain in precision results from use of a split-plot design. But we do have an increase in the precision of estimation of the effects of B and AB coming from a corresponding loss of precision of A. Therefore, a split-plot design should be used if B and AB are of greater interest than A or if the A effect cannot be measured on small amounts of materials. The analysis of variance for a split-plot design from a randomized complete block with r replications is shown in Table 11.14.

The sums of squares indicated in Table 11.14 are straightforward to compute and are given by

$$SS_R = \frac{\sum_{j=1}^{r} y_{\cdot j \cdot}^2}{ab} - \frac{y_{\cdots}^2}{rab},$$

$$SS_A = \frac{\sum_{i=1}^{a} y_{i \cdot \cdot}^2}{rb} - \frac{y_{\cdots}^2}{rab},$$

$$SS_{WPE} = \frac{\sum_{i=1}^{a} \sum_{j=1}^{r} y_{ij \cdot}^2}{b} - \frac{y_{\cdots}^2}{rab} - SS_R - SS_A,$$

$$SS_B = \frac{\sum_{k=1}^{b} y_{\cdot \cdot k}^2}{ra} - \frac{y_{\cdots}^2}{rab},$$

$$SS_{AB} = \frac{\sum_{j=1}^{r} \sum_{k=1}^{b} y_{\cdot jk}^2}{a} - \frac{y_{\cdots}^2}{rab} - SS_A - SS_B,$$

$$SS_{SPE} = SS_{TC} - SS_R - SS_A - SS_{WPE} - SS_B - SS_{AB}.$$

If the α effects and the τ effects are fixed, the F-ratios needed for testing H_0: all $\alpha_i = 0$, H_0: all $\tau_k = 0$, and H_0: all $(\alpha\tau)_{ik} = 0$ are $F = MS_A/MS_{WPE}$, $F = MS_B/MS_{SPE}$, and $F = MS_{AB}/MS_{SPE}$, respectively. Each F-statistic has the degrees of freedom corresponding to the mean squares used.

It is worth noting that the concept of splitting plots may be carried on for several stages, obtaining split-split-plot designs, and so on. Furthermore, split plots can be used in relation to other designs. For example, Table 11.15 gives a breakdown of the degrees of freedom of a split plot in a Latin square experiment with r replications. The corresponding sums of square are omitted but are easily obtained.

TABLE 11.14 Analysis of Variance for Split Plots in Random-ized Complete Block Experiment of r Replications

Source	d.f.	SS	MS
Whole plots			
Replicates (blocks)	$r - 1$	SS_R	MS_R
A	$a - 1$	SS_A	MS_A
Whole-plot error	$(r - 1)(a - 1)$	SS_{WPE}	MS_{WPE}
Subplots			
B	$(b - 1)$	SS_B	MS_B
AB	$(a - 1)(b - 1)$	SS_{AB}	MS_{AB}
Subplot error	$a(r - 1)(b - 1)$	SS_{SPE}	MS_{SPE}
Total (corrected)	$rab - 1$	SS_{TC}	

TABLE 11.15 Analysis of Variance for Split Plots in Latin Square Experiments with r Replications

Source	d.f.
Whole plots	
Rows	$a - 1$
Columns	$a - 1$
A	$a - 1$
Whole-plot error	$(a - 1)(a - 2)$
Subplots	
B	$b - 1$
AB	$(a - 1)(b - 1)$
Subplot error	$a(a - 1)(b - 1)$
Total	$a^2(b - 1)$

11.9.2 Incomplete Block Designs

In using a factorial design it may happen that the number of factors is so great that the size of the experiment becomes excessively large. Furthermore, the precision of the experiment, in terms of experimental error, may get out of hand. One way to get around the problem is to use fractional replication or to use what are called *incomplete block designs*. Our aim here is not to present a comprehensive treatment of such designs. Instead, we present a brief discussion of incomplete block designs and conclude with an example of a balanced incomplete block design.

Incomplete block designs are those designs arranged in groups smaller than a whole replication in order to eliminate heterogeneity to a greater extent than is possible, say, with a randomized complete block or with a Latin square design. The term *incomplete block* means that not all treatment combinations are present in every block. Incomplete block designs may be balanced or unbalanced. If they are balanced, every pair of treatments occurs once and only once in the same block. Thus all treatment pairs are compared with approximately the same precision. If the blocks cannot be balanced in separate replications, we have unbalanced incomplete blocks. If every treatment pair occurs once in the same row and once in the same column, the design is called a *lattice square*. For details as to the use and calculation methods pertaining to incomplete block designs, lattice squares, and other less frequently used experimental designs, interested students are advised to consult the text by Cochran and Cox [3], Chapters 9 through 13.

A balanced incomplete block can be viewed as an arrangement of treatments and blocks such that:

1. Every block contains k experimental units.
2. The number of treatments, t, exceeds the number of plots, k, in a block.
3. Every treatment appears in r blocks out of the possible b blocks.
4. Every pair of treatments appears in the same number of blocks (λ denotes this number).

Example 11.20 Because of the disappointing results obtained as described in Example 11.16, a second set of tests of that system were carried out in an attempt to determine the effect of water pressure and airflow rate on the scrubber-impinger's operating efficiency at higher inlet particulate concentrations. Table 11.16 tabulates the individual results from these tests. This series of tests was performed as a balanced incomplete block as shown in Table 11.17. In the complete block layout of Table 11.18, we see that there are $b = 3$ blocks, corresponding to 6400, 5120, and 4160, and $t = 3$ treatments, labeled 1, 2, and 3. Furthermore, $k = 2$, $r = 2$, and $\lambda = 1$. The entries in Table 11.17 are the averages over three replications.

As indicated in Table 11.18 the errors were pooled before testing the adjusted water pressure and airflow rate effects. Since the F values calculated were all less than the corresponding tabular values, the null hypotheses of no significant effects coming from airflow rates and water pressure changes on collection efficiency were accepted for each variable over the range studied.

TABLE 11.16   Results for the Analysis of the Effect of
Airflow Rate and Water Pressure on Efficiency at High
Inlet Particulate Concentration Levels

Airflow, ft³/min (block)	Water pressure level (treatment)	Efficiency ($\% \, \eta$)[a]
4160	2	53
6400	2	83
5120	3	96
4160	1	87
6400	3	83
5120	3	86
4160	2	97
6400	3	91
5120	3	92
4160	1	92
6400	2	97
5120	1	95
4160	2	90
6400	3	98
5120	1	92
4160	1	93
6400	2	98
5120	1	92

[a] η, collection efficiency.

TABLE 11.17   Balanced Incomplete Block Design for the Effect of
Airflow and Water Pressure on the Nozzle Taps on Efficiency[a]

Level of water pressure on each nozzle tap	Collection efficiency, η, for airflow rate (ft³/min):			$\sum_j \eta_j$
	6400	5120	4160	
3	91	91	--	182
2	93	—	80	173
1	—	93	91	184
$\sum_i \eta_i$	184	184	171	539

[a] Efficiencies shown are averages of three observations at each set of
conditions.

TABLE 11.18   Analysis of Variance for the Second Series of Air
Pollution Control Tests

Source of variation	d.f.	SS	MS	F
Mean	1	144,901.38	—	—
Air rate (crude)	—	174.12	—	—
Water pressure (adjusted)	2	131.44	65.72	0.535
Water pressure (crude)	— —	107.45	— —	— .
Air rate (adjusted)	2	198.11	99.055	0.806
Replication error	2	42.02	21.01	0.147
Experimental error	10	1,432.05	143.21	—
Pooled error	12	1,474.07	122.84	—
Total	18	146,705		

11.10   DESIGN EFFICIENCY

One of the greatest problems faced by the practicing engineer is that of
selecting the single design or designs for a particular application which are
most efficient and at the same time most effective for the particular needs.
As mentioned earlier, *effectiveness* refers to the ability of a design to answer
the questions that you have regarding a process, operation, effect, and so
on. *Efficiency* refers to how speedily they are answered with regard to
resource utilization and with what accuracy and precision. Most of all,
efficiency describes how well the existing data are utilized.

   Let us define the *variance* of a treatment mean in general as the mean
square for experimental error divided by the number of observations per
treatment. We can, by comparing the variance of a treatment mean for any
two designs, compare them with regard to their relative efficiency. Consider.
for example, the randomized complete block design where we have t
observations per treatment and b blocks. For these conditions the variance
of a treatment mean can be calculated as V_1. Now let us change the layout
of the entire RCB experiment. Instead of having t observations per treatment,
let us have w observations per treatment and let us have d blocks instead
of b blocks. Under these new conditions the variance of a treatment mean
is V_2. The relative efficiency of the original design compared to the revised
design is equal to $100V_2/V_1$. To compare the efficiency of the randomized
complete block design to the efficiency of a completely randomized design,
we must, at least on paper, manipulate the data involved so that it appears
that the treatments were assigned at random to the experimental units and

that there were no blocks involved at all. We would then estimate the experimental error mean square under these conditions of complete randomization and from it calculate the relative efficiency of the randomized complete block design with respect to the simulated completely randomized design. This is $100V_{CR}/V_{RCB}$. Comparisons of this type can be made for any two designs.

You will find that as the designs become more complex, the effectiveness of the design increases rapidly. We are able to find out more things from the results of the experiment. However, while this is going on, the relative efficiencies are changing in a manner that reflects the increase in accuracy of the estimate of experimental error and its corresponding decrease in sensitivity. With this goes a decreasing probability of obtaining a significant calculated F-value. A convenient calculational formula ([3], p. 112) to enable you to estimate the mean square for experimental error for a simulated completely random design from known randomized complete block data is

$$V_{CR} = \frac{(b - 1)MS_B + b(t - 1)MS_E}{bt - 1}. \tag{11.35}$$

The estimation ([3], p. 127) of the mean square for experimental error for a simulated randomized complete block design from known Latin square data is found from

$$V_{RCB} = \frac{(m - 1)MS_R + (m^2 + 1)MS_E}{m^2 - 1}. \tag{11.36}$$

Of course, the calculation of relative efficiencies after you have performed the experiment is foolish: It profits you absolutely nothing! What one should realize is that some designs are more effective and more efficient than others in certain applications. Learn them well and calculation of efficiency will be a problem of minor concern.

11.11 ANALYSIS OF COVARIANCE

In discussing analysis of variance and experimental designs, we have seen how, by incorporating an additional factor or by the use of blocks, effects that are of no interest in an experiment can be eliminated. This results in a smaller denominator in the F-ratio, thus resulting in a test that is more sensitive to the effects which are of interest. That in turn increases the power of the test (i.e., the probability of rejecting a false hypothesis). The analysis of covariance is another technique that may be used to accomplish the same objective.

As an example, suppose that one is interested in comparing several teaching methods. A number of students are randomly assigned to each teaching method. The dependent variable is Y, the final score obtained by the students on a final examination. The effect of student IQ on the final score can be eliminated by using IQ as a covariate, that is, by adjusting the final score for IQ.

As a second example, suppose that one is studying the carbon content Y (in grams) in four types of steel. There are five batches, not all of the same weight, for each type of steel. The determinations of carbon content are to be made by one analyst. The weight of each batch, X, can be used as a covariate, and differences in carbon content can be studied by an analysis of covariance. (Note that in this case, however, one also could do a one-way AOV on weight percent; see Problem 8.8.)

Consider a one-way AOV where the dependent (response) variable is Y and the independent (concomitant) variable is X. The usual AOV model of the Y values is

$$Y_{ij} = \mu + \alpha_i + \varepsilon_{ij}, \qquad i = 1,2,\ldots,t, \quad j = 1,2,\ldots,n_i.$$

The analysis-of-covariance model in a completely randomized design is

$$Y_{ij} = \mu + \alpha_i + \beta(X_{ij} - \bar{X}) + \varepsilon_{ij}, \quad i = 1,2,\ldots,t, \ j = 1,2,\ldots,n_i. \quad (11.37)$$

The model in (11.37) can be written as $Y_{ij} = v + \alpha_i + \beta X_{ij} + \varepsilon_{ij}$, with $v = \mu - \beta\bar{X}$. The form in (11.37) is more customary and is the one discussed here. It is assumed that the ε_{ij} are NID$(0,\sigma^2)$.

From (11.37) it is seen that $E(Y_{ij}) = \mu_{ij} = \mu + \alpha_i + \beta(X_{ij} - \bar{X})$. Thus the means of the Y values for each fixed i lie on a straight line. The slope of the line is constant for each of the t groups; there may be t different lines due to different values for the α_i, but they are parallel.

The usual assumptions as used for the completely randomized design are made. An additional and important assumption is that the values of X are not affected by the treatments. This may not always hold in practice, so the analyst should be aware of this assumption and take care in interpreting the results of a covariance analysis.

The results of a covariance analysis in the one-way AOV are summarized in Table 11.19. The notation in the table is as follows:

$$n = \sum_{i=1}^{t} n_i,$$

$$T_{xx} = \sum_{i=1}^{t}\sum_{j=1}^{n_i} (\bar{X}_{i.} - \bar{X}_{..})^2 = \sum_{i=1}^{t} \frac{X_{i.}^2}{n_i} - \frac{X_{..}^2}{n},$$

$$T_{yy} = \sum_{i=1}^{t} \frac{Y_{i.}^2}{n_i} - \frac{Y_{..}^2}{n}, \qquad T_{xy} = \sum_{i=1}^{t} \frac{X_{i.}\,Y_{i.}}{n_i} - \frac{X_{..}\,Y_{..}}{n},$$

TABLE 11.19 Results of Covariance Analysis in the One-Way AOV

Source	d.f.	SS_{x^2}	SS_{xy}	SS_{y^2}	SS'_y	d.f.	MS'_y	F
Treatment	$t-1$	T_{xx}	T_{xy}	T_{yy}	SS'_{yTr}	$t-1$	MS'_{yTr}	$\dfrac{MS'_{yTr}}{MS'_{yE}}$
Error	$n-t$	E_{xx}	E_{xy}	E_{yy}	SS'_{yE}	$n-t-1$	MS'_{yE}	
Total	$n-1$	$S_{xx}=T_{xx}+E_{xx}$	$S_{xy}=T_{xy}+E_{xy}$	$S_{yy}=T_{yy}+E_{yy}$	SS'_{yT}			

$$E_{xx} = \sum_{i=1}^{t} \sum_{j=1}^{n_i} (X_{ij} - \bar{X}_{i\cdot})^2 = \sum_{i=1}^{t} \sum_{j=1}^{n_i} X_{ij}^2 - \sum_{i=1}^{t} \frac{X_{i\cdot}^2}{n_i},$$

$$E_{yy} = \sum_{i=1}^{t} \sum_{j=1}^{n_i} Y_{ij}^2 - \sum_{i=1}^{t} \frac{Y_{i\cdot}^2}{n_i}, \qquad E_{xy} = \sum_{i=1}^{t} \sum_{j=1}^{n_i} X_{ij} Y_{ij} - \sum_{i=1}^{t} \frac{X_{i\cdot} Y_{i\cdot}}{n_i},$$

$$SS'_{yT} = S_{yy} - \frac{S_{xy}^2}{S_{xx}},$$

$$SS'_{yE} = E_{yy} - \frac{E_{xy}^2}{E_{xx}},$$

$$SS'_{yTr} = SS'_{yTr} - SS'_{yE} = T_{yy} - \frac{S_{xy}^2}{S_{xx}} + \frac{E_{xy}^2}{E_{xx}}.$$

The F-ratio to use in testing $H_0: \alpha_1 = \alpha_2 = \cdots = \alpha_t = 0$, after adjusting for the X values, is

$$F = \frac{SS'_{yTr}/(t-1)}{SS'_{yE}/(n-t-1)} = \frac{MS'_{yTr}}{MS'_{yE}}.$$

The hypothesis $H_0: \alpha_1 = \alpha_2 = \cdots = \alpha_t = 0$ is rejected at the significance level α if $F \geq F_{t-1,n-t-1,1-\alpha}$. This is a test of the hypothesis that there are no differences among the effects of the t treatments after adjusting for the concomitant variable (covariate) X.

The estimate of β is

$$\hat{\beta} = \frac{E_{xy}}{E_{xx}}$$

$$= \frac{\sum_{i=1}^{t} \sum_{j=1}^{n_i} (X_{ij} - \bar{X}_{i\cdot})(Y_{ij} - \bar{Y}_{i\cdot})}{\sum_{i=1}^{t} \sum_{j=1}^{n_i} (X_{ij} - \bar{X}_{i\cdot})^2}.$$

When doing a covariance analysis, it is assumed that $\beta \neq 0$. If $\beta = 0$, it is not necessary to analyze the data by covariance. Often, the experimenter will want to test the null hypothesis $H_0: \beta = 0$. This may be done by computing

$$F = \frac{E_{xy}^2/E_{xx}}{SS'_{yE}/(n-t-1)} = \frac{E_{xy}^2/E_{xx}}{MS'_{yE}}$$

and rejecting $H_0: \beta = 0$ if $F \geq F_{1,n-t-1,1-\alpha}$.

In addition to the F-test on treatment differences adjusted for X, one is often interested in the adjusted treatment means and their standard errors. The adjusted treatment means are

$$\text{adj.} \ \bar{Y}_{i\cdot} = \bar{Y}_{i\cdot} - \hat{\beta}(\bar{X}_{i\cdot} - \bar{X}_{\cdot\cdot}), \qquad i = 1,2,\ldots,t,$$

and the estimated variances are

$$\hat{V}(\text{adj. } \bar{Y}_{i.}) = MS'_{yE}\left[\frac{1}{n_i} + \frac{(\bar{X}_{i.} - \bar{X}_{..})^2}{E_{xx}}\right], \qquad i = 1, 2, \ldots, t.$$

The estimate of the variance of the difference of two adjusted means is

$$\hat{V}(\text{adj. } \bar{Y}_{i.} - \text{adj. } \bar{Y}_{j.}) = MS'_{yE}\left[\frac{1}{n_i} + \frac{1}{n_j} + \frac{(\bar{X}_{i.} - \bar{X}_{j.})^2}{E_{xx}}\right].$$

As has been seen, analysis of covariance combines analysis of variance with regression analysis. Analysis of covariance may be used with many other designs besides the one-way AOV; however, we will not elaborate further. We now present an example of analysis of covariance in a one-way AOV.

Example 11.21 Three lab groups varied the tube-side water rate X in a laboratory-scale, two-pass, double-pipe heat exchanger to determine the local inside heat-transfer film coefficient Y as a function of water rate. The three groups conducting this experiment did not use the same steam rates (STEAM) for this experiment. To evaluate the effect of different steam rates, we perform an analysis of covariance for these data using water rate as the covariate. The SAS program (*SAS User's Guide: Statistics, Version 5* [6]) is presented below, followed by the program output.

(put initial JCL cards/statements here)

DATA DP;

INPUT STEAM   X   Y;

CARDS;

0.5	749	241
0.5	749	259
0.5	1498	335
0.5	1498	317
0.5	1498	344
0.5	2246	380
0.5	2246	369
0.5	2246	370
0.5	2995	453
0.5	2995	433
0.5	3744	497
2.0	1470	593
2.0	4410	550
2.0	5880	670
2.0	7350	700

(continued)

2.0	8090	753
2.0	8820	766
2.0	1470	321
2.0	2940	442
2.0	4410	594
2.0	5880	647
2.0	5880	662
5.0	1470	305
5.0	4410	540
5.0	5880	690
5.0	7350	710
5.0	8090	771
5.0	8820	794
5.0	1470	300
5.0	2940	412
5.0	4410	641
5.0	5880	719
5.0	5880	722

PROC GLM;

CLASS STEAM;

MODEL Y = STEAM X/SOLUTION;

(put final JCL cards/statements here)

GENERAL LINEAR MODELS PROCEDURE

DEPENDENT VARIABLE: Y

SOURCE	DF	SUM OF SQUARES	MEAN SQUARE	F VALUE
MODEL	3	898732.53734567	299577.51244856	87.99
ERROR	29	98733.52326040	3404.60425036	PR > F
CORRECTED TOTAL	32	997466.06060606		0.0001

SOURCE	DF	TYPE III SS	F VALUE	PR > F
STEAM	2	15326.16040168	2.25	0.1234
X	1	471760.65855779	138.57	0.0001

PARAMETER		ESTIMATE	T FOR H0: PARAMETER=0	PR > |T|	STD ERROR OF ESTIMATE
INTERCEPT		241.50623113 B	11.83	0.0001	20.41649812
STEAM	2	60.14348519 B	2.04	0.0503	29.44223788
	5	51.59803065 B	1.75	0.0903	29.44223788
	0.5	0.00000000 B	.	.	.
X		0.05971472	11.77	0.0001	0.00507287

Note: The $\mathbf{X'X}$ matrix has been defined as singular and a generalized inverse has been employed to solve the normal equations. The estimates above represent only one of many possible solutions to the normal equations.

Estimates followed by the letter B are biased and do not estimate the parameter but are "the best linear unbiased estimates" for some linear combination of parameters (or are zero). The expected value of the biased estimators may be obtained from the general form of estimable functions. For the biased estimators, the STD ERR is that of the biased estimator and the T value tests H_0: E(biased estimator) $= 0$. Estimates not followed by the letter B are "the best linear unbiased estimates" for the parameter.

To test for differences among levels of the steam rate adjusted for the covariate (water rate) we use the type III SS to obtain

$$F = \frac{MS'_{Tr}}{MS'_{yE}} = \frac{15,326.16040/2}{98,733.52326/29} = 2.25.$$

As $F < F_{2,29,0.95} \cong 3.33$, we accept the hypothesis of no treatment (steam rate) effect as adjusted for the covariate (water rate). To test the null hypothesis of no effect of water rate on the heat-transfer coefficient, we can use either the type III SS to calculate

$$F = \frac{E_{xy}^2/E_{xx}}{MS'_{yE}} = \frac{471,760.65856/1}{98,733.52326/29} \cong 138.57 \text{ vs. } F_{1,29,0.95} \cong 4.19$$

or

$$T = 11.77 \text{ vs. } t_{29,0.975} = 2.045.$$

By either test, the null hypothesis that the water rate does not affect the heat-transfer coefficient (H_0: $\beta = 0$) is rejected, as it should be. Regardless of flow regime (laminar, transition, or turbulent), the local inside heat-transfer coefficient is known to be affected by the inside fluid (water in this case) rate. This can be shown by the Seider–Tate, Hausen, and Dittus–Boelter equations, respectively.

PROBLEMS

11.1 An experiment is planned that will measure the effect of applying various amounts of a chemical which will partially close the stomata on plant leaves on the transpiration rate. As the research engineer responsible for the pilot plant producing this new material, you are working closely with the plant scientist, who is running the field tests of your materials. The usual transpiration rate for sorghum is 5 g/dm^2 h for the wind velocity, degree of cloudiness, relative humidity, and dry-bulb temperature on the South Plains of Texas. This rate may fluctuate by 0.5 g/dm^2 h under normal daytime conditions. How many replicates will be required to detect, 9 times out of 10, a difference of 1.5 g/dm^2 h in transpiration rate? The significance level you and the plant physiologist have agreed to use is 5%.

11.2 Another experiment related to water conservation is concerned with the application of naturally occurring lipids to the soil between irrigated plant rows to cut down on surface evaporation. The effect of the control agent can be determined as a function of water content 6 in. below the surface. The average moisture content of an untreated soil is $25 \pm 1.4\%$. If a difference in water content of 2% moisture is to be detected four out of five times, how many replications will be needed? The significance level is 5%.

11.3 Using the data of Problem 7.29, how many observations are needed for a 5% test with a 80% probability of finding a significant result?

11.4 Using the data of Problem 7.7, if $\alpha = 0.01$ and a 95% probability of finding a significant result is desired, how many observations are needed? For $\alpha = 0.05$? What do you conclude regarding this experiment?

11.5 A large oil company has a water-flooded oilfield with a highly porous formation. The average overall ratio of oil to produced water is 0.05. This ratio varies greatly from well to well. In the current series of tests on well 16, the average value of the oil/water ratio was 0.0639 with a variance of 0.0035. Ten tests were performed. The required values of α and β were 0.05 and 0.20, respectively. Were an adequate number of tests conducted? Prove your work by showing all calculations.

11.6 Use the following data to determine whether surface tension as measured by the capillary rise method for 1-butanol is a function of concentration (mol/L) for aqueous solutions. The capillary rises below are given in centimeters for four replicates.

0.8	0.6	0.45	0.338	0.253	0.190
2.2	2.7	3.2	3.6	3.6	4.3
2.1	2.8	3.2	3.4	3.7	4.3
2.3	2.6	3.1	3.5	3.8	4.2
2.3	2.7	3.1	3.4	3.8	4.3

11.7 A study was conducted to determine the effect of catalyst pore size on gasoline yield. Seven independent observations were made for each of four pore sizes (μm). The yield data are given as fractions of the maximum possible yield for seven replicates.

3	7	9	12
0.70	0.65	0.59	0.55
0.69	0.66	0.60	0.54
0.68	0.64	0.61	0.53
0.71	0.63	0.58	0.55
0.71	0.67	0.57	0.56
0.70	0.68	0.60	0.53
0.68	0.65	0.61	0.53

Does the size of the pores in the zeolite catalysts affect the gasoline yield?

11.8 Heat-transfer calculations were performed on a steam-jacketed reactor in a polypropylene plant to determine the effect of viscosity on the overall heat-transfer coefficient U (But/hr ft^2 °F). The following heat-transfer coefficients were measured at four different viscosities, measured in centipoise (cP).

2100	1800	900	100
9	13	22	30
10	12	23	36
9	15	22	36
11	19	23	38
12	17	24	30

Does the viscosity significantly affect the overall heat-transfer coefficient? Use $\alpha = 0.05$.

11.9 The surface tension of aqueous 1-butanol solutions was measured in a 0.35-mm-ID capillary tube at 22.7°C. The height of the meniscus was measured (cm) in four replications for each concentration (mol/L).

0.8	0.6	0.45	0.34	0.25	0.19	0.14	0.11
3.00	4.15	4.05	5.07	5.11	5.59	5.50	6.00
3.25	4.16	4.09	5.05	5.60	5.60	5.49	6.00
3.30	4.18	4.09	5.09	5.11	5.60	5.51	6.01
3.30	4.19	4.08	5.06	5.11	5.61	5.51	6.00

Does concentration affect surface tension? Use $\alpha = 0.05$.

11.10   In a study of invertase activity in the developing embryonic chicken, duodenal loop samples were obtained from silver laced Winedot-Eastern Ray (*A*) and silver laced Winedot-AVA Conda (*B*) embryonic chicks from age 13 to 17 days. The duodenal tissue was homogenized and assayed for specific invertase activity.

	Age (days)				
Variety	13	14	15	16	17
A	56.3	60.5	64.3	86.1	97.5
B	53.4	61.2	64.5	87.0	94.2

Does age influence invertase activity? Is there any significant variation in invertase activity between the two varieties of chicken?

11.11   Results of a laboratory determination for the cotton dust weight (μg) present on air sampling filters at a yarn mill are given below using five vertical elutriators during shift *A* and shift *B*. The data were collected at two different sites in the mill: the warehouse and the bale opening area.

	Shift	
Site	*A*	*B*
Warehouse	610	350
	830	130
	630	380
	660	460
	490	400
Bale opening	430	170
	380	250
	690	270
	500	360
	330	370

Is there a significant ($\alpha = 0.05$) difference in the results between shifts? Does the site location make a significant difference?

11.12 The sampling program described in Problem 11.11 was continued for the other two shifts, with the results (in µg) shown below. It should be noted that the samples from all four shifts were obtained over a 6-h period and that the elutriator flow rates ranged from 7.43 to 7.59 L/min, as specified by 29CFR1910.1043.

	Shift	
Site	C	D
Warehouse	515	635
	635	355
	485	855
	465	655
	405	685
Bale opening	845	870
	765	815
	605	495
	610	530
	670	460

Prepare the analysis of variance table for these new data and interpret the results.

11.13 Using all the data of Problems 11.11 and 11.12, prepare the analysis of variance table and determine, at the $\alpha = 0.05$ level, the significance of site, shift, and any possible interaction between the two.

11.14 Ten different paints are to be evaluated for use in a chemical plant. To determine the effectiveness of each paint as an atmospheric corrosion inhibitor, 40 pieces of 18WF60 structural framework are selected on the plant site for experimental purposes. Ten test sections are located near the methanol oxidizers, 10 near the coal-fired powerhouse, 10 near the HNO_3 chamber process, and the last 10 near the synthetic NH_3 compressors. The following data were obtained when the paints were assigned at random within each of the 10 locations (data represent a function of discoloration, weight loss, and cracking after 10 months).

Location	Paint									
	1	2	3	4	5	6	7	8	9	10
1	2	5	8	6	1	3	8	6	4	4
2	3	4	7	5	2	5	8	12	5	4
3	3	5	10	5	1	7	7	2	6	2
4	5	5	9	2	2	8	8	5	3	3

(a) What does this design accomplish?

(b) Write the AOV showing expected mean squares for this experiment.

(c) What is the standard error of a treatment mean for these data?

(d) Are the paints equally effective?

(e) Based on your results in part (d), what would you do next?

11.15 A new engineer would like to investigate the effect of four temperatures on the yield of a certain product. Since he would like his experiment to be complete, he used four different pressures as blocks in an RCB design. In the time available, he could make 16 replications, so he assigned all four temperatures to each block. Give the model for this experiment and define all terms in it. What modifications, if any, would you make in the design of this experiment? If you recommend any changes, give the resulting model and tell specifically how each change is beneficial.

11.16 A plant-scale experiment is being designed to compare the effects of a number of changes in reaction conditions (temperature, pressure, catalyst concentration, regeneration frequency, and throughput) on the yield of ethylene glycol via the du Pont process. Because of the various process lags involved, considerable quantities of raw materials will be used. Give two general design approaches to this problem and discuss them as you would with the production superintendent.

11.17 A randomized block experiment was conducted to investigate the effects of the concentration of detergent (D) and sodium carbonate (S) on the working and suspending power of a solution. The following data refer to performance on working tests, with the higher figure indicating better performance. Each figure is the total of three observations. The levels were equally spaced.

	D_1	D_2	D_3	Sum
S_1	437	673	925	2035
S_2	711	1082	1157	2950
S_3	814	1146	1123	3083
	1962	2901	3205	8068

Compute the analysis of variance and test the significance of the D and S terms using the appropriate error terms. The total sum of squares is 2,875,729.

11.18 Given the following analysis of variance from a randomized complete block experiment:

Source	d.f.	MS
Blocks	5	81
Treatments	6	190
Error	30	40

(a) What is the variance of a treatment mean?
(b) What is the variance of the difference between any two treatment means?

11.19 A certain pilot-plant operation involves the saponification of a particular fatty alcohol. This reaction can be carried out in either of two different reactors. As the engineer in charge of this project, you feel that the raw material from one supplier is contaminated. Before you left for vacation. you told your technicians to run six batches from each of the two possible suppliers through the process under normal conditions. This will give you six replications on each of the two treatments (fatty alcohol sources) for the t-test you have in mind. Your summer technicans decided to speed things up and run three batches through each reactor. Now that you are back from your trip, how are you going to analyze the data that your technicians have obtained? Include in your answer the following: model, hypotheses to be tested and the necessary F-ratios, and the types of effects (fixed, random, mixed) being evaluated.

11.20   A test of paint primers used for corrosion inhibition gave the following results in a RCB experiment:

Source	d.f.	MS
Blocks	9	241.59
Primers	4	269.67
Error	36	97.41

Are differences in primers significant at the 5% level?

11.21   Data were obtained in a RCB experiment in which the treatments were a feeding arrangement in a triple-effect evaporator system. The effects of feed direction [forward, backward, mixed (2,3,1), mixed (2,1,3)] were measured in terms of steam economy, lb evaporated/lb steam used. The results were as follows:

	Treatment			
Block	T_1	T_2	T_3	T_4
B_1	3.6	4.1	3.0	—
B_2	3.8	3.7	3.3	4.5
B_3	3.0	3.9	3.4	4.0
B_4	4.0	4.2	3.8	4.8

For this experiment, the total Δt allowed was used to group the data into blocks. Prepare the analysis-of-variance table for tese data and test the hypothesis that the true treatment means are equal at the 5% significance level.

11.22   In an experiment to determine the effect of acid pickling in the galvanizing process, 10 samples were used. Four samples were dipped in the pickling solution once, three samples were dipped twice, two samples were dipped three times, and the other sample was dipped four times. The number of dippings was randomly assigned to the samples. The following data were

obtained, where the quality index is a measure of adherence of the zinc in
the next step in the process.

Sample (arbitrary numbering)	Number of dippings	Quality index
1	1	2
2	1	2
3	1	3
4	1	5
5	2	5
6	2	6
7	2	7
8	3	9
9	3	11
10	4	10

The uncorrected sum of squares for the quality index is 454. $SS_M = 360$.

(a) Set up the analysis of variance for this experiment.
(b) Compute the relation between number of dippings and quality index and test the regression coefficients.

11.23 Discussions with the design engineer revealed that half the peel strength values presented in Problem 8.26 (those marked with an asterisk) were obtained when using samples of one thickness. The other values (unmarked) were obtained using samples of a different thickness. You are required to include this new variable (sample thickness) in your analysis of variance.

(a) Give the appropriate model for this three-way classification.
(b) Prepare the complete analysis-of-variance table.
(c) State and test the null hypotheses involved.

11.24 As part of your senior project, you wish to study the effects of temperature, pressure, and molar feed rate ratios (liquid to gas) on the efficiency of a formaldehyde absorber. Three pressures, two temperatures, and three feed rate ratios are to be evaluated. Six small bench-scale absorbers are available and suitable for operation at any of the pressures desired. Temperature can be controlled at either level on each. Evaluation of any particular feed rate ratio will require one afternoon, allowing for startup and shutdown. In this semester you estimate having available an average of three afternoons per week for 10 weeks. Then you must stop all experimental results, correlate and analyze your data, and write up the report. Design an experiment to obtain the maximum amount of information in the time available. Discuss the reasons for your choice of designs. Describe how you would set up the experiment and analyze the data.

11.25 Given the following analysis of variance for fixed treatment levels:

Source	d.f.	MS
Among treatments	9	400
Among experimental units within treatments	70	80
Among samples within experimental units	320	5

(a) How many samples were used for each experimental unit?
(b) What are the expected mean squares for these sources of variation?
(c) Test the hypothesis of equality of treatment effects at $\alpha = 0.05$.
(d) What is your estimate of the difference between any two treatment means?
(e) Estimate the efficiency of this design relative to one in which 10 experimental units per treatment and three samples per experimental unit are used.

11.26 You designed an experiment to measure the wearing quality of sweaters as a function of the amount of drawing and crimping used in the process where the tow was originally made. This experiment started out as a Latin square with the amount of crimp as columns and the degree of drawing as rows. The treatments applied are surface finishes on the finished yarn. The original design is shown below:

Draw		Crimp		
	C_1	C_2	C_3	C_4
R_1	A	B	C	D
R_2	D	A	B	C
R_3	C	D	A	B
R_4	B	C	D	A

When the experiment was over, you realized from the lab notebook that your technician had applied treatment A in the R4C2 position and treatment C in the R4C4 position. This is just the reverse of what you had intended. To further complicate matters, you expect a much stronger gradient among rows than among columns. Indicate the various possible methods of data analysis. Discuss the relative merits of each.

11.27 In an industrial experiment investigating the effects of four different processes (I, II, III, IV), you believe you can eliminate any technician and/or day bias by using a Latin square design. Weekdays are rows and technicians are used as columns. Your design is laid out as below.

Day		Technician		
	1	2	3	4
M	I	II	III	IV
Tu	IV	I	II	III
Th	III	IV	I	II
F	II	III	IV	I

You showed this design to your supervisor. He though that a day–technician interaction exists. Does your design take this interaction into consideration? If so, how? For the design given, show the model used. defining all terms. Also show the source and corresponding degrees of freedom for all error sources in an AOV table.

11.28 Give at least four methods of increasing the precision of an experiment. Explain briefly the principle on which each is based.

11.29 In a *nested* sampling arrangement for the determination of percent solids from the viscosity of a polyester spinning solution, duplicate determinations were made on two samples from each of three Marco mixers. A partial analysis of variance is given below for the data obtained.

Source	Sum of squares
Between mixers	25.9
Between samples from mixers	0.42
Between determinations per sample	0.66

(a) Complete the analysis of variance showing the degrees of freedom, mean square, and expected mean square for each treatment.

(b) Compute the variance of the overall mean on a per determination basis.

(c) If the cost of the experiment were the same, what would have been the effect of taking four samples from each mixer and making only one determination of solids content per sample?

11.30 An experiment was conducted to determine the relative effectiveness of three different solvents in a vapor degreaser. Since cleaning efficiency is a function not only of the solvent used but also of the depth of immersion of the parts in the vapor and the amount of stripped grease already in the vat, the experiment was conducted as a $3 \times 3 \times 3$ factorial with two observations per factorial combination. The resulting sums of squares from the analysis of degreasing efficiency are given:

Source	Sum of squares
Solvent (S)	400
Immersion (I)	640
Stripped grease concentration (C)	1660
$S \times I$	620
$S \times C$	720
$I \times C$	420
$S \times I \times C$	1160
Error	2700
Total	8320

(a) Complete the analysis-of-variance table.
(b) Discuss the factors involved (S, I, C) as to whether they are fixed, random, or either and present the appropriate models and expected mean squares.
(c) This experiment could have been performed as a Latin square with stripped grease concentration as rows, immersion depth as columns, and solvent type as treatments. Discuss the advantages and disadvantages of such a design relative to the one actually used.

11.31 The following experiment was conducted in the University Hospital at Iowa City. The problem was to investigate the influence of an anesthetic on the consumption of oxygen by various brain tissues. Eight dogs were used in the experiment with a two-stage operation being performed, samples of the three brain tissues being removed before and after administration of the anesthetic. Each tissue sample was used to make up two $1 - g$ subsamples, and these were placed in separate flasks for the measurement of O_2 consumption. Readings of O_2 consumptions were taken every 30 min over a 4.5-h period, giving, in all, eight readings per subsample. Thus we have an experiment including 8 dogs, 3 brain tissues, 2 treatments (with and without anesthetic), 2 samples per treatment, 8 observations in time, and a total of 768 observations. Write out your version of the analysis of variance for this experiment showing the appropriate degrees of freedom; consider dogs as replicates. Indicate the proper F-tests for brain tissues, treatments, and the interaction of the two factors.

11.32 In an experiment to measure the variability in braking strengths of tire cords made at two different plants, Akutowicz and Truax [1] obtained the data given in the table.* (The numbers are $\frac{1}{10}$-lb$_f$ deflections from 21.5 lb$_f$.)

	0 yd		500 yd		1000 yd		1500 yd		2000 yd		2500 yd	
Adjacent breaks	1	2	1	2	1	2	1	2	1	2	1	2
Bobbin plant I												
1	−1	−5	−2	−8	−2	3	−3	−4	0	−1	−12	4
2	1	10	1	2	2	2	10	−4	−4	3	4	8

(continued)

* Data reprinted by permission; copyright American Society for Quality Control, Inc.

(Continued)

	0 yd		500 yd		1000 yd		1500 yd		2000 yd		2500 yd	
3	2	−3	5	−5	1	−1	−6	1	2	5	7	5
4	6	10	1	5	0	5	−2	−2	1	1	5	9
5	−1	−8	5	−10	1	−5	1	−4	5	−5	3	6
6	−1	−10	−9	−8	−2	2	0	−3	−8	−1	−2	−4
7	−9	−2	5	−2	7	−2	−2	−2	−1	2	10	5
8	0	2	−5	−2	5	3	10	−1	4	1	7	−1
Bobbin												
plant II												
1	10	8	−5	6	2	13	7	15	17	14	18	11
2	9	12	6	15	15	12	18	16	13	10	9	11
3	0	8	12	6	2	0	5	4	18	8	6	8
4	5	9	2	16	15	5	21	18	15	11	18	15
5	−1	−1	11	19	12	10	1	20	13	9	4	6
6	7	16	15	11	12	12	8	12	22	11	12	21
7	−5	1	−2	10	12	15	2	13	10	10	7	5
8	10	9	10	15	9	16	12	11	13	20	11	15

In this experiment, bobbins of tire cord were randomly selected in two groups of eight from regular production runs at two different plants. The breaking strength was recorded for adjacent pairs of cord at 500-yd intervals over the length of each bobbin to obtain these data.

(a) Write the model involved and the corresponding expected mean squares.

(b) Perform the complete analysis of variance to determine the effects of bobbins, spacing, position, and plant on the quality of tire cord as measured by braking strength. Interpret the results.

11.33 As a result of an economic analysis of finishing methods for automotive transmission gears, Yokota [9]* showed that the grinding technique was less expensive than the older honing method. Qualities of the finished products were equal. In an experiment to determine the optimum operating conditions for grinding, grinding worm movement, depth of cut, feed rate, and stock for finishing were varied over two levels each. The results of this experiment are shown below as they affected the noise abatement rating. Higher ratings (coded given) indicate lower transmission gear noise levels.

---

* Data reprinted by permission; copyright American Society for Quality Control, Inc.

		Factor			Coded rating	
Experi- ment	Grinding worm movement	Depth of cut (mm/rev)	Feed rate (mm/rev)	Stock for finishing (mm)	x_1	x_2
1	Fixed	0.020	1.8	0.020	6	8
2	Fixed	0.020	2.2	0.040	7	11
3	Fixed	0.040	1.8	0.020	8	8
4	Fixed	0.040	2.2	0.040	9	13
5	Shifted	0.020	1.8	0.040	11	13
6	Shifted	0.020	2.2	0.020	9	13
7	Shifted	0.040	1.8	0.040	15	19
8	Shifted	0.040	2.2	0.020	14	16

Which factor was of greatest significance in noise reduction? Present the complete analysis of variance table in support of your answer.

11.34 You would like to investigate the effect of the amount and quality of ammonium chloride on the yield of an organic chemical from a batch of base material. Interest is in the comparison between the use of finely ground and coarse ammonium chloride, the effect of a 10% increase in the charge of ammonium chloride to the batch, and the effect of two apparently identical units of the plant in which the actual conversion of base to the organic chemical product takes place.

Each batch of the base material was sufficient to make only two batches of the product. Facilities are available to do a fair job of mixing two batches of the base material to obtain a fairly uniform base material that will make four batches of the product. From chemical considerations there is unlikely to be any interaction between amount or quality of ammonium chloride and unit of plant; that is, if fine-grade ammonium chloride produced a given increase in yield in one unit, it should, apart from the experimental error, produce an equivalent increase in yield in the other unit. The three-factor interaction between quality of ammonium chloride, amount of ammonium chloride, and plant unit is even less likely to be appreciable.

Give the design you would use, including the physical layout, the model, and AOV. Four replications are needed.

11.35 One of your associates is working on the adsorption of Cl_2 in nuclear submarine atmospheres by various materials: activated charcoal, MnO_2, and so on. A total of six adsorbents are to be examined. Facilities are available for performing tests on up to eight adsorbents simultaneously.

The rescarch program allows time for testing 48 materials. Although four tests per material will give the required precision, he decides to use eight tests each in an attempt to improve the accuracy and precision of his results. Lay out the possible designs that this young engineer can use. Give and explain the model for each. Discuss each with regard to efficiency, ease of data handling, and reliability of results.

11.36 Describe the use of a Greco-Latin square for the tires–car–driver setup in Example 11.15. Give the model, the AOV, and make the appropriate statistical tests for the major variables at the 5% level.

11.37 Apply a Greco-Latin square to the experiment in Example 11.15 by adding a car and driver and cycling in the spare tire in regular rotation. Write out the model and show how to compute the analysis-of-variance table.

11.38 In a pilot-scale gas absorber (6 ft long, 3 in. ID, $\frac{1}{2}$-in. ceramic berl saddles), the number of transfer units (NTUs) was investigated for varying ammonia concentrations in air (NH_3). Various combinations of the ratio of the molar liquid-to-gas phase rates L/G were used.

NH_3	L/G	NTU
15	5.8	0.91
	14.6	0.77
	20.4	0.77
	29.2	0.76
5	22.3	3.60
	16.8	4.70
	13.4	6.60
	8.9	5.00
10	6.8	1.38
	13.6	1.40
	20.4	1.52
	34.0	1.65

Does ammonia concentration affect performance as measured by the number of transfer units? Does the phase-rate ratio affect the results? Use analysis of covariance for this problem.

11.39 In an experiment to measure the film coefficient HF for steam condensing on the outside surface of a single horizontal tube, various steam rates (MS) were used. The covariate is the cooling water flow rate V.

MS	V	HF
0.80	1.00	785
	0.90	1010
	0.80	1121
	0.70	1293
	0.60	1505
	0.80	1149
	0.50	1853
	0.40	2310
	0.80	1211
	0.30	3120
	0.20	4562
0.60	0.80	792
	1.00	946
	0.60	731
	0.40	684
	0.40	693
0.30	0.95	847
	0.75	800
	0.63	738
	0.45	715
	0.20	429
0.45	1.00	4049
	0.78	4000
	0.59	4883
	0.39	5053
	0.18	5054
0.70	0.20	4530
	0.30	3670
	0.50	3510
	0.70	2900
	0.80	3050
	1.00	2690
	1.10	2510

Do the steam and/or water rates affect HF?

11.40 In a full-scale determination of the airborne cotton-dust concentration in the cleaning, yarn preparation, dyeing, weaving, inspection, shipping, and associated "break" and shop areas, five independent samples are taken on each shift to satisfy the OSHA compliance program. From each sample, only a single value of the cotton-dust concentration ($\mu g/m^3$) can be obtained. The resulting data are used to determine whether the mill meets the standard as published in 29 CFR 1910.1043.

(a) Can these data also be used to determine which areas are not in compliance?
(b) Describe and justify the experimental design used.
(c) What F-tests must be done to carry out the AOV?

11.41 An experiment must be designed to determine the effects of varying temperature on the heat capacity ratio C_p/C_v for nitrogen. Four different physical chemistry lab groups (out of 21 available) will conduct this experiment over the predetermined temperature range 70 to 150°F. Each group can make eight determinations. Describe the best experimental design for this situation.

11.42 The effect of concentration on the conductance of an aqueous solution is to measured for various types of materials, such as sodium acetate, acetic acid, KCl, NaOH, Na_2SO_4, $KHCO_3$, $AlCl_3$, $(NH_4)_3PO_4$, Na_2CO_3, and so on. Describe a suitable experiment to determine whether or not salt concentration affects the dissociation constant and thus the conductance of the solutions. The chemicals are inexpensive. The range of concentrations is from 0.001 M to 10 M or the saturation limit, whichever is greater. All tests are conducted at 25°C. You can prepare a solution in about 12 min. Determination of the conductance takes about 1 min. This project must be complete by 4 P.M. tomorrow. Propose and justify a suitable experimental design.

11.43 Cooling tower performance, as measured by the number of transfer units (NTUs), is affected by feed-water temperature and the fluxes of the feed water and inlet air. A Latin square design was selected for an experiment using pilot-scale cooling-tower data. The rows were the feed-water flux, 100, 200, 300, and 400 $lb_m/hr\ ft^2$. The columns were the inlet air flux, 5000, 6000, 7000, and 8000 ft^3/hr per square foot. The treatment, inlet water temperature, was evaluated at 120, 130, 140, and 150°F. Why is the Latin square an inappropriate choice for an experimental design?

11.44 The effects of temperature and pressure on reaction rates are being studied using the following reaction:

$$CO + \tfrac{1}{2}O_2 \rightleftharpoons CO_2.$$

The desired temperatures and pressures are completely randomized and then the experiment is conducted. Samples are taken from each reaction mixture and are analyzed to determine the composition at equilibrium. Using the mole fractions and total pressures, values of the equilibrium constant are then calculated. A separate trial was also run after rerandomization of order. The results have been tabulated as follows:

Pressure	Reaction rate at temperature (K)				
(atm)	298	350	400	450	500
1	1.24	2.48	3.39	7.28	7.52
	1.22	2.44	3.43	7.25	7.55
5	1.35	2.61	3.52	7.43	7.61
	1.41	2.58	3.56	7.40	7.59
10	1.58	2.89	3.73	8.14	8.62
	1.52	2.82	3.68	8.10	8.76
15	1.92	3.43	4.26	9.45	9.73
	1.81	3.36	4.35	9.55	9.81
20	2.11	4.01	4.96	10.23	10.91
	2.15	3.98	5.02	10.30	11.01

Is the value of the equilibrium constant dependent on temperature, pressure, or temperature and pressure combined?

11.45 A sizable pressure drop across a fixed-bed catalytic reactor was desired for preliminary studies on the effect of a given antifouling agent on a sample of fouled catalyst removed from the first reactor of a hydrofinishing train. A pilot plant was used for the study. The flow rate and temperature were randomly increased and decreased, respectively, in hope of obtaining a considerable pressure drop. The technician was instructed to discontinue the increase in flow rate when a pressure drop ranging from 80 to 90 psig or when the maximum flow rate for the pilot plant was reached. The temperature was not to be lowered below 300°F. At a temperature below 300°F, a plug could easily form, which would in turn lead to technical problems. The pressure drop data are given below.

Conditions	Pressure drop (psi) at temperature (°F)	
	600	300
$H_2(g)$ 1.71 scf/hr	75	83
	77	82
$H_2(g)$ 1.2 scf/hr	53	65
	50	61
$H_2(g)$ 0.2 scf/hr	20	30
	19	33

(a)   What conclusions can be drawn from the data?
(b)   Did temperature affect the pressure drop?
(c)   Did flow rate affect the pressure drop?
(d)   Did both temperature and pressure affect the pressure drop?

11.46   Sixteen gaskets, four of each of four types, were tested at four different temperatures with four different fluids. The objective of this Latin square design was to determine which gasket type would last the longest under service conditions. In the experimental layout, the gaskets (treatments) are represented by letters, and the numerical values are the time to failure measured in weeks.

Fluid	Time to failure (weeks) at temperature (C)			
	50	60	70	80
Air	A	B	C	D
	30.8	36.9	24.2	27.9
Water	D	A	B	C
	17.3	18.7	20.8	16.3
Acid	C	D	A	B
	6.7	8.2	9.7	—
Base	B	C	D	A
	11.0	6.5	8.4	10.3

Is there a significant ($\alpha = 0.05$) difference in gasket materials?

11.47   The stainless steel vessels at your plant need to be replaced. You have the option of buying only one type of a particular vessel, so you are interested in getting the one that resists corrosion the best for all types of acids that will be encountered. The vessels that are to be replaced are exposed to concentrated hydrochloric, nitric, and sulfuric acids. The manufacturer can supply all different types of vessels made from any type of stainless steel that meets your specifications. To test the effectiveness of each material, you have ordered small ingots of each type of stainless steel and have subjected an ingot of each type to each corrosive environment and measured the amount of materials lost to corrosion (mg) per unit of area (dm^2) per day. The following data were collected from this series of accelerated corrosion tests.

Stainless steel type	HCl	HNO$_3$	H$_2$SO$_4$
1	0.64	0.30	0.70
2	0.44	0.78	0.48
3	0.85	0.80	0.22
4	0.19	0.42	0.46
5	0.18	0.32	0.37

Your solution to this problem must contain all of the following items:

1. Type of experimental design
2. Justification for design chosen
3. Mathematical statement of the model with all terms identified
4. Identification of each independent variable as to "random" or "fixed"
5. Explicit statements or equations showing how the AOV will be carried out (what F-tests will be done and in what order)
6. Discussion of how the results of the F-tests will be interpreted

11.48 The effect of different reaction times has been evaluated during determination of the optimal conditions for the oxidation of Texas lignite to soluble acids by oxygen in aqueous alkaline media. Reaction times of 0, 4, 8, and 12 h were used. The fraction conversion to alkaline-soluble acids was measured at three different oxygen partial pressures at two temperatures.

T (K)	pO$_2$ (psig)	Conversion
373	220	0.7243
	220	0.7244
	450	0.8211
	450	0.8212
	680	0.8219
	680	0.8218
433	220	0.8436
	220	0.8437
	450	0.8771
	450	0.8770
	680	0.8494
	680	0.8495

For the 8-h reaction time data shown, conduct the appropriate analysis of variance and interpret the results at $\alpha = 0.05$. Begin your solution with a statement and discussion of the appropriate statistical model for this part of the data.

11.49 A certain phenomenon is to be investigated in the laboratory. It is felt that four variables $(A, B, C,$ and $D)$ may affect the dependent response. Each experiment takes 45 min to set up and 2 h to reach equilibrium. Two complete experiments may thus be completed each day by a two-person team. Two weeks (10 workdays) have been allotted for your team to complete the experimental work, starting next Monday. You must design the experiment now so that the requisite supplies may be drawn from stock and the apparatus set up tomorrow.

State the model you assume to be followed by your design, writing out all appropriate terms. State any necessary assumptions. Describe your experimental plan in detail. Be sure to obtain the maximum amount of information available with the highest precision possible.

11.50 In a heat-transfer laboratory experiment, data were collected at several steam rates (0.5, 2, and 5 lb_m/min) to determine the effect of water rate (expressed as Reynolds number) on the inside heat-transfer coefficient, h_i (Btu/hr ft^2 °F). These data were collected in the transitional flow regime. A total of 33 observations were obtained.

Steam rate	N_{Re}	h_i
0.5	749	241, 259
	1498	335, 317, 344
	2246	380, 369, 370
	2995	453, 433
	3744	497
2	1470	593, 321
	2940	442
	4410	550, 594
	5880	670, 647, 662
	7350	700
	8090	753
	8820	766
5	1470	305, 300
	2940	412
	4410	540, 641
	5880	690, 719, 722
	7350	710
	8090	771
	8820	794

Use analysis of covariance to eliminate the effect of steam rate and determine the relationship desired.

REFERENCES

1. Akutowicz, F., and H. M. Truax, Establishing control of tire cord testing laboratories, *Ind. Qual. Control* **13**(2): 4–5 (1956).
2. Box, G. E. P., W. G. Hunter, and J. S. Hunter, *Statistics for Experimenters*, Wiley, New York (1978).
3. Cochran, W. G., and G. M. Cox, *Experimental Designs*, 2nd ed., Wiley, New York (1957), pp. 20–27.
4. Conover, W. J., *Practical Nonparametric Statistics*, 2nd ed., Wiley, New York (1980).
5. Ostle, B., and R. W. Mensing, *Statistics in Research*, 3rd ed., Iowa State University Press, Ames, Iowa (1975), pp. 395–408.
6. SAS Institute, *SAS User's Guide: Statistics*, Version 5 edition, SAS Institute, Cary, NC (1985).
7. Snedecor, G. W., and W. G. Cochran, *Statistical Methods*, 6th ed., Iowa State University Press, Ames, Iowa (1967).
8. Stillitto, G. F., *Research* **1**: 520–525 (1948). Copyright Butterworth & Co. (Publishers) Limited.
9. Yokota, Y., Quality control of gear grinding, *Ind. Qual. Control* **20**: 18–19 (1964).

Appendix A

Introduction to SAS

A.1 DEFINITIONS

Data Value: a single measurement (i.e., boiling point, resistivity, tensile strength).

Observation: a set of data values for the same item, process, individual (e.g., temperature, ohms, psi).

Variable: a set of data values for the same measurement (i.e., the average boiling points of all distillate samples collected during a shift, the resistance rating of heaters of a given type, the tensile strengths of all specimens from a mill run of steel strip stock).

Variable name: a name chosen for each variable, containing from one to eight characters. If alphameric, the variable name must begin with a letter. Some examples are AVGTEMP, RESIST, TENSILE.

Data set: a collection of observations, such as the temperature at 10° intervals along the boiling point curves for distillate samples, the actual resistances of each heater with a specific rating in a shipment from a single supplier, or the physical properties of rolled steel strip.

A.2 GENERAL INFORMATION

Jobs within the SAS system are made up of statements and procedures that provide information or ask the system to perform an activity. The first word of an SAS statement tells the system what activity you want to perform, while the rest of the statement (if required) gives the system more information about how you want the activity performed. Every SAS statement ends with a semicolon.

A.3 CREATING AN SAS DATA SET

To get your data into an SAS data set, you need:

A DATA statement
An INPUT statement
Either CARDS or INFILE statement

The DATA statement begins with the word DATA and then gives the name you chose for the data set: for example,

DATA PHYSPROP;

If you leave out the data set name, the system will create one for you. This may create a problem, especially if more than one activity is performed in an SAS job. In that event, the system will use the most recently named data set for a subsequent procedure. You may really want the activity performed on the original or some other data set instead. Always label data sets and specify the one the system is to use. Note that data set names are limited to eight characters.

The INPUT statement is used to describe your data to the system. Therefore, you need to know how your data are arranged. Begin with the word INPUT followed by the variable name for the first variable to be read. If the variable contains nonnumeric characters, follow the variable name with a dollar sign. Then give the column(s) in which the data values occur. Repeat for additional variables:

INPUT ASTM $ 10-14 TENSILE 16-18 BREAK 20-22 ROCK 24-27;

Variables that you do not wish to read in for a particular job may be skipped:

INPUT ASTM $ 10-14 BREAK 20-22;

Note that to avoid reading in all variables, you must know in exactly which columns your data are located.

If all variables are to be read in and the values are separated by at least one blank space, column designations may be omitted. However, any missing data values must be indicated by a period. Otherwise, the data entries are advanced into the wrong columns. If the breaking strength were not available for a specimen and no period were inserted, the Rockwell hardness would be read in for the breaking strength. If the data set consists of data values for only one observation, you may want to put more than one data value on a card. To tell the system to read them all instead of the first one only, use

INPUT name @@;

If the data are part of your job, the CARDS statement follows the DATA and INPUT statements in your program. Simply use

CARDS;

This tells the system that the data come next. Our mainframe operating system interprets data as part of the program as though they are cards. Your system may be different, so check with the computer center personnel for exact instructions. If the data values are on disk, you must tell both the computer's operating system using JCL and the SAS system where to find the data. The data set name in this example is **PHYSPROP** and must be used by the INFILE statement, thus:

INFILE PHYSPROP;

which goes before the INPUT statement. Entering data in an SAS job, then, could be illustrated by the following three examples:

(initial JCL cards/statements here)

```
DATA PHYSPROP;
INPUT ASTM $ 10-14 TENSILE 16-18 BREAK 20-22 ROCK 24-27;
CARDS
bbbbbbbbbbD1022b121b175b62.4
bbbbbbbbbbA0147b072b094b12.1
```
or

(JCL cards/statements here)

```
DATA PHYSPROP;
INFILE PHYSPROP;
INPUT ASTM $ 10-14 TENSILE 16-18;
```
or

(JCL cards/statements here)
```
DATA TENSTR;
INPUT TENSILE @@;
CARDS;
047   062   072   091   058   107
101   121   091   116   082   086
```

(note that in this example, all the data values belong to the same variable TENSILE which is the only variable in the data set TENSTR).

A.4 HOW TO USE SAS PROCEDURES

SAS procedures are used to analyze and process an SAS data set that you have created. The SAS statement is started with the word PROC to tell the system that you want a procedure executed. You may string together several procedures that use the same data set. After PROC, you must give the name of the procedure you want used, such as PRINT to print the data set or MEANS to calculate means. Unless you follow this by DATA = data set name, the system will use the most recently created data set.

The system will perform the indicated procedure on all variables unless otherwise specified. To indicate which variables are to be processed, you must follow the PROC statement with a variable statement VAR which lists the variables to use. For example:

```
PROC PRINT DATA=PHYSPROP;
VAR ASTM TENSILE;
PROC MEANS DATA=PHYSPROP;
VAR TENSILE BREAK ROCK;
```

will print the values for ASTM and TENSILE in PHYSPROP and will calculate means, standard deviation, and so on, for TENSILE, BREAK, and ROCK.

A.5 HISTOGRAMS AND FREQUENCY PLOTS

The procedure PROC FREQ can be used to obtain frequency tables. To get frequency tables for all variables in the most recently created data set, use

```
PROC FREQ;
```

If you want to use a previous data set and specify only some variables, specify the data set name and use a TABLES statement:

```
PROC FREQ DATA=PHYSPROP;
TABLES TENSILE;
```

The SAS system will go to data set PHYSPROP and make frequency tables for the variable TENSILE.

To create a bar chart histogram using frequency, use

```
PROC CHART DATA=data set name;
VBAR variable name/TYPE=FREQ;
```

SAS will create a vertical bar chart for the designated variable. It will create its own scales. If you want to specify the scale for the abscissa, you can indicate the range to be used and the distance between classes using the midpoints option, as in this example:

```
VBAR TENSILE/TYPE=FREQ MIDPOINTS=35 TO 155 BY 10;
```

This statement tells the system to make a frequency bar chart for the variable TENSILE using an abscissa scale of 35, 45, . . . , 145, 155. To get an estimate of the range and size of the interval to use, you can either let the system do a plot and then guess at modifications or you can use PROC FREQ results to get an idea of the distribution of your data.

To obtain the data you need to plot frequency polygons, you must use the frequency table that you obtained from PROC FREQ. Prepare a program to execute PROC FREQ and hold the output by using the command appropriate for your system. After the program has been executed, call a copy of the output into your active file with an appropriate command. Locate the frequency table in your active file and delete all other lines. You may also have to delete any page header information if the table runs over to a new page. Now your active file should contain only the numbers in the frequency table.

Renumber the remaining lines. List the last line of the frequency table and you will note the presence of a large empty space before the first column. Count these spaces and eliminate them as needed. Now relist the last line followed by a column number display command so you can find out exactly where your data are stored. Write down the columns where each variable is located so you can tell the SAS system where to find the data for PROC PLOT. Save this entire data set (table). Now clear your active file and start writing your new SAS program. When you have reached the place in the SAS program where the data are to be inserted, call in the frequency table and then proceed to finish the SAS program. Save the program on disk or tape, edit it as necessary, and then run it. Note that the cumulative relative frequency (CRF) will be plotted on the ordinate vs. the named variable (VAR) as the abscissa. You must already have named the variable in your SAS program.

Appendix B

Tables of Statistical Functions

TABLE I: BINOMIAL CUMULATIVE DISTRIBUTION

This table gives values of the cumulative binomial distribution function,

$$\mathscr{P}(X \le x) = \sum_{k=0}^{x} \binom{n}{k} p^k (1 - p)^{n-k}.$$

$\mathscr{P}(X \le x)$ is the probability of x or less successes in n independent binomial trials with probability p of success on a single trial. The table gives values of the distribution function for $x = 0,1,\ldots,n$; $n = 2,3,\ldots,25$; $p = 0.1$ through 0.9. All probabilities entered as 0.000 in the table are actually larger than 0 but less than 0.0005. All probabilities entered as 1.000, except for those when $x = n$, are actually larger than 0.9995 but less than 1.

BINOMIAL CUMULATIVE DISTRIBUTION

n	x	.1	.2	.3	.4	.5	.6	.7	.8	.9
2	0	0.810	C.640	C.490	0.360	0.250	0.160	0.090	0.040	0.010
2	1	0.990	C.960	0.910	0.840	0.750	0.640	0.510	0.360	0.190
2	2	1.000	1.000	1.000	1.000	1.000	1.000	1.000	1.000	1.000
3	0	0.729	0.512	0.343	0.216	0.125	0.064	0.027	0.008	0.001
3	1	0.972	0.896	0.784	0.648	0.500	0.352	0.216	0.104	0.028
3	2	0.999	C.992	C.973	0.936	0.875	0.784	0.657	0.488	0.271
3	3	1.000	1.000	1.000	1.000	1.000	1.000	1.000	1.000	1.000
4	0	0.656	C.410	C.240	0.130	0.C63	0.026	0.008	0.002	0.000
4	1	0.948	C.819	0.652	0.475	0.313	0.179	0.084	0.027	0.004
4	2	0.996	C.973	C.916	0.821	0.687	0.525	0.348	0.181	0.052
4	3	1.000	0.998	0.992	0.974	0.937	0.870	0.760	0.590	0.344
4	4	1.000	1.000	1.000	1.000	1.000	1.000	1.000	1.000	1.000
5	0	0.590	C.328	0.168	0.078	0.031	0.010	C.002	0.000	0.000
5	1	0.919	C.737	0.528	0.337	0.188	0.087	0.031	0.007	0.000
5	2	0.991	C.942	C.837	0.683	C.500	0.317	0.163	0.058	0.009
5	3	1.000	C.993	0.969	0.913	0.812	0.663	0.472	0.263	0.081
5	4	1.00C	1.CC0	C.998	C.990	C.969	0.922	0.832	0.672	0.410
5	5	1.000	1.000	1.000	1.000	1.0C0	1.000	1.000	1.000	1.000
6	C	0.531	C.262	0.118	0.047	C.016	0.004	0.001	0.000	0.000
6	1	0.886	0.655	0.420	0.233	0.1C9	0.041	0.011	0.002	0.000
6	2	0.984	C.901	0.744	0.544	0.344	0.179	0.070	0.017	0.001
6	3	0.999	C.983	C.930	0.821	0.656	0.456	0.256	0.099	0.016
6	4	1.000	C.998	0.989	0.959	0.891	0.767	0.580	0.345	0.114
6	5	1.000	1.C00	C.959	0.996	0.984	0.953	0.882	0.738	0.469
6	6	1.000	1.000	1.000	1.000	1.0C0	1.000	1.000	1.000	1.000
7	0	0.478	C.210	0.082	0.028	C.0C8	0.002	0.000	0.000	0.000
7	1	0.850	0.577	0.329	0.159	0.062	0.019	0.004	0.000	0.000
7	2	0.974	C.852	0.647	0.420	0.227	0.096	0.029	0.005	0.000
7	3	0.997	C.967	C.874	C.710	0.5C0	0.290	0.126	0.033	0.003
7	4	1.000	C.995	0.971	0.904	0.773	0.580	0.353	C.148	0.026
7	5	1.000	1.C00	C.956	0.981	0.938	0.841	0.671	0.423	0.150
7	6	1.000	1.000	1.000	0.998	0.992	0.972	0.918	0.790	0.522
7	7	1.000	1.C00	1.0C0	1.000	1.000	1.000	1.000	1.000	1.000
8	0	0.430	0.168	0.058	0.017	0.004	0.001	0.000	0.000	C.000
8	1	0.813	C.503	0.255	0.106	0.035	0.009	0.001	0.000	0.000
8	2	0.962	C.797	C.552	0.315	0.145	0.050	0.011	0.001	0.000
8	3	0.995	0.944	0.806	0.594	0.363	0.174	0.058	0.010	C.000
8	4	1.0C0	C.990	C.942	0.826	0.637	0.406	0.194	0.056	0.005
8	5	1.000	0.999	0.989	0.950	0.855	0.685	0.448	0.203	0.038
8	6	1.000	1.CC0	0.999	0.991	0.965	0.894	0.745	0.497	0.187
8	7	1.000	1.C00	1.0C0	0.999	C.996	0.983	0.942	0.832	0.570
8	8	1.000	1.000	1.000	1.000	1.000	1.000	1.000	1.000	1.000

BINOMIAL CUMULATIVE DISTRIBUTION

n	x	.1	.2	.3	.4	.5	.6	.7	.8	.9
9	0	0.387	C.134	0.040	0.010	0.002	0.000	0.000	0.000	0.000
9	1	0.775	C.436	0.196	0.071	0.020	0.004	0.000	0.000	0.000
9	2	0.947	C.738	0.463	0.232	0.090	0.025	0.004	0.000	0.000
9	3	0.992	C.914	C.730	0.483	0.254	0.099	0.025	0.003	0.000
9	4	0.999	C.980	0.901	0.733	0.5C0	0.267	0.099	0.020	0.001
9	5	1.000	C.997	0.975	0.901	0.746	0.517	0.270	0.086	0.008
9	6	1.000	1.CC0	C.956	0.975	C.910	0.768	0.537	0.262	0.053
9	7	1.000	1.000	1.000	0.996	0.980	0.929	0.804	0.564	0.225
9	8	1.000	1.C00	1.000	1.00C	0.958	0.990	0.960	0.866	0.613
9	9	1.000	1.000	1.000	1.000	1.000	1.000	1.000	1.000	1.000
10	0	0.349	C.1C7	0.028	0.006	C.0C1	0.000	0.000	0.000	0.000
10	1	0.736	0.376	0.149	0.046	0.011	0.002	0.000	0.000	0.000
10	2	0.930	C.678	0.383	0.167	0.055	0.012	0.002	0.000	0.000
10	3	0.987	0.879	0.650	0.382	0.172	0.055	C.011	0.001	0.000
10	4	0.998	C.967	0.850	0.633	0.377	0.166	0.047	0.006	0.000
1C	5	1.000	C.994	C.953	C.834	0.623	0.367	0.150	0.033	0.002
10	6	1.000	0.999	0.989	0.945	0.828	0.618	0.350	0.121	0.013
10	7	1.000	1.C0C	C.958	0.988	0.945	0.833	0.617	0.322	0.070
10	8	1.000	1.000	1.000	0.998	C.989	0.954	C.851	0.624	0.264
1C	9	1.000	1.000	1.000	1.000	0.999	0.994	0.972	0.893	0.651
1C	1C	1.000	1.C00	1.CC0	1.0CC	1.000	1.000	1.000	1.000	1.000
11	0	0.314	C.C86	0.020	0.004	0.000	0.000	0.000	0.000	0.000
11	1	0.697	C.322	0.113	0.030	0.0C6	0.001	0.000	0.000	0.000
11	2	0.910	0.617	0.313	0.119	0.033	0.006	0.001	0.000	C.000
11	3	0.981	C.E39	0.570	0.296	0.113	0.029	0.004	0.000	0.000
11	4	0.997	C.950	0.790	0.533	0.274	0.099	0.022	0.002	0.000
11	5	1.000	C.988	0.922	0.753	0.500	0.247	0.078	0.012	0.000
11	6	1.000	C.998	C.978	0.901	0.726	0.467	0.210	0.050	0.003
11	7	1.000	1.000	0.996	0.971	0.887	0.704	0.430	0.161	0.019
11	8	1.000	1.CC0	C.999	0.994	0.967	0.881	0.687	0.383	0.090
11	9	1.000	1.C00	1.000	0.999	0.994	0.970	0.887	0.678	0.303
11	10	1.000	1.000	1.000	1.000	1.000	0.996	0.980	0.914	0.686
11	11	1.000	1.C00	1.0C0	1.0CC	1.000	1.000	1.000	1.000	1.000
12	0	0.282	C.069	0.014	0.002	0.000	0.000	0.000	0.000	0.000
12	1	0.659	C.275	0.085	0.020	0.003	0.000	0.000	0.000	0.000
12	2	0.889	C.558	0.253	0.083	0.019	0.003	0.000	0.0C0	C.000
12	3	0.974	C.795	0.493	0.225	0.073	0.015	0.002	0.000	0.000
12	4	0.996	C.927	0.724	0.438	0.194	0.057	0.0C9	0.001	0.000
12	5	0.999	C.981	0.882	0.665	0.387	0.158	0.039	0.004	0.000
12	6	1.000	C.996	C.961	0.842	C.613	0.335	0.118	0.019	0.001
12	7	1.000	0.999	0.991	0.943	0.8C6	0.562	0.276	0.073	0.004
12	8	1.000	1.CC0	0.958	0.985	0.927	0.775	0.507	0.205	0.026
12	9	1.000	1.000	1.CC0	0.997	0.981	0.917	0.747	0.442	0.111
12	1C	1.000	1.000	1.000	1.000	0.997	0.980	0.915	0.725	C.341

BINOMIAL CUMULATIVE DISTRIBUTION

n	x	.1	.2	.3	.4	.5	.6	.7	.8	.9
12	11	1.000	1.000	1.000	1.000	1.000	0.998	0.986	0.931	0.718
12	12	1.000	1.000	1.000	1.000	1.000	1.000	1.000	1.000	1.000
13	0	0.254	0.055	0.010	0.001	0.000	0.000	0.000	0.000	0.000
13	1	0.621	0.234	0.064	0.013	0.002	0.000	0.000	0.000	0.000
13	2	0.866	0.502	0.202	0.058	0.011	0.001	0.000	0.000	0.000
13	3	0.966	0.747	0.421	0.169	0.046	0.008	0.001	0.000	0.000
13	4	0.994	0.901	0.654	0.353	0.133	0.032	0.004	0.000	0.000
13	5	0.999	0.970	0.835	0.574	0.291	0.098	0.018	0.001	0.000
13	6	1.000	0.993	0.938	0.771	0.500	0.229	0.062	0.007	0.000
13	7	1.000	0.999	0.982	0.902	0.709	0.426	0.165	0.030	0.001
13	8	1.000	1.000	0.996	0.968	0.867	0.647	0.346	0.099	0.006
13	9	1.000	1.000	0.999	0.992	0.954	0.831	0.579	0.253	0.034
13	10	1.000	1.000	1.000	0.999	0.989	0.942	0.798	0.498	0.134
13	11	1.000	1.000	1.000	1.000	0.998	0.987	0.936	0.766	0.379
13	12	1.000	1.000	1.000	1.000	1.000	0.999	0.990	0.945	0.746
13	13	1.000	1.000	1.000	1.000	1.000	1.000	1.000	1.000	1.000
14	0	0.229	0.044	0.007	0.001	0.000	0.000	0.000	0.000	0.000
14	1	0.585	0.198	0.047	0.008	0.001	0.000	0.000	0.000	0.000
14	2	0.842	0.448	0.161	0.040	0.006	0.001	0.000	0.000	0.000
14	3	0.956	0.698	0.355	0.124	0.029	0.004	0.000	0.000	0.000
14	4	0.991	0.870	0.584	0.279	0.090	0.018	0.002	0.000	0.000
14	5	0.999	0.956	0.781	0.486	0.212	0.058	0.008	0.000	0.000
14	6	1.000	0.988	0.907	0.692	0.395	0.150	0.031	0.002	0.000
14	7	1.000	0.998	0.969	0.850	0.605	0.308	0.093	0.012	0.000
14	8	1.000	1.000	0.992	0.942	0.788	0.514	0.219	0.044	0.001
14	9	1.000	1.000	0.998	0.982	0.910	0.721	0.416	0.130	0.009
14	10	1.000	1.000	1.000	0.996	0.971	0.876	0.645	0.302	0.044
14	11	1.000	1.000	1.000	0.999	0.994	0.960	0.839	0.552	0.158
14	12	1.000	1.000	1.000	1.000	0.999	0.992	0.953	0.802	0.415
14	13	1.000	1.000	1.000	1.000	1.000	0.999	0.993	0.956	0.771
14	14	1.000	1.000	1.000	1.000	1.000	1.000	1.000	1.000	1.000
15	0	0.206	0.035	0.005	0.000	0.000	0.000	0.000	0.000	0.000
15	1	0.549	0.167	0.035	0.005	0.000	0.000	0.000	0.000	0.000
15	2	0.816	0.398	0.127	0.027	0.004	0.000	0.000	0.000	0.000
15	3	0.944	0.648	0.297	0.091	0.018	0.002	0.000	0.000	0.000
15	4	0.987	0.836	0.515	0.217	0.059	0.009	0.001	0.000	0.000
15	5	0.998	0.939	0.722	0.403	0.151	0.034	0.004	0.000	0.000
15	6	1.000	0.982	0.869	0.610	0.304	0.095	0.015	0.001	0.000
15	7	1.000	0.996	0.950	0.787	0.500	0.213	0.050	0.004	0.000
15	8	1.000	0.999	0.985	0.905	0.696	0.390	0.131	0.018	0.000
15	9	1.000	1.000	0.996	0.966	0.849	0.597	0.278	0.061	0.002
15	10	1.000	1.000	0.999	0.991	0.941	0.783	0.485	0.164	0.013
15	11	1.000	1.000	1.000	0.998	0.982	0.909	0.703	0.352	0.056
15	12	1.000	1.000	1.000	1.000	0.996	0.973	0.873	0.602	0.184
15	13	1.000	1.000	1.000	1.000	1.000	0.995	0.965	0.833	0.451

BINOMIAL CUMULATIVE DISTRIBUTION

n	x	.1	.2	.3	.4	.5	.6	.7	.8	.9
						p				
15	14	1.000	1.000	1.000	1.000	1.000	1.000	0.995	0.965	0.794
15	15	1.000	1.000	1.000	1.000	1.000	1.000	1.000	1.000	1.000
16	0	0.185	0.028	0.003	0.000	0.000	0.000	0.000	0.000	0.000
16	1	0.515	0.141	0.026	0.003	0.000	0.000	0.000	0.000	0.000
16	2	0.789	0.352	0.099	0.018	0.002	0.000	0.000	0.000	0.000
16	3	0.932	0.558	0.246	0.065	0.011	0.001	0.000	0.000	0.000
16	4	0.983	0.798	0.450	0.167	0.038	0.005	0.000	0.000	0.000
16	5	0.997	0.918	0.660	0.329	0.105	0.019	0.002	0.000	0.000
16	6	0.999	0.973	0.825	0.527	0.227	0.058	0.007	0.000	0.000
16	7	1.000	0.993	0.926	0.716	0.402	0.142	0.026	0.001	0.000
16	8	1.000	0.999	0.974	0.858	0.598	0.284	0.074	0.007	0.000
16	9	1.000	1.000	0.993	0.942	0.773	0.473	0.175	0.027	0.001
16	10	1.000	1.000	0.998	0.981	0.895	0.671	0.340	0.082	0.003
16	11	1.000	1.000	1.000	0.995	0.962	0.833	0.550	0.202	0.017
16	12	1.000	1.000	1.000	0.999	0.989	0.935	0.754	0.402	0.068
16	13	1.000	1.000	1.000	1.000	0.998	0.982	0.901	0.648	0.211
16	14	1.000	1.000	1.000	1.000	1.000	0.997	0.974	0.859	0.485
16	15	1.000	1.000	1.000	1.000	1.000	1.000	0.997	0.972	0.815
16	16	1.000	1.000	1.000	1.000	1.000	1.000	1.000	1.000	1.000
17	0	0.167	0.023	0.002	0.000	0.000	0.000	0.000	0.000	0.000
17	1	0.482	0.118	0.019	0.002	0.000	0.000	0.000	0.000	0.000
17	2	0.762	0.310	0.077	0.012	0.001	0.000	0.000	0.000	0.000
17	3	0.917	0.549	0.202	0.046	0.006	0.000	0.000	0.000	0.000
17	4	0.978	0.758	0.389	0.126	0.025	0.003	0.000	0.000	0.000
17	5	0.995	0.894	0.597	0.264	0.072	0.011	0.001	0.000	0.000
17	6	0.999	0.962	0.775	0.448	0.166	0.035	0.003	0.000	0.000
17	7	1.000	0.989	0.895	0.641	0.315	0.092	0.013	0.000	0.000
17	8	1.000	0.997	0.960	0.801	0.500	0.199	0.040	0.003	0.000
17	9	1.000	1.000	0.987	0.908	0.685	0.359	0.105	0.011	0.000
17	10	1.000	1.000	0.997	0.965	0.834	0.552	0.225	0.038	0.001
17	11	1.000	1.000	0.999	0.989	0.928	0.736	0.403	0.106	0.005
17	12	1.000	1.000	1.000	0.997	0.975	0.874	0.611	0.242	0.022
17	13	1.000	1.000	1.000	1.000	0.994	0.954	0.798	0.451	0.083
17	14	1.000	1.000	1.000	1.000	0.999	0.988	0.923	0.690	0.238
17	15	1.000	1.000	1.000	1.000	1.000	0.998	0.981	0.882	0.518
17	16	1.000	1.000	1.000	1.000	1.000	1.000	0.998	0.977	0.833
17	17	1.000	1.000	1.000	1.000	1.000	1.000	1.000	1.000	1.000
18	0	0.150	0.018	0.002	0.000	0.000	0.000	0.000	0.000	0.000
18	1	0.450	0.099	0.014	0.001	0.000	0.000	0.000	0.000	0.000
18	2	0.734	0.271	0.060	0.008	0.001	0.000	0.000	0.000	0.000
18	3	0.902	0.501	0.165	0.033	0.004	0.000	0.000	0.000	0.000
18	4	0.972	0.716	0.333	0.094	0.015	0.001	0.000	0.000	0.000
18	5	0.994	0.867	0.534	0.209	0.048	0.006	0.000	0.000	0.000
18	6	0.999	0.949	0.722	0.374	0.119	0.020	0.001	0.000	0.000
18	7	1.000	0.984	0.859	0.563	0.240	0.058	0.006	0.000	0.000

BINOMIAL CUMULATIVE DISTRIBUTION

n	x	.1	.2	.3	.4	.5	.6	.7	.8	.9
18	8	1.000	0.996	0.940	0.737	0.407	0.135	0.021	0.001	0.000
18	9	1.000	0.999	0.979	0.865	0.593	0.263	0.060	0.004	0.000
18	10	1.000	1.000	0.994	0.942	0.760	0.437	0.141	0.016	0.000
18	11	1.000	1.000	0.999	0.980	0.881	0.626	0.278	0.051	0.001
18	12	1.000	1.000	1.000	0.994	0.952	0.791	0.466	0.133	0.006
18	13	1.000	1.000	1.000	0.999	0.985	0.906	0.667	0.284	0.028
18	14	1.000	1.000	1.000	1.000	0.996	0.967	0.835	0.499	0.098
18	15	1.000	1.000	1.000	1.000	0.999	0.992	0.940	0.729	0.266
18	16	1.000	1.000	1.000	1.000	1.000	0.999	0.986	0.901	0.550
18	17	1.000	1.000	1.000	1.000	1.000	1.000	0.998	0.982	0.850
18	18	1.000	1.000	1.000	1.000	1.000	1.000	1.000	1.000	1.000
19	0	0.135	0.014	0.001	0.000	0.000	0.000	0.000	0.000	0.000
19	1	0.420	0.083	0.010	0.001	0.000	0.000	0.000	0.000	0.000
19	2	0.705	0.237	0.046	0.005	0.000	0.000	0.000	0.000	0.000
19	3	0.885	0.455	0.133	0.023	0.002	0.000	0.000	0.000	0.000
19	4	0.965	0.673	0.282	0.070	0.010	0.001	0.000	0.000	0.000
19	5	0.991	0.837	0.474	0.163	0.032	0.003	0.000	0.000	0.000
19	6	0.998	0.932	0.666	0.308	0.084	0.012	0.001	0.000	0.000
19	7	1.000	0.977	0.818	0.488	0.180	0.035	0.003	0.000	0.000
19	8	1.000	0.993	0.916	0.667	0.324	0.088	0.011	0.000	0.000
19	9	1.000	0.998	0.967	0.814	0.500	0.186	0.033	0.002	0.000
19	10	1.000	1.000	0.989	0.912	0.676	0.333	0.084	0.007	0.000
19	11	1.000	1.000	0.997	0.965	0.820	0.512	0.182	0.023	0.000
19	12	1.000	1.000	0.999	0.988	0.916	0.692	0.334	0.068	0.002
19	13	1.000	1.000	1.000	0.997	0.968	0.837	0.526	0.163	0.009
19	14	1.000	1.000	1.000	0.999	0.990	0.930	0.718	0.327	0.035
19	15	1.000	1.000	1.000	1.000	0.998	0.977	0.867	0.545	0.115
19	16	1.000	1.000	1.000	1.000	1.000	0.995	0.954	0.763	0.295
19	17	1.000	1.000	1.000	1.000	1.000	0.999	0.990	0.917	0.580
19	18	1.000	1.000	1.000	1.000	1.000	1.000	0.999	0.986	0.865
19	19	1.000	1.000	1.000	1.000	1.000	1.000	1.000	1.000	1.000
20	0	0.122	0.012	0.001	0.000	0.000	0.000	0.000	0.000	0.000
20	1	0.392	0.069	0.008	0.001	0.000	0.000	0.000	0.000	0.000
20	2	0.677	0.206	0.035	0.004	0.000	0.000	0.000	0.000	0.000
20	3	0.867	0.411	0.107	0.016	0.001	0.000	0.000	0.000	0.000
20	4	0.957	0.630	0.238	0.051	0.006	0.000	0.000	0.000	0.000
20	5	0.989	0.804	0.416	0.126	0.021	0.002	0.000	0.000	0.000
20	6	0.998	0.913	0.608	0.250	0.058	0.006	0.000	0.000	0.000
20	7	1.000	0.968	0.772	0.416	0.132	0.021	0.001	0.000	0.000
20	8	1.000	0.990	0.887	0.596	0.252	0.057	0.005	0.000	0.000
20	9	1.000	0.997	0.952	0.755	0.412	0.128	0.017	0.001	0.000
20	10	1.000	0.999	0.983	0.872	0.588	0.245	0.048	0.003	0.000
20	11	1.000	1.000	0.995	0.943	0.748	0.404	0.113	0.010	0.000
20	12	1.000	1.000	0.999	0.979	0.868	0.584	0.228	0.032	0.000
20	13	1.000	1.000	1.000	0.994	0.942	0.750	0.392	0.087	0.002
20	14	1.000	1.000	1.000	0.998	0.979	0.874	0.584	0.196	0.011

BINOMIAL CUMULATIVE DISTRIBUTION

n	x	.1	.2	.3	.4	.5	.6	.7	.8	.9
20	15	1.000	1.000	1.000	1.000	0.994	0.949	0.762	0.370	0.043
20	16	1.000	1.000	1.000	1.000	0.999	0.984	0.893	0.589	0.133
20	17	1.000	1.000	1.000	1.000	1.000	0.996	0.965	0.794	0.323
20	18	1.000	1.000	1.000	1.000	1.000	0.999	0.992	0.931	0.608
20	19	1.000	1.000	1.000	1.000	1.000	1.000	0.999	0.988	0.878
20	20	1.000	1.000	1.000	1.000	1.000	1.000	1.000	1.000	1.000
21	0	0.109	0.009	0.001	0.000	0.000	0.000	0.000	0.000	0.000
21	1	0.365	0.058	0.006	0.000	0.000	0.000	0.000	0.000	0.000
21	2	0.648	0.179	0.027	0.002	0.000	0.000	0.000	0.000	0.000
21	3	0.848	0.370	0.086	0.011	0.001	0.000	0.000	0.000	0.000
21	4	0.943	0.586	0.198	0.037	0.004	0.000	0.000	0.000	0.000
21	5	0.986	0.769	0.363	0.096	0.013	0.001	0.000	0.000	0.000
21	6	0.997	0.891	0.551	0.200	0.039	0.004	0.000	0.000	0.000
21	7	0.999	0.957	0.723	0.350	0.095	0.012	0.001	0.000	0.000
21	8	1.000	0.986	0.852	0.524	0.192	0.035	0.002	0.000	0.000
21	9	1.000	0.996	0.932	0.691	0.332	0.085	0.009	0.000	0.000
21	10	1.000	0.999	0.974	0.826	0.500	0.174	0.026	0.001	0.000
21	11	1.000	1.000	0.991	0.915	0.668	0.309	0.068	0.004	0.000
21	12	1.000	1.000	0.998	0.965	0.808	0.476	0.148	0.014	0.000
21	13	1.000	1.000	0.999	0.988	0.905	0.650	0.277	0.043	0.001
21	14	1.000	1.000	1.000	0.996	0.961	0.800	0.449	0.109	0.003
21	15	1.000	1.000	1.000	0.999	0.987	0.904	0.637	0.231	0.014
21	16	1.000	1.000	1.000	1.000	0.996	0.963	0.802	0.414	0.052
21	17	1.000	1.000	1.000	1.000	0.999	0.989	0.914	0.630	0.152
21	18	1.000	1.000	1.000	1.000	1.000	0.998	0.973	0.821	0.352
21	19	1.000	1.000	1.000	1.000	1.000	1.000	0.994	0.942	0.635
21	20	1.000	1.000	1.000	1.000	1.000	1.000	0.999	0.991	0.891
21	21	1.000	1.000	1.000	1.000	1.000	1.000	1.000	1.000	1.000
22	0	0.098	0.007	0.000	0.000	0.000	0.000	0.000	0.000	0.000
22	1	0.339	0.048	0.004	0.000	0.000	0.000	0.000	0.000	0.000
22	2	0.620	0.154	0.021	0.002	0.000	0.000	0.000	0.000	0.000
22	3	0.828	0.332	0.068	0.008	0.000	0.000	0.000	0.000	0.000
22	4	0.938	0.543	0.165	0.027	0.002	0.000	0.000	0.000	0.000
22	5	0.982	0.733	0.313	0.072	0.008	0.000	0.000	0.000	0.000
22	6	0.996	0.867	0.494	0.158	0.026	0.002	0.000	0.000	0.000
22	7	0.999	0.944	0.671	0.290	0.067	0.007	0.000	0.000	0.000
22	8	1.000	0.980	0.814	0.454	0.143	0.021	0.001	0.000	0.000
22	9	1.000	0.994	0.908	0.624	0.262	0.055	0.004	0.000	0.000
22	10	1.000	0.998	0.961	0.772	0.416	0.121	0.014	0.000	0.000
22	11	1.000	1.000	0.986	0.879	0.584	0.228	0.039	0.002	0.000
22	12	1.000	1.000	0.996	0.945	0.738	0.376	0.092	0.006	0.000
22	13	1.000	1.000	0.999	0.979	0.857	0.546	0.186	0.020	0.000
22	14	1.000	1.000	1.000	0.993	0.933	0.710	0.329	0.056	0.001
22	15	1.000	1.000	1.000	0.998	0.974	0.842	0.506	0.133	0.004
22	16	1.000	1.000	1.000	1.000	0.992	0.928	0.687	0.267	0.018
22	17	1.000	1.000	1.000	1.000	0.998	0.973	0.835	0.457	0.062

BINOMIAL CUMULATIVE DISTRIBUTION

n	x					p				
		.1	.2	.3	.4	.5	.6	.7	.8	.9
22	18	1.000	1.000	1.000	1.000	1.000	0.992	0.932	0.668	0.172
22	19	1.000	1.000	1.000	1.000	1.000	0.998	0.979	0.846	0.380
22	20	1.000	1.000	1.000	1.000	1.000	1.000	0.996	0.952	0.661
22	21	1.000	1.000	1.000	1.000	1.000	1.000	1.000	0.993	0.902
22	22	1.000	1.000	1.000	1.000	1.000	1.000	1.000	1.000	1.000
23	0	0.089	0.006	0.000	0.000	0.000	0.000	0.000	0.000	0.000
23	1	0.315	0.040	0.003	0.000	0.000	0.000	0.000	0.000	0.000
23	2	0.592	0.133	0.016	0.001	0.000	0.000	0.000	0.000	0.000
23	3	0.807	0.297	0.054	0.005	0.000	0.000	0.000	0.000	0.000
23	4	0.927	0.501	0.136	0.019	0.001	0.000	0.000	0.000	0.000
23	5	0.977	0.695	0.269	0.054	0.005	0.000	0.000	0.000	0.000
23	6	0.994	0.840	0.440	0.124	0.017	0.001	0.000	0.000	0.000
23	7	0.999	0.928	0.618	0.237	0.047	0.004	0.000	0.000	0.000
23	8	1.000	0.973	0.771	0.388	0.105	0.013	0.001	0.000	0.000
23	9	1.000	0.991	0.880	0.556	0.202	0.035	0.002	0.000	0.000
23	10	1.000	0.997	0.945	0.713	0.339	0.081	0.007	0.000	0.000
23	11	1.000	0.999	0.979	0.836	0.500	0.164	0.021	0.001	0.000
23	12	1.000	1.000	0.993	0.919	0.661	0.287	0.055	0.003	0.000
23	13	1.000	1.000	0.998	0.965	0.798	0.444	0.120	0.009	0.000
23	14	1.000	1.000	0.999	0.987	0.895	0.612	0.229	0.027	0.000
23	15	1.000	1.000	1.000	0.996	0.953	0.763	0.382	0.072	0.001
23	16	1.000	1.000	1.000	0.999	0.983	0.876	0.560	0.160	0.006
23	17	1.000	1.000	1.000	1.000	0.995	0.946	0.731	0.305	0.023
23	18	1.000	1.000	1.000	1.000	0.999	0.981	0.864	0.499	0.073
23	19	1.000	1.000	1.000	1.000	1.000	0.995	0.946	0.703	0.193
23	20	1.000	1.000	1.000	1.000	1.000	0.999	0.984	0.867	0.408
23	21	1.000	1.000	1.000	1.000	1.000	1.000	0.997	0.960	0.685
23	22	1.000	1.000	1.000	1.000	1.000	1.000	1.000	0.994	0.911
23	23	1.000	1.000	1.000	1.000	1.000	1.000	1.000	1.000	1.000
24	0	0.080	0.005	0.000	0.000	0.000	0.000	0.000	0.000	0.000
24	1	0.292	0.033	0.002	0.000	0.000	0.000	0.000	0.000	0.000
24	2	0.564	0.115	0.012	0.001	0.000	0.000	0.000	0.000	0.000
24	3	0.786	0.264	0.042	0.004	0.000	0.000	0.000	0.000	0.000
24	4	0.915	0.460	0.111	0.013	0.001	0.000	0.000	0.000	0.000
24	5	0.972	0.656	0.229	0.040	0.003	0.000	0.000	0.000	0.000
24	6	0.993	0.811	0.389	0.096	0.011	0.001	0.000	0.000	0.000
24	7	0.998	0.911	0.565	0.192	0.032	0.002	0.000	0.000	0.000
24	8	1.000	0.964	0.725	0.328	0.076	0.008	0.000	0.000	0.000
24	9	1.000	0.987	0.847	0.489	0.154	0.022	0.001	0.000	0.000
24	10	1.000	0.996	0.926	0.650	0.271	0.053	0.004	0.000	0.000
24	11	1.000	0.999	0.969	0.787	0.419	0.114	0.012	0.000	0.000
24	12	1.000	1.000	0.988	0.886	0.581	0.213	0.031	0.001	0.000
24	13	1.000	1.000	0.996	0.947	0.729	0.350	0.074	0.004	0.000
24	14	1.000	1.000	0.999	0.978	0.846	0.511	0.153	0.013	0.000
24	15	1.000	1.000	1.000	0.992	0.924	0.672	0.275	0.036	0.000
24	16	1.000	1.000	1.000	0.998	0.968	0.808	0.435	0.089	0.002

BINOMIAL CUMULATIVE DISTRIBUTION

n	x					p				
		.1	.2	.3	.4	.5	.6	.7	.8	.9
24	17	1.000	1.000	1.000	0.999	0.989	0.904	0.611	0.189	0.007
24	18	1.000	1.000	1.000	1.000	0.997	0.960	0.771	0.344	0.028
24	19	1.000	1.000	1.000	1.000	0.999	0.987	0.889	0.540	0.085
24	20	1.000	1.000	1.000	1.000	1.000	0.996	0.958	0.736	0.214
24	21	1.000	1.000	1.000	1.000	1.000	0.999	0.988	0.885	0.436
24	22	1.000	1.000	1.000	1.000	1.000	1.000	0.998	0.967	0.708
24	23	1.000	1.000	1.000	1.000	1.000	1.000	1.000	0.995	0.920
24	24	1.000	1.000	1.000	1.000	1.000	1.000	1.000	1.000	1.000
25	0	0.072	0.004	0.000	0.000	0.000	0.000	0.000	0.000	0.000
25	1	0.271	0.027	0.002	0.000	0.000	0.000	0.000	0.000	0.000
25	2	0.537	0.098	0.009	0.000	0.000	0.000	0.000	0.000	0.000
25	3	0.764	0.234	0.033	0.002	0.000	0.000	0.000	0.000	0.000
25	4	0.902	0.421	0.090	0.009	0.000	0.000	0.000	0.000	0.000
25	5	0.967	0.617	0.193	0.029	0.002	0.000	0.000	0.000	0.000
25	6	0.991	0.780	0.341	0.074	0.007	0.000	0.000	0.000	0.000
25	7	0.998	0.891	0.512	0.154	0.022	0.001	0.000	0.000	0.000
25	8	1.000	0.953	0.677	0.274	0.054	0.004	0.000	0.000	0.000
25	9	1.000	0.983	0.811	0.425	0.115	0.013	0.000	0.000	0.000
25	10	1.000	0.994	0.902	0.586	0.212	0.034	0.002	0.000	0.000
25	11	1.000	0.998	0.956	0.732	0.345	0.078	0.006	0.000	0.000
25	12	1.000	1.000	0.983	0.846	0.500	0.154	0.017	0.000	0.000
25	13	1.000	1.000	0.994	0.922	0.655	0.268	0.044	0.002	0.000
25	14	1.000	1.000	0.998	0.966	0.788	0.414	0.098	0.006	0.000
25	15	1.000	1.000	1.000	0.987	0.885	0.575	0.189	0.017	0.000
25	16	1.000	1.000	1.000	0.996	0.946	0.726	0.323	0.047	0.000
25	17	1.000	1.000	1.000	0.999	0.978	0.846	0.488	0.109	0.002
25	18	1.000	1.000	1.000	1.000	0.993	0.926	0.659	0.220	0.009
25	19	1.000	1.000	1.000	1.000	0.998	0.971	0.807	0.383	0.033
25	20	1.000	1.000	1.000	1.000	1.000	0.991	0.910	0.579	0.098
25	21	1.000	1.000	1.000	1.000	1.000	0.998	0.967	0.766	0.236
25	22	1.000	1.000	1.000	1.000	1.000	1.000	0.991	0.902	0.463
25	23	1.000	1.000	1.000	1.000	1.000	1.000	0.998	0.973	0.729
25	24	1.000	1.000	1.000	1.000	1.000	1.000	1.000	0.996	0.928
25	25	1.000	1.000	1.000	1.000	1.000	1.000	1.000	1.000	1.000

TABLE II: POISSON CUMULATIVE DISTRIBUTION

This table gives values of $F(x)$ where

$$F(x) = \mathscr{P}(X \leq x) = \frac{\sum_{k=0}^{x} e^{-\lambda}\lambda^k}{k!}$$

All entries of 0.000 in the table are actually larger than 0.0 but less than 0.0005. All entries of 1.000 are actually larger than 0.9995 but less than 1.

POISSON CUMULATIVE DISTRIBUTION

λ \ x	C	1	2	3	4	5	6	7
0.01	C.99C	1.000	1.000	1.000	1.000	1.000	1.000	1.000
0.02	0.980	1.000	1.C00	1.0C0	1.000	1.000	1.000	1.000
0.03	C.970	1.000	1.000	1.000	1.000	1.000	1.000	1.0C0
0.04	C.961	0.999	1.000	1.CCC	1.000	1.000	1.000	1.000
0.05	0.951	0.999	1.000	1.000	1.000	1.000	1.000	1.000
0.06	C.942	C.998	1.000	1.000	1.000	1.000	1.000	1.000
0.07	0.932	0.998	1.000	1.000	1.000	1.000	1.000	1.000
0.08	0.923	0.997	1.000	1.000	1.000	1.000	1.000	1.0C0
0.09	C.914	0.996	1.000	1.0CC	1.000	1.000	1.000	1.000
0.10	0.905	0.995	1.000	1.000	1.000	1.000	1.000	1.0C0
0.15	C.861	0.990	C.999	1.000	1.000	1.000	1.000	1.000
0.20	0.819	0.982	0.999	1.CCC	1.000	1.000	1.000	1.000
0.25	0.779	0.974	0.998	1.000	1.000	1.000	1.000	1.0C0
0.30	C.741	0.963	0.996	1.000	1.000	1.000	1.000	1.000
0.35	0.705	0.951	0.994	1.0C0	1.000	1.000	1.000	1.000
0.40	0.670	0.938	0.992	0.999	1.000	1.000	1.000	1.000
0.45	0.638	0.925	C.989	C.999	1.000	1.000	1.000	1.000
0.50	C.607	0.910	0.986	0.998	1.000	1.000	1.000	1.0C0
0.55	0.577	0.894	0.982	0.958	1.000	1.000	1.000	1.000
0.60	0.549	0.878	0.977	C.997	1.000	1.000	1.000	1.000
0.65	0.522	0.861	0.972	0.996	0.999	1.000	1.000	1.000
0.70	0.497	0.844	C.966	0.994	0.999	1.000	1.000	1.000
0.75	C.472	0.827	0.959	0.993	0.999	1.000	1.000	1.CC0
0.80	C.449	0.809	0.953	0.991	0.999	1.000	1.000	1.000
0.85	0.427	0.791	0.945	C.989	C.998	1.000	1.000	1.0C0
0.90	0.407	0.772	0.937	0.987	0.998	1.000	1.000	1.000
0.95	0.387	0.754	C.929	0.984	C.997	1.000	1.000	1.000
1.00	0.368	0.736	0.920	0.981	0.996	0.999	1.000	1.0C0
1.10	0.333	0.699	0.900	0.974	0.995	0.999	1.000	1.000
1.20	0.301	0.663	0.879	C.966	C.992	0.998	1.000	1.000
1.30	0.273	0.627	0.857	0.957	0.989	0.998	1.000	1.000
1.40	0.247	C.592	C.833	C.946	0.986	0.997	0.999	1.000
1.50	0.223	0.558	0.809	0.934	0.981	0.996	0.999	1.0C0
1.60	0.202	0.525	C.783	0.921	0.976	0.994	0.999	1.000
1.70	0.183	0.493	0.757	0.907	C.970	C.992	0.998	1.000
1.80	0.165	0.463	0.731	0.891	0.964	0.990	0.997	0.999
1.90	0.150	0.434	0.704	0.875	C.956	C.987	0.997	0.999
2.00	0.135	0.406	0.677	0.857	C.947	0.983	0.995	0.999
2.10	C.122	0.380	C.650	0.839	0.938	0.980	0.994	0.999
2.20	0.111	0.355	0.623	0.819	C.928	0.975	0.993	0.998
2.30	0.100	0.331	0.596	0.799	0.916	0.970	0.991	C.997
2.40	0.091	0.308	C.570	C.779	C.904	0.964	0.988	0.997
2.50	0.082	0.287	0.544	0.758	C.891	0.958	0.986	0.996
2.60	C.074	0.267	C.518	0.736	0.877	0.951	0.983	0.995
2.70	0.067	0.249	0.494	0.714	0.863	0.943	0.979	0.993
2.80	C.061	0.231	0.469	0.692	0.848	0.935	0.976	C.992
2.90	C.055	0.215	0.446	C.670	C.832	0.926	0.971	0.990
3.00	0.050	0.199	0.423	0.647	0.815	C.916	0.966	0.988

POISSON CUMULATIVE DISTRIBUTION

λ \ x	8	9	10	11	12	13	14	15
2.20	1.000	1.000	1.000	1.000	1.000	1.000	1.000	1.000
2.30	0.999	1.000	1.000	1.000	1.000	1.000	1.000	1.000
2.40	0.999	1.000	1.000	1.000	1.000	1.000	1.000	1.000
2.50	0.999	1.000	1.000	1.000	1.000	1.000	1.000	1.000
2.60	0.999	1.000	1.000	1.000	1.000	1.000	1.000	1.000
2.70	0.998	0.999	1.000	1.000	1.000	1.000	1.000	1.000
2.80	0.998	0.999	1.000	1.000	1.000	1.000	1.000	1.000
2.90	0.997	0.999	1.000	1.000	1.000	1.000	1.000	1.000
3.00	0.996	0.999	1.000	1.000	1.000	1.000	1.000	1.000

λ \ x	0	1	2	3	4	5	6	7
3.20	0.041	0.171	0.380	0.603	0.781	0.895	0.955	0.983
3.40	0.033	0.147	0.340	0.558	0.744	0.871	0.942	0.977
3.60	0.027	0.126	0.303	0.515	0.706	0.844	0.927	0.969
3.80	0.022	0.107	0.269	0.473	0.668	0.816	0.909	0.960
4.00	0.018	0.092	0.238	0.433	0.629	0.785	0.889	0.949
4.20	0.015	0.078	0.210	0.395	0.590	0.753	0.867	0.936
4.40	0.012	0.066	0.185	0.359	0.551	0.720	0.844	0.921
4.60	0.010	0.056	0.163	0.326	0.513	0.686	0.818	0.905
4.80	0.008	0.048	0.143	0.294	0.476	0.651	0.791	0.887
5.00	0.007	0.040	0.125	0.265	0.440	0.616	0.762	0.867
5.20	0.006	0.034	0.109	0.238	0.406	0.581	0.732	0.845
5.40	0.005	0.029	0.095	0.213	0.373	0.546	0.702	0.822
5.60	0.004	0.024	0.082	0.191	0.342	0.512	0.670	0.797
5.80	0.003	0.021	0.072	0.170	0.313	0.478	0.638	0.771
6.00	0.002	0.017	0.062	0.151	0.285	0.446	0.606	0.744
6.20	0.002	0.015	0.054	0.134	0.259	0.414	0.574	0.716
6.40	0.002	0.012	0.046	0.119	0.235	0.384	0.542	0.687
6.60	0.001	0.010	0.040	0.105	0.213	0.355	0.511	0.658
6.80	0.001	0.009	0.034	0.093	0.192	0.327	0.480	0.628
7.00	0.001	0.007	0.030	0.082	0.173	0.301	0.450	0.599
7.20	0.001	0.006	0.025	0.072	0.156	0.276	0.420	0.569
7.40	0.001	0.005	0.022	0.063	0.140	0.253	0.392	0.539
7.60	0.001	0.004	0.019	0.055	0.125	0.231	0.365	0.510
7.80	0.000	0.004	0.016	0.048	0.112	0.210	0.338	0.481
8.00	0.000	0.003	0.014	0.042	0.100	0.191	0.313	0.453
8.50	0.000	0.002	0.009	0.030	0.074	0.150	0.256	0.386
9.00	0.000	0.001	0.006	0.021	0.055	0.116	0.207	0.324
9.50	0.000	0.001	0.004	0.015	0.040	0.089	0.165	0.269
10.00	0.000	0.000	0.003	0.010	0.029	0.067	0.130	0.220
10.50	0.000	0.000	0.002	0.007	0.021	0.050	0.102	0.179
11.00	0.000	0.000	0.001	0.005	0.015	0.038	0.079	0.143
11.50	0.000	0.000	0.001	0.003	0.011	0.028	0.060	0.114
12.00	0.000	0.000	0.001	0.002	0.008	0.020	0.046	0.090
12.50	0.000	0.000	0.000	0.002	0.005	0.015	0.035	0.070
13.00	0.000	0.000	0.000	0.001	0.004	0.011	0.026	0.054

POISSON CUMULATIVE DISTRIBUTION

λ \ x	8	9	10	11	12	13	14	15
3.20	0.994	0.998	1.000	1.000	1.000	1.000	1.000	1.000
3.40	0.992	0.997	0.999	1.000	1.000	1.000	1.000	1.000
3.60	0.988	0.996	0.999	1.000	1.000	1.000	1.000	1.000
3.80	0.984	0.994	0.998	0.999	1.000	1.000	1.000	1.000
4.00	0.979	0.992	0.997	0.999	1.000	1.000	1.000	1.000
4.20	0.972	0.989	0.996	0.999	1.000	1.000	1.000	1.000
4.40	0.964	0.985	0.994	0.998	0.999	1.000	1.000	1.000
4.60	0.955	0.980	0.992	0.997	0.999	1.000	1.000	1.000
4.80	0.944	0.975	0.990	0.996	0.999	1.000	1.000	1.000
5.00	0.932	0.968	0.986	0.995	0.998	0.999	1.000	1.000
5.20	0.918	0.960	0.982	0.993	0.997	0.999	1.000	1.000
5.40	0.903	0.951	0.977	0.990	0.996	0.999	1.000	1.000
5.60	0.886	0.941	0.972	0.988	0.995	0.998	0.999	1.000
5.80	0.867	0.929	0.965	0.984	0.993	0.997	0.999	1.000
6.00	0.847	0.916	0.957	0.980	0.991	0.996	0.999	0.999
6.20	0.826	0.902	0.949	0.975	0.989	0.995	0.998	0.999
6.40	0.803	0.886	0.939	0.969	0.986	0.994	0.997	0.999
6.60	0.780	0.869	0.927	0.963	0.982	0.992	0.997	0.999
6.80	0.755	0.850	0.915	0.955	0.978	0.990	0.996	0.998
7.00	0.729	0.830	0.901	0.947	0.973	0.987	0.994	0.998
7.20	0.703	0.810	0.887	0.937	0.967	0.984	0.993	0.997
7.40	0.676	0.788	0.871	0.926	0.961	0.980	0.991	0.996
7.60	0.648	0.765	0.854	0.915	0.954	0.976	0.989	0.995
7.80	0.620	0.741	0.835	0.902	0.945	0.971	0.986	0.993
8.00	0.593	0.717	0.816	0.888	0.936	0.966	0.983	0.992
8.50	0.523	0.653	0.763	0.849	0.909	0.949	0.973	0.986
9.00	0.456	0.587	0.706	0.803	0.876	0.926	0.959	0.978
9.50	0.392	0.522	0.645	0.752	0.836	0.898	0.940	0.967
10.00	0.333	0.458	0.583	0.697	0.792	0.864	0.917	0.951
10.50	0.279	0.397	0.521	0.639	0.742	0.825	0.888	0.932
11.00	0.232	0.341	0.460	0.579	0.689	0.781	0.854	0.907
11.50	0.191	0.289	0.402	0.520	0.633	0.733	0.815	0.878
12.00	0.155	0.242	0.347	0.462	0.576	0.682	0.772	0.844
12.50	0.125	0.201	0.297	0.406	0.519	0.628	0.725	0.806
13.00	0.100	0.166	0.252	0.353	0.463	0.573	0.675	0.764

λ \ x	16	17	18	19	20	21	22	23
6.00	1.000	1.000	1.000	1.000	1.000	1.000	1.000	1.000
6.20	1.000	1.000	1.000	1.000	1.000	1.000	1.000	1.000
6.40	1.000	1.000	1.000	1.000	1.000	1.000	1.000	1.000
6.60	0.999	1.000	1.000	1.000	1.000	1.000	1.000	1.000
6.80	0.999	1.000	1.000	1.000	1.000	1.000	1.000	1.000
7.00	0.999	1.000	1.000	1.000	1.000	1.000	1.000	1.000
7.20	0.999	0.999	1.000	1.000	1.000	1.000	1.000	1.000
7.40	0.998	0.999	1.000	1.000	1.000	1.000	1.000	1.000
7.60	0.998	0.999	1.000	1.000	1.000	1.000	1.000	1.000
7.80	0.997	0.999	1.000	1.000	1.000	1.000	1.000	1.000
8.00	0.996	0.998	0.999	1.000	1.000	1.000	1.000	1.000

POISSON CUMULATIVE DISTRIBUTION

λ \ x	16	17	18	19	20	21	22	23
8.50	0.993	0.997	0.999	0.999	1.000	1.000	1.000	1.000
9.00	0.989	0.995	0.998	0.999	1.000	1.000	1.000	1.000
9.50	0.982	0.991	0.996	0.998	0.999	1.000	1.000	1.000
10.00	0.973	0.986	0.993	0.997	0.998	0.999	1.000	1.000
10.50	0.960	0.978	0.988	0.994	0.997	0.999	0.999	1.000
11.00	0.944	0.968	0.982	0.991	0.995	0.998	0.999	1.000
11.50	0.924	0.954	0.974	0.986	0.992	0.996	0.998	0.999
12.00	0.899	0.937	0.963	0.979	0.988	0.994	0.997	0.999
12.50	0.869	0.916	0.948	0.969	0.983	0.991	0.995	0.998
13.00	0.835	0.890	0.930	0.957	0.975	0.986	0.992	0.996

λ \ x	24	25	26	27	28	29	30	31
12.00	0.999	1.000	1.000	1.000	1.000	1.000	1.000	1.000
12.50	0.999	0.999	1.000	1.000	1.000	1.000	1.000	1.000
13.00	0.998	0.999	1.000	1.000	1.000	1.000	1.000	1.000

λ \ x	2	3	4	5	6	7	8	9
13.50	0.000	0.001	0.003	0.008	0.019	0.041	0.079	0.135
14.00	0.000	0.000	0.002	0.006	0.014	0.032	0.062	0.109
14.50	0.000	0.000	0.001	0.004	0.010	0.024	0.048	0.088
15.00	0.000	0.000	0.001	0.003	0.008	0.018	0.037	0.070
16.00	0.000	0.000	0.000	0.001	0.004	0.010	0.022	0.043
17.00	0.000	0.000	0.000	0.001	0.002	0.005	0.013	0.026
18.00	0.000	0.000	0.000	0.000	0.001	0.003	0.007	0.015
19.00	0.000	0.000	0.000	0.000	0.001	0.002	0.004	0.009
20.00	0.000	0.000	0.000	0.000	0.000	0.001	0.002	0.005
21.00	0.000	0.000	0.000	0.000	0.000	0.000	0.001	0.003
22.00	0.000	0.000	0.000	0.000	0.000	0.000	0.001	0.002
23.00	0.000	0.000	0.000	0.000	0.000	0.000	0.000	0.001

λ \ x	10	11	12	13	14	15	16	17
13.50	0.211	0.304	0.409	0.518	0.623	0.718	0.798	0.861
14.00	0.176	0.260	0.358	0.464	0.570	0.669	0.756	0.827
14.50	0.145	0.220	0.311	0.413	0.518	0.619	0.711	0.790
15.00	0.118	0.185	0.268	0.363	0.466	0.568	0.664	0.749
16.00	0.077	0.127	0.193	0.275	0.368	0.467	0.566	0.659
17.00	0.049	0.085	0.135	0.201	0.281	0.371	0.468	0.564
18.00	0.030	0.055	0.092	0.143	0.208	0.287	0.375	0.469
19.00	0.018	0.035	0.061	0.098	0.150	0.215	0.292	0.378
20.00	0.011	0.021	0.039	0.066	0.105	0.157	0.221	0.297
21.00	0.006	0.013	0.025	0.043	0.072	0.111	0.163	0.227
22.00	0.004	0.008	0.015	0.028	0.048	0.077	0.117	0.169
23.00	0.002	0.004	0.009	0.017	0.031	0.052	0.082	0.123
24.00	0.001	0.003	0.005	0.011	0.020	0.034	0.056	0.087
25.00	0.001	0.001	0.003	0.006	0.012	0.022	0.038	0.060

POISSON CUMULATIVE DISTRIBUTION

λ \ x	18	19	20	21	22	23	24	25
13.50	0.908	0.942	0.965	0.980	0.989	0.994	0.997	0.998
14.00	0.883	0.923	0.952	0.971	0.983	0.991	0.995	0.997
14.50	0.853	0.901	0.936	0.960	0.976	0.986	0.992	0.996
15.00	0.819	0.875	0.917	0.947	0.967	0.981	0.989	0.994
16.00	0.742	0.812	0.868	0.911	0.942	0.963	0.978	0.987
17.00	0.655	0.736	0.805	0.861	0.905	0.937	0.959	0.975
18.00	0.562	0.651	0.731	0.799	0.855	0.899	0.932	0.955
19.00	0.469	0.561	0.647	0.725	0.793	0.849	0.893	0.927
20.00	0.381	0.470	0.559	0.644	0.721	0.787	0.843	0.888
21.00	0.302	0.384	0.471	0.558	0.640	0.716	0.782	0.838
22.00	0.232	0.306	0.387	0.472	0.556	0.637	0.712	0.777
23.00	0.175	0.238	0.310	0.389	0.472	0.555	0.635	0.708
24.00	0.128	0.180	0.243	0.314	0.392	0.473	0.554	0.632
25.00	0.092	0.134	0.185	0.247	0.318	0.394	0.473	0.553

λ \ x	26	27	28	29	30	31	32	33
13.50	0.999	1.000	1.000	1.000	1.000	1.000	1.000	1.000
14.00	0.999	0.999	1.000	1.000	1.000	1.000	1.000	1.000
14.50	0.998	0.999	0.999	1.000	1.000	1.000	1.000	1.000
15.00	0.997	0.998	0.999	1.000	1.000	1.000	1.000	1.000
16.00	0.993	0.996	0.998	0.999	0.999	1.000	1.000	1.000
17.00	0.985	0.991	0.995	0.997	0.999	0.999	1.000	1.000
18.00	0.972	0.983	0.990	0.994	0.997	0.998	0.999	0.999
19.00	0.951	0.969	0.980	0.988	0.993	0.996	0.998	0.999
20.00	0.922	0.948	0.966	0.978	0.987	0.992	0.995	0.997
21.00	0.883	0.917	0.944	0.963	0.976	0.985	0.991	0.994
22.00	0.832	0.877	0.913	0.940	0.959	0.973	0.983	0.989
23.00	0.772	0.827	0.873	0.908	0.936	0.956	0.971	0.981
24.00	0.704	0.768	0.823	0.868	0.904	0.932	0.953	0.969
25.00	0.629	0.700	0.763	0.818	0.863	0.900	0.929	0.950

λ \ x	34	35	36	37	38	39	40	41
18.00	1.000	1.000	1.000	1.000	1.000	1.000	1.000	1.000
19.00	0.999	1.000	1.000	1.000	1.000	1.000	1.000	1.000
20.00	0.998	0.999	1.000	1.000	1.000	1.000	1.000	1.000
21.00	0.997	0.998	0.999	0.999	1.000	1.000	1.000	1.000
22.00	0.994	0.996	0.998	0.999	0.999	1.000	1.000	1.000
23.00	0.988	0.993	0.996	0.997	0.999	0.999	1.000	1.000
24.00	0.979	0.987	0.992	0.995	0.997	0.998	0.999	0.999
25.00	0.966	0.978	0.985	0.991	0.994	0.997	0.998	0.999

λ \ x	42	43	44	45	46	47	48	49
24.00	1.000	1.000	1.000	1.000	1.000	1.000	1.000	1.000
25.00	0.999	1.000	1.000	1.000	1.000	1.000	1.000	1.000

TABLE III: STANDARD NORMAL CUMULATIVE DISTRIBUTION

This table gives values of the standard normal cumulative distribution function

$$F(z) = \mathscr{P}(Z \le z) = \int_{-\infty}^{z} \frac{1}{\sqrt{2\pi}} e^{-t^2/2}\, dt$$

for values of $z = -4.00$ to 4.00 by 0.01 increments.

STANDARD NORMAL CUMULATIVE DISTRIBUTION

z	F(z)	z	F(z)	z	F(z)
-4.00	0.00003	-3.60	0.00016	-3.20	0.00069
-3.99	0.00003	-3.59	0.00017	-3.19	0.00071
-3.98	0.00003	-3.58	0.00017	-3.18	0.00074
-3.97	0.00004	-3.57	0.00018	-3.17	0.00076
-3.96	0.00004	-3.56	0.00019	-3.16	0.00079
-3.95	0.00004	-3.55	0.00019	-3.15	0.00082
-3.94	0.00004	-3.54	0.00020	-3.14	0.00084
-3.93	0.00004	-3.53	0.00021	-3.13	0.00087
-3.92	0.00004	-3.52	0.00022	-3.12	0.00090
-3.91	0.00005	-3.51	0.00022	-3.11	0.00094
-3.90	0.00005	-3.50	0.00023	-3.10	0.00097
-3.89	0.00005	-3.49	0.00024	-3.09	0.00100
-3.88	0.00005	-3.48	0.00025	-3.08	0.00104
-3.87	0.00005	-3.47	0.00026	-3.07	0.00107
-3.86	0.00006	-3.46	0.00027	-3.06	0.00111
-3.85	0.00006	-3.45	0.00028	-3.05	0.00114
-3.84	0.00006	-3.44	0.00029	-3.04	0.00118
-3.83	0.00006	-3.43	0.00030	-3.03	0.00122
-3.82	0.00007	-3.42	0.00031	-3.02	0.00126
-3.81	0.00007	-3.41	0.00032	-3.01	0.00131
-3.80	0.00007	-3.40	0.00034	-3.00	0.00135
-3.79	0.00008	-3.39	0.00035	-2.99	0.00139
-3.78	0.00008	-3.38	0.00036	-2.98	0.00144
-3.77	0.00008	-3.37	0.00038	-2.97	0.00149
-3.76	0.00008	-3.36	0.00039	-2.96	0.00154
-3.75	0.00009	-3.35	0.00040	-2.95	0.00159
-3.74	0.00009	-3.34	0.00042	-2.94	0.00164
-3.73	0.00010	-3.33	0.00043	-2.93	0.00169
-3.72	0.00010	-3.32	0.00045	-2.92	0.00175
-3.71	0.00010	-3.31	0.00047	-2.91	0.00181
-3.70	0.00011	-3.30	0.00048	-2.90	0.00187
-3.69	0.00011	-3.29	0.00050	-2.89	0.00193
-3.68	0.00012	-3.28	0.00052	-2.88	0.00199
-3.67	0.00012	-3.27	0.00054	-2.87	0.00205
-3.66	0.00013	-3.26	0.00056	-2.86	0.00212
-3.65	0.00013	-3.25	0.00058	-2.85	0.00219
-3.64	0.00014	-3.24	0.00060	-2.84	0.00226
-3.63	0.00014	-3.23	0.00062	-2.83	0.00233
-3.62	0.00015	-3.22	0.00064	-2.82	0.00240
-3.61	0.00015	-3.21	0.00066	-2.81	0.00248

STANDARD NORMAL CUMULATIVE DISTRIBUTION

z	F(z)	z	F(z)	z	F(z)
-2.80	0.00256	-2.40	0.00820	-2.00	0.02275
-2.79	0.00264	-2.39	0.00842	-1.99	0.02330
-2.78	0.00272	-2.38	0.00866	-1.98	0.02385
-2.77	0.00280	-2.37	0.00889	-1.97	0.02442
-2.76	0.00289	-2.36	0.00914	-1.96	0.02500
-2.75	0.00298	-2.35	0.00939	-1.95	0.02559
-2.74	0.00307	-2.34	0.00964	-1.94	0.02619
-2.73	0.00317	-2.33	0.00990	-1.93	0.02680
-2.72	0.00326	-2.32	0.01017	-1.92	0.02743
-2.71	0.00336	-2.31	0.01044	-1.91	0.02807
-2.70	0.00347	-2.30	0.01072	-1.90	0.02872
-2.69	0.00357	-2.29	0.01101	-1.89	0.02938
-2.68	0.00368	-2.28	0.01130	-1.88	0.03005
-2.67	0.00379	-2.27	0.01160	-1.87	0.03074
-2.66	0.00391	-2.26	0.01191	-1.86	0.03144
-2.65	0.00402	-2.25	0.01222	-1.85	0.03216
-2.64	0.00415	-2.24	0.01255	-1.84	0.03288
-2.63	0.00427	-2.23	0.01287	-1.83	0.03362
-2.62	0.00440	-2.22	0.01321	-1.82	0.03438
-2.61	0.00453	-2.21	0.01355	-1.81	0.03515
-2.60	0.00466	-2.20	0.01390	-1.80	0.03593
-2.59	0.00480	-2.19	0.01426	-1.79	0.03673
-2.58	0.00494	-2.18	0.01463	-1.78	0.03754
-2.57	0.00508	-2.17	0.01500	-1.77	0.03836
-2.56	0.00523	-2.16	0.01539	-1.76	0.03920
-2.55	0.00539	-2.15	0.01578	-1.75	0.04006
-2.54	0.00554	-2.14	0.01618	-1.74	0.04093
-2.53	0.00570	-2.13	0.01659	-1.73	0.04182
-2.52	0.00587	-2.12	0.01700	-1.72	0.04272
-2.51	0.00604	-2.11	0.01743	-1.71	0.04363
-2.50	0.00621	-2.10	0.01786	-1.70	0.04457
-2.49	0.00639	-2.09	0.01831	-1.69	0.04551
-2.48	0.00657	-2.08	0.01876	-1.68	0.04648
-2.47	0.00676	-2.07	0.01923	-1.67	0.04746
-2.46	0.00695	-2.06	0.01970	-1.66	0.04846
-2.45	0.00714	-2.05	0.02018	-1.65	0.04947
-2.44	0.00734	-2.04	0.02068	-1.64	0.05050
-2.43	0.00755	-2.03	0.02118	-1.63	0.05155
-2.42	0.00776	-2.02	0.02169	-1.62	0.05262
-2.41	0.00798	-2.01	0.02222	-1.61	0.05370

STANDARD NORMAL CUMULATIVE DISTRIBUTION

z	F(z)	z	F(z)	z	F(z)
-1.60	0.05480	-1.20	0.11507	-0.80	0.21186
-1.59	0.05592	-1.19	0.11702	-0.79	0.21476
-1.58	0.05705	-1.18	0.11900	-0.78	0.21770
-1.57	0.05821	-1.17	0.12100	-0.77	0.22065
-1.56	0.05938	-1.16	0.12302	-0.76	0.22363
-1.55	0.06057	-1.15	0.12507	-0.75	0.22663
-1.54	0.06178	-1.14	0.12714	-0.74	0.22965
-1.53	0.06301	-1.13	0.12924	-0.73	0.23270
-1.52	0.06426	-1.12	0.13136	-0.72	0.23576
-1.51	0.06552	-1.11	0.13350	-0.71	0.23885
-1.50	0.06681	-1.10	0.13567	-0.70	0.24196
-1.49	0.06811	-1.09	0.13786	-0.69	0.24510
-1.48	0.06944	-1.08	0.14007	-0.68	0.24825
-1.47	0.07078	-1.07	0.14231	-0.67	0.25143
-1.46	0.07215	-1.06	0.14457	-0.66	0.25463
-1.45	0.07353	-1.05	0.14686	-0.65	0.25785
-1.44	0.07493	-1.04	0.14917	-0.64	0.26109
-1.43	0.07636	-1.03	0.15150	-0.63	0.26435
-1.42	0.07780	-1.02	0.15386	-0.62	0.26763
-1.41	0.07927	-1.01	0.15625	-0.61	0.27093
-1.40	0.08076	-1.00	0.15866	-0.60	0.27425
-1.39	0.08226	-0.99	0.16109	-0.59	0.27760
-1.38	0.08379	-0.98	0.16354	-0.58	0.28096
-1.37	0.08534	-0.97	0.16602	-0.57	0.28434
-1.36	0.08691	-0.96	0.16853	-0.56	0.28774
-1.35	0.08851	-0.95	0.17106	-0.55	0.29116
-1.34	0.09012	-0.94	0.17361	-0.54	0.29460
-1.33	0.09176	-0.93	0.17619	-0.53	0.29806
-1.32	0.09342	-0.92	0.17879	-0.52	0.30153
-1.31	0.09510	-0.91	0.18141	-0.51	0.30503
-1.30	0.09680	-0.90	0.18406	-0.50	0.30854
-1.29	0.09853	-0.89	0.18673	-0.49	0.31207
-1.28	0.10027	-0.88	0.18943	-0.48	0.31561
-1.27	0.10204	-0.87	0.19215	-0.47	0.31918
-1.26	0.10383	-0.86	0.19489	-0.46	0.32276
-1.25	0.10565	-0.85	0.19766	-0.45	0.32636
-1.24	0.10749	-0.84	0.20045	-0.44	0.32997
-1.23	0.10935	-0.83	0.20327	-0.43	0.33360
-1.22	0.11123	-0.82	0.20611	-0.42	0.33724
-1.21	0.11314	-0.81	0.20897	-0.41	0.34090

STANDARD NORMAL CUMULATIVE DISTRIBUTION

z	F(z)	z	F(z)	z	F(z)
-0.40	C.34458	0.0	0.50000	0.40	0.65542
-0.39	C.34927	C.01	0.50399	0.41	0.65910
-0.38	0.35197	0.02	0.50798	0.42	0.66276
-0.37	0.35569	0.03	0.51197	0.43	0.66640
-0.36	0.35942	C.04	0.51595	0.44	0.67003
-0.35	0.36317	0.05	0.51994	0.45	0.67364
-0.34	0.36693	C.06	0.52392	0.46	0.67724
-0.33	C.37070	0.07	0.52790	0.47	C.68082
-0.32	C.37448	0.08	0.53188	0.48	0.68439
-0.31	0.37828	0.09	0.53586	0.49	0.68793
-0.30	C.38209	C.10	0.53983	0.50	0.69146
-0.29	0.38591	0.11	0.54380	0.51	0.69497
-0.28	C.38974	0.12	0.54776	0.52	0.69847
-0.27	0.39358	0.13	0.55172	0.53	0.70194
-0.26	C.39743	0.14	0.55567	0.54	0.70540
-0.25	C.40129	C.15	0.55962	0.55	C.70884
-0.24	C.40517	0.16	0.56356	0.56	0.71226
-0.23	C.40905	0.17	0.56749	0.57	0.71566
-0.22	0.41294	0.18	0.57142	C.58	0.71904
-0.21	C.41683	0.19	0.57535	0.59	0.72240
-0.20	0.42074	0.20	0.57926	0.60	0.72575
-0.19	C.42465	0.21	0.58317	0.61	0.72907
-0.18	C.42858	C.22	0.58706	0.62	C.73237
-0.17	0.43251	0.23	0.59095	0.63	0.73565
-0.16	C.43644	0.24	0.59483	0.64	0.73891
-0.15	C.44038	0.25	0.59871	0.65	C.74215
-0.14	C.44433	C.26	0.60257	0.66	0.74537
-0.13	0.44828	0.27	0.60642	0.67	C.74857
-0.12	C.45224	0.28	0.61026	0.68	0.75175
-0.11	C.45620	C.29	0.61409	0.69	0.75490
-0.10	C.46017	0.30	0.61791	0.70	0.75804
-0.09	0.46414	0.31	0.62172	C.71	0.76115
-0.08	0.46812	0.32	0.62552	0.72	0.76424
-0.07	C.47210	C.33	0.62930	0.73	0.76730
-0.06	0.47608	0.34	0.63307	0.74	0.77035
-0.05	C.48006	0.35	0.63683	0.75	0.77337
-0.04	0.48405	0.36	0.64058	0.76	0.77637
-0.03	C.48803	0.37	0.64431	0.77	0.77935
-0.02	0.49202	0.38	0.64803	C.78	C.78230
-0.01	0.49601	0.39	0.65173	0.79	0.78524

STANDARD NORMAL CUMULATIVE DISTRIBUTION

z	F(z)	z	F(z)	z	F(z)
0.80	0.78314	1.20	0.88493	1.60	0.94520
0.81	0.79103	1.21	0.88686	1.61	0.94630
0.82	0.79389	1.22	0.88877	1.62	0.94738
0.83	0.79673	1.23	0.89065	1.63	0.94845
0.84	0.79955	1.24	0.89251	1.64	0.94950
0.85	0.80234	1.25	0.89435	1.65	0.95053
0.86	0.80511	1.26	0.89617	1.66	0.95154
0.87	0.80785	1.27	0.89796	1.67	0.95254
0.88	0.81057	1.28	0.89973	1.68	0.95352
0.89	0.81327	1.29	0.90147	1.69	0.95449
0.90	0.81594	1.30	0.90320	1.70	0.95543
0.91	0.81859	1.31	0.90490	1.71	0.95637
0.92	0.82121	1.32	0.90658	1.72	0.95728
0.93	0.82381	1.33	0.90824	1.73	0.95818
0.94	0.82639	1.34	0.90988	1.74	0.95907
0.95	0.82894	1.35	0.91149	1.75	0.95994
0.96	0.83147	1.36	0.91309	1.76	0.96080
0.97	0.83398	1.37	0.91466	1.77	0.96164
0.98	0.83646	1.38	0.91621	1.78	0.96246
0.99	0.83891	1.39	0.91774	1.79	0.96327
1.00	0.84134	1.40	0.91924	1.80	0.96407
1.01	0.84375	1.41	0.92073	1.81	0.96485
1.02	0.84614	1.42	0.92220	1.82	0.96562
1.03	0.84850	1.43	0.92364	1.83	0.96638
1.04	0.85083	1.44	0.92507	1.84	0.96712
1.05	0.85314	1.45	0.92647	1.85	0.96784
1.06	0.85543	1.46	0.92785	1.86	0.96856
1.07	0.85769	1.47	0.92922	1.87	0.96926
1.08	0.85993	1.48	0.93056	1.88	0.96995
1.09	0.86214	1.49	0.93189	1.89	0.97062
1.10	0.86433	1.50	0.93319	1.90	0.97128
1.11	0.86650	1.51	0.93448	1.91	0.97193
1.12	0.86864	1.52	0.93574	1.92	0.97257
1.13	0.87076	1.53	0.93699	1.93	0.97320
1.14	0.87286	1.54	0.93822	1.94	0.97381
1.15	0.87493	1.55	0.93943	1.95	0.97441
1.16	0.87698	1.56	0.94062	1.96	0.97500
1.17	0.87900	1.57	0.94179	1.97	0.97558
1.18	0.88100	1.58	0.94295	1.98	0.97615
1.19	0.88298	1.59	0.94408	1.99	0.97670

STANDARD NORMAL CUMULATIVE DISTRIBUTION

z	F(z)	z	F(z)	z	F(z)
2.00	0.97725	2.40	0.99180	2.80	0.99744
2.01	0.97778	2.41	0.99202	2.81	0.99752
2.02	0.97831	2.42	0.99224	2.82	0.99760
2.03	0.97882	2.43	0.99245	2.83	0.99767
2.04	0.97932	2.44	0.99266	2.84	0.99774
2.05	0.97982	2.45	0.99286	2.85	0.99781
2.06	0.98030	2.46	0.99305	2.86	0.99788
2.07	0.98077	2.47	0.99324	2.87	0.99795
2.08	0.98124	2.48	0.99343	2.88	0.99801
2.09	0.98169	2.49	0.99361	2.89	0.99807
2.10	0.98214	2.50	0.99379	2.90	0.99813
2.11	0.98257	2.51	0.99396	2.91	0.99819
2.12	0.98300	2.52	0.99413	2.92	0.99825
2.13	0.98341	2.53	0.99430	2.93	0.99831
2.14	0.98382	2.54	0.99446	2.94	0.99836
2.15	0.98422	2.55	0.99461	2.95	0.99841
2.16	0.98461	2.56	0.99477	2.96	0.99846
2.17	0.98500	2.57	0.99492	2.97	0.99851
2.18	0.98537	2.58	0.99506	2.98	0.99856
2.19	0.98574	2.59	0.99520	2.99	0.99861
2.20	0.98610	2.60	0.99534	3.00	0.99865
2.21	0.98645	2.61	0.99547	3.01	0.99869
2.22	0.98679	2.62	0.99560	3.02	0.99874
2.23	0.98713	2.63	0.99573	3.03	0.99878
2.24	0.98745	2.64	0.99585	3.04	0.99882
2.25	0.98778	2.65	0.99598	3.05	0.99886
2.26	0.98809	2.66	0.99609	3.06	0.99889
2.27	0.98840	2.67	0.99621	3.07	0.99893
2.28	0.98870	2.68	0.99632	3.08	0.99896
2.29	0.98899	2.69	0.99643	3.09	0.99900
2.30	0.98928	2.70	0.99653	3.10	0.99903
2.31	0.98956	2.71	0.99664	3.11	0.99906
2.32	0.98983	2.72	0.99674	3.12	0.99910
2.33	0.99010	2.73	0.99683	3.13	0.99913
2.34	0.99036	2.74	0.99693	3.14	0.99916
2.35	0.99061	2.75	0.99702	3.15	0.99918
2.36	0.99086	2.76	0.99711	3.16	0.99921
2.37	0.99111	2.77	0.99720	3.17	0.99924
2.38	0.99134	2.78	0.99728	3.18	0.99926
2.39	0.99158	2.79	0.99736	3.19	0.99929

STANDARD NORMAL CUMULATIVE DISTRIBUTION

z	F(z)	z	F(z)	z	F(z)
3.20	0.99931	3.50	0.99977	3.80	0.99993
3.21	0.99934	3.51	0.99978	3.81	0.99993
3.22	0.99936	3.52	0.99978	3.82	0.99993
3.23	0.99938	3.53	0.99979	3.83	0.99994
3.24	0.99940	3.54	0.99980	3.84	0.99994
3.25	0.99942	3.55	0.99981	3.85	0.99994
3.26	0.99944	3.56	0.99981	3.86	0.99994
3.27	0.99946	3.57	0.99982	3.87	0.99995
3.28	0.99948	3.58	0.99983	3.88	0.99995
3.29	0.99950	3.59	0.99983	3.89	0.99995
3.30	0.99952	3.60	0.99984	3.90	0.99995
3.31	0.99953	3.61	0.99985	3.91	0.99995
3.32	0.99955	3.62	0.99985	3.92	0.99996
3.33	0.99957	3.63	0.99986	3.93	0.99996
3.34	0.99958	3.64	0.99986	3.94	0.99996
3.35	0.99960	3.65	0.99987	3.95	0.99996
3.36	0.99961	3.66	0.99987	3.96	0.99996
3.37	0.99962	3.67	0.99988	3.97	0.99996
3.38	0.99964	3.68	0.99988	3.98	0.99997
3.39	0.99965	3.69	0.99989	3.99	0.99997
3.40	0.99966	3.70	0.99989	4.00	0.99997
3.41	0.99968	3.71	0.99990		
3.42	0.99969	3.72	0.99990		
3.43	0.99970	3.73	0.99990		
3.44	0.99971	3.74	0.99991		
3.45	0.99972	3.75	0.99991		
3.46	0.99973	3.76	0.99992		
3.47	0.99974	3.77	0.99992		
3.48	0.99975	3.78	0.99992		
3.49	0.99976	3.79	0.99992		

TABLE IV: CUMULATIVE *t*-DISTRIBUTION

This table gives selected critical values of Student's t-distribution. Entries in the table are values of t_p where

$$p = \mathscr{P}(T \leq t_p),$$

where T has Student's t-distribution with v degrees of freedom. The table covers values of $p = 0.9, 0.95, 0.97, 0.99,$ and 0.995 and $v = 1(1)30(10)100,$ $120, 200, \infty$. For $v = \infty$ the critical values are actually those of the standard normal distribution.

CUMULATIVE *t*-DISTRIBUTION

df, ν	p				
	0.900	0.950	0.975	0.990	0.995
1	3.078	6.314	12.706	31.821	63.657
2	1.886	2.920	4.303	6.965	9.925
3	1.638	2.353	3.182	4.541	5.841
4	1.533	2.132	2.776	3.747	4.604
5	1.476	2.015	2.571	3.365	4.032
6	1.440	1.943	2.447	3.143	3.707
7	1.415	1.895	2.365	2.998	3.499
8	1.397	1.860	2.306	2.896	3.355
9	1.383	1.833	2.262	2.821	3.250
10	1.372	1.812	2.228	2.764	3.169
11	1.363	1.796	2.201	2.718	3.106
12	1.356	1.782	2.179	2.681	3.055
13	1.350	1.771	2.160	2.650	3.012
14	1.345	1.761	2.145	2.624	2.977
15	1.341	1.753	2.131	2.602	2.947
16	1.337	1.746	2.120	2.583	2.921
17	1.333	1.740	2.110	2.567	2.898
18	1.330	1.734	2.101	2.552	2.878
19	1.328	1.729	2.093	2.539	2.861
20	1.325	1.725	2.086	2.528	2.845
21	1.323	1.721	2.080	2.518	2.831
22	1.321	1.717	2.074	2.508	2.819
23	1.319	1.714	2.069	2.500	2.807
24	1.318	1.711	2.064	2.492	2.797
25	1.316	1.708	2.060	2.485	2.787
26	1.315	1.706	2.056	2.479	2.779
27	1.314	1.703	2.052	2.473	2.771
28	1.313	1.701	2.048	2.467	2.763
29	1.311	1.699	2.045	2.462	2.756
30	1.310	1.697	2.042	2.457	2.750
40	1.303	1.684	2.021	2.423	2.704
50	1.299	1.676	2.008	2.403	2.678
60	1.296	1.671	2.000	2.390	2.660
70	1.294	1.667	1.994	2.381	2.648
80	1.293	1.665	1.990	2.374	2.638
90	1.291	1.662	1.987	2.369	2.632
100	1.290	1.661	1.984	2.364	2.626
120	1.289	1.658	1.980	2.358	2.617
200	1.286	1.653	1.972	2.345	2.601
∞	1.282	1.645	1.960	2.326	2.576

TABLE V: CUMULATIVE CHI-SQUARE (χ^2) DISTRIBUTION

This table gives critical values of the χ^2-distribution. Entries in the table are values of χ_p^2 where

$$p = \mathscr{P}(\chi^2 \le \chi_p^2)$$

and χ^2 has a chi-square distribution with v degrees of freedom.

The table covers values of $p = 0.005, 0.01, 0.025, 0.05, 0.10, 0.90, 0.95, 0.975, 0.99,$ and $0.995,$ and $v = 1(1)100.$

CUMULATIVE χ^2 -DISTRIBUTION

df, v	p				
	0.005	0.010	0.025	0.050	0.100
1	.000039	.000157	.000982	.00393	.0158
2	.0100	.0201	.0506	.103	.211
3	.0717	.115	.216	.352	.584
4	.207	.297	.484	.711	1.06
5	.412	.554	.831	1.15	1.61
6	.676	.872	1.24	1.64	2.20
7	.989	1.24	1.69	2.17	2.83
8	1.34	1.65	2.18	2.73	3.49
9	1.73	2.09	2.70	3.33	4.17
10	2.16	2.56	3.25	3.94	4.87
11	2.60	3.05	3.82	4.57	5.58
12	3.07	3.57	4.40	5.23	6.30
13	3.57	4.11	5.01	5.89	7.04
14	4.07	4.66	5.63	6.57	7.79
15	4.60	5.23	6.26	7.26	8.55
16	5.14	5.81	6.91	7.96	9.31
17	5.70	6.41	7.56	8.67	10.1
18	6.26	7.01	8.23	9.39	10.9
19	6.84	7.63	8.91	10.1	11.7
20	7.43	8.26	9.59	10.9	12.4
21	8.03	8.90	10.3	11.6	13.2
22	8.64	9.54	11.0	12.3	14.0
23	9.26	10.2	11.7	13.1	14.8
24	9.89	10.9	12.4	13.8	15.7

CUMULATIVE χ^2 -DISTRIBUTION

df, ν	0.005	0.010	0.025	0.050	0.100
			p		
25	10.5	11.5	13.1	14.6	16.5
26	11.2	12.2	13.8	15.4	17.3
27	11.8	12.9	14.6	16.2	18.1
28	12.5	13.6	15.3	16.9	18.9
29	13.1	14.3	16.0	17.7	19.8
30	13.8	15.0	16.8	18.5	20.6
31	14.5	15.7	17.5	19.3	21.4
32	15.1	16.4	18.3	20.1	22.3
33	15.8	17.1	19.0	20.9	23.1
34	16.5	17.8	19.8	21.7	24.0
35	17.2	18.5	20.6	22.5	24.8
36	17.9	19.2	21.3	23.3	25.6
37	18.6	20.0	22.1	24.1	26.5
38	19.3	20.7	22.9	24.9	27.3
39	20.0	21.4	23.7	25.7	28.2
40	20.7	22.2	24.4	26.5	29.1
41	21.4	22.9	25.2	27.3	29.9
42	22.1	23.7	26.0	28.1	30.8
43	22.9	24.4	26.8	29.0	31.6
44	23.6	25.1	27.6	29.8	32.5
45	24.3	25.9	28.4	30.6	33.4
46	25.0	26.7	29.2	31.4	34.2
47	25.8	27.4	30.0	32.3	35.1
48	26.5	28.2	30.8	33.1	35.9
49	27.2	28.9	31.6	33.9	36.8
50	28.0	29.7	32.4	34.8	37.7
51	28.7	30.5	33.2	35.6	38.6
52	29.5	31.2	34.0	36.4	39.4
53	30.2	32.0	34.8	37.3	40.3
54	31.0	32.8	35.6	38.1	41.2
55	31.7	33.6	36.4	39.0	42.1
56	32.5	34.3	37.2	39.8	42.9
57	33.2	35.1	38.0	40.6	43.8
58	34.0	35.9	38.8	41.5	44.7
59	34.8	36.7	39.7	42.3	45.6
60	35.5	37.5	40.5	43.2	46.5
61	36.3	38.3	41.3	44.0	47.3
62	37.1	39.1	42.1	44.9	48.2

CUMULATIVE χ^2 -DISTRIBUTION

df, ν	0.005	0.010	0.025	0.050	0.100
63	37.8	39.9	43.0	45.7	49.1
64	38.6	40.6	43.8	46.6	50.0
65	39.4	41.4	44.6	47.4	50.9
66	40.2	42.2	45.4	48.3	51.8
67	40.9	43.0	46.3	49.2	52.7
68	41.7	43.8	47.1	50.0	53.5
69	42.5	44.6	47.9	50.9	54.4
70	43.3	45.4	48.8	51.7	55.3
71	44.1	46.2	49.6	52.6	56.2
72	44.8	47.1	50.4	53.5	57.1
73	45.6	47.9	51.3	54.3	58.0
74	46.4	48.7	52.1	55.2	58.9
75	47.2	49.5	52.9	56.1	59.8
76	48.0	50.3	53.8	56.9	60.7
77	48.8	51.1	54.6	57.8	61.6
78	49.6	51.9	55.5	58.7	62.5
79	50.4	52.7	56.3	59.5	63.4
80	51.2	53.5	57.2	60.4	64.3
81	52.0	54.4	58.0	61.3	65.2
82	52.8	55.2	58.8	62.1	66.1
83	53.6	56.0	59.7	63.0	67.0
84	54.4	56.8	60.5	63.9	67.9
85	55.2	57.6	61.4	64.7	68.8
86	56.0	58.5	62.2	65.6	69.7
87	56.8	59.3	63.1	66.5	70.6
88	57.6	60.1	63.9	67.4	71.5
89	58.4	60.9	64.8	68.2	72.4
90	59.2	61.8	65.6	69.1	73.3
91	60.0	62.6	66.5	70.0	74.2
92	60.8	63.4	67.4	70.9	75.1
93	61.6	64.2	68.2	71.8	76.0
94	62.4	65.1	69.1	72.6	76.9
95	63.2	65.9	69.9	73.5	77.8
96	64.1	66.7	70.8	74.4	78.7
97	64.9	67.6	71.6	75.3	79.6
98	65.7	68.4	72.5	76.2	80.5
99	66.5	69.2	73.4	77.0	81.4
100	67.3	70.1	74.2	77.9	82.4

CUMULATIVE χ^2 -DISTRIBUTION

df, ν	p 0.900	0.950	0.975	0.990	0.995
1	2.71	3.84	5.02	6.63	7.88
2	4.61	5.99	7.38	9.21	10.6
3	6.25	7.81	9.35	11.3	12.8
4	7.78	9.49	11.1	13.3	14.9
5	9.24	11.1	12.8	15.1	16.7
6	10.6	12.6	14.4	16.8	18.5
7	12.0	14.1	16.0	18.5	20.3
8	13.4	15.5	17.5	20.1	22.0
9	14.7	16.9	19.0	21.7	23.6
10	16.0	18.3	20.5	23.2	25.2
11	17.3	19.7	21.9	24.7	26.8
12	18.5	21.0	23.3	26.2	28.3
13	19.8	22.4	24.7	27.7	29.8
14	21.1	23.7	26.1	29.1	31.3
15	22.3	25.0	27.5	30.6	32.8
16	23.5	26.3	28.8	32.0	34.3
17	24.8	27.6	30.2	33.4	35.7
18	26.0	28.9	31.5	34.8	37.2
19	27.2	30.1	32.9	36.2	38.6
20	28.4	31.4	34.2	37.6	40.0
21	29.6	32.7	35.5	38.9	41.4
22	30.8	33.9	36.8	40.3	42.8
23	32.0	35.2	38.1	41.6	44.2
24	33.2	36.4	39.4	43.0	45.6
25	34.4	37.7	40.6	44.3	46.9
26	35.6	38.9	41.9	45.6	48.3
27	36.7	40.1	43.2	47.0	49.6
28	37.9	41.3	44.5	48.3	51.0
29	39.1	42.6	45.7	49.6	52.3
30	40.3	43.8	47.0	50.9	53.7
31	41.4	45.0	48.2	52.2	55.0
32	42.6	46.2	49.5	53.5	56.3
33	43.7	47.4	50.7	54.8	57.6
34	44.9	48.6	52.0	56.1	59.0
35	46.1	49.8	53.2	57.3	60.3
36	47.2	51.0	54.4	58.6	61.6
37	48.4	52.2	55.7	59.9	62.9
38	49.5	53.4	56.9	61.2	64.2

CUMULATIVE χ^2 -DISTRIBUTION

df, ν	p				
	0.900	0.950	0.975	0.990	0.995
39	50.7	54.6	58.1	62.4	65.5
40	51.8	55.8	59.3	63.7	66.8
41	52.9	56.9	60.6	65.0	68.1
42	54.1	58.1	61.8	66.2	69.3
43	55.2	59.3	63.0	67.5	70.6
44	56.4	60.5	64.2	68.7	71.9
45	57.5	61.7	65.4	70.0	73.2
46	58.6	62.8	66.6	71.2	74.4
47	59.8	64.0	67.8	72.4	75.7
48	60.9	65.2	69.0	73.7	77.0
49	62.0	66.3	70.2	74.9	78.2
50	63.2	67.5	71.4	76.2	79.5
51	64.3	68.7	72.6	77.4	80.7
52	65.4	69.8	73.8	78.6	82.0
53	66.5	71.0	75.0	79.8	83.3
54	67.7	72.2	76.2	81.1	84.5
55	68.8	73.3	77.4	82.3	85.7
56	69.9	74.5	78.6	83.5	87.0
57	71.0	75.6	79.8	84.7	88.2
58	72.2	76.8	80.9	86.0	89.5
59	73.3	77.9	82.1	87.2	90.7
60	74.4	79.1	83.3	88.4	92.0
61	75.5	80.2	84.5	89.6	93.2
62	76.6	81.4	85.7	90.8	94.4
63	77.7	82.5	86.8	92.0	95.6
64	78.9	83.7	88.0	93.2	96.9
65	80.0	84.8	89.2	94.4	98.1
66	81.1	86.0	90.3	95.6	99.3
67	82.2	87.1	91.5	96.8	100.6
68	83.3	88.3	92.7	98.0	101.8
69	84.4	89.4	93.9	99.2	103.0
70	85.5	90.5	95.0	100.4	104.2
71	86.6	91.7	96.2	101.6	105.4
72	87.7	92.8	97.4	102.8	106.6
73	88.8	93.9	98.5	104.0	107.9
74	90.0	95.1	99.7	105.2	109.1
75	91.1	96.2	100.8	106.4	110.3
76	92.2	97.4	102.0	107.6	111.5

CUMULATIVE χ^2 -DISTRIBUTION

df, ν	p				
	0.900	0.950	0.975	0.990	0.995
77	93.3	98.5	103.2	108.8	112.7
78	94.4	99.6	104.3	110.0	113.9
79	95.5	100.7	105.5	111.1	115.1
80	96.6	101.9	106.6	112.3	116.3
81	97.7	103.0	107.8	113.5	117.5
82	98.8	104.1	108.9	114.7	118.7
83	99.9	105.3	110.1	115.9	119.9
84	101.0	106.4	111.2	117.1	121.1
85	102.1	107.5	112.4	118.2	122.3
86	103.2	108.6	113.5	119.4	123.5
87	104.3	109.8	114.7	120.6	124.7
88	105.4	110.9	115.8	121.8	125.9
89	106.5	112.0	117.0	122.9	127.1
90	107.6	113.1	118.1	124.1	128.3
91	108.7	114.3	119.3	125.3	129.5
92	109.8	115.4	120.4	126.5	130.7
93	110.9	116.5	121.6	127.6	131.9
94	111.9	117.6	122.7	128.8	133.1
95	113.0	118.8	123.9	130.0	134.2
96	114.1	119.9	125.0	131.1	135.4
97	115.2	121.0	126.1	132.3	136.6
98	116.3	122.1	127.3	133.5	137.8
99	117.4	123.2	128.4	134.6	139.0
100	118.5	124.3	129.6	135.8	140.2

TABLE VI: CUMULATIVE *F*-DISTRIBUTION

This table gives critical values of the F-distribution. Entries in the table are values of F_p where

$$p = \mathscr{P}(F \le F_p)$$

and F has an F-distribution with v_1 (numerator) and v_2 (denominator) degrees of freedom. The table covers values v_1 (and v_2) = 1(1)15, 20, 24, 30, 40, 50, 60, 100, 120, ∞ and p = 0.005, 0.01, 0.025, 0.05, 0.10, 0.90, 0.95, 0.975, 0.99, 0.995.

CUMULATIVE F-DISTRIBUTION

ν_2	p	ν_1					
		1	2	3	4	5	6
1	.005	.C00062	.0050	.C18	.032	.044	.054
	.010	.00025	.010	.029	.047	.062	.073
	.025	.00154	.026	.057	.082	.100	.113
	.050	.00619	.054	.C99	.130	.151	.167
	.100	.025	.117	.181	.220	.246	.265
	.900	39.9	49.5	53.6	55.8	57.2	58.2
	.950	161.	200.	216.	225.	230.	234.
	.975	648.	799.	864.	900.	922.	937.
	.990	4052.	5000.	5403.	5625.	5764.	5859.
	.995	16211.	20000.	21615.	22500.	23056.	23437.
2	.005	.000050	.0050	.020	.038	.055	.069
	.010	.00020	.010	.032	.056	.075	.092
	.025	.00125	.C26	.062	.094	.119	.138
	.050	.0050	.053	.105	.144	.173	.194
	.100	.020	.111	.183	.231	.265	.289
	.900	8.53	9.00	9.16	9.24	9.29	9.33
	.950	18.5	19.0	19.2	19.2	19.3	19.3
	.975	38.5	39.0	39.2	39.2	39.3	39.3
	.990	98.5	99.0	99.2	99.3	99.3	99.3
	.995	198.	199.	199.	199.	199.	199.
3	.005	.000046	.0050	.021	.041	.060	.077
	.010	.C0019	.010	.034	.060	.083	.102
	.025	.0C116	.026	.065	.100	.129	.152
	.050	.0046	.052	.108	.152	.185	.210
	.100	.019	.1C9	.185	.239	.276	.304
	.900	5.54	5.46	5.39	5.34	5.31	5.28
	.950	10.1	9.55	9.28	9.12	9.01	8.94
	.975	17.4	16.0	15.4	15.1	14.9	14.7
	.990	34.1	30.8	29.5	28.7	28.2	27.9
	.995	55.6	49.8	47.5	46.2	45.4	44.8
4	.005	.000044	.0050	.022	.043	.064	.083
	.010	.00018	.C10	.035	.063	.088	.109
	.025	.00111	.025	.066	.104	.135	.161
	.C50	.0045	.052	.110	.157	.193	.221
	.100	.018	.1C8	.187	.243	.284	.314
	.900	4.54	4.32	4.19	4.11	4.05	4.01
	.950	7.71	6.94	6.59	6.39	6.26	6.16
	.975	12.2	10.6	9.98	9.60	9.36	9.20
	.99C	21.2	18.0	16.7	16.0	15.5	15.2
	.995	31.3	26.3	24.3	23.2	22.5	22.0

CUMULATIVE F-DISTRIBUTION

ν_2	p	ν_1					
		1	2	3	4	5	6
5	.005	.000043	.0050	.022	.045	.067	.087
	.010	.00017	.010	.035	.064	.091	.114
	.025	.00108	.025	.067	.107	.140	.167
	.050	.0043	.052	.111	.160	.198	.228
	.100	.017	.108	.188	.247	.290	.322
	.900	4.06	3.78	3.62	3.52	3.45	3.40
	.950	6.61	5.79	5.41	5.19	5.05	4.95
	.975	10.0	8.43	7.76	7.39	7.15	6.98
	.990	16.3	13.3	12.1	11.4	11.0	10.7
	.995	22.8	18.3	16.5	15.6	14.9	14.5
6	.005	.CC0043	.C050	.022	.046	.069	.090
	.010	.00017	.C10	.036	.066	.094	.118
	.025	.00107	.025	.068	.109	.143	.172
	.050	.0043	.052	.112	.162	.202	.233
	.100	.017	.107	.189	.249	.294	.327
	.900	3.78	3.46	3.29	3.18	3.11	3.05
	.950	5.99	5.14	4.76	4.53	4.39	4.28
	.975	8.81	7.26	6.60	6.23	5.99	5.82
	.990	13.7	10.9	9.78	9.15	8.75	8.47
	.995	18.6	14.5	12.9	12.0	11.5	11.1
7	.005	.CC0042	.C050	.023	.046	.070	.093
	.010	.00017	.010	.036	.067	.096	.121
	.025	.00105	.025	.068	.110	.146	.176
	.050	.0042	.C52	.113	.164	.205	.238
	.100	.017	.107	.190	.251	.297	.332
	.900	3.59	3.26	3.07	2.96	2.88	2.83
	.950	5.59	4.74	4.35	4.12	3.97	3.87
	.975	8.07	6.54	5.89	5.52	5.29	5.12
	.990	12.2	9.55	8.45	7.85	7.46	7.19
	.995	16.2	12.4	10.9	10.1	9.52	9.16
8	.005	.000042	.CC50	.C23	.047	.072	.095
	.010	.00017	.010	.036	.068	.097	.123
	.025	.CC105	.025	.069	.111	.148	.179
	.050	.0042	.052	.113	.166	.208	.241
	.100	.017	.107	.190	.253	.299	.335
	.900	3.46	3.11	2.92	2.81	2.73	2.67
	.950	5.32	4.46	4.07	3.84	3.69	3.58
	.975	7.57	6.06	5.42	5.05	4.82	4.65
	.990	11.3	8.65	7.59	7.01	6.63	6.37
	.995	14.7	11.0	9.60	8.81	8.30	7.95

CUMULATIVE F-DISTRIBUTION

ν_2	p	ν_1					
		1	2	3	4	5	6
9	.005	.C00042	.C050	.023	.047	.073	.096
	.010	.00017	.010	.037	.068	.098	.125
	.025	.C0104	.025	.069	.112	.150	.181
	.050	.0042	.052	.113	.167	.210	.244
	.100	.017	.107	.191	.254	.302	.338
	.900	3.36	3.C1	2.81	2.69	2.61	2.55
	.950	5.12	4.26	3.86	3.63	3.48	3.37
	.975	7.21	5.71	5.08	4.72	4.48	4.32
	.990	10.6	8.02	6.99	6.42	6.C6	5.80
	.995	13.6	10.1	8.72	7.96	7.47	7.13
10	.005	.000041	.0050	.023	.048	.073	.098
	.010	.00017	.010	.037	.069	.099	.127
	.025	.C0103	.025	.C69	.113	.151	.183
	.050	.0041	.052	.114	.168	.211	.246
	.100	.017	.106	.191	.255	.303	.340
	.900	3.28	2.92	2.73	2.61	2.52	2.46
	.950	4.96	4.10	3.71	3.48	3.33	3.22
	.975	6.94	5.46	4.83	4.47	4.24	4.07
	.990	10.0	7.56	6.55	5.99	5.64	5.39
	.995	12.8	9.43	8.08	7.34	6.87	6.54
11	.005	.000041	.0050	.023	.048	.074	.099
	.010	.00016	.010	.037	.069	.100	.128
	.025	.00103	.025	.07C	.114	.152	.185
	.050	.0041	.052	.114	.168	.213	.248
	.100	.017	.106	.191	.256	.305	.343
	.900	3.23	2.86	2.66	2.54	2.45	2.39
	.950	4.84	3.98	3.59	3.36	3.20	3.09
	.975	6.72	5.26	4.63	4.28	4.C4	3.88
	.990	9.65	7.21	6.22	5.67	5.32	5.07
	.995	12.2	8.91	7.60	6.88	6.42	6.10
12	.005	.000041	.0050	.023	.048	.075	.100
	.C10	.00016	.010	.037	.070	.101	.130
	.025	.00102	.025	.07C	.114	.153	.186
	.C50	.0041	.052	.114	.169	.214	.250
	.100	.016	.106	.192	.257	.306	.344
	.900	3.18	2.81	2.61	2.48	2.39	2.33
	.950	4.75	3.89	3.49	3.26	3.11	3.00
	.975	6.55	5.10	4.47	4.12	3.89	3.73
	.990	9.33	6.93	5.95	5.41	5.06	4.82
	.995	11.8	8.51	7.23	6.52	6.07	5.76

CUMLLATIVE F-DISTRIBUTION

ν_2	P	ν_1					
		1	2	3	4	5	6
13	.005	.000041	.0050	.023	.049	.075	.100
	.010	.00016	.010	.037	.070	.102	.131
	.025	.00102	.025	.070	.115	.154	.188
	.050	.0041	.051	.115	.170	.215	.251
	.100	.016	.106	.192	.257	.307	.346
	.900	3.14	2.76	2.56	2.43	2.35	2.28
	.950	4.67	3.81	3.41	3.18	3.03	2.92
	.975	6.41	4.97	4.35	4.00	3.77	3.60
	.990	9.07	6.70	5.74	5.21	4.86	4.62
	.995	11.4	8.19	6.93	6.23	5.79	5.48
14	.005	.000041	.0050	.023	.049	.076	.101
	.010	.00016	.010	.037	.070	.102	.131
	.025	.00102	.025	.070	.115	.155	.189
	.050	.0041	.051	.115	.170	.216	.253
	.100	.016	.106	.192	.258	.308	.347
	.900	3.10	2.73	2.52	2.39	2.31	2.24
	.950	4.60	3.74	3.34	3.11	2.96	2.85
	.975	6.30	4.86	4.24	3.89	3.66	3.50
	.990	8.86	6.51	5.56	5.04	4.69	4.46
	.995	11.1	7.92	6.68	6.00	5.56	5.26
15	.005	.000041	.0050	.023	.049	.076	.102
	.010	.00016	.010	.037	.070	.103	.132
	.025	.00102	.025	.070	.116	.156	.190
	.050	.0041	.051	.115	.171	.217	.254
	.100	.016	.106	.192	.258	.309	.348
	.900	3.07	2.70	2.49	2.36	2.27	2.21
	.950	4.54	3.68	3.29	3.06	2.90	2.79
	.975	6.20	4.77	4.15	3.80	3.58	3.41
	.990	8.68	6.36	5.42	4.89	4.56	4.32
	.995	10.8	7.70	6.48	5.80	5.37	5.07
20	.005	.000040	.0050	.023	.050	.077	.104
	.010	.00016	.010	.037	.071	.105	.135
	.025	.00101	.025	.071	.117	.158	.193
	.050	.0040	.051	.115	.172	.219	.258
	.100	.016	.106	.193	.260	.312	.353
	.900	2.97	2.59	2.38	2.25	2.16	2.09
	.950	4.35	3.49	3.10	2.87	2.71	2.60
	.975	5.87	4.46	3.86	3.51	3.29	3.13
	.990	8.10	5.85	4.94	4.43	4.10	3.87
	.995	9.94	6.99	5.82	5.17	4.76	4.47

CUMULATIVE F-DISTRIBUTION

ν_2	p	ν_1 1	2	3	4	5	6
24	.005	.000040	.0050	.023	.050	.078	.106
	.010	.00016	.010	.038	.072	.106	.137
	.025	.00100	.025	.071	.117	.159	.195
	.050	.0040	.051	.116	.173	.221	.260
	.100	.016	.106	.193	.261	.313	.355
	.900	2.93	2.54	2.33	2.19	2.10	2.04
	.950	4.26	3.40	3.01	2.78	2.62	2.51
	.975	5.72	4.32	3.72	3.38	3.15	2.99
	.990	7.82	5.61	4.72	4.22	3.90	3.67
	.995	9.55	6.66	5.52	4.89	4.49	4.20
30	.005	.000040	.0050	.024	.050	.079	.107
	.010	.00016	.010	.038	.072	.107	.138
	.025	.00100	.025	.071	.118	.161	.197
	.050	.0040	.051	.116	.174	.222	.263
	.100	.016	.106	.193	.262	.315	.357
	.900	2.88	2.49	2.28	2.14	2.05	1.98
	.950	4.17	3.32	2.92	2.69	2.53	2.42
	.975	5.57	4.18	3.59	3.25	3.03	2.87
	.990	7.56	5.39	4.51	4.02	3.70	3.47
	.995	9.18	6.35	5.24	4.62	4.23	3.95
40	.005	.000040	.0050	.024	.051	.080	.108
	.010	.00016	.010	.038	.073	.108	.140
	.025	.00099	.025	.071	.119	.162	.199
	.050	.0040	.051	.116	.175	.224	.265
	.100	.016	.106	.194	.263	.317	.360
	.900	2.84	2.44	2.23	2.09	2.00	1.93
	.950	4.08	3.23	2.84	2.61	2.45	2.34
	.975	5.42	4.05	3.46	3.13	2.90	2.74
	.990	7.31	5.18	4.31	3.83	3.51	3.29
	.995	8.83	6.07	4.98	4.37	3.99	3.71
50	.005	.000040	.0050	.024	.051	.080	.109
	.010	.00016	.010	.038	.073	.108	.141
	.025	.00099	.025	.071	.119	.163	.201
	.050	.0040	.051	.117	.175	.225	.266
	.100	.016	.106	.194	.263	.318	.361
	.900	2.81	2.41	2.20	2.06	1.97	1.90
	.950	4.03	3.18	2.79	2.56	2.40	2.29
	.975	5.34	3.97	3.39	3.05	2.83	2.67
	.990	7.17	5.06	4.20	3.72	3.41	3.19
	.995	8.63	5.90	4.83	4.23	3.85	3.58

CUMMLATIVE F-DISTRIBUTION

ν_2	p	ν_1 1	2	3	4	5	6
6C	.005	.000040	.0050	.024	.051	.081	.110
	.010	.00016	.C10	.038	.073	.109	.142
	.025	.00099	.025	.071	.120	.163	.202
	.C50	.0040	.051	.117	.176	.226	.267
	.100	.016	.106	.194	.264	.318	.362
	.900	2.79	2.39	2.18	2.04	1.95	1.87
	.950	4.00	3.15	2.76	2.53	2.37	2.25
	.975	5.29	3.93	3.34	3.01	2.79	2.63
	.990	7.08	4.98	4.13	3.65	3.34	3.12
	.995	8.49	5.79	4.73	4.14	3.76	3.49
1C0	.005	.CC0035	.C050	.024	.051	.081	.111
	.010	.00016	.010	.C38	.074	.110	.143
	.025	.00099	.025	.072	.120	.164	.203
	.050	.0040	.C51	.117	.177	.227	.269
	.100	.016	.105	.194	.265	.320	.364
	.900	2.76	2.36	2.14	2.00	1.91	1.83
	.950	3.94	3.09	2.70	2.46	2.31	2.19
	.975	5.18	3.83	3.25	2.92	2.70	2.54
	.990	6.90	4.82	3.98	3.51	3.21	2.99
	.995	8.24	5.59	4.54	3.96	3.59	3.33
120	.005	.0C0039	.C050	.024	.051	.081	.111
	.010	.00016	.010	.038	.074	.110	.143
	.025	.C0099	.025	.072	.120	.165	.204
	.050	.0039	.C51	.117	.177	.227	.270
	.100	.016	.105	.194	.265	.320	.365
	.900	2.75	2.35	2.13	1.99	1.90	1.82
	.950	3.92	3.07	2.68	2.45	2.29	2.17
	.975	5.15	3.80	3.23	2.89	2.67	2.52
	.990	6.85	4.79	3.95	3.48	3.17	2.96
	.995	8.18	5.54	4.50	3.92	3.55	3.28
∞	.005	.CC0C39	.C050	.024	.052	.082	.113
	.010	.00016	.010	.038	.074	.111	.145
	.025	.C0098	.C25	.072	.121	.166	.206
	.050	.0039	.051	.117	.178	.229	.273
	.100	.016	.105	.195	.266	.322	.367
	.900	2.71	2.30	2.C8	1.94	1.85	1.77
	.950	3.84	3.00	2.60	2.37	2.21	2.10
	.975	5.02	3.69	3.12	2.79	2.57	2.41
	.990	6.63	4.61	3.78	3.32	3.C2	2.80
	.995	7.88	5.30	4.28	3.72	3.35	3.09

CUMULATIVE F-DISTRIBUTION

ν_2	p	ν_1 7	8	9	10	11	12
1	.005	.062	.068	.073	.078	.082	.085
	.010	.082	.089	.095	.100	.104	.107
	.025	.124	.132	.139	.144	.149	.153
	.050	.179	.188	.195	.201	.206	.211
	.100	.279	.289	.298	.304	.310	.315
	.900	58.9	59.4	59.9	60.2	60.5	60.7
	.950	237.	239.	241.	242.	243.	244.
	.975	948.	957.	963.	969.	973.	977.
	.990	5928.	5981.	6023.	6056.	6083.	6106.
	.995	23715.	23925.	24091.	24224.	24334.	24426.
2	.005	.081	.091	.099	.106	.112	.118
	.010	.105	.116	.125	.132	.139	.144
	.025	.153	.165	.175	.183	.190	.196
	.050	.211	.224	.235	.244	.251	.257
	.100	.307	.321	.333	.342	.350	.356
	.900	9.35	9.37	9.38	9.39	9.40	9.41
	.950	19.4	19.4	19.4	19.4	19.4	19.4
	.975	39.4	39.4	39.4	39.4	39.4	39.4
	.990	99.4	99.4	99.4	99.4	99.4	99.4
	.995	199.	199.	199.	199.	199.	199.
3	.005	.092	.104	.115	.124	.132	0.14
	.010	.118	.132	.143	.153	.161	.168
	.025	.170	.185	.197	.207	.216	.223
	.C50	.230	.246	.259	.270	.279	.287
	.100	.325	.342	.356	.367	.376	.384
	.900	5.27	5.25	5.24	5.23	5.22	5.22
	.950	8.89	8.85	8.81	8.79	8.76	8.74
	.975	14.6	14.5	14.5	14.4	14.4	14.3
	.990	27.7	27.5	27.3	27.2	27.1	27.1
	.995	44.4	44.1	43.9	43.7	43.5	43.4
4	.005	.C99	.114	.126	.136	.145	0.15
	.010	.127	.143	.156	.167	.176	.185
	.025	.181	.198	.212	.224	.234	.243
	.050	.243	.261	.275	.288	.298	.307
	.100	.338	.356	.371	.384	.394	.403
	.900	3.98	3.95	3.94	3.92	3.91	3.90
	.950	6.09	6.04	6.00	5.96	5.94	5.91
	.975	9.07	8.98	8.90	8.84	8.79	8.75
	.990	15.0	14.8	14.7	14.5	14.5	14.4
	.995	21.6	21.4	21.1	21.0	20.8	20.7

CUMULATIVE F-DISTRIBUTION

ν_2	p	ν_1					
		7	8	9	10	11	12
5	.005	.105	.120	.134	.146	.156	0.16
	.010	.134	.151	.165	.177	.188	.197
	.025	.189	.208	.223	.236	.247	.257
	.050	.252	.271	.287	.301	.312	.322
	.100	.347	.367	.383	.397	.408	.418
	.900	3.37	3.34	3.32	3.30	3.28	3.27
	.950	4.88	4.82	4.77	4.74	4.70	4.68
	.975	6.85	6.76	6.68	6.62	6.57	6.52
	.990	10.5	10.3	10.2	10.1	9.96	9.89
	.995	14.2	14.0	13.8	13.6	13.5	13.4
6	.005	.109	.126	.140	.153	.164	0.17
	.010	.139	.157	.172	.186	.197	.207
	.025	.195	.215	.231	.246	.258	.268
	.050	.259	.279	.296	.311	.323	.334
	.100	.354	.375	.392	.406	.419	.429
	.900	3.01	2.98	2.96	2.94	2.92	2.90
	.950	4.21	4.15	4.10	4.06	4.03	4.00
	.975	5.70	5.60	5.52	5.46	5.41	5.37
	.990	8.26	8.10	7.98	7.87	7.79	7.72
	.995	10.8	1C.6	10.4	10.3	10.1	10.0
7	.005	.113	.130	.145	.159	.171	0.18
	.010	.143	.162	.179	.192	.205	.216
	.025	.200	.221	.238	.253	.266	.277
	.050	.264	.286	.304	.319	.332	.343
	.100	.359	.381	.395	.414	.427	.438
	.900	2.78	2.75	2.72	2.70	2.68	2.67
	.950	3.79	3.73	3.68	3.64	3.60	3.57
	.975	4.99	4.90	4.82	4.76	4.71	4.67
	.990	6.99	6.84	6.72	6.62	6.54	6.47
	.995	8.89	8.68	8.51	8.38	8.27	8.18
8	.005	.115	.133	.149	.164	.176	0.19
	.010	.146	.166	.183	.198	.211	.222
	.025	.204	.226	.244	.259	.273	.285
	.050	.268	.291	.310	.326	.339	.351
	.100	.363	.386	.405	.421	.434	.446
	.900	2.62	2.59	2.56	2.54	2.52	2.50
	.950	3.5C	3.44	3.39	3.35	3.31	3.28
	.975	4.53	4.43	4.36	4.30	4.24	4.20
	.990	6.18	6.03	5.91	5.81	5.73	5.67
	.995	7.69	7.50	7.34	7.21	7.10	7.01

CUMULATIVE F-DISTRIBUTION

ν_2	p	ν_1					
		7	8	9	10	11	12
9	.005	.117	.136	.153	.168	.181	0.19
	.010	.149	.169	.187	.202	.216	.228
	.025	.207	.229	.248	.265	.279	.291
	.050	.272	.295	.315	.331	.345	.358
	.100	.367	.390	.410	.426	.440	.452
	.900	2.51	2.47	2.44	2.42	2.40	2.38
	.950	3.29	3.23	3.18	3.14	3.10	3.07
	.975	4.20	4.10	4.03	3.96	3.91	3.87
	.990	5.61	5.47	5.35	5.26	5.18	5.11
	.995	6.88	6.69	6.54	6.42	6.31	6.23
10	.005	.119	.139	.156	.171	.185	0.20
	.010	.151	.172	.190	.206	.220	.233
	.025	.210	.233	.252	.269	.284	.296
	.050	.275	.299	.319	.336	.350	.363
	.100	.370	.394	.414	.431	.445	.457
	.900	2.41	2.38	2.35	2.32	2.30	2.28
	.950	3.14	3.07	3.02	2.98	2.94	2.91
	.975	3.95	3.85	3.78	3.72	3.66	3.62
	.990	5.20	5.06	4.94	4.85	4.77	4.71
	.995	6.30	6.12	5.97	5.85	5.75	5.66
11	.005	.121	.141	.158	.174	.188	0.20
	.010	.153	.174	.193	.210	.224	.237
	.025	.212	.236	.256	.273	.288	.301
	.050	.278	.302	.322	.340	.355	.368
	.100	.373	.397	.417	.434	.449	.462
	.900	2.34	2.30	2.27	2.25	2.23	2.21
	.950	3.01	2.95	2.90	2.85	2.82	2.79
	.975	3.76	3.66	3.59	3.53	3.47	3.43
	.990	4.89	4.74	4.63	4.54	4.46	4.40
	.995	5.86	5.68	5.54	5.42	5.32	5.24
12	.005	.122	.143	.161	.177	.191	0.20
	.010	.155	.176	.196	.212	.227	.241
	.025	.214	.238	.259	.276	.292	.305
	.050	.280	.305	.325	.343	.359	.372
	.100	.375	.400	.420	.438	.453	.466
	.900	2.28	2.24	2.21	2.19	2.17	2.15
	.950	2.91	2.85	2.80	2.75	2.72	2.69
	.975	3.61	3.51	3.44	3.37	3.32	3.28
	.990	4.64	4.50	4.39	4.30	4.22	4.16
	.995	5.52	5.35	5.20	5.09	4.99	4.91

CUMULATIVE F-DISTRIBUTION

| ν_2 | p | \multicolumn{6}{c}{ν_1} |
		7	8	9	10	11	12
13	.005	.124	.144	.163	.179	.194	0.21
	.010	.156	.178	.198	.215	.230	.244
	.025	.216	.240	.261	.279	.295	.309
	.050	.282	.307	.328	.346	.362	.376
	.100	.377	.402	.423	.441	.456	.469
	.900	2.23	2.20	2.16	2.14	2.12	2.10
	.950	2.83	2.77	2.71	2.67	2.63	2.60
	.975	3.48	3.39	3.31	3.25	3.20	3.15
	.990	4.44	4.30	4.19	4.10	4.02	3.96
	.995	5.25	5.08	4.94	4.82	4.72	4.64
14	.005	.125	.146	.164	.181	.196	0.21
	.010	.157	.180	.200	.217	.233	.247
	.025	.218	.242	.263	.282	.298	.312
	.050	.283	.309	.331	.349	.365	.379
	.100	.378	.404	.425	.443	.459	.472
	.900	2.19	2.15	2.12	2.10	2.07	2.05
	.950	2.76	2.70	2.65	2.60	2.57	2.53
	.975	3.38	3.29	3.21	3.15	3.09	3.05
	.990	4.28	4.14	4.03	3.94	3.86	3.80
	.995	5.03	4.86	4.72	4.60	4.51	4.43
15	.C05	.126	.147	.166	.183	.198	0.21
	.010	.158	.181	.202	.219	.235	.249
	.025	.219	.244	.265	.284	.300	.315
	.050	.285	.311	.333	.351	.368	.382
	.100	.380	.406	.427	.446	.461	.475
	.900	2.16	2.12	2.09	2.06	2.04	2.02
	.950	2.71	2.64	2.59	2.54	2.51	2.48
	.975	3.29	3.20	3.12	3.06	3.01	2.96
	.990	4.14	4.00	3.89	3.80	3.73	3.67
	.995	4.85	4.67	4.54	4.42	4.33	4.25
20	.C05	.129	.151	.171	.190	.206	0.22
	.010	.162	.187	.208	.227	.244	.259
	.025	.224	.250	.273	.293	.310	.325
	.050	.290	.317	.341	.360	.378	.393
	.100	.385	.412	.435	.454	.471	.486
	.900	2.04	2.00	1.96	1.94	1.91	1.89
	.950	2.51	2.45	2.39	2.35	2.31	2.28
	.975	3.01	2.91	2.84	2.77	2.72	2.68
	.990	3.70	3.56	3.46	3.37	3.29	3.23
	.995	4.26	4.09	3.96	3.85	3.76	3.68

CUMULATIVE F-DISTRIBUTION

ν_2	p	ν_1					
		7	8	9	10	11	12
24	.005	.131	.154	.175	.193	.210	0.23
	.010	.165	.189	.211	.231	.249	.265
	.025	.226	.253	.277	.297	.315	.331
	.050	.293	.321	.345	.365	.383	.399
	.100	.388	.416	.439	.459	.476	.491
	.900	1.98	1.94	1.91	1.88	1.85	1.83
	.950	2.42	2.36	2.30	2.25	2.22	2.18
	.975	2.87	2.78	2.70	2.64	2.59	2.54
	.990	3.50	3.36	3.26	3.17	3.09	3.03
	.995	3.99	3.83	3.69	3.59	3.50	3.42
30	.005	.133	.156	.178	.197	.215	0.23
	.010	.167	.192	.215	.235	.254	.270
	.025	.229	.257	.281	.302	.321	.337
	.050	.296	.325	.349	.370	.389	.405
	.100	.391	.420	.444	.464	.482	.497
	.900	1.93	1.88	1.85	1.82	1.79	1.77
	.950	2.33	2.27	2.21	2.16	2.13	2.09
	.975	2.75	2.65	2.57	2.51	2.46	2.41
	.990	3.30	3.17	3.07	2.98	2.91	2.84
	.995	3.74	3.58	3.45	3.34	3.25	3.18
40	.005	.135	.159	.181	.201	.220	0.24
	.010	.169	.195	.219	.240	.259	.276
	.025	.232	.260	.285	.307	.327	.344
	.050	.299	.329	.354	.376	.395	.412
	.100	.394	.423	.448	.469	.487	.503
	.900	1.87	1.83	1.79	1.76	1.74	1.71
	.950	2.25	2.18	2.12	2.08	2.04	2.00
	.975	2.62	2.53	2.45	2.39	2.33	2.29
	.990	3.12	2.99	2.89	2.80	2.73	2.66
	.995	3.51	3.35	3.22	3.12	3.03	2.95
50	.005	.136	.161	.183	.204	.223	0.24
	.010	.171	.197	.221	.243	.262	.280
	.025	.234	.263	.288	.310	.330	.348
	.050	.301	.331	.357	.379	.399	.416
	.100	.396	.426	.451	.472	.491	.508
	.900	1.84	1.80	1.76	1.73	1.70	1.68
	.950	2.20	2.13	2.07	2.03	1.99	1.95
	.975	2.55	2.46	2.38	2.32	2.26	2.22
	.990	3.02	2.89	2.78	2.70	2.63	2.56
	.995	3.38	3.22	3.09	2.99	2.90	2.82

CUMULATIVE F-DISTRIBUTION

ν_2	p	ν_1 7	8	9	10	11	12
60	.005	.137	.162	.185	.206	.225	0.24
	.010	.172	.199	.223	.245	.265	.283
	.025	.235	.264	.290	.313	.333	.351
	.050	.303	.333	.359	.382	.402	.419
	.100	.398	.428	.453	.475	.494	.510
	.900	1.82	1.77	1.74	1.71	1.68	1.66
	.950	2.17	2.10	2.04	1.99	1.95	1.92
	.975	2.51	2.41	2.33	2.27	2.22	2.17
	.990	2.95	2.82	2.72	2.63	2.56	2.50
	.995	3.29	3.13	3.01	2.90	2.82	2.74
10C	.005	.139	.164	.188	.210	.229	0.25
	.010	.174	.201	.226	.249	.270	.288
	.025	.238	.267	.294	.317	.338	.357
	.050	.305	.336	.363	.386	.407	.426
	.100	.400	.431	.457	.479	.499	.516
	.900	1.78	1.73	1.69	1.66	1.64	1.61
	.950	2.10	2.03	1.97	1.93	1.89	1.85
	.975	2.42	2.32	2.24	2.18	2.12	2.08
	.990	2.82	2.69	2.59	2.50	2.43	2.37
	.995	3.13	2.97	2.85	2.74	2.66	2.58
120	.005	.139	.165	.189	.211	.231	0.25
	.010	.174	.202	.227	.250	.271	.290
	.025	.238	.268	.295	.318	.340	.359
	.050	.306	.337	.364	.388	.408	.427
	.100	.401	.432	.458	.480	.500	.518
	.900	1.77	1.72	1.68	1.65	1.63	1.60
	.950	2.09	2.02	1.96	1.91	1.87	1.83
	.975	2.39	2.30	2.22	2.16	2.10	2.05
	.990	2.79	2.66	2.56	2.47	2.40	2.34
	.995	3.09	2.93	2.81	2.71	2.62	2.54
∞	.005	.141	.168	.193	.216	.237	0.26
	.010	.177	.206	.232	.256	.278	.298
	.025	.241	.272	.300	.325	.347	.367
	.050	.310	.342	.369	.394	.416	.435
	.100	.405	.436	.463	.487	.507	.525
	.900	1.72	1.67	1.63	1.60	1.57	1.55
	.950	2.01	1.94	1.88	1.83	1.79	1.75
	.975	2.29	2.19	2.11	2.05	1.99	1.94
	.990	2.64	2.51	2.41	2.32	2.25	1.18
	.995	2.90	2.74	2.62	2.52	2.43	2.36

CUMULATIVE F-DISTRIBUTION

ν_2	p	ν_1 13	14	15	20	24	30
1	.005	.088	.090	.093	.101	.105	0.11
	.010	.110	.113	.115	.124	.128	.132
	.025	.156	.159	.161	.170	.175	.180
	.050	.214	.217	.220	.230	.235	.240
	.100	.319	.322	.325	.336	.342	.347
	.900	60.9	61.1	61.2	61.7	62.0	62.3
	.950	245.	245.	246.	248.	249.	250.
	.975	980.	983.	985.	993.	997.	1001.
	.990	6126.	6143.	6157.	6209.	6235.	6261.
	.995	24504.	24572.	24630.	24836.	24940.	25044.
2	.005	.122	.126	.130	.143	.150	0.16
	.010	.149	.153	.157	.171	.178	.186
	.025	.201	.206	.210	.224	.232	.239
	.050	.263	.267	.272	.286	.294	.302
	.100	.362	.367	.371	.386	.394	.402
	.900	9.41	9.42	9.42	9.44	9.45	9.46
	.950	19.4	19.4	19.4	19.4	19.5	19.5
	.975	39.4	39.4	39.4	39.4	39.5	39.5
	.990	99.4	99.4	99.4	99.4	99.5	99.5
	.995	199.	199.	199.	199.	199.	199.
3	.005	.144	.150	.154	.172	.181	0.19
	.010	.174	.180	.185	.202	.212	.222
	.025	.230	.236	.241	.259	.269	.279
	.050	.293	.299	.304	.323	.332	.342
	.100	.391	.396	.402	.420	.430	.439
	.900	5.21	5.20	5.20	5.18	5.18	5.17
	.950	8.73	8.71	8.70	8.66	8.64	8.62
	.975	14.3	14.3	14.3	14.2	14.1	14.1
	.990	27.0	26.9	26.9	26.7	26.6	26.5
	.995	43.3	43.2	43.1	42.8	42.6	42.5
4	.005	.160	.167	.172	.193	.205	0.22
	.010	.192	.199	.204	.226	.237	.249
	.025	.250	.257	.263	.285	.296	.308
	.050	.315	.321	.327	.349	.360	.372
	.100	.411	.418	.423	.445	.456	.467
	.900	3.89	3.88	3.87	3.84	3.83	3.82
	.950	5.89	5.87	5.86	5.80	5.77	5.75
	.975	8.71	8.68	8.66	8.56	8.51	8.46
	.990	14.3	14.2	14.2	14.0	13.9	13.8
	.995	20.6	20.5	20.4	20.2	20.0	19.9

CUMULATIVE F-DISTRIBUTION

ν_2	p	ν_1					
		13	14	15	20	24	30
5	.005	.173	.180	.186	.210	.223	0.24
	.010	.206	.213	.220	.244	.257	.270
	.025	.265	.273	.280	.304	.317	.330
	.050	.331	.338	.345	.369	.382	.395
	.100	.426	.433	.440	.463	.475	.488
	.900	3.26	3.25	3.24	3.21	3.19	3.17
	.950	4.66	4.64	4.62	4.56	4.53	4.50
	.975	6.49	6.46	6.43	6.33	6.28	6.23
	.990	9.82	9.77	9.72	9.55	9.47	9.38
	.995	13.3	13.2	13.1	12.9	12.8	12.7
6	.005	.182	.190	.197	.224	.238	0.25
	.010	.216	.224	.232	.258	.273	.288
	.025	.277	.286	.293	.320	.334	.349
	.050	.343	.351	.358	.385	.399	.413
	.100	.438	.446	.453	.478	.491	.505
	.900	2.89	2.88	2.87	2.84	2.82	2.80
	.950	3.98	3.96	3.94	3.87	3.84	3.81
	.975	5.33	5.30	5.27	5.17	5.12	5.07
	.990	7.66	7.60	7.56	7.40	7.31	7.23
	.995	9.95	9.88	9.81	9.59	9.47	9.36
7	.005	.190	.199	.206	.235	.251	0.27
	.010	.225	.234	.241	.270	.286	.303
	.025	.287	.296	.304	.333	.348	.364
	.050	.353	.362	.369	.398	.413	.428
	.100	.448	.456	.463	.490	.504	.519
	.900	2.65	2.64	2.63	2.59	2.58	2.56
	.950	3.55	3.53	3.51	3.44	3.41	3.38
	.975	4.63	4.60	4.57	4.47	4.41	4.36
	.990	6.41	6.36	6.31	6.16	6.07	5.99
	.995	8.10	8.03	7.97	7.75	7.64	7.53
8	.005	.197	.206	.214	.244	.261	0.28
	.010	.232	.242	.250	.281	.297	.315
	.025	.295	.304	.313	.343	.360	.377
	.050	.361	.371	.379	.409	.425	.441
	.100	.456	.464	.472	.500	.515	.531
	.900	2.49	2.48	2.46	2.42	2.40	2.38
	.950	3.26	3.24	3.22	3.15	3.12	3.08
	.975	4.16	4.13	4.10	4.00	3.95	3.89
	.990	5.61	5.56	5.52	5.36	5.28	5.20
	.995	6.94	6.87	6.81	6.61	6.50	6.40

CUMULATIVE F-DISTRIBUTION

ν_2	p	ν_1					
		13	14	15	20	24	30
9	.005	.203	.212	.220	.253	.271	0.29
	.010	.239	.248	.257	.289	.307	.326
	.025	.302	.312	.320	.353	.370	.388
	.050	.368	.378	.386	.418	.435	.452
	.100	.462	.471	.479	.509	.525	.541
	.900	2.36	2.35	2.34	2.30	2.28	2.25
	.950	3.05	3.03	3.01	2.94	2.90	2.86
	.975	3.83	3.80	3.77	3.67	3.61	3.56
	.990	5.05	5.01	4.96	4.81	4.73	4.65
	.995	6.15	6.09	6.03	5.83	5.73	5.62
10	.005	.207	.217	.226	.260	.279	0.30
	.010	.244	.254	.263	.297	.316	.336
	.025	.308	.318	.327	.361	.379	.398
	.050	.374	.384	.393	.426	.444	.462
	.100	.468	.477	.486	.516	.533	.550
	.900	2.27	2.26	2.24	2.20	2.18	2.16
	.950	2.89	2.86	2.84	2.77	2.74	2.70
	.975	3.58	3.55	3.52	3.42	3.37	3.31
	.990	4.65	4.60	4.56	4.41	4.33	4.25
	.995	5.59	5.53	5.47	5.27	5.17	5.07
11	.005	.212	.222	.231	.266	.286	0.31
	.010	.248	.259	.268	.304	.323	.344
	.025	.313	.323	.332	.368	.387	.407
	.050	.380	.390	.399	.433	.451	.470
	.100	.473	.482	.491	.523	.540	.557
	.900	2.19	2.18	2.17	2.12	2.10	2.08
	.950	2.76	2.74	2.72	2.65	2.61	2.57
	.975	3.39	3.36	3.33	3.23	3.17	3.12
	.990	4.34	4.29	4.25	4.10	4.02	3.94
	.995	5.16	5.10	5.05	4.86	4.76	4.65
12	.005	.215	.226	.235	.272	.292	0.31
	.010	.252	.263	.273	.309	.330	.352
	.025	.317	.328	.337	.374	.394	.415
	.050	.384	.395	.404	.439	.458	.478
	.100	.477	.487	.496	.528	.546	.564
	.900	2.13	2.12	2.10	2.06	2.04	2.01
	.950	2.66	2.64	2.62	2.54	2.51	2.47
	.975	3.24	3.21	3.18	3.07	3.02	2.96
	.990	4.10	4.05	4.01	3.86	3.78	3.70
	.995	4.84	4.77	4.72	4.53	4.43	4.33

CUMLLATIVE F-DISTRIBUTION

ν_1

ν_2	p	13	14	15	20	24	30
13	.005	.219	.229	.239	.277	.298	0.32
	.010	.256	.267	.277	.315	.336	.359
	.025	.321	.332	.342	.379	.400	.422
	.050	.388	.399	.408	.445	.464	.485
	.100	.481	.491	.500	.533	.551	.570
	.900	2.08	2.07	2.05	2.01	1.98	1.96
	.950	2.58	2.55	2.53	2.46	2.42	2.38
	.975	3.11	3.08	3.05	2.95	2.89	2.84
	.990	3.91	3.86	3.82	3.66	3.59	3.51
	.995	4.57	4.51	4.46	4.27	4.17	4.07
14	.005	.222	.233	.243	.281	.303	0.33
	.010	.259	.270	.281	.320	.341	.365
	.025	.324	.336	.346	.384	.405	.428
	.050	.392	.403	.412	.449	.470	.491
	.100	.484	.494	.504	.538	.556	.576
	.900	2.04	2.02	2.01	1.96	1.94	1.91
	.950	2.51	2.48	2.46	2.39	2.35	2.31
	.975	3.01	2.98	2.95	2.84	2.79	2.73
	.990	3.75	3.70	3.66	3.51	3.43	3.35
	.995	4.36	4.30	4.25	4.06	3.96	3.86
15	.005	.224	.235	.246	.286	.308	0.33
	.010	.262	.274	.284	.324	.346	.370
	.025	.328	.339	.349	.389	.410	.433
	.050	.395	.406	.416	.454	.474	.496
	.100	.487	.498	.507	.542	.561	.581
	.900	2.00	1.99	1.97	1.92	1.90	1.87
	.950	2.45	2.42	2.40	2.33	2.29	2.25
	.975	2.92	2.89	2.86	2.76	2.70	2.64
	.990	3.61	3.56	3.52	3.37	3.29	3.21
	.995	4.18	4.12	4.07	3.88	3.79	3.69
20	.005	.234	.246	.258	.301	.327	0.35
	.010	.273	.285	.297	.340	.365	.392
	.025	.339	.352	.363	.406	.430	.456
	.050	.407	.419	.430	.471	.493	.518
	.100	.498	.510	.520	.557	.578	.600
	.900	1.87	1.86	1.84	1.79	1.77	1.74
	.950	2.25	2.22	2.20	2.12	2.08	2.04
	.975	2.64	2.60	2.57	2.46	2.41	2.35
	.990	3.18	3.13	3.09	2.94	2.86	2.78
	.995	3.61	3.55	3.50	3.32	3.22	3.12

CUMULATIVE F-DISTRIBUTION

$$\nu_1$$

ν_2	p	13	14	15	20	24	30
24	.005	.240	.252	.264	.310	.337	0.37
	.010	.279	.292	.304	.350	.376	.405
	.025	.346	.359	.370	.415	.441	.468
	.050	.413	.426	.437	.480	.504	.530
	.100	.504	.516	.527	.566	.588	.611
	.900	1.81	1.80	1.78	1.73	1.70	1.67
	.950	2.15	2.13	2.11	2.03	1.98	1.94
	.975	2.50	2.47	2.44	2.33	2.27	2.21
	.990	2.98	2.93	2.89	2.74	2.66	2.58
	.995	3.35	3.30	3.25	3.06	2.97	2.87
30	.005	.246	.259	.271	.320	.349	0.38
	.010	.285	.299	.311	.360	.388	.419
	.025	.352	.366	.378	.426	.453	.482
	.050	.420	.433	.445	.490	.516	.543
	.100	.511	.523	.534	.575	.598	.622
	.900	1.75	1.74	1.72	1.67	1.64	1.61
	.950	2.06	2.04	2.01	1.93	1.89	1.84
	.975	2.37	2.34	2.31	2.20	2.14	2.07
	.990	2.79	2.74	2.70	2.55	2.47	2.39
	.995	3.11	3.06	3.01	2.82	2.73	2.63
40	.005	.252	.266	.279	.331	.362	0.40
	.010	.292	.306	.319	.371	.401	.435
	.025	.360	.374	.387	.437	.466	.498
	.050	.427	.441	.454	.502	.529	.558
	.100	.518	.530	.542	.585	.610	.636
	.900	1.69	1.68	1.66	1.61	1.57	1.54
	.950	1.97	1.95	1.92	1.84	1.79	1.74
	.975	2.25	2.21	2.18	2.07	2.01	1.94
	.990	2.61	2.56	2.52	2.37	2.29	2.20
	.995	2.89	2.83	2.78	2.60	2.50	2.40
50	.005	.256	.270	.284	.338	.370	0.41
	.010	.296	.311	.325	.378	.410	.445
	.025	.364	.379	.392	.445	.475	.508
	.050	.432	.446	.459	.509	.537	.568
	.100	.522	.535	.547	.592	.617	.644
	.900	1.66	1.64	1.63	1.57	1.54	1.50
	.950	1.92	1.89	1.87	1.78	1.74	1.69
	.975	2.18	2.14	2.11	1.99	1.93	1.87
	.990	2.51	2.46	2.42	2.27	2.18	2.10
	.995	2.76	2.70	2.65	2.47	2.37	2.27

CUMULATIVE F-DISTRIBUTION

ν_2	p	ν_1					
		13	14	15	20	24	30
60	.005	.259	.274	.287	.343	.376	0.41
	.010	.299	.314	.328	.383	.416	.453
	.025	.368	.383	.396	.450	.481	.515
	.050	.435	.450	.463	.514	.543	.575
	.100	.525	.538	.550	.596	.622	.650
	.900	1.64	1.62	1.60	1.54	1.51	1.48
	.950	1.89	1.86	1.84	1.75	1.70	1.65
	.975	2.13	2.09	2.06	1.94	1.88	1.82
	.990	2.44	2.39	2.35	2.20	2.12	2.03
	.995	2.68	2.62	2.57	2.39	2.29	2.19
100	.005	.265	.280	.295	.354	.389	0.43
	.010	.306	.321	.336	.394	.429	.469
	.025	.374	.390	.404	.461	.494	.531
	.050	.442	.457	.471	.525	.555	.590
	.100	.531	.545	.558	.606	.633	.664
	.900	1.59	1.57	1.56	1.49	1.46	1.42
	.950	1.82	1.79	1.77	1.68	1.63	1.57
	.975	2.04	2.00	1.97	1.85	1.78	1.71
	.990	2.31	2.27	2.22	2.07	1.98	1.89
	.995	2.52	2.46	2.41	2.23	2.13	2.02
120	.005	.266	.282	.297	.356	.393	0.43
	.010	.307	.323	.338	.397	.433	.474
	.025	.376	.392	.406	.464	.498	.536
	.050	.444	.459	.473	.527	.559	.594
	.100	.533	.547	.560	.609	.636	.667
	.900	1.58	1.56	1.54	1.48	1.45	1.41
	.950	1.80	1.77	1.75	1.66	1.61	1.55
	.975	2.01	1.98	1.94	1.82	1.76	1.69
	.990	2.28	2.23	2.19	2.03	1.95	1.86
	.995	2.48	2.42	2.37	2.19	2.09	1.98
∞	.005	.274	.291	.307	.372	.412	0.46
	.010	.316	.333	.349	.413	.452	.498
	.025	.385	.402	.417	.480	.517	.560
	.050	.453	.469	.484	.543	.577	.616
	.100	.542	.556	.570	.622	.652	.687
	.900	1.52	1.50	1.49	1.42	1.38	1.34
	.950	1.72	1.69	1.67	1.57	1.52	1.46
	.975	1.90	1.86	1.83	1.71	1.64	1.57
	.990	2.13	2.08	2.04	1.88	1.79	1.70
	.995	2.29	2.24	2.19	2.00	1.90	1.79

CUMULATIVE F-DISTRIBUTION

ν_2	p	ν_1 40	50	60	100	120	∞
1	.005	.113	.116	.118	.121	.122	.127
	.010	.137	.139	.141	.145	.146	.151
	.025	.184	.187	.189	.193	.194	.199
	.050	.245	.248	.250	.254	.255	.260
	.100	.353	.356	.358	.363	.364	.370
	.900	62.5	62.7	62.8	63.0	63.1	63.3
	.950	251.	252.	252.	253.	253.	254.
	.975	1006.	1008.	1010.	1013.	1014.	1018.
	.990	6287.	6303.	6313.	6334.	6339.	6366.
	.995	25148.	25211.	25253.	25337.	25359.	25465.
2	.005	.165	.169	.173	.179	.181	0.19
	.010	.193	.198	.201	.207	.209	.217
	.025	.247	.252	.255	.261	.263	.271
	.050	.309	.314	.317	.324	.326	.334
	.100	.410	.415	.418	.424	.426	.434
	.900	9.47	9.47	9.47	9.48	9.48	9.49
	.950	19.5	19.5	19.5	19.5	19.5	19.5
	.975	39.5	39.5	39.5	39.5	39.5	39.5
	.990	99.5	99.5	99.5	99.5	99.5	99.5
	.995	199.	199.	199.	199.	199.	200.
3	.005	.201	.207	.211	.220	.222	0.23
	.010	.232	.238	.242	.251	.253	.264
	.025	.289	.295	.299	.308	.310	.321
	.050	.352	.358	.363	.371	.373	.384
	.100	.449	.455	.459	.467	.469	.480
	.900	5.16	5.15	5.15	5.14	5.14	5.13
	.950	8.59	8.58	8.57	8.55	8.55	8.53
	.975	14.0	14.0	14.0	14.0	13.9	13.9
	.990	26.4	26.4	26.3	26.2	26.2	26.1
	.995	42.3	42.2	42.1	42.0	42.0	41.8
4	.005	.229	.236	.242	.252	.255	0.27
	.010	.261	.269	.274	.285	.287	.301
	.025	.320	.327	.332	.343	.346	.359
	.050	.384	.391	.396	.406	.409	.422
	.100	.478	.485	.490	.500	.502	.514
	.900	3.80	3.80	3.79	3.78	3.78	3.76
	.950	5.72	5.70	5.69	5.66	5.66	5.63
	.975	8.41	8.38	8.36	8.32	8.31	8.26
	.990	13.7	13.7	13.7	13.6	13.6	13.5
	.995	19.8	19.7	19.6	19.5	19.5	19.3

CUMULATIVE F-DISTRIBUTION

ν_2	P	ν_1					
		40	50	60	100	120	∞
5	.005	.251	.260	.266	.279	.282	0.30
	.010	.285	.293	.299	.312	.315	.331
	.025	.344	.353	.359	.371	.374	.390
	.050	.408	.417	.422	.434	.437	.452
	.100	.501	.509	.514	.525	.527	.541
	.900	3.16	3.15	3.14	3.13	3.12	3.10
	.950	4.46	4.44	4.43	4.41	4.40	4.36
	.975	6.18	6.14	6.12	6.08	6.07	6.02
	.990	9.29	9.24	9.20	9.13	9.11	9.02
	.995	12.5	12.5	12.4	12.3	12.3	12.1
6	.005	.269	.279	.286	.301	.304	0.32
	.010	.304	.314	.321	.335	.338	.357
	.025	.364	.374	.381	.394	.398	.415
	.050	.428	.437	.444	.456	.460	.477
	.100	.519	.528	.533	.545	.548	.564
	.900	2.78	2.77	·2.76	2.75	2.74	2.72
	.950	3.77	3.75	3.74	3.71	3.70	3.67
	.975	5.01	4.98	4.96	4.92	4.90	4.85
	.990	7.14	7.09	7.06	6.99	6.97	6.88
	.995	9.24	9.17	9.12	9.03	9.00	8.88
7	.005	.285	.296	.304	.320	.324	0.35
	.010	.320	.331	.339	.354	.358	.379
	.025	.381	.392	.399	.414	.418	.437
	.050	.445	.455	.462	.476	.479	.498
	.100	.534	.543	.550	.563	.566	.583
	.900	2.54	2.52	2.51	2.50	2.49	2.47
	.950	3.34	3.32	3.30	3.27	3.27	3.23
	.975	4.31	4.28	4.25	4.21	4.20	4.14
	.990	5.91	5.86	5.82	5.75	5.74	5.65
	.995	7.42	7.35	7.31	7.22	7.19	7.08
8	.005	.299	.311	.319	.336	.341	0.36
	.010	.334	.346	.354	.371	.376	.398
	.025	.395	.407	.415	.431	.435	.456
	.050	.459	.469	.477	.492	.496	.516
	.100	.547	.557	.563	.577	.581	.599
	.900	2.36	2.35	2.34	2.32	2.32	2.29
	.950	3.04	3.02	3.01	2.97	2.97	2.93
	.975	3.84	3.81	3.78	3.74	3.73	3.67
	.990	5.12	5.07	5.03	4.96	4.95	4.86
	.995	6.29	6.22	6.18	6.09	6.06	5.95

CUMULATIVE F-DISTRIBUTION

ν_1

ν_2	p	40	50	60	100	120	∞
9	.005	.310	.323	.332	.351	.356	0.38
	.010	.346	.359	.368	.386	.391	.415
	.025	.408	.420	.428	.446	.450	.473
	.050	.471	.482	.490	.506	.511	.532
	.100	.558	.568	.575	.590	.594	.613
	.900	2.23	2.22	2.21	2.19	2.18	2.16
	.950	2.83	2.80	2.79	2.76	2.75	2.71
	.975	3.51	3.47	3.45	3.40	3.39	3.33
	.990	4.57	4.52	4.48	4.41	4.40	4.31
	.995	5.52	5.45	5.41	5.32	5.30	5.19
10	.005	.321	.335	.344	.364	.370	0.40
	.010	.357	.371	.380	.399	.405	.431
	.025	.419	.432	.440	.459	.464	.488
	.050	.481	.494	.502	.519	.523	.546
	.100	.567	.578	.586	.601	.605	.626
	.900	2.13	2.12	2.11	2.09	2.08	2.06
	.950	2.66	2.64	2.62	2.59	2.58	2.54
	.975	3.26	3.22	3.20	3.15	3.14	3.08
	.990	4.17	4.12	4.08	4.01	4.00	3.91
	.995	4.97	4.90	4.86	4.77	4.75	4.64
11	.005	.330	.345	.355	.376	.382	0.41
	.010	.367	.381	.391	.411	.417	.445
	.025	.428	.442	.451	.471	.476	.502
	.050	.491	.504	.512	.530	.535	.559
	.100	.576	.587	.595	.611	.615	.637
	.900	2.05	2.04	2.03	2.00	2.00	1.97
	.950	2.53	2.51	2.49	2.46	2.45	2.40
	.975	3.06	3.03	3.00	2.96	2.94	2.88
	.990	3.86	3.81	3.78	3.71	3.69	3.60
	.995	4.55	4.49	4.44	4.36	4.34	4.23
12	.005	.339	.354	.365	.387	.393	0.42
	.010	.375	.390	.401	.422	.428	.844
	.025	.437	.451	.461	.481	.487	.514
	.050	.499	.512	.522	.540	.545	.571
	.100	.583	.595	.603	.620	.625	.647
	.900	1.99	1.97	1.96	1.94	1.93	1.90
	.950	2.43	2.40	2.38	2.35	2.34	2.30
	.975	2.91	2.87	2.85	2.80	2.79	2.72
	.990	3.62	3.57	3.54	3.47	3.45	3.36
	.995	4.23	4.17	4.12	4.04	4.01	3.90

CUMULATIVE F-DISTRIBUTION

ν_2	p	ν_1					
		40	50	60	100	120	∞
13	.005	.346	.362	.374	.397	.403	0.44
	.010	.383	.399	.410	.432	.438	.470
	.025	.445	.460	.470	.491	.497	.526
	.050	.507	.520	.530	.550	.555	.582
	.100	.590	.602	.611	.628	.633	.657
	.900	1.93	1.92	1.90	1.88	1.88	1.85
	.950	2.34	2.31	2.30	2.26	2.25	2.21
	.975	2.78	2.74	2.72	2.67	2.66	2.60
	.990	3.43	3.38	3.34	3.27	3.25	3.17
	.995	3.97	3.91	3.87	3.78	3.76	3.65
14	.005	.353	.370	.382	.406	.413	0.45
	.010	.390	.406	.418	.441	.448	.481
	.025	.452	.467	.478	.500	.506	.536
	.050	.513	.528	.538	.558	.563	.591
	.100	.596	.609	.618	.636	.640	.665
	.900	1.89	1.87	1.86	1.83	1.83	1.80
	.950	2.27	2.24	2.22	2.19	2.18	2.13
	.975	2.67	2.64	2.61	2.56	2.55	2.49
	.990	3.27	3.22	3.18	3.11	3.09	3.00
	.995	3.76	3.70	3.66	3.57	3.55	3.44
15	.005	.360	.377	.389	.415	.421	0.46
	.010	.397	.413	.425	.450	.456	.491
	.025	.458	.474	.485	.508	.514	.546
	.050	.520	.534	.545	.566	.571	.600
	.100	.602	.615	.624	.642	.647	.672
	.900	1.85	1.83	1.82	1.79	1.79	1.76
	.950	2.20	2.18	2.16	2.12	2.11	2.07
	.975	2.58	2.55	2.52	2.47	2.46	2.40
	.990	3.13	3.08	3.05	2.98	2.96	2.87
	.995	3.58	3.52	3.48	3.39	3.37	3.26
20	.005	.385	.405	.419	.449	.457	0.50
	.010	.422	.441	.455	.484	.491	.532
	.025	.484	.502	.514	.541	.548	.585
	.050	.544	.560	.572	.596	.603	.637
	.100	.623	.638	.648	.669	.675	.704
	.900	1.71	1.69	1.68	1.65	1.64	1.61
	.950	1.99	1.97	1.95	1.91	1.90	1.84
	.975	2.29	2.25	2.22	2.17	2.16	2.09
	.990	2.69	2.64	2.61	2.54	2.52	2.42
	.995	3.02	2.96	2.92	2.83	2.81	2.69

CUMULATIVE F-DISTRIBUTION

ν_2	p	ν_1					
		40	50	60	100	120	∞
24	.005	.400	.421	.437	.470	.479	0.53
	.010	.437	.458	.473	.504	.513	.558
	.025	.498	.518	.531	.561	.568	.610
	.050	.558	.576	.588	.615	.622	.659
	.100	.635	.651	.662	.685	.691	.723
	.900	1.64	1.62	1.61	1.58	1.57	1.53
	.950	1.89	1.86	1.84	1.80	1.79	1.73
	.975	2.15	2.11	2.08	2.02	2.01	1.94
	.990	2.49	2.44	2.40	2.33	2.31	2.21
	.995	2.77	2.70	2.66	2.57	2.55	2.43
30	.005	.416	.440	.457	.494	.504	0.56
	.010	.454	.477	.493	.528	.538	.589
	.025	.515	.536	.551	.583	.592	.639
	.050	.573	.593	.606	.636	.643	.685
	.100	.649	.666	.678	.703	.710	.745
	.900	1.57	1.55	1.54	1.51	1.50	1.46
	.950	1.79	1.76	1.74	1.69	1.68	1.62
	.975	2.01	1.97	1.94	1.88	1.87	1.79
	.990	2.30	2.24	2.21	2.13	2.11	2.01
	.995	2.52	2.46	2.42	2.32	2.30	2.18
40	.005	.436	.462	.481	.523	.534	0.60
	.010	.473	.498	.517	.556	.567	.628
	.025	.533	.557	.573	.610	.620	.674
	.050	.591	.612	.627	.660	.669	.717
	.100	.664	.683	.696	.724	.731	.772
	.900	1.51	1.48	1.47	1.43	1.42	1.38
	.950	1.69	1.66	1.64	1.59	1.58	1.51
	.975	1.88	1.83	1.80	1.74	1.72	1.64
	.990	2.11	2.06	2.02	1.94	1.92	1.80
	.995	2.30	2.23	2.18	2.09	2.06	1.93
50	.005	.449	.477	.498	.543	.556	0.63
	.010	.486	.513	.533	.576	.588	.657
	.025	.546	.571	.589	.628	.639	.700
	.050	.602	.625	.641	.677	.687	.741
	.100	.674	.694	.708	.738	.746	.792
	.900	1.46	1.44	1.42	1.39	1.38	1.33
	.950	1.63	1.60	1.58	1.52	1.51	1.44
	.975	1.80	1.75	1.72	1.66	1.64	1.55
	.990	2.01	1.95	1.91	1.82	1.80	1.68
	.995	2.16	2.10	2.05	1.95	1.93	1.79

CUMULATIVE F-DISTRIBUTION

ν_2	p	ν_1					
		40	50	60	100	120	∞
60	.005	.458	.488	.510	.559	.572	0.65
	.010	.495	.524	.545	.591	.604	.679
	.025	.555	.581	.600	.642	.654	.720
	.050	.611	.635	.652	.689	.700	.759
	.100	.682	.702	.717	.749	.757	.806
	.900	1.44	1.41	1.40	1.36	1.35	1.29
	.950	1.59	1.56	1.53	1.48	1.47	1.39
	.975	1.74	1.70	1.67	1.60	1.58	1.48
	.990	1.94	1.88	1.84	1.75	1.73	1.60
	.995	2.08	2.01	1.96	1.86	1.83	1.69
100	.005	.479	.512	.537	.595	.611	0.71
	.010	.516	.548	.572	.626	.641	.736
	.025	.575	.604	.625	.674	.688	.772
	.050	.629	.656	.675	.719	.731	.804
	.100	.698	.720	.737	.773	.783	.844
	.900	1.38	1.35	1.34	1.29	1.28	1.21
	.950	1.52	1.48	1.45	1.39	1.38	1.28
	.975	1.64	1.59	1.56	1.48	1.46	1.35
	.990	1.80	1.74	1.69	1.60	1.57	1.43
	.995	1.91	1.84	1.79	1.68	1.65	1.49
120	.005	.485	.519	.545	.605	.623	0.73
	.010	.522	.555	.579	.636	.652	.755
	.025	.580	.610	.632	.683	.698	.788
	.050	.634	.662	.682	.727	.740	.819
	.100	.702	.725	.742	.780	.791	.856
	.900	1.37	1.34	1.32	1.28	1.26	1.19
	.950	1.50	1.46	1.43	1.37	1.35	1.25
	.975	1.61	1.56	1.53	1.45	1.43	1.31
	.990	1.76	1.70	1.66	1.56	1.53	1.38
	.995	1.87	1.80	1.75	1.64	1.61	1.43
∞	.005	.518	.560	.592	.673	.699	1.00
	.010	.554	.594	.625	.701	.724	1.00
	.025	.611	.647	.675	.742	.763	1.00
	.050	.663	.695	.720	.779	.798	1.00
	.100	.726	.754	.774	.824	.838	1.00
	.900	1.30	1.26	1.24	1.18	1.17	1.00
	.950	1.39	1.35	1.32	1.24	1.22	1.00
	.975	1.48	1.43	1.39	1.30	1.27	1.00
	.990	1.59	1.52	1.47	1.36	1.32	1.00
	.995	1.67	1.59	1.53	1.40	1.36	1.00

TABLE VII Percentiles of the Wilcoxon Signed-Ranks Test Statistic[a]

n =	$W_{0.005}$	$W_{0.01}$	$W_{0.025}$	$W_{0.05}$	$W_{0.10}$	$W_{0.20}$	$W_{0.30}$	$W_{0.40}$	$W_{0.50}$	$\dfrac{n(n+1)}{2}$
4	0	0	0	0	1	3	3	4	5	10
5	0	0	0	1	3	4	5	6	7.5	15
6	0	0	1	3	4	6	8	9	10.5	21
7	0	1	3	4	6	9	11	12	14	28
8	1	2	4	6	9	12	14	16	18	36
9	2	4	6	9	11	15	18	20	22.5	45
10	4	6	9	11	15	19	22	25	27.5	55
11	6	8	11	14	18	23	27	30	33	66
12	8	10	14	18	22	28	32	36	39	78
13	10	13	18	22	27	33	38	42	45.5	91
14	13	16	22	26	32	39	44	48	52.5	105
15	16	20	26	31	37	45	51	55	60	120
16	20	24	30	36	43	51	58	63	68	136
17	24	28	35	42	49	58	65	71	76.5	153
18	28	33	41	48	56	66	73	80	85.5	171
19	33	38	47	54	63	74	82	89	95	190
20	38	44	53	61	70	83	91	98	105	210
21	44	50	59	68	78	91	100	108	115.5	131
22	49	56	67	76	87	100	110	119	126.5	153
23	55	63	74	84	95	110	120	130	138	176
24	62	70	82	92	105	120	131	141	150	300
25	69	77	90	101	114	131	143	153	162.5	325
26	76	85	99	111	125	142	155	165	175.5	351
27	84	94	108	120	135	154	167	178	189	378
28	92	102	117	131	146	166	180	192	203	406
29	101	111	127	141	158	178	193	206	217.5	435
30	110	121	138	152	170	191	207	220	232.5	465
31	119	131	148	164	182	205	221	235	248	496
32	129	141	160	176	195	219	236	250	264	528
33	139	152	171	188	208	233	251	266	280.5	561
34	149	163	183	201	222	248	266	282	297.5	595
35	160	175	196	214	236	263	283	299	315	630
36	172	187	209	228	251	279	299	317	333	666
37	184	199	222	242	266	295	316	335	351.5	703
38	196	212	236	257	282	312	334	353	370.5	741
39	208	225	250	272	298	329	352	372	390	780
40	221	239	265	287	314	347	371	391	410	820
41	235	253	280	303	331	365	390	411	430.5	861
42	248	267	295	320	349	384	409	431	451.5	903
43	263	282	311	337	366	403	429	452	473	946
44	277	297	328	354	385	422	450	473	495	990

$n =$	$w_{0.005}$	$w_{0.01}$	$w_{0.025}$	$w_{0.05}$	$w_{0.10}$	$w_{0.20}$	$w_{0.30}$	$w_{0.40}$	$w_{0.50}$	$\dfrac{n(n+1)}{2}$
45	292	313	344	372	403	442	471	495	517.5	1035
46	308	329	362	390	423	463	492	517	540.5	1081
47	324	346	379	408	442	484	514	540	564	1128
48	340	363	397	428	463	505	536	563	588	1176
49	357	381	416	447	483	527	559	587	612.5	1225
50	374	398	435	467	504	550	583	611	637.5	1275

For n larger than 50, the pth quantile w_p of the Wilcoxon signed ranks test statistic may be approximated by $w_p = [n(n+1)/4] + x_p\sqrt{n(n+1)(2n+1)/24}$, where x_p is the pth quantile of a standard normal random variable, obtained from Table B.III.

SOURCE: Adapted from Harter and Owen, *Selected Tables in Mathematical Statistics* (1970), with permission from the American Mathematical Society.
[a] The entries in this table are percentiles w_p of the Wilcoxon signed ranks test statistic W_n^-, given by Equation (7.5), for selected values of $p \leq 0.50$. Percentiles for $p > 0.50$ may be computed from the equation

$$w_p = n(n+1)2 - w_{1-p}$$

where $n(n+1)$ is given in the right hand column in the table. Note that $P(W_n^- < w_p) \leq p$ and $P(W_n^- > w_p) \leq 1 - p$ if H_0 is true. Critical regions correspond to values of W_n^- less than (or greater than) but not including the appropriate percentile.

TABLE VIII · (Continued)

n	p	m=2	3	4	5	6	7	8	9	10	11	12	13	14	15	16	17	18	19	20
11	.001	66	66	67	69	71	73	75	77	79	82	84	87	89	91	94	96	99	101	104
	.005	66	67	69	72	74	77	80	83	85	88	91	94	97	100	103	106	109	112	115
	.01	66	68	71	74	76	79	82	85	89	92	95	98	101	104	108	111	114	117	120
	.025	67	70	73	76	80	83	86	90	93	97	100	104	107	111	114	118	122	125	129
	.05	68	72	75	79	83	86	90	94	98	101	105	109	113	117	121	124	128	132	136
	.10	70	74	78	82	86	90	94	98	103	107	111	115	119	124	128	132	136	140	145
12	.001	78	78	79	81	83	86	88	91	93	96	98	102	104	106	110	113	116	118	121
	.005	78	80	82	85	88	91	94	97	100	103	106	110	113	116	120	123	126	130	133
	.01	78	81	84	87	90	93	96	100	103	107	110	114	117	121	125	128	132	135	139
	.025	80	83	86	90	93	97	101	105	108	112	116	120	124	128	132	136	140	144	148
	.05	81	84	88	92	96	100	105	109	111	117	121	126	130	134	139	143	147	151	156
	.10	83	87	91	96	100	105	109	114	118	123	128	132	137	142	146	151	156	160	165
13	.001	91	91	93	95	97	100	103	106	109	112	115	118	121	124	127	130	134	137	140
	.005	91	93	95	99	102	105	109	112	116	119	123	126	130	134	137	141	145	149	152
	.01	92	94	97	101	104	108	112	115	119	123	127	131	135	139	143	147	151	155	159
	.025	93	96	100	104	108	112	116	120	125	129	133	137	142	146	151	155	159	164	168
	.05	94	98	102	107	111	116	120	125	129	134	139	143	148	153	157	162	167	172	176
	.10	96	101	105	110	115	120	125	130	135	140	145	150	155	160	166	171	176	181	186
14	.001	105	105	107	109	112	115	118	121	125	128	131	135	138	142	145	149	152	156	160
	.005	105	107	110	113	117	121	124	128	132	136	140	144	148	152	156	160	164	169	173
	.01	106	108	112	116	119	123	128	132	136	140	144	149	153	157	162	166	171	175	179
	.025	107	111	115	119	123	128	132	137	142	146	151	156	161	165	170	175	180	184	189
	.05	109	113	117	122	127	132	137	142	147	152	157	162	167	172	177	183	188	193	198
	.10	110	116	121	126	131	137	142	147	153	158	164	169	175	180	186	191	197	203	208

	α																				
15	.001	120	120	122	125	128	133	135	138	142	145	149	153	157	161	164	168	172	176	180
	.005	120	123	126	129	133	137	141	145	150	154	158	163	167	172	176	181	185	190	194
	.01	121	124	128	132	136	140	145	149	154	158	163	168	172	177	182	187	191	196	201
	.025	122	126	131	135	140	145	150	155	160	165	170	175	180	185	191	196	201	206	211
	.05	124	128	133	139	144	149	154	160	165	171	176	182	187	193	198	204	209	215	221
	.10	126	131	137	143	148	154	160	166	172	178	184	189	195	201	207	213	219	225	231
16	.001	136	136	139	142	145	148	152	156	160	164	168	172	176	180	185	189	193	197	202
	.005	136	139	142	146	150	155	159	164	168	173	178	182	187	192	197	202	207	211	216
	.01	137	140	144	149	153	158	163	168	173	178	183	188	193	198	203	208	213	219	224
	.025	138	143	148	152	158	163	168	174	179	184	190	196	201	207	212	218	223	229	235
	.05	140	145	151	156	162	167	173	179	185	191	197	202	208	214	220	226	232	238	244
	.10	142	148	154	160	166	173	179	185	191	198	204	211	217	223	230	236	243	249	256
17	.001	153	154	156	159	163	167	171	175	179	183	188	192	197	201	206	211	215	220	224
	.005	153	156	160	164	169	173	178	183	188	193	198	203	208	214	219	224	229	235	240
	.01	154	158	162	167	172	177	182	187	192	198	203	209	214	220	225	231	236	242	247
	.025	156	160	165	171	176	182	188	193	199	205	211	217	223	229	235	241	247	253	259
	.05	157	163	169	174	180	187	193	199	205	211	218	224	231	237	243	250	256	263	269
	.10	160	166	172	179	185	192	199	206	212	219	226	233	239	246	253	260	267	274	281
18	.001	171	172	175	178	182	186	190	195	199	204	209	214	218	223	228	233	238	243	248
	.005	171	174	178	183	188	193	198	203	209	214	219	225	230	236	242	247	253	259	264
	.01	172	176	181	186	191	196	202	208	213	219	225	231	237	242	248	254	260	266	272
	.025	174	179	184	190	196	202	208	214	220	227	233	239	246	252	258	265	271	278	284
	.05	176	181	188	194	200	207	213	220	227	233	240	247	254	260	267	274	281	288	295
	.10	178	185	192	199	206	213	220	227	234	241	249	256	263	270	278	285	292	300	307
19	.001	190	191	194	198	202	206	211	216	220	225	231	236	241	246	251	257	262	268	273
	.005	191	194	198	203	208	213	219	224	230	236	242	248	254	260	265	272	278	284	290
	.01	192	195	200	206	211	217	223	229	235	241	247	254	260	266	273	279	285	292	298
	.025	193	198	204	210	216	223	229	236	243	249	256	263	269	276	283	290	297	304	310
	.05	195	201	208	214	221	228	235	242	249	256	263	271	278	285	292	300	307	314	321
	.10	198	205	212	219	227	234	242	249	257	264	272	280	288	295	303	311	319	326	334

(Continued)

TABLE VIII   *(Continued)*

n	p	m=2	3	4	5	6	7	8	9	10	11	12	13	14	15	16	17	18	19	20
20	.001	210	211	214	218	223	227	232	237	243	248	253	259	265	270	276	281	287	293	299
	.005	211	214	219	224	229	235	241	247	253	259	265	271	278	284	290	297	303	310	316
	.01	212	216	221	227	233	239	245	251	258	264	271	278	284	291	298	304	311	318	325
	.025	213	219	225	231	238	245	251	259	266	273	280	287	294	301	309	316	323	330	338
	.05	215	222	229	236	243	250	258	265	273	280	288	295	303	311	318	326	334	341	349
	.10	218	226	233	241	249	257	265	273	281	289	297	305	313	321	330	338	346	354	362

For n or m greater than 20, the pth percentile w_p of the Wilcoxon rank-sum statistic may be approximated by

$$w_p = n(N-1)/2 + x_p \sqrt{nm(N+1)/12}$$

where x_p is the pth percentile of a standard normal random variable, obtained from Table B.III and where $N = m + n$.

[a] The entries in this table are quantiles w_p of the Wilcoxon rank-sum test statistic w_N, given by Equation (7.10), for selected values of p. Note that $\mathscr{P}(W_N < w_p) \leq p$. Upper quantiles may be found from the equation

$$w_p = n(n+m+1) - w_{1-p}$$

Critical regions correspond to values of W_N less than (or greater than) but not equal to the appropriate quantile.

TABLE IX Percentiles of the Kruskal–Wallis Test Statistic for Small Sample Sizes[a]

Sample Sizes	$w_{0.90}$	$w_{0.95}$	$w_{0.99}$
2, 2, 2	3.7143	4.5714	4.5714
3, 2, 1	3.8571	4.2857	4.2857
3, 2, 2	4.4643	4.5000	5.3571
3, 3, 1	4.0000	4.5714	5.1429
3, 3, 2	4.2500	5.1389	6.2500
3, 3, 3	4.6000	5.0667	6.4889
4, 2, 1	4.0179	4.8214	4.8214
4, 2, 2	4.1667	5.1250	6.0000
4, 3, 1	3.8889	5.0000	5.8333
4, 3, 2	4.4444	5.4000	6.3000
4, 3, 3	4.7000	5.7273	6.7091
4, 4, 1	4.0667	4.8667	6.1667
4, 4, 2	4.4455	5.2364	6.8727
4, 4, 3	4.773	5.5758	7.1364
4, 4, 4	4.5000	5.6538	7.5385
5, 2, 1	4.0500	4.4500	5.2500
5, 2, 2	4.2933	5.0400	6.1333
5, 3, 1	3.8400	4.8711	6.4000
5, 3, 2	4.4946	5.1055	6.8218
5, 3, 3	4.4121	5.5152	6.9818
5, 4, 1	3.9600	4.8600	6.8400
5, 4, 2	4.5182	5.2682	7.1182
5, 4, 3	4.5231	5.6308	7.3949
5, 4, 4	4.6187	5.6176	7.7440
5, 5, 1	4.0364	4.9091	6.8364
5, 5, 2	4.5077	5.2462	7.2692
5, 5, 3	4.5363	5.6264	7.5429
5, 5, 4	4.5200	5.6429	7.7914
5, 5, 5	4.5000	5.6600	7.9800

Source. Adapted from Iman, Quade, and Alexander (1975), with permission from the American Mathematical Society.

[a] The null hypothesis may be rejected at the level α if the Kruskal–Wallis test statistic, given by Equation 8.46 exceeds the $1 - \alpha$ percentile given in the table.

Appendix C

Answers to Selected Problems

Chapter 2

2.1. 35,910   **2.2.** 165   **2.3.** 56   **2.4.** 56   **2.5.** 5040   **2.6.** $\frac{1}{3}$ **2.7.** $\frac{2}{3}$ **2.8.** $\frac{11}{36}$

2.9. (a) $\frac{1}{36}$ **2.10.** (a) $\mathscr{P}(A) = \frac{1}{2}$, $\mathscr{P}(B) = \frac{1}{2}$, $\mathscr{P}(E) = \frac{1}{6}$;   (b) A and B;

(d) $\mathscr{P}(C \text{ or } D) = \frac{5}{6}$ **2.11.** $\frac{1}{69}$, 300   **2.12.** (a) $\frac{7}{12}$ (assuming no "curve" on the grades)

2.13. (a) $(\frac{1}{12})^4$;   (b) $(\frac{1}{3})^4$;   (c) $(\frac{7}{12})^4$

2.14. $\mathscr{P}(1) = 12!(0.05)(0.95)^{11}/11!\ 1!$, $\mathscr{P}(5) = 12!(0.05)^5(0.95)^7/7!\ 5!$

2.15. (b) $\mathscr{P}(\text{one needs service call}) = 0.64$ **2.16.** 0.0182   **2.17.** 0.0144

2.18. (b) 0.9182   **2.20.** $\mathscr{P}(\text{accepting}) = 0.7659$ **2.22.** 0.657

2.23. (a) $\frac{18}{35}$;   (b) $\frac{1}{7}$;   (c) 0.414   **2.24** (b) $\mathscr{P}(2 \text{ alike}) = \frac{10}{36}$

2.25. (a) 0.43046;   (b) 0.18689;   (c) 3;   (d) $\mathscr{P}(30 \text{ good}) = 0.04239$

2.26. (b) 0.6676;   (c) 0.4420   **2.27.** (a) 0.09693;   (b) 0.0494;   (c) 1.48 cents/lb

2.28. (a) 0.2589;   (b) 0.3150;   (c) 0.3665; $\mathscr{P}(= 12 \text{ points}) = 0.2013$

Chapter 3

3.15. (a) 0.88;   (b) 0.128;   (c) 0.829   **3.18.** $s_x = 1.27$, $s_{\bar{x}} = 0.366$

3.19. $\mathscr{P}(r \geq 1) = \dfrac{12!(0.84)^{11}(0.16)}{11!\ 1!} + 12!(0.84)^{12}(0.16)^0/12!\ 0!$

3.20. $m = 10.08$, $\mathscr{P}(11 \leq r \leq 12) = e^{-10.08}\left[\dfrac{(10.08)^{11}}{11!} + \dfrac{(10.08)^{12}}{12!}\right]$

3.21. 0.168   **3.22.** 0.000519

3.24. (a) $\mathscr{P}(1) = 0.23423$: (b) $\mathscr{P}(2) = 0.001909$
3.25. $\mathscr{P}(6 \text{ or } 7 \text{ or } 8) = 0.0064$
3.27. $m = 0.48$, $\mathscr{P}(0) = 0.619$, $\mathscr{P}(1) = 0.297$, $\mathscr{P}(2) = 0.071$, $\mathscr{P}(3) = 0.011$
3.28. $\mathscr{P}(0) = 0.1377$
3.29. (a) 4 broken at 95% CI; (b) 6 broken at 99% CI **3.30.** $m = 0.85$

Chapter 4

4.1. (a) $\bar{x} = \bar{z}(250)$; (b) $\bar{x} = \bar{w} + 25$ **4.2.** $\bar{x} = 44.9833$, $s^2 = 3518.7896$
4.3. $\bar{x} = 144.5$, $s^2 = 2788.4$ **4.4.** $\bar{x} = 82.782$, $s^2 = 56.648$
4.5. $\bar{x} = 2.79$, $s^2 = 8.4706$ **4.6.** $\bar{x} = 0.92085$, $s^2 = 0.50503$
4.7. $\bar{x} = 79.04$, $s^2 = 2.7944$ **4.8.** $\bar{x} = 14.63$, $s^2 = 16.8037$
4.9. $\bar{x} = 0.9358$, $s^2 = 0.0026$ **4.10.** $\bar{x} = 1449.3$, $s^2 = 42,053.07$
4.11. $\bar{x} = 51.19$, $s^2 = 193.854$ **4.12.** $x = 0.350$, $s^2 = 0.113306$ (evap/day)
4.13. $\bar{x} = 7.505416$, $s^2 = 0.00329344$ (data from Problem 3.17)

Chapter 5

5.1. (a) $k = \sqrt{2\theta}$; (b) $f(x) = x/\theta$ **5.2.** $a = 3$
5.3. (a) $A = \frac{1}{18}$; (b) $\mu = 3.07$, $\sigma^2 = 0.328$; (c) $\mathscr{P}(3 \le x \le 4) = 0.556$;
(d) $\mathscr{P}(x < 3) = 0.444$ **5.6.** $\mu'_1 = \frac{1}{2}$, $\mu_2 = \frac{1}{4}$ **5.7.** $\mu'_1 = m$, $\mu_2 = m$ **5.8.** 0.0256
5.13. $\text{var}(x) = 276$

Chapter 6

6.1. $\bar{x} = 2.501 \times 10^{11}$ ft/lb, $s_x^2 = 1.11 \times 10^{20}$, range on α is 2.446×10^{11} to 2.556×10^{11} ft/lb.
6.3. $\bar{x} = 11.411$, $s_x^2 = 3.5551$, $z_{0.025} = -1.96$, range on $SO_4{}^{2-}$ is 10.24 to 12.58 $\mu g/m^3$.
6.4. $\bar{x} = \$516.62$, $s_x = \$8.551$, $z_{0.025} = -1.96$; range on cost = $500.06 to $533.58. (*Note*: Disregard all units if either cost or capacity specification is not met.
6.5. (a) $\bar{x} = 1.09375$, $\sigma_{\bar{x}}^2 = 0.00071257$, $z = 2.7178$; (b) confidence level = 99.3%
6.6. $\bar{x} = 50.46$ psig, $s_{\bar{x}} = 0.45318$ psig/sample, $t_{0.99} = 3.250$, range = 48.98 to 51.93 psig
6.7. $\bar{x} = 12.75$, $s_{\bar{x}} = 0.98107$, $z_{\text{calc}} = 1.4131$, confidence level = 84.2%
6.8. (a) $\bar{x} = 0.91$, $s_x = 2.9277 \times 10^{-3}$, $F(z) = 0.833$;
(b) instructor was not fair as 61% is passing and he failed at 66.6%.
6.9. $s_X = 1.1686$, 95% C.I. is $58.18 < \mu < 63.32$ **6.10.** (a) $\bar{x} = 17.8$, $s_{\bar{x}} = 2.1908902$
6.11. $s_{\bar{X}}^2 = 0.033928$, 99% C.I. is 0.745% Si $< \mu < 0.931$% Si
6.17. $f = 12.55$, $t_f = \pm 3.0356$, $s_{\bar{x}_A - \bar{x}_B} = 2.073$
6.18. (a) $s_{\bar{D}} = 1.823$, $\bar{d} = 5.5$, 95% CI is $1.191 < \mu_d < 9.809$ **6.22.** $\sigma_{\bar{x}} = 1.55$%
6.23. 0.1915 **6.24.** $-0.0496 < P_1 - P_2 < 0.0208$ **6.25.** $s_{\bar{P}_1 - \bar{P}_2} = 0.047191$, $-0.0826 < P_1 - P_2 < 0.1605$

Chapter 7

7.1. $z = 1.725$, accept H_0: $\mu_1 = \mu_2$

7.2. $\bar{x} = 60.75$, $s_X^2 = 16.386$, $T = -3.6368$; at $\alpha = 0.05$, accept H_0: $\mu < \mu_0 = 65$.

7.4. $\bar{x} = 17.8$, $s_X^2 = 480$; 95% CI is $13.46 < \mu < 21.14$.

7.5. $\bar{x} = 39$, $s_X = 2.4$, $T = -1.8633$, accept H_0: $\mu = \mu_0 = 40$;
CI is $37.474 < \mu < 40.526$

7.6. $\bar{x} = 0.838$, $s_{\bar{x}} = 0.0336$, $T = -0.357$; accept H_0: $\mu_{new} \geq \mu = 0.85$

7.7. $\bar{x} = 0.197$ in., $s_X = 0.005$ in.; $t_{99,0.025} = 1.9824$; reject H_0: $\mu_1 = \mu_2 = 0.1875$
at both levels of α

7.8. $s_A^2 = 17.95$, $s_B^2 = 15.61$, $T = -1.615$; $s_{\bar{X}_A - \bar{X}_B} = 2.0542$; accept
H_0: $\mu_A = \mu_B$ at $\alpha = 0.02$

7.10. (a) $\bar{x} = 69.07$, $s_{\bar{x}} = 0.0789$, $T = 2.914$; $\alpha \leq 0.072$ to accept H_0

7.14. $s_{\Delta P}^2 = 2.06105$, $s_F^2 = 0.060475$; to test H_0: $\mu_{\Delta P_{exp}} = \mu_{\Delta P_{theor}}$, $T = -2.815$,
which is outside the acceptance region for $\alpha = 0.05$

7.16. $\bar{x}_1 = 4.69$, $\bar{x}_2 = 4.63$, $s_{X_1} = 29.15 \times 10^{-3}$, $s_{X_2} = 5.612 \times 10^{-2}$,
$F_{4,4,0.025} = 0.104$; $F_{4,4,0.975} = 9.60$. The 95% CI on σ_1/σ_2 is $0.2326 < \sigma_1/\sigma_2 < 2.233$.
From F-test using s_1/s_2 and $\sqrt{}$ of the foregoing F values, accept H_0: $\sigma_1 = \sigma_2$.
Alternative solution: $T_f = 2.122$, $f \cong 6.01$, since $t_{6.01,0.025} = -2.447$, accept
H_0: $\mu_1 = \mu_2$ by t-test.

7.17. (a) $s_{\bar{x}} = 37.4$, $\bar{X}_2 = 2365$; $T = 7.674$, so reject H_0 and conclude that the
modification is helpful; (b) $2260 < \mu < 2470$ lb/day wastage

7.18. $SS_T = 14,754$, $SS_M = 14,045$, $SS_E = 563.2$; reject H_0: $\alpha_i = 0$ at 5% level;
(b) $18.67 < \mu_1 < 28.93$; $25.41 < \mu_2 < 32.99$

7.19. $s_p^2 = 13.9194$, $s_X^2 = 67.71$. As $F_{1,18,0.05} = 4.41$, we reject H_0: cleaning had no
effect. H_0 would be accepted at 99% level.

7.20. $\bar{d} = 0$, $s_{\bar{D}} = 33.28$, $T = 0$; $t_{3,0.975} = 4.176$, so accept H_0: $\mu_{\bar{D}} = 0$. Alternative
solution: Calculate $\chi^2 = 27.459$; $\chi^2_{3,0.95} = 7.815$. As $\chi^2_{calc} > \chi^2_{tab}$, reject H_0: $0 = E$

7.21. $T_f = -3.0593$, $f = 12.17$, accept H_0: $\mu_{140} < \mu_{120}$

7.22. $\bar{x}_A = 76.2$, $\bar{x}_B = 78.286$, $s_A^2 = 1.075$, $s_B^2 = 3.5714$; (a) $T_{f,0.05} = -1.822$,
$f = 9.57$. Accept H_0: $\mu_B > \mu_A$ by one-tailed test as $T = 2.449$. Alternative solution:
$\chi^2_{calc} = 19.9468$; $\chi^2_{6,0.05} = 12.592$, so accept H_0: $\sigma_B > \sigma_A$; (b) 0.953 chance of being
wrong by t-test; 0.5% chance of being wrong according to χ^2 test

7.23. $s_{x_1}^2 = 8.75$; $s_{x_2}^2 = 6.625$; $F_{8,8,0.95} = 3.44$, so we accept H_0: $\sigma_1^2 = \sigma_2^2$. As the
variances are thus shown to be statistically equal, pool them to get $s_p^2 = 7.6875$ and
$T_p = 2.21$, which does fall in the 99% acceptance region for the difference in means.

7.25. $s_X^2 = 0.027461$, $s_Y^2 = 0.0048073$

7.26. $\chi^2 = 12.32$, $\chi^2_{3,0.95} = 7.81$, so reject hypothesis of uniform population.

7.27. $\chi^2 = 6.881$, $\chi^2_{6,0.95} = 11.1$, so accept hypotheis.

7.28. $\chi^2 = 3.408$, $\chi^2_{2,0.95} = 5.99$, so accept hypothesis of normality.

7.29. $\chi^2_{calc} = 6.1667$, $\chi^2_{3 \times 1,0.95} = 7.815$, so accept H_0: no regional differences.

7.30. $\chi^2 = 32.6048$ vs. $\chi^2_{17,0.95} = 27.6$, so reject H_0: predictive method is accurate.

7.31. $\chi^2 = 0.00772$; compare to $\chi^2_{6,0.025}$ and $\chi^2_{6,0.975}$ and reject the null hypothesis.

7.32. $\chi^2 = 9$ vs. $\chi^2_{9,0.99} = 6.635$, so the coin is probably biased.

7.33. $s_{\bar{d}} = 6.05923$, $-14.8269 < \mu_D < 14.8269$, $\bar{d} = 0$, dye probably has no effect;
$\chi^2 = 14.46$, which is not in the 95% CI for χ^2_6.

7.34. $\chi^2 = 17.38 > \chi^2_{5,0.95} = 11.1$, so we must conclude that the system was not installed properly.

Chapter 8

8.1. (a) $s_p^2 = 1.808$, $s_Y^2 = 24.596$, reject H_0: $\alpha_i = 0$ at 5% level;
(b) $t_{20,0.95} = 2.086$; $36.522 < \mu_{III} < 38.812$ for highest yield
8.2. (Coke yield): $s_p^2 = 13.7923$, $s_Y^2 = 5.6205$; accept H_0: hydrogenation of feed does not affect coke yield; (conversion): $s_p^2 = 96.248$, $s_Y^2 = 54.908$; accept H_0: no significant difference in conversion.
8.3. (a) $s_p^2 = 0.89391$, $s_Y^2 = 5.1566$; accept H_0: there is a significant difference between populations; to compare $(A + B)$ vs. $(C + D)$, $s_p^2 = 0.85286$, $s_Y^2 = 15.2420$ and we accept H_0: $\mu_{A+B} \neq \mu_{C+D}$.
8.6. $s_p^2 = 0.075$, $s_Y^2 = 0.08988$; $F_{5,12,0.99} = 5.06$, so accept H_0: $\alpha_i = 0$.
8.7. $s_Y^2 = 0.0014167$, $s_p^2 = 0.0007458$, $s_{\bar{p}} = 0.01362$, $t_{12,0.995} = 3.055$; accept H_0: $\alpha_i = 0$ at both 95% and 99% levels; $\mu_{.1} = 0.385 \pm 3.055(0.01362)$.
8.9. $s_p^2 = 0.1678$, $s_Y^2 = 0.357$, accept H_0: different gases have no effect.
8.10. $s_p^2 = 0.008085$, $s_Y^2 = 0.81334$; reject H_0 and conclude that flow rate does affect CO_2 leakage.
8.11. $s_p^2 = 66.083$, $f = 1.411$; accept H_0: no significant difference in lab techniques.
8.12. $SS_T = 2.3236$, $SS_M = 2.3104$, $SS_C = 0.00425$, $SS_E = 0.0084$; accept H_0: $\alpha_i = 0$, $\beta_j = 0$ and conclude no difference between batches or analysts.
8.13. $SS_T = 119{,}047.54$, $SS_C = 15{,}449.66$, $SS_R = 1.6366$, $SS_E = 31.27334$; at $\alpha = 0.05$, inlet concentration affects outlet gas composition but lab groups do not.
8.14. $SS_T = 760.430$, $SS_C = 189.393$, $SS_R = 0.433333$, $SS_F = 0.186667$; both flow rate and lab group affect the HTU (at $\alpha = 0.05$).
8.15. (a) $SS_T = 310{,}933{,}275$, $SS_{col} = 5{,}653{,}437.5$, $SS_R = 170{,}343{,}939.6$, $SS_E = 7{,}306{,}879.2$; accept H_{0_1}: supply pressure does not affect output; reject H_{0_2}: model has a significant effect on output.
8.16. (a) $SS_{col} = 95.7905$, $SS_{row} = 556.402$, $SS_{AC} = 959.305$, $SS_{EE} = 307.1125$, $SS_{SE} = 272.225$. By F-test, we find sampling errors to be insignificant so we pool the error sources to get $MS_{pooled\,error} = 19.9772$. As the resulting $F > F$ at $\alpha = 0.05$, we accept H_{0_1}: $\alpha_i = 0$ and H_{0_2}: $\beta_j = 0$; (b) $SS_{level} = 17.2925$, $SS_E = 562.045$, column and row effects are unchanged. $F_{col} = 4.772$, $F_{row} = 3.080$, $F_{level} = 0.861$. Reject insignificance of row and column effects; accept insignificance of level effect. All tests at $\alpha = 0.05$.
8.18. $SS_M = 7.196$, $SS_{flow} = 2.148$, $SS_{CONC} = 1.492$, $MS_E = 0.007662$. By 5% F-tests, both null hypotheses are rejected and we conclude that flow rate and amine concentration affect performance.
8.19. $SS_M = 2{,}601.907$, $SS_{col} = 685.815$, $SS_{row} = 33.163$, $SS_E = 9.525$. Reject both null hypotheses and conclude that both temperature and pipe diameter influence flow rate, the former possibly more strongly than the latter.
8.20. $F_{feed\,rate} = 165.707$, $F_{conc} = 7.607$, reject both null hypotheses at $\alpha = 0.05$ and conclude that feed rate and amount of catalyst affect the % conversion. $MS_E = 1.90276$.

8.21. $SS_M = 2,596,126.5625$, $SS_{col} = 39,934.1875$, $SS_{row} = 324,082.1875$, $SS_E = 9,232.0625$. Reject the null hypotheses that liquid and gas rates have no effect at $\alpha = 0.01$.
8.22. $SS_{col} = 24.97$, $SS_{row} = 1.62$, $SS_E = 0.16$; reject H_{0_1}: blue colorant is insignificant; reject H_{0_2}: yellow colorant is insignificant.
8.23. $SS_T = 2164.06$, $SS_{col} = 12.8833$, $SS_{row} = 1.8067$, $SS_E = 0.0067$. Reject the null hypotheses of no effects on greenness of the yellow and blue colorants.
8.24. $SS_T = 572.07$, $SS_M = 460.041$, $SS_{col} = 47.33$, $SS_{row} = 27.96$. Accept H_{0_1}: $\alpha_i = 0$ and H_{0_2}: $\beta_j = 0$.
8.25. $SS_{col} = 246.966$, $SS_{row} = 58.157$, $SS_E = 91.006$. Accept H_{0_1}: pull angle does not affect force. Reject H_{0_2}: different connectors do not affect force. $\alpha = 0.05$ in both tests.
8.26. $SS_M = 101,798.13$, $SS_{col} = 6274.307$, $SS_{row} = 4.380$, $SS_{AC} = 7083.911$, $F_{SE} = 0.636$, so pool errors. Based on pooled errors, $F_{cols} = 11.771$, $F_{rows} = 0.024$. Surface pretreatment effects are found insignificant; adhesive effect is significant.
8.27. Method I: no interaction. $SS_{TC} = 14,481.344$, $SS_{row} = 9.276$, $SS_{col} = 13,761.23$. At $\alpha = 0.05$, accept H_{0_1}: primer has no effect; reject H_{0_2}: adhesive system does not contribute significantly to variation in data; method II: interactions exist as measured by repetition effect. $SS_M = 169,183.126$, $SS_{rep} = 19.2777$; $SS_{primer \times adhesive} = 14,036.7154$. At $\alpha = 0.05$, replication and primer effects are insignificant but adhesive and adhesive $\times$ primer interaction are significant with F values of 232.15 and 236.8, respectively.
8.28. $SS_T = 105,788$, $SS_C = 105.25$, $SS_R = 326.25$; at $\alpha = 0.05$, both students and sleep time affect test-taking ability.
8.30. $SS_T = 156.5279$, $SS_M = 133.0061$, $SS_R = 10.92024$, $SS_C = 11.6725$; at $\alpha = 0.05$, conclude that catalyst effectiveness is affected by both space velocity and surface area.
8.31. $s_p^2 = 19.7599$, $s_Y^2 = 18.344$; accept H_0: there is no significant difference in areas.
8.32. $SS_{Tr} = 21,851.1$, $SS_{EE} = 1.74$, $SS_{SE} = 0.08$; reject H_0: experimental error is indistinguishable from sampling error; reject H_0: sample source is not significant for this model I situation.
8.33. $MS_{Tr} = 22,699.4$, $MS_{EE} = 133.18$, $MS_{SE} = 18.0138$, $F_{EE} = 7.393$; reject H_0: experimental error and sampling error measure the same thing $F_{Tr} = 170.44$, so reject H_0: $\alpha_j = 0$.
8.34. $MS_{EE} = 3.9333$, $MS_{SE} = 12.6055$, $F_{EE} = MS_{EE}/MS_{PE} = 0.312$ but no need to pool the error terms as $v_{SE} = 45$ is sufficient. $F_{Tr} = 356.67$, so reject H_0: $\alpha_j = 0$.
8.36. $MS_{EE} = 39.95$, $MS_{SE} = 0.1083$, $F_1 = 368.8$, we reject H_0: experimental error is insignificant. Comparing MS_{Tr} to MS_{EE} gives $F_2 = 6.326$, so reject H_0: age does not affect invertase activity.

Chapter 9

9.1. (a) $\sum xy = 14.626$, $\sum y^2 = 0.005331$, $\sum d^2 = 0.000442$, $s_{\beta_1} = 5.025 \times 10^{-5}$, $s_{\beta_0} = 0.9785 \times 10^{-2}$, $C_p = 0.353332 + 3.34292 \times 10^{-4}t$.
9.2. (a) $\hat{Y} = 6.00 + 0.8X$, $s_{\beta_1} = 0.2$
9.3. $\sum 0^2 = 279.9997$, $\sum F = 872.0988$, $(\sum F)^2 = 760,556.4$, $F = 41.4239 + 0.7270230$

9.4. At 1.1% yellow, $s_Y = 0.0516$, $L = 81.49709 - 1.93353$ (% blue); at 1.9% yellow, $s_{\bar Y} = 0.12187$, $L = 81.64635 - 2.18323$ (% blue)

9.5. $\sum x_1^2 = 46.8125$, $\sum y^2 = 61 \times 10^{-6}$, $\sum x_1 y = 5.561005$, $\hat C = -89.5017 \times 10^{-3} + 11.2732 \times 10^{-4}$ (% T)

9.6. For loading zone, $12.64 < L/G < 21.7$: $-\Delta P = 0.3344 - 0.01111(L/G)$; for safe operating region, $12.64 \le L/G < 5.557$: $-\Delta P = 1.186 - 0.07853(L/G)$; for flooding zone, $5.557 \le L/G < 7.6$, $-\Delta P = 2.391 - 0.2953(L/G)$; loading point is at intersection of first two equations: $(L/G) = 12.637$, $-\Delta P = 0.1940$ in H_2O; flooding point is at intersection of second and third equations: $(L/G) = 5.5571$, $-\Delta P = 0.7500$ in H_2O

9.7. $\hat Y = 3.2666 + 1.98789X$, $r^2 = 0.99516$, $\sum x^2 = 82.49951$, $\sum xy = 164$

9.8. $s_{E_1}^2 = 1.86222 \times 10^{-5}$, $\sum x_1^2 = 0.717779$, $\bar x_1 = 0.509559$, $T_{\beta_1} = -25.8626$, $s_{E_2}^2 = 1.30333 \times 10^{-5}$, $s_{\beta_0} = 0.0024289$, $T_{\beta_0} = 617.085$; for testing H_0: $\beta_{0_1} = \beta_{0_2}$, $T_f = 0.6132$, $f = 21.41$

9.9. (a) $\hat Y = -0.26672 + 3.73417X$, $s_{\beta_1} = 0.40765$, $T_{\text{calc}} = 9.1602$;
(b) $\hat Y = 3.48945 + 2.44597X_w + 0.82089X_a$, $SS_E = 48{,}380.21$, $F_{\text{calc}} = 49.138$; reject H_0: regression is not significant.

9.10. $\hat Y = 45.1972 - 2.68408X_1 + 4.20910X_2$, $R^2 = 0.94808$

9.16. $\hat Y = -0.8868835 + 0.76595168X$, $s_{\beta_0} = 0.02767$, $s_{\beta_1} = 0.03725$, $r^2 = 0.97246$

9.18. $\hat Y_1 = 3.09204 + 0.193694X_1 + 0.0362872X_1^2$, $R_1^2 = 0.99956$, $\hat Y_2 = 2.41588 + 0.180067X_2 + 0.034122X_2^2$, $R_2^2 = 0.998707$; $F_{\text{calc},1} = 398.37$, $F_{\text{calc},2} = 68.567$

9.20. (100% gasoline): $\hat Y = -72.6072 + 0.807145X - 0.944949 \times 10^{-3}X^2$, $R^2 = 0.996678$; (100% acetone, 142°F endpoint): $\hat Y = -2507.72 + 27.5217X - 0.0641013X^2$, $R^2 = 0.89897$; (50% gasoline, 50% acetone): $\hat Y = 148.651 - 3.0168X + 0.05172X^2$, $R^2 = 0.94478$; (70% gasoline, 30% acetone): $\hat Y = 152.229 - 2.1634X + 0.0442202X^2$, $R^2 = 0.98786$

9.21. $MV = 43.123445 - 40.784467 \log_{10}(NH_3 \text{ concentration})$; $T_{\beta_0} = 38.68$; $T_{\beta_1} = -72.48$

9.24. $\log \widehat{N_{St} N_{Pr}^{0.2}} = -0.0166505 - 0.625178 \log N_{Pr}$, $\sum y^2 = 0.552948$, $\sum x^2 = 1.384811$, $\sum xy = -0.865753$, $r^2 = 0.978844$

9.25. $\log \hat U = -0.0853233 + 0.390123 \log v_{\max}$, $\sum x^2 = 0.821564$, $\sum xy = 0.320511$, $\sum y^2 = 0.139633$

9.26. (Run 12-18-1): $\log \hat Y = -2.54230 + 0.690076 \log X$, where $Y = $ fractional weight gain, $X = $ time; $s_{\beta_0} = 0.02129$, $s_{\beta_1} = 0.006197$, $T_{\beta_1} = 111.47$, so reject H_0: $\beta_1 = 0$. For run 12-18-2, $\log \hat Y = -2.51246 + 0.687049 \log X$, $s_{\beta_1} = 0.001632$, $s_{\beta_0} = 0.005671$, $t_{\beta_0} = 442$, so reject H_0: $\beta_0 = 0$. To compare intercepts, H_{0_1}: $\beta_0 = \beta_0'$, for which $T_f = -8.011383$, $f = 39.078$. For slopes, H_{0_2}: $\beta_1 = \beta_1'$, for which $T_f = 3.441$ and $f = 38.966$. Both hypotheses are rejected.

9.27. (Run 1-13-3): $\hat Y = 0.00315X^{0.70405}$ where $Y = $ fractional weight gain,

X = time. For run 1-13-6, $Y = 2.116X^{0.70842}$. To compare runs, calculate the sum of squares due to regression, $SS_R = b_1 \sum xy$ and compare runs by $F = SS_{R_1}/SS_{R_2}$, etc. $SS_R(1\text{-}13\text{-}3) = -58.35$; $SS_R(1\text{-}13\text{-}6) = -59.18$, $SS_R(12\text{-}18\text{-}2) = -59.98$.

9.28. $\widehat{\log k} = -8.96025 + 0.0483922T$, where $T = {}^{\circ}C$; $s_E^2 = 0.0381656$, $T_{\beta_1} = 13.107$, $T_{\beta_0} = 14.772$; 95% CI on β_0 is -10.9626 to -6.9579, 95% CI on β_1 is 0.03890 to 0.05788. **9.29.** 0.916938 **9.31.** 0.13341
9.32. 1.1% yellow, 0.99925; 1.5% yellow, 0.99655; 1.9% yellow, 0.99879
9.33. 0.97535 **9.34.** 0.99287 **9.35.** 0.99516 **9.36.** Run 1, 0.98653; run 2, 0.99052
9.38. 0.94808 **9.39.** 0.82525 **9.40.** 0.993818
9.45. $\hat{Y} = 9.5663 + 0.452X$, $r^2 = 0.9708$ **9.52.** 0.995 **9.56.** 0.99902
9.59. 0.995474 **9.60.** 0.971 **9.61.** 0.97582 **9.62.** 0.94898 **9.63.** 9.99034
9.62. $U = \hat{\beta}_0 + \hat{\beta}_1 \log_{10} \Delta T = -1022.4986 + 626.0556 \log_{10} \Delta T$; $F = 398$ vs. $F_{1.3.0.95} = 10.1$, so accept H_A: regression is significant; $R^2 = 0.9925$; $T_{\beta_1} = 19.962$, so reject H_0: slope = 0.

Chapter 10

10.2. (b) $LCL_{\bar{x}} = 9.56$, $UCL_{\bar{x}} = 11.38$, UCL for range is 3.333. Sample 10, 18, 20 are out of control limits for $\bar{x}$. R always is control; **(d)** yes
10.3. (a) Control limits for $\bar{x}$ are 89.43, 171.2, for range 0 to 102.9;
(c) more consistent, not very accurate; **(d)** yes, no;
(e) 103.5 to 199.1 for $\bar{x}$, 0 to 120.3 for R; **(f)** No, no; **(g) (g)** should be used;
(h) week 12; **(i)** week 15
10.4. (a) $LCL_{\bar{x}} = 1.93$, $UCL_{\bar{x}} = 2.104$, $UCL_R = 0.7658$ using all 6 days. Since R is out of control on days 1 and 2. The $\bar{x}$ chart should not be used. Reconstruct using only days 3 to 6 where control is maintained. For these days $UCL_{\bar{x}} = 2.001$, $LCL_{\bar{x}} = 2$, $LCL_R = 0.001$, $UCL_R = 0.00411$; **(c)** no; **(d)** on day 39
10.5. (a) $LCL_{\bar{x}} = 2.219$, $UCL_x = 5.895$, $UCL_R = 7.62$;
(b) A.M.: $LCL_{\bar{x}} = 1.068$, $UCL_x = 5.038$, $UCL_R = 4.993$; P.M.: $LCL_{\bar{x}} = 3.08$, $UCL_{\bar{x}} = 7.04$, $UCL_R = 4.981$
10.6. (b) Probably not since 2 points above UCL and 2 points below LCL
10.7. Lot 5 out of control
10.8. (a) $OC(0.1) = 0.95$, $OC(0.48) = 0.05$; **(b)** $OC(0.24) = 0.95$, $OC(0.64) = 0.08$;
(c) Producer may object to plan (a) since probability of rejection for small p is larger than plan (b). Consumer may object to plan (b) since probability of acceptance for large p is higher than plan (a).
10.9. (a) Use binomial approx.: $OC(0.05) = 0.988$, $OC(0.2) = 0.6778$,
$OC(0.4) = 0.167$; **(b)** $OC(0.05) = 0.9999$, $OC(0.2) = 0.9672$, $OC(0.4) = 0.633$
10.10. (a) $E(T) = \beta$, $\text{var}(T) = \beta^2$; **(c)** $R(t) = e^{-t\beta}$; **(e)** $h(3) = 1/\beta$
10.11. (c) $\alpha^{-2 \cdot \beta}[\Gamma(1 + 2/\beta) - \Gamma^2(1 + 1/\beta)]$; **(e)** e^{-x}, $x > 0$
10.12. (a) 0.9976, 0.97; **(b)** 0.92926

Chapter 11

11.1. 4 **11.2.** 12 for $\alpha = 0.05$ **11.3.** $\hat{\sigma} = 1.9925$, n is about 50

11.4. $n = 8$ for $\alpha = 0.05$, $n = 11$ for $\alpha = 0.01$

11.10. $SS_{Tr} = 2510.96$, $SS_B = 1.936$, $SS_E = 8.384$, $f_{Tr} = 299.5$; reject H_0: age does not affect invertase activity.

11.11. $F_{shift} = 1.2644$, $F_{site} = 0.3083$, $F_{shift \times site} = 0.0758$, so we accept all model I hypotheses.

11.14. $F_H = 1.785$, $F_M = 5.193$, $F_{H \times M} = 1 \times 10^{-5}$, so accept all model I hypotheses.

11.17. $F_D = 2.877$, $F_S = 2.230$, $F_{DS} = 0.113$

11.22. $s_{\hat{\beta}_0} = 0.990$, $s_{\hat{\beta}_1} = 0.443$, $1.78 \leq \beta_1 \leq 3.822$

11.23. *Method I*: no interaction, no sampling error. $SS_M = 266,725.25$; $SS_{col} = 17,721.175$; $SS_{row} = 13.202$; $SS_{level} = 4256.007$; $SS_E = 4386.9$. Reject hypotheses of no effect of thickness and adhesive type. Accept null hypothesis of no effect due to primer. *Method II*: no interaction, sampling error exists. $SS_{AC} = 25,381.63$; $SS_{SE} = 1000.57$; $SS_{EE} = 3386.16$. $F_{calc} = MS_{SE}/MS_{EE} \ll 1$, the error sources are pooled to give $MS_P = 48.73$. Results of F-tests at $\alpha = 0.05$ are the same as in Method I. *Method III*: all interactions exist. Calculated F values are: $F_P < 1$; $F_T = 71.38$; $F_A = 99.08$; $F_{PT} < 1$; $F_{PA} = 5.32$, $F_{TA} = 12.93$; $F_{PTA} = 18.93$. Degrees of freedom for error = 75.

11.25. (c) $F = 5$;   (d) 2;   (e) 133%   **11.29.** (b) 1.1

11.30. Each main factor has 2 degrees of freedom. The second-order interaction has 8 d.f. and there are 27 d.f. for error.

11.33. $F_{movement} = 41.67$, $F_{depth} = 15.00$, $F_{stock} = 6.67$, $F_{movement \times feed\,rate} = 6.67$, $F_{movement \times depth} = 3.75$

11.38. F(adjusted) $= 34.05$, so reject H_0: ammonia concentration has no effect on performance, $T = -0.51$, so accept H_0: phase rate ratio is not important.

11.39. F(adjusted) $= 44.37$, $T = -4.89$, so reject both null hypotheses.

Index

Milton Keynes UK
Ingram Content Group UK Ltd.
UKHW021934071024
449327UK00022B/1805